AF342643

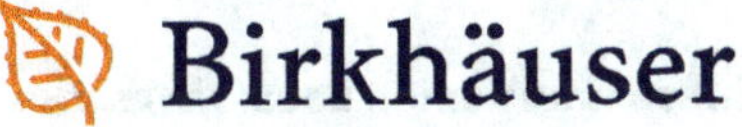

Progress in Nonlinear Differential Equations and Their Applications

Volume 91

More information about this series at http://www.springer.com/series/4889

Nicholas D. Alikakos • Giorgio Fusco •
Panayotis Smyrnelis

Elliptic Systems of Phase Transition Type

Birkhäuser

Nicholas D. Alikakos
Department of Mathematics
National and Kapodistrian University
Athens, Greece

Giorgio Fusco
Department of Mathematics
University of L'Aquila
Coppito, Italy

Panayotis Smyrnelis
Center for Mathematical Modeling
University of Chile
Santiago, Chile

ISSN 1421-1750   ISSN 2374-0280   (electronic)
Progress in Nonlinear Differential Equations and Their Applications
ISBN 978-3-319-90571-6   ISBN 978-3-319-90572-3   (eBook)
https://doi.org/10.1007/978-3-319-90572-3

Library of Congress Control Number: 2018962004

Mathematics Subject Classification (2010): 35J02, 49J02, 34C02

This book is published under the imprint Birkhäuser, www.birkhauser-science.com by the registered company Springer Nature Switzerland AG
The registered company address is: Gewerbestrasse 11, 6330 Cham, Switzerland

To Haim Brezis

Acknowledgements

N.D.A, is grateful to Rafe Mazzeo for inviting him to Stanford in the Spring of 2009 and 2012, where much of the point of view permeating this work was shaped. G.F. is indebted to Michał Kowalczyk for many comments and suggestions on several parts of this work. P.S. would like to thank Michał Kowalczyk, and the Center for Mathematical Modeling as a whole, for their support with this project, and the wonderful collaboration during the last three years. We owe considerably more than a formal acknowledgement to the program Aristeia (PDEGE, co-financed by the European Union (ESF) and national resources) through which the third author (P.S.) was supported as a postdoc, and the second author (G.F.) visited Athens for one semester. P.S. also enjoyed the support of Fondo Basal CMM-Chile, and Fondecyt postdoctoral grant 3160055.

The book is dedicated to Haim Brezis, for his profound influence on nonlinear PDE, for his steady interest and encouragement, and for his friendship.

Contents

Contents

Chapter 1
Introduction

Abstract In this chapter we give an overview of the book. We state and motivate the main theorems and refer the reader to the appropriate sections.

Let $f : \mathbb{R}^m \to \mathbb{R}^m$ be a map and $\Omega \subset \mathbb{R}^n$ be a domain that may be unbounded, with the case $\Omega = \mathbb{R}^n$ being particularly significant in our considerations. The problem of describing the structure of all bounded solutions $u : \Omega \to \mathbb{R}^m$ for the system

$$\Delta u - f(u) = 0, \quad x \in \Omega, \tag{1.1}$$

is a very challenging one.

For the scalar case $m = 1$, De Giorgi in 1978 [30] suggested a striking analogy with minimal surface theory that led to significant developments in P.D.E. and the Calculus of Variations, by stating the following conjecture about bounded solutions on $\mathbb{R}^n$:

Conjecture 1.1 (De Giorgi) Let $u \in C^2(\mathbb{R}^n)$ be a solution to $\Delta u - (u^3 - u) = 0$ such that

1. $|u| < 1$,
2. $\dfrac{\partial u}{\partial x_n} > 0$ for all $x \in \mathbb{R}^n$.

Is it true that all the level sets of u are hyperplanes, at least for $n \leq 8$?

The relationship with the Bernstein problem for minimal graphs is the reason why $n \leq 8$ appears in the conjecture.

Among the major contributions to this direction of research are those by Ghoussoub and Gui [40], Ambrosio and Cabré [13], Savin [58], del Pino et al. [31], Modica and Mortola [51], Modica [49, 50], Barlow et al. [17], Jerisson and Monneau

© Springer Nature Switzerland AG 2018
N. D. Alikakos et al., *Elliptic Systems of Phase Transition Type*,
Progress in Nonlinear Differential Equations and Their Applications 91,
https://doi.org/10.1007/978-3-319-90572-3_1

[45], Caffarelli and Córdoba [24, 25], Berestycki et al. [19], Farina [32], Berestycki et al. [20], Cabré and Terra [23], Sternberg [66]. We refer to the expository papers of Farina and Valdinoci [34] and of Savin [58] for a detailed account.

In the vector case $m > 1$, the mathematical phenomena are considerably richer. Moreover, the lack of the maximum principle is a further obstruction.

An important special case is when (1.1) is of gradient type, that is, $f(u) = W_u(u)$ is the (transposed) gradient of a potential $W : \mathbb{R}^m \to \mathbb{R}$, $W \geq 0$, and

$$\Delta u - W_u(u) = 0, \quad x \in \Omega \tag{1.2}$$

is the Euler-Lagrange equation for the functional

$$J_\Omega(v) = \int_\Omega \left(\frac{1}{2}|\nabla v|^2 + W(v) \right) dx. \tag{1.3}$$

The complementary class of interest is when f is of Hamiltonian type, which however we do not address in this book (cf. [29, 61, 63]). In the scalar case $m = 1$, (1.1) is automatically of gradient type, and the model nonlinearity is the double-well potential $W(u) = \frac{1}{4}(u^2 - 1)^2$. On the contrary, in the vector case $m \geq 2$, the behavior of the solutions to (1.2) depends strongly on the degree of connectedness of the zero set of the potential, $\{W = 0\} \neq \varnothing$ (Fig. 1.1).

With the exception of Chap. 3, where all potentials and all solutions are allowed, we restrict ourselves to phase transition potentials (i.e., $\{W = 0\}$ is a finite set of points representing the phases), and to minimal solutions to be defined later. The scalar problem models the coexistence of two phases and is linked to the fact that hyperplanes partition optimally the space into two parts, from the point of view of minimizing the interface perimeter, at least in low dimensions. On the other hand, in the vector case, when the number of coexisting phases is three or more, singular minimal cones are the appropriate objects (cf. Fig. 1.2 below). These are minimal, and nonorientable. They can have a cylindrical structure and they may be stratified into lower dimensional cones. In this book we pursue this analogy for symmetric potentials and equivariant solutions. We note that a vector order parameter is necessary for describing coexistence of three or more phases [54, §1.7]. In the case where $\{W = 0\}$ is a connected set, the problem is less geometrical

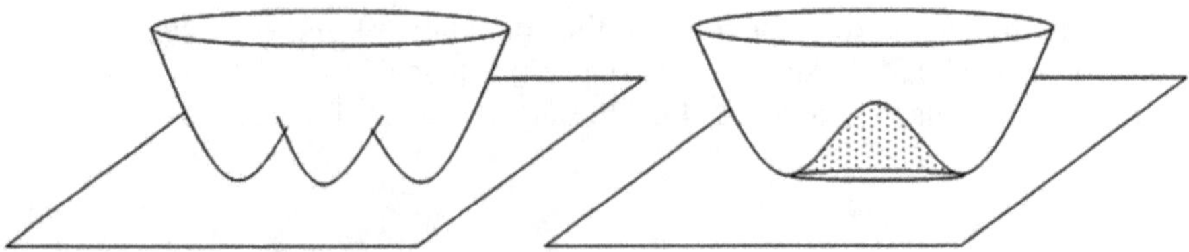

Fig. 1.1 Phase transition potential $\{W = 0\} = \{a_1, \ldots, a_N\}$ on the left, and Ginzburg-Landau potential $\{W = 0\} = \mathbb{S}^{m-1}$ on the right

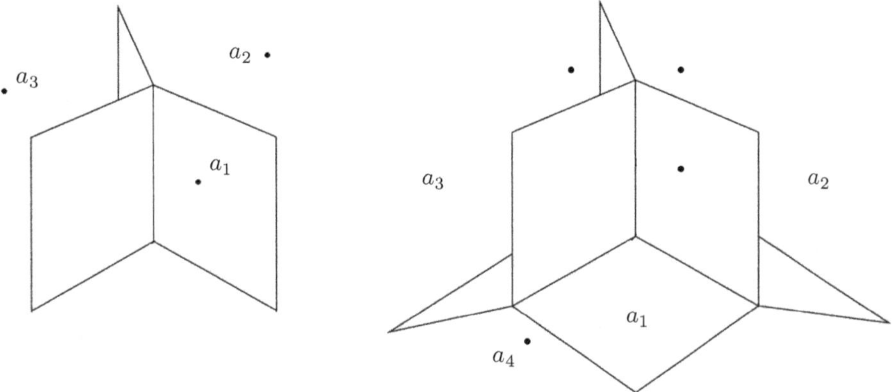

Fig. 1.2 The triod and the tetrahedral cone are the two singular minimal cones in $\mathbb{R}^3$ [69]

and instead is related to harmonic maps. Among the major contributions here are the monograph of Bethuel et al. [21], Sandier and Serfaty [57] and the papers of Caffarelli and Lin [26], and Caffarelli et al. [27].

Next we give a detailed account of the contents of the book.

Chapter 2 is rather special, as it addresses the $n = 1$ case, where (1.2) reduces to a Hamiltonian O.D.E. system. First we consider bistable potentials with two global minima and establish existence of *connections*, that is solutions to

$$u'' - W_u(u) = 0, \quad u : \mathbb{R} \to \mathbb{R}^m, \tag{1.4}$$

connecting the minima at infinity, $u(\pm\infty) = a^{\pm}$.

One-dimensional solutions are important because they capture the behavior of the transition near and across the interface. This fact is well established and plays a fundamental role, for example, in the context of Γ-convergence [1].

The first existence proofs of a heteroclinic connection in the vector case for a double-well potential were given by Rabinowitz [55] via minimization of the action functional and by Sternberg [67], who utilized the Jacobi principle under somewhat restrictive hypotheses on the behavior of W at the minima. In [55] aside from smoothness no other assumption is required on the behavior of W at the minima. Sternberg's approach was recently revisited in Sternberg and Zuniga [68] and in Monteil and Santambrogio [52]. These papers prove existence under minimal assumptions on W. In particular, as already observed by Sourdis [64], no conditions need to be imposed on W in a neighborhood of $a^{\pm}$. In Chap. 2, under the assumptions in [52], we give a proof of existence of connections by direct minimization of the *action functional*

$$J_{\mathbb{R}}(u) = \int_{\mathbb{R}} \left(\frac{1}{2}|u'|^2 + W(u)\right)dx.$$

Our presentation is based on Fusco et al. [38, 39]. In Sect. 2.4, in preparation for the traveling wave problem (cf. (1.5)), and for introducing a few concepts that we develop later in the P.D.E. context, we give an alternative proof of the existence of connections. This proof however requires that W satisfies a very mild non-degeneracy condition at $a^{\pm}$. We follow the presentation in [4]. Section 2.4 serves as an introduction to the rest of the book. Theorem 2.2 is invoked in various places later on, and its method of proof is extended to higher dimensions in Chap. 4. The connection problem has also been studied by André and Shafrir [14], and by Lin et al. [47] for special classes of potentials, but on the other hand for connections between more general sets. In this vein is also the work of Antonopoulos and Smyrnelis [15], where homoclinic and periodic connections between more general sets are also considered.

Next, we consider unbalanced bistable potentials $W(a^{-}) < 0 = W(a^{+})$, and examine with the method in Sect. 2.4 the connection problem for the minima at infinity:

$$u'' - W_u(u) = -cu', \quad c > 0, \quad u(\pm\infty) = a^{\pm}, \tag{1.5}$$

and under a convexity hypothesis on the a^{-} well, we establish existence, together with a variational characterization for c and u. Our presentation is a major simplification and improvement of the material in Alikakos and Katzourakis [12]. We note that the topological argument of Heinze [44] is one of the major ingredients in our argument. Risler [56], and Lucia et al. [48], at about the same time with [12], established with different variational arguments related results. Terman [70] already in 1987 showed existence of infinitely many solutions to (1.5) via topological methods.

Finally, in the last section of the chapter, we examine the phenomena of nonexistence/nonuniqueness of connections for potentials with three or more global minima. Restricting to $m = 2$ and identifying $\mathbb{R}^2$ with $\mathbb{C}$, we employ the geodesic method for the metric $g_{ij} = W(u)\delta_{ij}$ (conformal, degenerate), and construct several examples of potentials with explicit solutions for (1.4) exhibiting a variety of behaviors. Our presentation follows Alikakos et al. [11].

In Chap. 3, we collect a few basic facts that hold for all solutions and all potentials $W \geq 0$, hence in particular not distinguishing between the two types of potential depicted in Fig. 1.1. The point of departure for all of these considerations is the property that (1.2) can be written as a divergence-free condition

$$\mathrm{div}\, T(u) = 0, \tag{1.6}$$

where $T_{ij} = u_{x_i} \cdot u_{x_j} - \delta_{ij}\left(\frac{1}{2}|\nabla u|^2 + W(u)\right)$ is the stress-energy tensor associated to (1.2) that goes back to Noether and is well known to physicists (cf. [46]).

From (1.6), one can derive several useful properties for the solutions $u : \mathbb{R}^n \to \mathbb{R}^m$ of (1.2). For example, following Schoen [60], we derive the monotonicity formula (cf. [2] and also Farina [33]):

$$\frac{d}{dr}\left(r^{-(n-2)} J_{B_r(x_0)}(u)\right) \geq 0, \quad \text{for } r > 0, \tag{1.7}$$

which implies the lower bound

$$\begin{cases} J_{B_r}(u) \geq cr^{n-2}, & \text{for } n \geq 3, \ r > 0 \\ J_{B_r}(u) \geq c \ln r, & \text{for } n = 2, \ r > 0. \end{cases} \tag{1.8}$$

From this it follows, in particular, that

$$J_{\mathbb{R}^n}(u) < \infty \iff u \equiv \text{Const.}, \quad \text{if } n \geq 2, \tag{1.9}$$

and therefore for constructing solutions over $\mathbb{R}^n$, $n \geq 2$, we can no longer employ the direct method of the Calculus of Variations on $J_{\mathbb{R}^n}$. Instead we are forced to minimize on a ball B_R to obtain a minimizer $u_R : B_R \to \mathbb{R}^m$ of J_{B_R}, and construct the solution $u : \mathbb{R}^n \to \mathbb{R}^n$ by taking the limit along a subsequence

$$u = \lim_{R \to \infty} u_R. \tag{1.10}$$

For carrying out this procedure successfully we need uniform estimates in R, first simply for securing the nontriviality of the limit, and second for establishing the desired asymptotic behavior. Also as a corollary to (1.6) we obtain Gui's Hamiltonian Identities [42]. These are P.D.E. analogs of the conservation relation $\frac{1}{2}|u'|^2 - W(u) = \text{Const.}$ satisfied by solutions $u : \mathbb{R} \to \mathbb{R}^m$ to (1.4). The simplest example is

$$\int_{\mathbb{R}} \left[\frac{1}{2}\left(|u_{x_1}|^2 - |u_{x_2}|^2\right) + W(u(x))\right] dx_1 = \text{Const.}, \quad \forall x_2 \in \mathbb{R},$$

which holds for solutions $u : \mathbb{R}^2 \to \mathbb{R}^m$ of (1.2) that have limits $a(x_2)$, $b(x_2)$, as $x_1 \to \pm\infty$, for all x_2.

From Chap. 4 and on, we restrict ourselves to *minimal solutions* (which we also call *minimizers*[1]), that is, solutions of (1.2) that minimize $J_{\Omega'}$ subject to their own Dirichlet data on $\partial\Omega'$, for any open, bounded, Lipschitz $\Omega' \subset \Omega$ (for example,

---

[1] The term *minimizing minimal* has also been adopted for surfaces by X. Cabré, to avoid confusion with (local) minimizers and with critical points, and it could have been adopted in our setting.

solutions obtained via (1.10)). For bounded minimal solutions $u : \mathbb{R}^n \to \mathbb{R}^m$, provided $W \geq 0$ vanishes at least at a point, we have the basic estimate

$$J_{B_r}(u) \leq C r^{n-1}, \quad r > 0. \tag{1.11}$$

Furthermore, we restrict ourselves to potentials W with the property that the zero set $\{W = 0\}$ contains an isolated point $a \in \mathbb{R}^m$, but quite arbitrary otherwise. This automatically excludes the Ginzburg-Landau and the segregation potentials. Based then on the minimality of u or u_R, we derive certain general estimates that in specific problems yield information on the structure of the map u or u_R. Most of the key estimates in this book are derived from minimality, and are developed in Chap. 5. More on this later.

In Chap. 4, we establish the following maximum principle type result for minimizers (cf. [6]). We introduce the following assumptions:

1. $W \in C(\mathbb{R}^m; \mathbb{R})$, $W(a) = 0$ for some $a \in \mathbb{R}^m$, $W \geq 0$.
2. There exists $r_0 > 0$ such that for *every* unit vector $\xi \in \mathbb{R}^m$, the function

$$(0, r_0] \ni r \mapsto W(a + r\xi) \text{ is nondecreasing, and } W(a + r_0\xi) > 0.$$

Note that this is a very weak condition that even allows for W to vanish on a ball centered at a.

Theorem 1.1 (Maximum Principle) *Let $A \subset \mathbb{R}^n$, open, bounded, and with Lipschitz boundary. Let $v(\cdot) \in W^{1,2}(A; \mathbb{R}^m) \cap L^\infty(A; \mathbb{R}^m)$ be a minimizer of $J_A(u)$ with respect to its own Dirichlet values on ∂A,*

$$J_A(v) = \min\{J_A(u), \ u = v \text{ on } \partial A\}. \tag{1.12}$$

Assume

$$|v(x) - a| \leq r \text{ on } \partial A, \quad 0 < 2r \leq r_0.$$

Then,

$$|v(x) - a| \leq r \quad \text{on } A. \tag{1.13}$$

Moreover, if $u \to W_u(u)$ is Lipschitz, then the attainment of equality in (1.13) at an interior point of A,

$$|v(\hat{x}) - a| = r, \quad \text{for some } \hat{x} \in A,$$

implies

$$v(x) \equiv \text{Const.} \quad \text{in the connected component of } \hat{x} \text{ in } A.$$

Note that the theorem above is not true if v is just a *local* minimizer in the sense that (1.12) is satisfied only for u in a neighborhood of v, for instance in the sense of the second variation. As a useful corollary we obtain the following replacement result.

Lemma 1.1 (Cut-off Lemma) *Let $u(\cdot) \in W^{1,2}(A; \mathbb{R}^m) \cap L^\infty(A; \mathbb{R}^m)$, and suppose that*

(I) $|u(x) - a| \le r$ *on* ∂A, $0 < 2r \le r_0$,
(II) $\mathscr{L}^n(A \cap \{|u(x) - a| > r\}) > 0$ ($\mathscr{L}^n(E)$, *the n-dimensional Lebesgue measure*),

then, there is $\tilde{u}(\cdot) \in W^{1,2}(A; \mathbb{R}^m) \cap L^\infty(A; \mathbb{R}^m)$ with

$$\begin{cases} \tilde{u} = u, & \text{on } \partial A, \\ |\tilde{u}(x) - a| \le r, & \text{in } A, \\ J_A(\tilde{u}) < J_A(u). \end{cases}$$

The proof of the theorem is based on the use of the *polar form* of a map u:

$$u(x) = a + q^u(x)\boldsymbol{n}^u(x), \quad \left(q^u(x) := |u(x) - a|, \ \boldsymbol{n}^u(x) := \frac{u(x) - a}{|u(x) - a|}\right), \quad (1.14)$$

for the construction of competing maps $\tilde{u}$ defined by varying only the *modulus $q^u(x)$* and keeping the *direction vector $\boldsymbol{n}^u(x)$* fixed. In these coordinates:

$$|\nabla u|^2 = |\nabla q^u|^2 + (q^u)^2 |\nabla \boldsymbol{n}^u|^2.$$

As it turns out, obtaining estimates on the scalar quantity q^u, and understanding its level sets is sufficient for controlling the vector quantity u. The polar form (1.14) is used extensively throughout this book. It appears first in the proof of the Lemmas 2.4 and 2.5, that are behind Theorem 2.2, the one-dimensional analogs of Lemma 1.1, and later in the proof of Theorem 1.2 below, and the proof of the density estimates (Theorem 1.5). Finally, the polar form is extended by replacing the (zero-dimensional solution) a with a more general lower-dimensional solution (cf. (1.43)).

We then focus on phase transition potentials $W \ge 0$, with a finite number $N \ge 1$ of distinct zeros $a_1, \ldots, a_N \in \mathbb{R}^m$, the phases. If $N > 1$, W is not convex, and a difficult problem in the analysis of the structure of the minimizers is the a priori characterization of the regions Ω_i where $u(x)$, for $|x| \gg 1$ is close to the phase a_i.

The presence of symmetries has been instrumental for handling such difficulties. Chapters 6 and 7 are dedicated to a systematic study of symmetric solutions of (1.2). In Chap. 6 we assume that a finite reflection group G acts on both the domain x-space, and the target u-space, both coinciding with $\mathbb{R}^n$. The action of G implies the existence of an open convex set $F \subset \mathbb{R}^n$, the fundamental domain [41], that satisfies:

$$\bigcup_{g \in G} g\overline{F} = \mathbb{R}^n, \quad gF \cap F = \emptyset, \quad g \in G \setminus \{I\}.$$

We assume that W is invariant under G:

$$W(gu) = W(u), \quad u \in \mathbb{R}^n, \quad g \in G,$$

and that W has a unique zero, say a_1, on $\overline{F}$. We denote by $\mathrm{Stab}(a_1) = \{g \in G : ga_1 = a_1\}$ the subgroup of G that leaves a_1 fixed and set

$$D_1 := \mathrm{Int} \left(\bigcup_{g \in \mathrm{Stab}(a_1)} g\overline{F} \right).$$

Chapter 6 is dedicated to the following existence result [5, 3, 36], in chronological order):

Theorem 1.2 (Point Group, $u : \mathbb{R}^n \to \mathbb{R}^n$) *Under the previous assumptions on W and a_1, and under smoothness and nondegeneracy hypotheses (cf. $\mathbf{H}_1$–$\mathbf{H}_3$ in Sect. 6.2), there exists a solution $u : \mathbb{R}^n \to \mathbb{R}^n$ of (1.2) which is equivariant:*

$$u(gx) = gu(x), \quad g \in G,$$

and positive in the sense that

$$u(\overline{F}) \subset \overline{F} \quad \text{(positivity).} \tag{1.15}$$

Moreover, there exist $c > 0$ and $C > 0$ such that

$$|u(x) - a_1| \leq C e^{-cd(x, \partial D_1)}, \quad x \in D_1, \tag{1.16}$$

The invariance of W implies that a_1 is the unique minimum of W in D_1 and that W has exactly $N = \frac{|G|}{|\mathrm{Stab}(a_1)|}$ minima ($|\Gamma|$ denote the order of the group Γ). The estimate (1.16) and equivariance imply that u connects the N minima of W. A comment is in order also on the positivity condition (1.15). The proof of Theorem 1.2 is variational and a key point is to show that minimizing under the positivity constraint does not affect the Euler-Lagrange equation. We show that the associated gradient flow with Neumann conditions on B_R preserves positivity (in the strong maximum principle sense), and since it reduces J_{B_R}, we conclude that positivity is a removable constraint. Once this is established, we can use equivariance and positivity to reduce the problem on D_1, where W has a single minimum. Also from $u(\overline{D}_1) \subset \overline{D}_1$ it follows that $d(u(\overline{D}_1), \{W = 0\} \setminus \{a_1\}) > 0$, which is the basis for deriving the exponential estimate (1.16) and for understanding the structure of u. We note that the equivariance requirement is not a constraint that can affect the Euler-Lagrange equation, as it is well known [53]. Finally, we mention that positivity does not seem to follow automatically from minimization, for general G and W.

An important example of Theorem 1.2 is obtained for the choice $n = m = 2$, W a triple-well potential, $\{W = 0\} = \{a_1, a_2, a_3\}$, and G the symmetry group of the equilateral triangle with vertices in a_1, a_2, a_3 (Fig. 1.3).

In the example at hand we obtain D_1, D_2, D_3, the three $120°$ sectors that partition $\mathbb{R}^2$. Moreover, the estimate in (1.16) implies that the restriction of $u(x)$ along rays emanating from the origin and contained in D_i, approaches a_i, as $|x| \to \infty$. Thus, positivity reduces the problem to a single D_i, where there is a unique minimum of W, converting it to a morally convex problem.

The other important example is obtained for $n = m = 3$, W a quadruple potential with $\{W = 0\} = \{a_1, a_2, a_3, a_4\}$, G the symmetry group of the tetrahedron with vertices in a_1, a_2, a_3, a_4. We note that the boundary of the partition in the first example is the unique singular minimal cone in $\mathbb{R}^2$, while the partition in the second example corresponds to one of the two singular minimal cones in $\mathbb{R}^3$ (see Fig. 1.4 below).

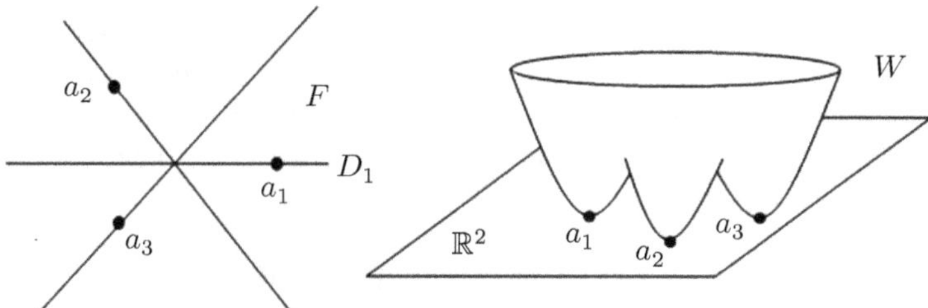

Fig. 1.3 The triple junction solution on $\mathbb{R}^2$. $|G| = 6$, $|\mathrm{Stab}(a_1)| = 2$, F is the $\frac{1}{3}\pi$ sector, D_1 the $\frac{2}{3}\pi$ sector that contains a_1

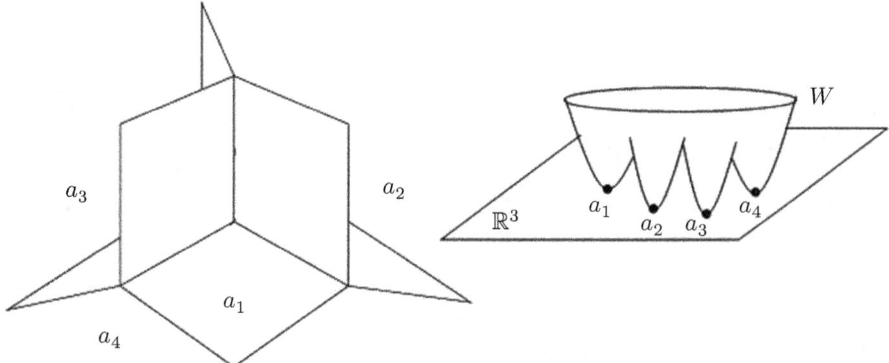

Fig. 1.4 The four junction solution on $\mathbb{R}^3$. $|G| = 24$, $|\mathrm{Stab}(a_1)| = 6$

These two examples have been worked out respectively by Bronsard et al. [22] in 1996, and by Gui and Schatzman [43] in 2008. The method of construction in these papers is based on a systematic assembling of the solution out of lower-dimensional pieces, along the lines of the asymptotic expansion. Our approach on the other hand is different. We first obtain an object via a not very hard integral estimate, and subsequently we dissect it and analyze its structure. For this purpose, pointwise estimates like (1.16) are needed, and are deduced from the general estimates in Chap. 5. In particular, the geometrical details of the symmetry group do not enter at all. These two solutions are in a sense the singular minimal cones in $\mathbb{R}^2$ and $\mathbb{R}^3$, respectively, in the diffused interface setting. We conjecture they are stable under general (not necessarily G-symmetric) perturbations.

Our purpose in Chap. 7 is to generalize Theorem 1.2, by extending the notion of equivariance and considering a different stucture on the domain and on the range. More precisely, two different reflection groups, related by a group homomorphism, may act on the domain space $\mathbb{R}^n$, and on the target space $\mathbb{R}^m$. By including also translations, we are led naturally to *discrete reflection groups*, and to the existence of *lattice solutions*. Due to the variety of choices for n, m, the reflection groups acting on the domain and the target space, and the corresponding homomorphism, we deduce the existence of various complex symmetric structures. The solutions of interest are those mapping a fundamental domain F in the domain space into a fundamental domain Φ in the target space. To ensure this property, which we call *positivity*, the homomorphism between the two reflection groups has to be chosen appropriately (cf. Sect. 7.3). We refer to these homomorphisms as *positive*. In the case of Theorem 1.2, we establish positivity by reducing the problem to a half-space determined by a reflection plane. Now, in our general set up including also discrete reflection groups, one has to deal with the fundamental domain all at once [9, 62]. According to the structure considered on the domain space (finite or discrete reflection group), we obtain the following existence results [18]:

Theorem 1.3 (Point Group, $u : \mathbb{R}^n \to \mathbb{R}^m$) *Assume that*

- *W is invariant with respect to a reflection point group Γ, and the closure of the fundamental domain $\overline{\Phi}$ contains a unique zero of W, say a_1;*
- *there exist: a finite reflection group G acting on $\mathbb{R}^n$, and a positive homomorphism $f : G \to \Gamma$ (cf. Definition 7.1) that associates Φ with the fundamental domain F of G,*

Then there exists an f-equivariant solution u of (1.2), $u(gx) = f(g)u(x)$, for $g \in G$, which is positive, and connects the phases at infinity:

$$u(\overline{F}) \subset \overline{\Phi} \quad \text{(positivity)}, \tag{1.17}$$

$$|u(x) - a_1| \leq Ce^{-cd(x,\partial D_1)}, \quad x \in D_1, \tag{1.18}$$

where $D_1 = \mathrm{Int}\left(\bigcup_{g \in f^{-1}(\mathrm{Stab}(a_1))} g\overline{F}\right)$.

In the case where a discrete reflection group acts on the domain space, we give a slightly different version of the theorem. Since the fundamental domain of a discrete reflection group is bounded or has a cylindrical structure, the exponential estimate applies when the corresponding lattice blows up. By rescaling, this is equivalent to multiplying the gradient of the potential in (1.2) by a factor R^2, and consider the lattice in the domain space as fixed.

Theorem 1.4 (Lattice) *Assume that*

- *W is invariant with respect to a reflection point group Γ, and that the closure of the fundamental domain $\overline{\Phi}$ contains a unique zero of W, say a_1;*
- *there exist: a discrete reflection group G acting on $\mathbb{R}^n$, and a positive homomorphism $f : G \to \Gamma$ (cf. Definition 7.1) that associates Φ with the fundamental domain F of G.*

Then there exists, for every $R > R_0$, an f-equivariant solution $u_R : \mathbb{R}^n \to \mathbb{R}^m$ $(u_R(gx) = f(g)u_R(x)$, for $g \in G)$ to system

$$\Delta u_R - R^2 W_u(u_R) = 0, \quad \text{for } x \in \mathbb{R}^n, \tag{1.19}$$

which is positive, and connects the phases:

$$u(\overline{F}) \subset \overline{\Phi} \quad \text{(positivity)}, \tag{1.20}$$

$$|u_R(x) - a_1| \le Ce^{-cRd(x,\partial D_1)}, \quad x \in D_1, \tag{1.21}$$

where $D_1 = \text{Int}\left(\bigcup_{g \in f^{-1}(\text{Stab}(a_1))} g\overline{F}\right)$.

However, to deal with a general, not necessarily symmetric potential W, there is no established general approach yet. A potentially useful tool should be the blow-down process via the change of variables (cf. Alberti [1])

$$x = \frac{y}{\epsilon} = Ry, \quad v_\epsilon(y) = u_\epsilon(x), \tag{1.22}$$

that provides a reformulation of (1.2) in the unit ball B_1 and relates the asymptotic behavior $|x| \to \infty$ to the limit of $v_\epsilon(y)$ as $\epsilon \to 0$. The minimization of J_{B_R} is replaced by the minimization of the rescaled functional

$$J_\epsilon(v) = \int_{B_1} \left(\frac{\epsilon}{2}|\nabla_y v|^2 + \frac{1}{\epsilon}W(v)\right)dy, \tag{1.23}$$

with an appropriate Dirichlet constraint. The additional benefit from this reformulation is that (by Baldo [16]), $J_\epsilon(v)$ Γ-converges to the weighted perimeter functional

$$\sum_{i \neq j}^{N} \sigma_{ij} \mathcal{H}^{n-1}(\partial D_i \cap \partial D_j), \tag{1.24}$$

where σ_{ij} are the so called *surface tension coefficients* determined by the *actions* of the *connections* between a_i and a_j (Chap. 2 deals exclusively with the action functional).

Moreover,

$$v_\epsilon(y) \to v_0(y) \text{ in } L^1(B_1; \mathbb{R}^m), \tag{1.25}$$

with v_0 being a minimizer of the perimeter functional (1.24), $v_0(y) \in \{a_1, \ldots, a_N\}$ a.e. in B_1,

$$v_0(y) = \sum_{j=1}^{N} a_j \mathbb{1}_{D_j^*}, \tag{1.26}$$

where $D_1^*, \ldots, D_N^*$ is a minimal partition of B_1, with prescribed Dirichlet conditions. This relationship between the diffuse and the sharp interface problem is sometimes called *linking*, and has been studied extensively since De Giorgi introduced Γ-convergence, beginning with Modica and Mortola [51], and also in the more general setting of gradient flows.

It is natural to try to pull back this information to v_ϵ and consequently to u_R, and finally to the limit map u. This is a delicate step, since L^1 convergence is too weak for establishing a quantified correspondence between v_0 and u_R. An important tool which could be useful also in this direction is provided by the *density* estimates derived first by Caffarelli and Cordoba [24] for the scalar case; these are based on the following analogy of surface and volume: $A(r) = \int_{B_r \cap \{|u-a| \leq \lambda\}} W(u)du$, $V(r) = \mathcal{L}^n(B_r \cap \{|u-a| > \lambda\})$. A large part of Chap. 5 is devoted to vector analogs of these estimates and their corollaries. We describe now some of its contents. A first estimate of this kind can be stated as follows for bounded minimal solutions of (1.2) [7]:

Theorem 1.5 (Density Estimate) *Assume that $u : \Omega \to \mathbb{R}^m$ $\Omega \subset \mathbb{R}^n$, open is a minimizer. Let $W : \mathbb{R}^m \to \mathbb{R}$ be continuous and nonnegative, with $a \in \{W = 0\}$ an isolated nondegenerate zero of W. Let $\lambda \in (0, \min_\xi\{|a - \xi| : \xi \neq a, \ W(\xi) = 0\}) = (0, d_0)$. Then the condition*

$$\mathcal{L}^n(B_{r_0}(x_0) \cap \{|u - a| > \lambda\}) \geq \mu_0 > 0, \tag{1.27}$$

for some $r_0 > 0$, implies that for all $r \geq r_0$ with $B_r(x_0) \subset \Omega$, the estimate

$$\mathcal{L}^n(B_r(x_0) \cap \{|u - a| > \lambda\}) \geq c^* r^n \tag{1.28}$$

holds for some constant $c^* = c^*(\mu_0, \lambda) > 0$.

The admissible W's are continuous and quite general otherwise, as in Fig. 1.5 below, with the important requirement that a is isolated in the zero set of W.

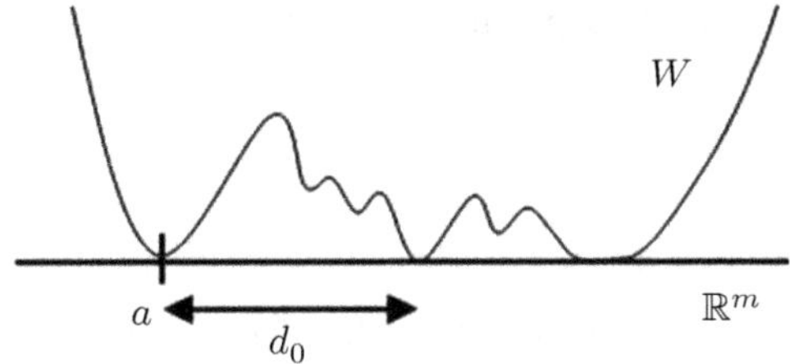

Fig. 1.5 The potential W in the density theorem

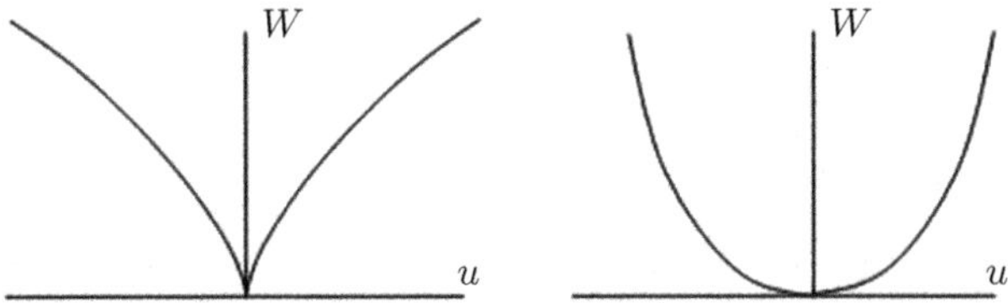

Fig. 1.6 Nondegenerate behavior near a: $W \in C^\alpha$, with $0 < \alpha \leq 1$ on the left, and $1 < \alpha \leq 2$ on the right

This condition excludes for example the Ginzburg-Landau potentials. We require W near a to be as in Fig. 1.6 ($W \in C^\alpha \alpha \in (0, 2)$, free boundary case), and in the smooth case $\alpha = 2$ (reaction-diffusion case) we assume that the Hessian matrix $W_{uu}(a)$ of W at a is positive definite.

The point of departure in our proof is the polar form. Otherwise, we follow quite closely as far as the essentials are concerned the argument in [24], including an improvement from Valdinoci [71].

For phase transition potentials, a consequence of the density estimate above is a *refinement of linking* (1.25), that gives a tighter relationship between the *sharp interface* ∂P, namely the jump set

$$\partial P = \bigcup_{i,j=1, i\neq j}^{N} (\partial D_i \cap \partial D_j) \tag{1.29}$$

of v_0, and the *diffuse interface* $I_{\gamma,\epsilon}$ of v_ϵ,

$$I_{\gamma,\epsilon} = \{y \in B_1 : \min_{j=1,\dots,N} |v_\epsilon(y) - a_j| \geq \gamma\}, \quad \gamma \in (0, \min_{i\neq j} |a_i - a_j|). \tag{1.30}$$

Indeed, from (1.28), we can derive the uniform convergence of the diffuse interface to the sharp interface (on compacts in B_1):

$$\lim_{\epsilon \to 0^+} \max\{d(y, \partial P) : y \in I_{\gamma,\epsilon}\} = 0,$$

together with an estimate. Thus, given $\delta > 0$ there is an $\epsilon_\delta > 0$ such that

$$0 < \epsilon \leq \epsilon_\delta \quad \Longrightarrow \quad |v_\epsilon - a_j| < \gamma \ \text{ on } \ D_{j,\delta} \qquad (1.31)$$

where

$$D_{j,\delta} = \{y \in D_j : d(y, \partial P) > \delta\}.$$

Under the assumption that ∂P is a minimal cone, which is true in certain important cases, this information on v_ϵ translates into an analogous statement on the structure of u_R:

$$R \geq \frac{1}{\epsilon_\delta} \quad \Longrightarrow \quad |u_R - a_j| < \lambda \ \text{ on } \ RD_{j,\delta}. \qquad (1.32)$$

This is the basis for further analysis of the minimizer u_R, aiming at establishing that u is a map tending at ∞ along different directions to different global minima a_i. The case of bistable W's is quite special since by (1.11), on most of the set where $u(x)$ is far from a_1, necessarily it has to be close to a_2. Thus (1.27), (1.28) hold with reversed inequalities $|u_R - a_i| \leq \lambda$ as well ($i = 1, 2$).

In certain situations it so happens that the restriction of a minimizer u to a certain region A is bounded away from all but one of the zeros of W, say a:

$$d(u(A), \{W = 0\} \setminus \{a\}) = \delta > 0. \qquad (1.33)$$

When this is the case we expect that minimality forces u near a in the center of any sufficiently large ball contained in A. This was proved in [35] for nonnegative C^2 potentials that satisfy a mild nondegeneracy condition at a: for some $r_0 > 0$ and for all unit vectors $\xi \in \mathbb{R}^m$, the map $(0, r_0) \ni r \mapsto W(a + r\xi)$ has a positive second derivative. This condition allows for C^∞ contact of W with 0 at a.

Theorem 1.6 (Pointwise Estimate) *Assume W and a are as before and let u a minimizer that satisfies (1.33). Then, given $\lambda > 0$, there is an $r(\lambda) > 0$ such that*

$$B_{r(\lambda)}(x_0) \subset A \quad \Longrightarrow \quad |u(x_0) - a| < \lambda. \qquad (1.34)$$

In other words, if A contains a sufficiently large ball, then at its center x_0 the distance of u from a is less than any preassigned $\lambda > 0$.

Theorem 1.6 follows from Theorem 1.5. Indeed, suppose for the sake of contradiction that $|u(x_0) - a| > \lambda$. Then, by continuity, $\mathscr{L}^n(B_{r_0}(x_0) \cap \{|u - a| > \lambda\}) > 0$, and so $\mathscr{L}^n(B_r(x_0) \cap \{|u - a| > \lambda\}) \geq c^* r^n, r \geq r_0$. This and (1.33) imply

$$\int_{B_r(x_0)} W(u)\mathrm{d}x \geq cr^n, \quad \text{ for some } c > 0,$$

as long as $B_r(x_0) \subset A$, which is in contradiction with the basic estimate (1.11), for r large enough. The basic estimate is established in this book for $W \in C^2$, and in Alikakos and Zarnescu [10] for $0 \leq \alpha < 2$.

The pointwise estimate (1.34) is very useful in various circumstances. For example, in the derivation of estimate (1.16) symmetry and positivity are utilized to show that (1.33) holds. Indeed, (1.15) implies $u(\overline{D}_1) \subset \overline{D}_1$, and since a_1 is the unique zero of W in $\overline{D}_1$, (1.34) applies with $A = \overline{D}_1$. The exponential estimate then follows by linear theory, since W is convex near its minima.

Statement (1.34) can produce also Liouville type theorems. For example, let u be an entire bounded minimal solution of (1.2), and suppose that $u(x)$ is bounded away from all the zeros of W, except possibly a. Then, by (1.34), necessarily $u \equiv a$. On the other hand, if u instead is defined on the upper half-space $\{x_n \geq 0\}$ and a is nondegenerate, it satisfies the estimate $|u(x) - a| \leq Ke^{-kx_n}$.

Another interesting application of (1.34) for phase transition potentials concerns the set

$$I_\gamma := \{x \in \mathbb{R}^n : \min_{j=1,\ldots,N} |u(x) - a_j| \geq \gamma\}$$

that, for small $\gamma > 0$, can be regarded as the diffuse interface for a minimizer $u : \mathbb{R}^n \to \mathbb{R}^m$. Combining the pointwise estimate (1.34) with the Maximum Principle we obtain that, for $n \geq 2$, I_γ is either unbounded or empty, in which case $u \equiv a$ for some $a \in \{a_1, \ldots, a_N\}$. Indeed, I_γ contained in a ball B_r implies that u is near some $a \in \{a_1, \ldots, a_N\}$ in $\mathbb{R}^n \setminus B_r$ and (1.34) implies that u converges to a as $|x| \to +\infty$. This and Theorem 1.1 yield $u \equiv a$.

We also present the original direct proof of (1.34) based on the idea that, if a minimizer u violates (1.34), then it is possible to deform u into a suitable comparison map v with less energy. We do this deformation in two steps. Set $\bar{q} = \frac{\lambda}{2}$ and observe that $q^u(x_0) = |u(x_0) - a| \geq \lambda$ implies $q^u(x) \geq \bar{q}$ on $B_{r_0}(x_0)$ for some $r_0 > 0$. With $\eta > 0$ a fixed number, we set $r_h = r_0 + 2h\eta$ for $h = 0, 1, 2, \ldots$. We first spend a certain amount of energy J_h^- to transform u into a map $\tilde{u}$ that satisfies the condition

$$q^{\tilde{u}} \leq \bar{q} \quad \text{on } B_{r_h+\eta}(x_0), \tag{1.35}$$

and coincides with u outside $B_{r_h+2\eta}(x_0)$. Then we gain a quantity J_h^+ of energy by exploiting (1.35), which allows to transform $\tilde{u}$ into a map v that satisfies

$$q^v(x_0) < \bar{q}$$

and coincides with $\tilde{u}$ outside $B_{r_h+\eta}(x_0)$. The point is to show that the inequality $J_h^+ \leq J_h^-$ can hold only for a finite number of values of h. For this we derive quantitative estimates for $J_h^\pm$:

$$k\mathscr{L}^n\left(B_{r_h} \cap \{q^u \geq \bar{q}\}\right) \leq J_h^+, \quad J_h^- \leq K\mathscr{L}^n\left((B_{r_h+2\eta} \setminus B_{r_h}) \cap \{q^u \geq \bar{q}\}\right). \tag{1.36}$$

If in (1.36) we replace the comma between J_h^+ and J_h^- with $\leq$, we obtain an inequality in which the left-hand side grows exponentially in h while the right-hand side only algebraically, and therefore we get $J_h^+ > J_h^-$ for h sufficiently large.

This is how originally the proof of Theorem 1.6 under optimal hypotheses was given in [35], evolving out of [5, 36] and [3]. The proof in [35] applies also to degenerate W's.

In Sect. 5.7 we derive the lower bound

$$\int_{B_r(x_0)} W(u)\mathrm{d}x \geq Cr^{n-1}, \quad \forall r \geq r(x_0), \tag{1.37}$$

with $C > 0$ independent of x_0. Inequality (1.37) is sharp in light of (1.11), and is derived under the hypotheses that u is an entire nonconstant minimizer in $W^{1,2}(\mathbb{R}^n; \mathbb{R}^m) \cap L^\infty(\mathbb{R}^n; \mathbb{R}^m)$, and W is a phase transition potential, $\{W = 0\} = \{a_1, \ldots, a_N\}$. For $n = 2$, it was derived in [65].

The results in Chap. 5, including the density estimates, are analogous to corresponding results in minimal surface theory. The relationship is suggested by the blow-down argument (1.23)–(1.26) above. However, generally, they are not derived by reduction to the geometric setting.

As mentioned above, estimate (1.16) implies that $u(x)$ converges to a_i along rays emanating from the center, and contained entirely in the ith compartment of the partition. The a_i's are zero-dimensional solutions of (1.2). In analogy with the structure theorems of singular minimal cones, particularly Federer's dimension reduction principle and Almgren's stratification theorem [28, 72], one expects that if instead we choose a direction parallel to one of the walls of the fundamental domain F, then $u(x)$ ought to converge, as $|x| \to \infty$, to a one-dimensional solution of (1.2) (a connection), and so on. That is, we expect the limits along certain special directions ξ:

$$\lim_{\lambda \to +\infty} u(x' + \lambda\xi) = \hat{u}(x'), \quad x' \perp \xi, \tag{1.38}$$

to produce a hierarchy of solutions of varying dimension. Such asymptotic information was built in the two constructed solutions in [22] and [43], already mentioned above. In Chap. 8 we take up this programme in the equivariant setting. We describe now some of its contents.

For describing solutions $u : \mathbb{R}^n \to \mathbb{R}^m$, which in a subspace $\mathbb{R}^d \subset \mathbb{R}^n = \mathbb{R}^d \times \mathbb{R}^{n-d}$ have a well-defined asymptotic behavior, it is natural to represent them as maps $\tilde{u} : \mathbb{R}^{n-d} \to W_\#^{1,2}(\mathbb{R}^d; \mathbb{R}^m)$, where $W_\#^{1,2}(\mathbb{R}^d; \mathbb{R}^m)$ is a suitable subspace of $W_{\mathrm{loc}}^{1,2}(\mathbb{R}^n; \mathbb{R}^m)$, and $\tilde{u}$ is defined by

$$\tilde{u}(y)(s) = u((s, y)), \quad x = (s, y), \ s \in \mathbb{R}^d, \ y \in \mathbb{R}^{n-d}.$$

The interest in this point of view is that solutions $e_1, \ldots, e_N \in W_\#^{1,2}(\mathbb{R}^d; \mathbb{R}^m)$ of (1.2) can play the role of the minima $a_1, \ldots, a_N$. For example, for $d = 1$, $e(\cdot)$

is a connection between two minima of W. To state the simplest result of this kind, take $d = 1$ and W symmetric with respect to a plane, and with two minima a^+ and a^-, and take u equivariant with respect to this symmetry. The analog of (1.11) in the present context is

$$0 \leq \int_{C_r(y_0)} \left(\frac{1}{2}|\nabla u|^2 + W(u) - J_{\mathbb{R}}(e)\right)dx \leq Cr^{n-2}, \tag{1.39}$$

where $C_r(y_0)$ is the cylinder $\mathbb{R} \times B_r^{n-1}(y_0)$ with center $y_0 \in \mathbb{R}^{n-1}$. Taking into account that the connection e minimizes the action $J_{\mathbb{R}}$ among a certain class of curves in $\mathbb{R}^m$ connecting the two minima, which we denote by E^{xp}, we define the *effective potential* $\mathscr{W} : \mathrm{E}^{\mathrm{xp}} \to \mathbb{R}$ by setting

$$\mathscr{W}(v(\cdot)) = J_{\mathbb{R}}(v(\cdot)) - J_{\mathbb{R}}(e(\cdot)) \geq 0. \tag{1.40}$$

In this context, under the assumption that $e(\cdot)$ is a nondegenerate minimizer of $J_{\mathbb{R}}(v)$ in E^{xp}, hence an isolated minimizer of $\mathscr{W}$, we establish the density estimate

$$\mathscr{L}^{n-1}(B_1^{n-1}(y_0) \cap \{y : \|u(\cdot, y) - e(\cdot)\|_{L^2(\mathbb{R})} \geq \lambda\}) \geq \mu_0 > 0$$

$$\implies \mathscr{L}^{n-1}(B_r^{n-1}(y_0) \cap \{y : \|u(\cdot, y) - e(\cdot)\|_{L^2(\mathbb{R})} \geq \lambda\}) \geq Cr^{n-1}, \tag{1.41}$$

for $r \geq 1$, $\lambda > 0$ appropriately small, and a constant $C = C(\mu_0, \lambda) > 0$. This is the simplest possible theorem of its kind. It is based on the (second-order) analogy of surface and volume

$$\mathscr{A}(r) = \int_{B_r^{n-1}(y^0) \cap \{y : \|u(\cdot,y)-e(\cdot)\|_{L^2(\mathbb{R})} \leq \lambda\}} \mathscr{W}(u(\cdot, y))dy, \tag{1.42a}$$

$$\mathscr{V}(r) = \mathscr{L}^{n-1}(B_r^{n-1}(y^0) \cap \{y : \|u(\cdot, y) - e(\cdot)\|_{L^2(\mathbb{R})} > \lambda\}). \tag{1.42b}$$

Its proof is based on the *polar form* of u with respect to e,

$$u(\cdot, y) = e(\cdot) + q^u(y)\boldsymbol{n}^u(\cdot, y),$$

$$q^u(y) = \|u(\cdot, y) - e(\cdot)\|_{L^2(\mathbb{R})}, \quad \boldsymbol{n}^u(\cdot, y) = \frac{u(\cdot, y) - e(\cdot)}{\|u(\cdot, y) - e(\cdot)\|_{L^2(\mathbb{R})}}, \tag{1.43}$$

where (1.39) takes the form

$$0 \leq \int_{B_r^{n-1}(y_0)} \left\{\frac{1}{2}\left(|\nabla q^u|^2 + (q^u)^2 \sum_{i=1}^{n-1} \|\boldsymbol{n}_{y_i}^u\|_{L^2(\mathbb{R})}^2\right) + \mathscr{W}(u)\right\}dy \leq Cr^{n-2}, \tag{1.44}$$

and minimality over cylinders plays the role of minimality over balls in the proof of Theorem 1.5. From this, under the hypothesis of uniqueness for e, one can establish that a minimizer $u : \mathbb{R}^n \to \mathbb{R}^m$ symmetric in the plane $\{x_1 = 0\}$ is necessarily one-dimensional:

$$u(\cdot, y) = e(\cdot), \quad y \in \mathbb{R}^{n-1}.$$

These results were obtained in [7].

To establish the limits in (1.38), we need to localize the density estimate (1.41) in a suitable way in domains obtained as intersections of half spaces. The cylinders in (1.39) now have finite height.

We now describe the simplest result which requires localization of the density estimate. We consider the upper half plane, $\Omega = \{x_n \geq 0\}$, and decompose it into two sets with disjoint interiors, $\Omega = \Omega_I \bigcup \Omega_{II}$, where $\Omega_I = \{x_n \geq |x_1|\}$ (cf. Fig. 8.3). Set $x = (s, y)$, $s = x_1$, $y = (x_2, \ldots, x_n)$ and for $r \in (0, y_{n-1})$ denote by $C_r(y)$ the cylinder with cross section $B_r^{n-1}(y)$ and height $2l_r^y$, $l_r^y = y_{n-1} - r$. Note that $C_r(y)$ is contained in Ω_I and touches its boundary.

Theorem 1.7 *Under the hypotheses of smoothness and nondegeneracy on W, symmetry with respect to the hyperplane $x_1 = 0$, minimality for $u : \Omega_I \to \mathbb{R}^m$, $n \geq 2$, and hyperbolicity for e, there is $\lambda^* > 0$ such that for any $\mu_0 > 0$ and $\lambda \in (0, \lambda^*)$, there exists $l^* = l^*(\lambda)$ so that the conditions $l_1^{y^0} > l^*$ and*

$$\mathscr{L}^{n-1}(B_1^{n-1}(y^0) \cap \{y : \|u(\cdot, y) - e(\cdot)\|_{l_1^{y^0}} \geq \lambda\}) \geq \mu_0 > 0,$$

imply

$$\mathscr{L}^{n-1}(B_r^{n-1}(y^0) \cap \{y : \|u(\cdot, y) - e(\cdot)\|_{l_r^{y^0}} \geq \lambda\}) \geq Cr^{n-1}, \quad 1 \leq r \leq y_{n-1}^0 - l^*,$$

where $C = C(\mu_0, \lambda) > 0$ is independent of y^0 [8].

Corollary 1.1 *Let $\Omega = \{x_n \geq 0\}$, and assume the hypotheses in Theorem 1.7 above, and moreover assume that the connection e is unique. Then, there exist $\bar{k}, \bar{K}$, positive constants, such that*

$$|u(x) - e(x_1)| \leq \bar{K}e^{-\bar{k}x_n}, \quad x = (x_1, \ldots, x_n) \in \Omega.$$

The proof of the corollary is inductive. In Ω_{II} it follows from (1.18), while in Ω_I it is based on Theorem 1.7.

Theorem 1.7 has not been published before, although it does utilize several ingredients from [8]. The replacement of the infinite cylinders with finite ones of increasing length introduces an exponentially small term that can be absorbed in the scheme of the proof of the exponential estimates. This fact appears to be useful also in handling mass constraints and other situations. By replacing $e(x_1)$

with $u_{\mathrm{tr}}(x_1, x_2)$, $u_{\mathrm{tr}} : \mathbb{R}^2 \to \mathbb{R}^3$, equivariant with respect to the symmetries of the equilateral triangle, minimal, and hyperbolic, we obtain a higher-dimensional analog of the results above:

$$|u(x_1, x_2, x_3) - u_{\mathrm{tr}}(x_1, x_2)| \leq K \mathrm{e}^{-kd(x, \partial \Omega)}. \tag{1.45}$$

Here $u : \mathbb{R}^3 \to \mathbb{R}^3$ is the tetrahedral solution provided by Theorem 1.2, and $u_{\mathrm{tr}} : \mathbb{R}^2 \to \mathbb{R}^3$ is the triod solution provided by Theorem 1.3. If u_{tr} is assumed also unique in its symmetry class, then we show that any $u : \mathbb{R}^3 \to \mathbb{R}^3$, equivariant with respect to the equilateral triangle symmetries, and minimal has to coincide with u_{tr}:

$$u(x_1, x_2, x_3) = u_{\mathrm{tr}}(x_1, x_2).$$

This is a reduction of variables result in the spirit of De Giorgi's conjecture [30]. The stratification results in Chap. 8 were obtained in [8]. Our presentation is different and is based on the localization of (1.41) and its higher dimensional generalization.

Chapter 9 is devoted to minimal solutions $u : \mathbb{R}^2 \to \mathbb{R}^m$ of (1.2), without symmetry hypotheses, and is based on [37]. We assume that $W : \mathbb{R}^m \to \mathbb{R}$ has two nondegenerate zeros $a^- \neq a^+$ and $W > 0$ on $\mathbb{R}^m \setminus \{a^-, a^+\}$. We assume the existence of $N \geq 1$ distinct minimal orbits connecting a^- to a^+, represented by maps $\bar{u}_j : \mathbb{R} \to \mathbb{R}^m$, $j = 1, \ldots, N$ which are nondegenerate in the sense that 0 is a simple eigenvalue of the operator $T : W^{2,2}(\mathbb{R}; \mathbb{R}^m) \to L^2(\mathbb{R}; \mathbb{R}^m)$

$$Tv = -v'' + W_{uu}(\bar{u})v, \quad \bar{u} = \bar{u}_j, \quad j = 1, \ldots, N.$$

For the case $N = 2$ we present a proof, see Theorem 9.3, of Schatzman's theorem [59] concerning the existence of a double heteroclinic solution of (1.2): that is a solution $u : \mathbb{R}^2 \to \mathbb{R}^m$ that satisfies

$$\lim_{y \to \pm\infty} u(x, y) = a^{\pm}, \tag{1.46}$$

$$\lim_{x \to \pm\infty} u(x, y) = \bar{u}_{\pm}(y - \eta_{\pm}),$$

where $\eta_{\pm} \in \mathbb{R}$ specify to what particular element of the manifold of the translates of $\bar{u}_{\pm}$ the map $u(x, \cdot)$ is converging to. The constants $\eta_{\pm}$ are extra unknowns of the problem that must be determined together with the map u itself. This is a marked difference with the problem considered in Chap. 8, where the restriction to the symmetry context automatically implies $\eta_{\pm} = 0$. To our knowledge, the double heteroclinic solution constructed by Schatzman is the only known non-symmetric vector minimizer for the case of phase transition potentials. Problem (1.46) occurs naturally in the context of phase transitions, and shows up when two kind of interfaces are present (cf. [11, Fig. 1]). Our proof of Theorem 9.3 differs substantially from the original proof.

We observe that the problem of the existence of solutions of (1.2) that satisfy (1.46) is somehow analogous to the connection problem discussed in Chap. 2. In this analogy, the minima $a^\pm$ of the potential $W : \mathbb{R}^m \to \mathbb{R}$ correspond to the minima $\bar{u}_\pm$ of the effective potential $\mathscr{W}(u) = J_\mathbb{R}(u) - J_\mathbb{R}(\bar{u}_\pm)$ defined on $\mathscr{H} = \bar{v} + W^{1,2}(\mathbb{R}; \mathbb{R}^m)$ (with $\bar{v} : \mathbb{R} \to \mathbb{R}^m$ a map with the same asymptotic behavior as $\bar{u}_\pm$). Continuing with the analogy, we reinterpret the double heteroclinic solution $u : \mathbb{R}^2 \to \mathbb{R}^m$ as a map $\mathbb{R} \ni x \mapsto u(x, \cdot) \in \mathscr{H}$ that satisfies

$$\lim_{x \to \pm\infty} u(x, \cdot) = \bar{u}_\pm(\cdot - \eta_\pm),$$

which corresponds to the condition

$$\lim_{x \to \pm\infty} u(x) = a^\pm$$

imposed to the connecting map $u : \mathbb{R} \to \mathbb{R}^m$ in Chap. 2. To stress this analogy, in Chap. 9, we assign to the variable x the role of independent variable, in contrast to our discussion in Chap. 8. By means of Lemma 1.1 and by exploiting the Hamiltonian identities discussed in Chap. 4, we derive a representation formula for the energy density which allows for an extension to the infinite-dimensional setting of some of the arguments in the proofs of Lemmas 2.4 and 2.5. Theorem 9.3 is complemented by Theorem 9.1 stating that, under the above assumptions, any minimizer $u : \mathbb{R}^2 \to \mathbb{R}^m$ that satisfies $(1.46)_1$ uniformly in x must be a double heteroclinic solution, and by Theorem 9.2 that concerns the case $N = 1$ where, as in the scalar case $m = 1$, there is a unique orbit connecting a^- to a^+. Theorem 9.2 states that in this case u is one-dimensional:

$$u(x, y) = \bar{u}(y - \eta),$$

for some $\eta \in \mathbb{R}$.

Chapters 2, 3, 4, 5, and 9 do not require any symmetry hypotheses. Also they are self-contained and can be read independently from the rest of the book. Chapters 6, 7, and 8 require symmetry and depend on Chap. 5.

Acknowledgements N. D. A. would like to thank Alex Freire for his drawing of the tetrahedral cone in Fig. 1.2.

References

1. Alberti, G.: Variational models for phase transitions, an approach via Gamma convergence. In: Ambrosio, L., Dancer, N. (eds.) Calculus of Variations and Partial Differential Equations, pp. 95–114. Springer, Berlin (2000)
2. Alikakos, N.D.: Some basic facts on the system $\Delta u - \nabla W(u) = 0$. Proc. Am. Math. Soc. **139**, 153–162 (2011)

3. Alikakos, N.D.: A new proof for the existence of an equivariant entire solution connecting the minima of the potential for the system $\Delta u - W_u(u) = 0$. Commun. Partial Differ. Equ. **37**(12), 2093–2115 (2012)

4. Alikakos, N.D., Fusco, G.: On the connection problem for potentials with several global minima. Indiana Univ. Math. J. **57**, 1871–1906 (2008)

5. Alikakos, N.D., Fusco, G.: Entire solutions to equivariant elliptic systems with variational structure. Arch. Ration. Mech. Anal. **202**(2), 567–597 (2011)

6. Alikakos, N.D., Fusco, G.: A maximum principle for systems with variational structure and an application to standing waves. J. Eur. Math. Soc. **17**(7), 1547–1567 (2015)

7. Alikakos, N.D., Fusco, G.: Density estimates for vector minimizers and application. Discrete Contin. Dyn. Syst. **35**(12), 5631–5663 (2015). Special issue edited by E. Valdinoci

8. Alikakos, N.D., Fusco, G.: Asymptotic behavior and rigidity results for symmetric solutions of the elliptic system $\Delta u = W_u(u)$. Annali della Scuola Normale Superiore di Pisa **XV**(special issue), 809–836 (2016)

9. Alikakos, N.D., Smyrnelis, P.: Existence of lattice solutions to semilinear elliptic systems with periodic potential. Electron. J. Differ. Equ. **15**, 1–15 (2012)

10. Alikakos, N.D., Zarnescu, A.: (In preparation)

11. Alikakos, N.D., Betelú, S.I., Chen, X.: Explicit stationary solutions in multiple well dynamics and non-uniqueness of interfacial energies. Eur. J. Appl. Math. **17**, 525–556 (2006)

12. Alikakos, N.D., Katzourakis, N.: Heteroclinic travelling waves of gradient diffusion systems. Trans. Am. Math. Soc. **363**, 1362–1397 (2011)

13. Ambrosio, L., Cabré, X.: Entire solutions of semilinear elliptic equations in $\mathbb{R}^3$ and a conjecture of De Giorgi. J. Am. Math. Soc. **13**, 725–739 (2000)

14. André, N., Shafrir, I.: On a vector-valued singular perturbation problem on the sphere. In: Proceedings of the International Conference on Nonlinear Analysis, Recent advances in nonlinear Analysis, pp. 11–42. World Scientific Publishing, Singapore (2008)

15. Antonopoulos, P., Smyrnelis, P.: On minimizers of the Hamiltonian system $u'' = \nabla W(u)$, and on the existence of heteroclinic, homoclinic and periodic orbits. Indiana Univ. Math. J. **65**(5), 1503–1524 (2016)

16. Baldo, S.: Minimal interface criterion for phase transitions in mixtures of Cahn-Hilliard fluids. Ann. Inst. Henri Poincaré **7**(2), 67–90 (1990)

17. Barlow, M.T., Bass, R.F, Gui, C.: The Liouville property and a conjecture of De Giorgi. Commun. Pure Appl. Math. **53**, 1007–1038 (2000)

18. Bates, P.W., Fusco, G., Smyrnelis, P.: Multiphase solutions to the vector Allen-Cahn equation: crystalline and other complex symmetric structures. Arch. Ration. Mech. Anal. **225**(2), 685–715 (2017)

19. Berestycki, H., Caffarelli, L., Nirenberg, L.: Further qualitative properties for elliptic equations in unbounded domains. Ann. Scuola Norm. Sup. Pisa Cl. Sci. **25**(1–2), 69–94 (1997)

20. Berestycki, H., Hamel, F., Monneau, R.: One dimensional symmetry of bounded entire solutions of some elliptic equations. Duke Math. J. **103**(3), 375–396 (2000)

21. Bethuel, F., Brezis, H., Helein, F.: Ginzburg-Landau Vortices. Progress in Nonlinear Differential Equations and Their Applications, vol. 13. Birkhäuser, Basel (1994)

22. Bronsard, L., Gui, C., Schatzman, M.: A three-layered minimizer in $\mathbb{R}^2$ for a variational problem with a symmetric three-well potential. Commun. Pure. Appl. Math. **49**(7), 677–715 (1996)

23. Cabré, X., Terra, J.: Saddle-shaped solutions of bistable diffusion equations in all of $\mathbb{R}^{2m}$. J. Eur. Math. Soc. **11**, 819–943 (2009)

24. Caffarelli, L., Córdoba, A.: Uniform convergence of a singular perturbation problem. Commun. Pure Appl. Math. **48**, 1–12 (1995)

25. Caffarelli, L., Córdoba, A.: Phase transitions: uniform regularity of the intermediate layers. J. Reine Angew. Math. **593**, 209–235 (2006)

26. Caffarelli, L., Lin, F.H.: Singularly perturbed elliptic systems and multi-valued harmonic functions with free boundaries. J. Am. Math. Soc. **21**(3), 847–862 (2008)

27. Caffarelli, L., Karakhanyan, A.L., Lin, F.H.: The geometry of solutions to a segregation problem for nondivergence systems. J. Fixed Point Theory Appl. **5**, 319–351 (2009)
28. Colding, T.H., Minicozzi, W.P.: A Course in Minimal Surfaces. Graduate Studies in Mathematics, vol. 121. AMS, Providence (2011)
29. De Figueirdo, D.G., Magalhaes, C.A.: On nonquadratic Hamiltonian elliptic systems. Adv. Differ. Equ. **1**(5), 881–898 (1996)
30. De Giorgi, E.: Convergence problems for functionals and operators. In: Proceedings of the International Meeting on Recent Methods in Nonlinear Analysis, Rome (1978), pp. 131–188. Pitagora, Bologna (1979)
31. del Pino, M., Kowalczyk, M., Wei, J.: On De Giorgi's conjecture in dimension $N \geq 9$. Ann. Math. **174**, 1485–1569 (2011)
32. Farina, A.: Symmetry for solutions of semilinear elliptic equations in $\mathbb{R}^N$ and related conjectures. Ricerche Mat. **10**(Suppl 48), 129–154 (1999)
33. Farina, A.: Two results on entire solutions of Ginzburg–Landau system in higher dimensions. J. Funct. Anal. **214**(2), 386–395 (2004)
34. Farina, A., Valdinoci, E.: The state of art for a conjecture of De Giorgi and related questions. Reaction-diffusion systems and viscosity solutions. In: Recent Progress on Rection-Diffusion Systems and Viscosity Solutions, pp. 74–96. World Scientific, Singapore (2008)
35. Fusco, G.: On some elementary properties of vector minimizers of the Allen-Cahn energy. Commun. Pure Appl. Anal. **13**(3), 1045–1060 (2014)
36. Fusco, G.: Equivariant entire solutions to the elliptic system $\Delta u - W_u(u) = 0$ for general G-invariant potentials. Calc. Var. Partial Differ. Equ. **49**(3), 963–985 (2014)
37. Fusco, G.: Layered solutions to the vector Allen-Cahn equation in $\mathbb{R}^2$, minimizers and heteroclinic connections. Commun. Pure Appl. Anal. **16**(5), 1807–1841 (2017)
38. Fusco, G., Gronchi, G.F., Novaga, M.: On the existence of connecting orbits for critical values of the energy. J. Differ. Equ. **263**, 8848–8872 (2017)
39. Fusco, G., Gronchi, G.F., Novaga, M.: On the existence of heteroclinic connections. Sao Paulo J. Math. Sci. **12**, 1–14 (2017)
40. Ghoussoub, N., Gui, C.: On a conjecture of De Giorgi and some related problems. Math. Ann. **311**, 481–491 (1998)
41. Grove, L.C., Benson, C.T.: Finite Reflection Groups. Graduate Texts in Mathematics, vol. 99, 2nd edn. Springer, Berlin (1985)
42. Gui, C.: Hamiltonian identities for partial differential equations. J. Funct. Anal. **254**(4), 904–933 (2008)
43. Gui, C., Schatzman, M.: Symmetric quadruple phase transitions. Indiana Univ. Math. J. **57**(2), 781–836 (2008)
44. Heinze, S.: Travelling waves for semilinear parabolic partial differential equations in cylindrical domains. PhD thesis, Heidelberg University (1988)
45. Jerison, D., Monneau, R.: Towards a counter-example to a conjecture of De Giorgi in high dimensions. Ann. Mat. Pura. Appl. **183**, 439–467 (2004)
46. Landau, L.D., Lifschitz, E.M.: Course of Theoretical Physics. Classical Field Theory, vol. 2, 4th edn. Butterworth-Heinemann, Oxford (1980)
47. Lin, F., Pan, X. B., Wang, C.: Phase transition for potentials of high-dimensional wells. Commun. Pure Appl. Math. **65**(6), 833–888 (2012)
48. Lucia, M., Muratov, C., Novaga, M.: Existence of traveling wave solutions for Ginzburg-Landau-type problems in infinite cylinders. Arch. Ration. Mech. Anal. **188**(3), 475–508 (2008)
49. Modica, L.: The gradient theory of phase transitions and its minimal interface criterion. Arch. Ration. Mech. Anal. **98**(2), 123–142 (1987)
50. Modica, L.: Monotonicity of the energy for entire solutions of semilinear elliptic equations. In: Colombini, F., Marino, A., Modica, L. (eds.) Partial Differential Equations and the Calculus of Variations, Essays in Honor of Ennio De Giorgi, vol. 2, pp. 843–850. Birkhäuser, Boston (1989)
51. Modica, L., Mortola, S.: Un esempio di Γ-convergenza. Boll. Unione. Mat. Ital. Sez B **14**, 285–299 (1977)

52. Monteil, A., Santambrogio, F.: Metric methods for heteroclinic connections. Math. Methods Appl. Sci. **41**(3), 1019–1024 (2018)
53. Palais, R.S.: The principle of symmetric criticality. Commun. Math. Phys. **69**(1), 19–30 (1979)
54. Porter, D.A., Easterling, K.E.: Phase Transformations in Metals and Alloys, 2nd edn. Chapman and Hall, London (1996)
55. Rabinowitz, P.H.: Periodic and heteroclinic orbits for a periodic hamiltonian system. Ann. Inst. Henri Poincaré **6**(5), 331–346 (1989)
56. Risler, R.E.: Global convergence towards travelling fronts in nonlinear parabolic systems with a gradient structure. Ann. Inst. Henri Poincaré Anal. Non Linear **25**(2), 381–424 (2008)
57. Sandier, E., Serfaty, S.: Vortices in the Magnetic Ginzburg-Landau Model. Progress in Nonlinear Differential Equations and Their Applications, vol. 70. Birkhäuser, Basel (2007)
58. Savin, O.: Minimal Surfaces and Minimizers of the Ginzburg Landau energy. In: Contemporary Mathematics Mechanical Analysis AMS, vol. 526, pp. 43–58. American Mathematical Society, Providence (2010)
59. Schatzman, M.: Asymmetric heteroclinic double layers. Control Optim. Calc. Var. **8**, 965–1005 (2002). A tribute to J. L. Lions (electronic)
60. Schoen, R.: Lecture Notes on General Relativity. Stanford University, Stanford (2009)
61. Serrin, J., Zou, H.: The existence of positive entire solutions of elliptic Hamiltonian systems. Commun. Partial Differ. Equ. **23**, 577–599 (1998)
62. Smyrnelis, P.: Solutions to elliptic systems with mixed boundary conditions. Phd thesis (2012)
63. Souplet, P.: The proof of the Lane-Emden conjecture in four space dimensions. Adv. Math. **221**, 1409–1427 (2009)
64. Sourdis, C.: The heteroclinic connection problem for general double-well potentials. Mediterr. J. Math. **13**, 4693–4710 (2016)
65. Sourdis, C.: Optimal energy growth lower bounds for a class of solutions to the vectorial Allen-Cahn Equation. Math. Methods Appl. Sci. **41**(3), 966–972 (2018)
66. Sternberg, P.: The effect of a singular perturbation on nonconvex variational problems. Arch. Ration. Mech. Anal. **101**(3), 209–260 (1988)
67. Sternberg, P.: Vector-valued local minimizers of nonconvex variational problems. Rocky Mountain J. Math. **21**, 799–807 (1991)
68. Sternberg, P., Zuniga, A.: On the heteroclinic problem for multi-well gradient systems. J. Differ. Equ. **261**, 3987–4007 (2016)
69. Taylor, J.E.: The structure of singularities in soap-bubble-like and soap-film-like minimal surfaces. Ann. Math. **103**, 489–539 (1976)
70. Terman, D.: Infinitely many traveling wave solutions of a gradient system. Trans. Am. Math. Soc. **301**(2), 537–556 (1987)
71. Valdinoci, E.: Plane-like minimizers in periodic media: jet flows and Ginzburg-Landau-type functionals. J. Reine Angew. Math. **574**, 147–185 (2004)
72. White, B.: Topics in geometric measure theory. Lecture notes (taken by O. Chodosh), Stanford (2012)

Chapter 2
Connections

Abstract We begin by giving a concise proof of the existence of a heteroclinic connection (Theorem 2.1). The experienced reader then can move on to Sect. 2.6. In Sect. 2.4 we develop an alternative approach via constrained minimization. Most readers will find this easier and also good preparation for the polar form and the cut-off lemma in Chap. 4. In Sect. 2.6 we consider the connection problem for an unbalanced double-well potential, and handle it via the constrained method. Finally in Sect. 2.7 we investigate the failure of the existence of a connection when three or more global minima are present.

2.1 Motivation

To motivate the study of connections and of other solutions of (1.2) that we consider in this and other chapters, we focus on what can be regarded as the simplest model for the free energy of a substance that can exist in $N \geq 1$ equally preferred *phases*: the Allen-Cahn functional

$$J_{\Omega}(u) = \int_{\Omega} \left(\frac{\epsilon^2}{2} |\nabla u|^2 + W(u) \right) dx,$$

where $\Omega \subset \mathbb{R}^n$ is the region containing the material and $W : \mathbb{R}^m \to \mathbb{R}$ is a potential that models the bulk free energy of the substance and is assumed to satisfy

$$0 = W(a_j) < W(u), \quad \text{for } u \in \mathbb{R}^m \setminus \{a_1, \ldots, a_N\}, \tag{2.1}$$

for some $a_1, \ldots, a_N \in \mathbb{R}^m$ that represent the pure phases: the N different equally preferred states in which the substance can exist at thermodynamical equilibrium. For $0 < \epsilon \ll 1$, the term $\frac{\epsilon^2}{2}|\nabla u|^2$ penalizes high gradients of the map $u : \Omega \to \mathbb{R}^m$ and models the interfacial energy. Given an initial state u_0, the simplest model for

© Springer Nature Switzerland AG 2018

N. D. Alikakos et al., *Elliptic Systems of Phase Transition Type*,
Progress in Nonlinear Differential Equations and Their Applications 91,
https://doi.org/10.1007/978-3-319-90572-3_2

the evolution of the system toward a final stationary state is the L^2 gradient system associated to (2.1), that is, the parabolic Allen-Cahn equation

$$\begin{cases} u_t = \epsilon^2 \Delta u - W_u(u), & x \in \Omega \\ \dfrac{\partial}{\partial \nu} u = 0, & x \in \partial \Omega, \\ u(0, \cdot) = u_0. \end{cases} \tag{2.2}$$

For small $\epsilon > 0$ we can distinguish different behaviors in the dynamics of (2.2). In an initial time interval of $O(1)$ the evolution is essentially dictated by the O.D.E. $u_t = -W_u(u)$, which, depending on the structure of u_0, evolves the solution $u^\epsilon(t, x, u_0)$ towards the set of the minima of W, and as a result Ω is partitioned into subregions where u^ϵ is approximately constant and equal to one of the a_j. These subregions are separated by an interface of thickness $O(\epsilon)$ across which u^ϵ makes a transition from the one to the other of the minima of W associated to the neighboring subregions. At the end of this first period, the so called separation stage, u^ϵ develops high gradients and the two terms on the right-hand side of (2.2) become comparable and a second, slower phase of dynamics begins, the so called coarsening stage, during which u^ϵ keeps its partitioned structure and the boundaries of the subregions evolve with speed of $O(\epsilon^2)$. The method of matched asymptotic expansion can be used to get some insight into the structure of u^ϵ during this second period of evolution. If we focus on a point of the interface that separates two subregions, we find that the profile of u^ϵ across the interface is approximately determined by

$$u^\epsilon(x) \approx \bar{u}\left(\frac{d(x)}{\epsilon}\right),$$

where $d(x)$ is the signed distance of x from the interface and $\bar{u} : \mathbb{R} \to \mathbb{R}^m$ is a solution of the problem

$$u'' = W_u(u),$$
$$\lim_{s \to \pm\infty} \bar{u}(s) = a^\pm, \tag{2.3}$$

with $a^\pm \in \{a_1, \ldots, a_N\}$ the values corresponding to the two neighboring subregions. The next step is the description of the structure of u^ϵ in a neighborhood of a point p of the interface where three or more subregions come together. Considering for simplicity the case of three subregions and assuming that $\Omega \subset \mathbb{R}^2$ and that $p = 0$, we find

$$u^\epsilon(x_1, x_2) \approx \tilde{u}\left(\frac{x_1}{\epsilon}, \frac{x_2}{\epsilon}\right),$$

where $\tilde{u} : \mathbb{R}^2 \to \mathbb{R}^m$ is a solution of (1.2) that satisfies

$$\lim_{r \to +\infty} \tilde{u}(r v_{1,2} + s v_{1,2}^\top) = \bar{u}_{1,2}(s),$$

$$\lim_{r \to +\infty} \tilde{u}(r v_{2,3} + s v_{2,3}^\top) = \bar{u}_{2,3}(s), \tag{2.4}$$

$$\lim_{r \to +\infty} \tilde{u}(r v_{3,1} + s v_{3,1}^\top) = \bar{u}_{3,1}(s),$$

uniformly in compacts of $s \in \mathbb{R}$. In (2.4) $v_{1,2}$, $v_{2,3}$, $v_{3,1}$ are suitable unit vectors and $\bar{u}_{i,j}$ is a solution of (2.3) with a_i and a_j in place of $a^\pm$ and a_1, a_2, a_3 are the minima of W associated to the subregions that meet at p.

In this chapter we present a systematic study of problem (2.3). The problem of determining a solution $\tilde{u}$ of (1.2) that satisfies (2.4) and other similar problems are discussed in a symmetry context in Chaps. 6, 7, and 8.

2.2 The Hamilton and Jacobi Principles

We begin by explaining the two variational principles that are the main tools for constructing connections, and the relationship between them. Consider Newton's equation for a unit mass and *potential energy* $-W$ (see Fig. 2.1):

$$u'' - W_u(u) = 0, \quad u : \mathbb{R} \to \mathbb{R}^m, \tag{2.5}$$

where $W : \mathbb{R}^m \to \mathbb{R}$ is a C^2 function, $W_u(u) := (\partial W/\partial u_1, \ldots, \partial W/\partial u_m)^\top$, $x \in \mathbb{R}$ stands for time, $T = \frac{1}{2}|u'|^2$ the *kinetic energy*, $H = T - W$ the *Hamiltonian*, or total mechanical energy, and $L = T + W$ the *Lagrangian*. The functional

$$J_{(x_1,x_2)}(u) := \int_{x_1}^{x_2} \left\{ \frac{1}{2}|u'|^2 + W(u) \right\} dx, \tag{2.6}$$

is called the *action*. Equation (2.5) preserves H along solutions.

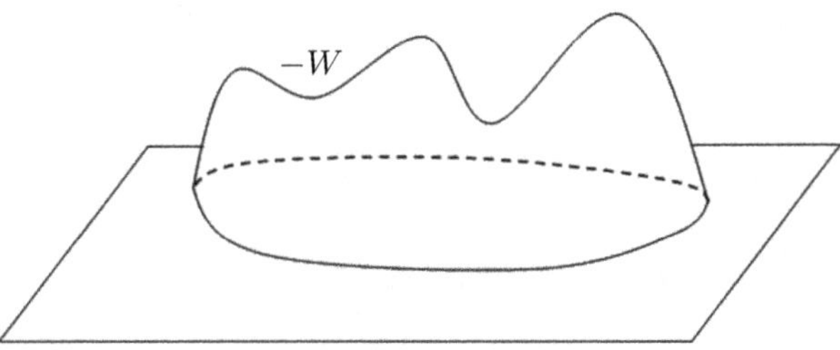

Fig. 2.1 The potential $-W$

In case there is friction, the equation is modified to

$$u'' - W_u(u) = -cu', \quad c > 0, \tag{2.7}$$

and now

$$\frac{\mathrm{d}}{\mathrm{d}x} H = -c|u'|^2 \leq 0. \tag{2.8}$$

Hamilton's principle or the least action principle stated loosely says that global minimizers of J in the class of u's satisfying $u(x_1) = u_1$, $u(x_2) = u_2$ solve (2.5). Connections are global minimizers that connect picks of $-W$ of equal height. Traveling wave solutions satisfy (2.7), and connect picks of $-W$ of different height.

Jacobi's principle deals with curves and detects geodesics. In the context of connections, it can be stated as follows. Given Γ, a C^1 curve in $\mathbb{R}^m$, $\Gamma = \{u(x) : x \in [x_1, x_2], \ u(x_i) = u_i, \ |u'(x)| \neq 0\}$, one minimizes the length functional

$$L(u) := \sqrt{2} \int_{x_1}^{x_2} \sqrt{W(u)} \, |u'(x)| \mathrm{d}x. \tag{2.9}$$

In this subsection, we will assume $W > 0$ on $\mathbb{R}^m \setminus \{a^+, a^-\}$. We are interested in connecting the states $u_1 = a^-$, $u_2 = a^+$, $W(a^\pm) = 0$ and take $(x_1, x_2) = \mathbb{R}$. We have the two functionals:

$$L(u) = \sqrt{2} \int_{\mathbb{R}} \sqrt{W(u(x))} \, |u'(x)| \mathrm{d}x, \tag{2.10}$$

$$J(u) = \int_{\mathbb{R}} \left\{ \frac{1}{2}|u'|^2 + W(u) \right\} \mathrm{d}x, \tag{2.11}$$

in the class of functions

$$X_{a^- a^+} = \{u \in W_{\mathrm{loc}}^{1,2}(\mathbb{R}; \mathbb{R}^m) : \lim_{x \to \pm\infty} u(x) = a^\pm\}. \tag{2.12}$$

We note that L is invariant under the group of orientation preserving diffeomorphisms $\psi : \mathbb{R} \to \mathbb{R}$, $\psi' > 0$, that is,

$$L(u \circ \psi) = L(u). \tag{2.13}$$

Lemma 2.1 $L(u) \leq J(u)$, *with equality if and only if* u *is* equipartitioned, *that is,*

$$\frac{1}{2}|u'|^2 = W(u). \tag{2.14}$$

Fig. 2.2 $E(u_t) \geq L(u_t)$

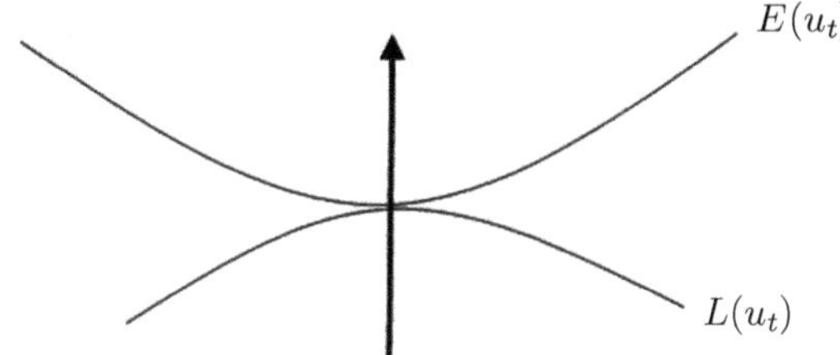

Proof $\sqrt{2W(u)}\,|u'| \leq \frac{1}{2}\big(\sqrt{2W(u)}\,\big)^2 + \frac{1}{2}|u'|^2.$ □

Corollary 2.1 *If $u_0 \in W^{1,2}_{\mathrm{loc}}(\mathbb{R};\mathbb{R}^m)$ is minimal in the sense that*

$$J_{(\alpha,\beta)}(u) \leq J_{(\alpha,\beta)}(u + v),$$

for every $v \in W^{1,2}_0([\alpha,\beta];\mathbb{R}^m)$ and every interval $[\alpha,\beta] \subset \mathbb{R}$, and if u_0 is equipartitioned,[1] then u_0 is a critical point of L (see Fig. 2.2).

Proof Let u_t be a perturbation of u_0 for $|t|$ small. Then, $\phi(t) := J(u_t) - L(u_t) \geq 0$, $\phi(0) = J(u_0) - L(u_0) = 0$ (since u_0 is equipartitioned), $\phi'(0) = 0$, and
$\left(\frac{\mathrm{d}}{\mathrm{d}t}\right)_{t=0} J(u_t) = 0 \Rightarrow \left(\frac{\mathrm{d}}{\mathrm{d}t}\right)_{t=0} L(u_t) = 0.$ □

Proposition 2.1 *Let $t \mapsto u(t)$ be a C^1 parametrization, with $|u'(t)| \neq 0$, $u(t) \neq a^{\pm}$, and $u \in X_{a^- a^+}$. Then there is an orientation-preserving diffeomorphism $t = \phi(x)$, such that $u \circ \phi$ is equipartitioned, and $u \circ \phi \in X_{a^- a^+}$. The diffeomorphism ϕ is called an equipartition parametrization.*

Proof Define

$$x(t) = \int_0^t \frac{|u'(\tau)|}{\sqrt{2W(u(\tau))}}\,\mathrm{d}\tau, \quad t \in \mathbb{R}. \tag{2.15}$$

Then, $\frac{\mathrm{d}x}{\mathrm{d}t} > 0$, hence $t \mapsto x(t)$ is invertible. Let $t = \phi(x)$ and set $v(x) = (u \circ \phi)(x)$. We have

$$\frac{1}{2}\left|\frac{\mathrm{d}v}{\mathrm{d}x}\right|^2 = \frac{1}{2}\left|\frac{\mathrm{d}u}{\mathrm{d}t}\frac{\mathrm{d}\phi}{\mathrm{d}x}\right|^2 = \frac{1}{2}\left|\frac{\mathrm{d}u}{\mathrm{d}t}\frac{\sqrt{2W(u(t))}}{|u'(t)|}\right|^2 = W(u(\phi(x))) = W(v(x)).$$

Hence v is equipartitioned. To show that $x \to \pm\infty$ as $t \to \pm\infty$ in (2.15), we use the inequality

$$W(u) \leq c|u - a|^2$$

---

[1] If u_0 is minimal and bounded, then it satisfies the equipartition relation (cf. Theorem 2.3).

which holds for u in a neighborhood of $a \in \{a^+, a^-\}$, and for some constant $c > 0$. We have for $t \gg 1$,

$$\frac{|u'(t)|}{\sqrt{2W(u(t))}} \geq \frac{|u'(t)|}{\sqrt{2c}|u(t) - a^+|} = \frac{|u'(t)||u(t) - a^+|}{\sqrt{2c}|u(t) - a^+|^2} \geq -\frac{\frac{d}{dt}|u(t) - a^+|^2}{2\sqrt{2c}|u(t) - a^+|^2}.$$

Thus for T large

$$\int_T^t \frac{|u'(\tau)|}{\sqrt{2W(u(\tau))}} d\tau \geq \frac{-1}{\sqrt{8c}} \int_T^t \frac{\frac{d}{d\tau}|u(\tau) - a^+|^2}{|u(\tau) - a^+|^2} d\tau = \frac{-1}{\sqrt{8c}} \ln\left(\frac{|u(t) - a^+|^2}{|u(T) - a^+|^2}\right) \to +\infty,$$

as $t \to +\infty$. $\qquad\qquad\qquad\qquad\qquad\qquad\qquad\qquad\qquad\qquad\qquad\qquad\qquad\qquad\qquad\qquad\square$

Proposition 2.2 *Let $L_{a^-a^+} = \inf\{L(u) : u \in X_{a^-a^+}\}$, and $J_{a^-a^+} = \inf\{J(u) : u \in X_{a^-a^+}\}$. Then we have $L_{a^-a^+} = J_{a^-a^+}$.*

Proof That $L_{a^-a^+} \leq J_{a^-a^+}$ is immediate from Lemma 2.1. Let $u \in X_{a^-a^+}$ such that $L(u) < L_{a^-a^+} + \epsilon$, with $\epsilon > 0$. Assuming that $u(\cdot)$ is C^1, $|u'(t)| \neq 0$, and $u(t) \neq a^\pm$, there is, by Proposition 2.1, an equipartition parametrization ϕ such that $u \circ \phi \in X_{a^-a^+}$. Hence $u \circ \phi$ is equipartitioned. On the one hand we have $L(u \circ \phi) = L(u)$, and on the other hand we have, by Lemma 2.1, $L(u \circ \phi) = J(u \circ \phi)$. Therefore,

$$J_{a^-a^+} \leq J(u \circ \phi) = L(u \circ \phi) = L(u) < L_{a^-a^+} + \epsilon.$$

In the general case $u \in X_{a^-a^+}$ may be nonsmooth and may have the set $\{u' = 0\}$ of positive measure. Given $\eta > 0$ small, let $l_+ := \min\{t : |u(t) - a^+| \leq \eta\}$, $l_- := \max\{t : |u(t) - a^-| \leq \eta\}$ and set

$$u_\eta(t) = \begin{cases} a^- & \text{for } t \in (-\infty, l_-], \\ (-t + l_-)a^- + (t - l_- + 1)u(l_-) & \text{for } t \in (l_- - 1, l_-), \\ u(t) & \text{for } t \in [l_-, l_+], \\ (1 - t + l_+)u(l_+) + (t - l_+)a^+ & \text{for } t \in (l_+, l_+ + 1), \\ a^+ & \text{for } t \in [l_+ + 1, \infty). \end{cases}$$

Then we have that $J_{(-\infty, l_-] \cup [l_+, \infty)}(u_\eta) \leq \eta^2 + 2\max_{|u - a^\pm| \leq \eta} W(u) \to 0$ as $\eta \to 0$. Thus we deduce the existence of a map $u_\epsilon = u_{\eta_\epsilon} \in X_{a^-a^+}$ satisfying

$$J_{(-\infty, l_-] \cup [l_+, \infty)}(u_\epsilon) \leq \epsilon.$$

Now, since $I = [l_-, l_+]$ is bounded, $C_c^\infty(I; \mathbb{R}^m)$ is dense in $W_0^{1,2}(I; \mathbb{R}^m)$. Moreover, since L is continuous in $W^{1,2}(I; \mathbb{R}^m)$, we can assume that u is smooth and $u \neq a^\pm$ on I. Consider the reparametrization $v : [\lambda_-, \lambda_+] \to \mathbb{R}^m$ of

$u : [l_-, l_+] \to \mathbb{R}^m$ defined by $t = \phi(x)$, where $\phi : [\lambda_-, \lambda_+] \to [l_-, l_+]$ is the inverse of the map $x = \psi(t)$ defined by

$$\psi(t) = \int_{\frac{l_-^u + l_+^u}{2}}^{t} \frac{\max\{|u'(\tau)|, \delta\}}{\sqrt{2W(u(\tau))}} d\tau, \quad t \in [l_-, l_+]. \tag{2.16}$$

Note that ϕ satisfies $\phi(0) = \frac{l_- + l_+}{2}$ and the equation

$$\phi' = \frac{\sqrt{2W(u(\phi))}}{\max\{|u'(\phi)|, \delta\}}, \quad x \in [\lambda_-, \lambda_+], \quad \lambda_\pm = \psi(l_\pm), \tag{2.17}$$

which is approximately the condition one must impose to ϕ in order to obtain an equipartitioned reparametrization v. In (2.16) and (2.17) we use the approximate expression $\max\{|u'|, \delta\}$ instead of $|u'|$ to have well-defined strictly increasing maps ψ and ϕ even when u' vanishes in a set of positive measure. From (2.17) we obtain

$$\frac{1}{2}|u'(\phi)|^2 (\phi')^2 + W(u(\phi)) - \sqrt{2W(u(\phi))}|u'(\phi)|\phi' = \gamma_\delta, \quad x \in [\lambda_-, \lambda_+], \tag{2.18}$$

where

$$\gamma_\delta = \begin{cases} 0, & \text{if } |u'| > \delta, \\ \left(\frac{\delta - |u'|}{\delta}\right)^2 W, & \text{if } |u'| \le \delta. \end{cases} \tag{2.19}$$

From (2.16) and (2.17) we obtain

$$|\{x \in [\lambda_-, \lambda_+] : |u'(\phi(x))| \le \delta\}| = \int_{\{t \in [l_-, l_+] : |u'(t)| \le \delta\}} \frac{\max\{|u'(t)|, \delta\}}{\sqrt{2W(u(t))}} dt \le C\delta, \tag{2.20}$$

where $|S|$ denotes the measure of S and $C = \frac{l_+ - l_-}{\min_{t \in [l_-, l_+]} \sqrt{2W(u(t))}}$. Therefore integrating (2.18) in $[\lambda_-, \lambda_+]$ and using that $\gamma_\delta \le 2 \max_{t \in [l_-, l_+]} W(u(t))$ yields

$$J_{[\lambda_-, \lambda_+]}(v) - L(v, [\lambda_-, \lambda_+]) = J_{[\lambda_-, \lambda_+]}(v) - L(u, [l_-, l_+]) \le C\delta, \tag{2.21}$$

with $C > 0$ independent of δ. Now extend $v = u_\epsilon(\phi)$ from $[\lambda_-, \lambda_+]$ to $\mathbb{R}$ by setting

$$\phi(x) = \begin{cases} x - \lambda_- + l_- & \text{for } x \le \lambda_-, \\ x - \lambda_+ + l_+ & \text{for } x \ge \lambda_+. \end{cases} \tag{2.22}$$

The map v so extended belong to $X_{a^- a^+}$, and satisfies:

$$J_{\mathbb{R}}(v) \le L(u, [l_-, l_+]) + C\delta + J_{(-\infty, l_-] \cup [l_+, \infty)}(u_\epsilon) \le L_{a^- a^+} + 3\epsilon,$$

for $\delta > 0$ small enough. Therefore we conclude that $J_{a^-a^+} \leq L_{a^-a^+} + 3\epsilon$, and the proof is complete. $\square$

Remark 2.1 It follows from Proposition 2.2 that if $t \mapsto u(t)$ is a minimizer of L in $X_{a^-a^+}$, then $v = u \circ \phi$ is a minimizer of J in $X_{a^-a^+}$, ϕ an equipartition parametrization hence it satisfies

$$v'' - W_u(v) = 0, \quad v(\pm\infty) = a^{\pm}, \tag{2.23}$$

that is, v is a *connection*. Now, given a diffeomorphism

$$\xi : \mathbb{R} \to \mathbb{R}, \quad \xi' > 0, \quad \lim_{\tau \to \pm\infty} \xi(\tau) = \pm\infty, \tag{2.24}$$

$\hat{u}(\tau) := (u \circ \xi)(\tau)$ (where $t = \xi(\tau)$), is also a minimizer of L. Then,

$$\hat{v}(x) := (\hat{u} \circ \hat{\phi})(x) \tag{2.25}$$

will also be a connection where $\hat{\phi}$ is defined below by inverting $t \mapsto \hat{x}(t)$. It is natural to ask about the relationship between $\hat{v}$ and v. It is not hard to see that $\hat{v}(x) = v(x - c)$ for some constant c. In other words, the group of diffeomorphisms for the length functional manifests itself in the action functional as the group of translations. Both are sources of noncompactness that cause difficulty in establishing that the infima in Proposition 2.2 above are actually minima. Here is the calculation relating the two groups:

$$\hat{x}(t) = \int_0^t \frac{|\hat{u}'(\tau)|}{\sqrt{2W(\hat{u}(\tau))}} d\tau = \int_0^t \frac{|u'(\xi(\tau))|}{\sqrt{2W(\hat{u}(\tau))}} \xi'(\tau) d\tau = \int_{\xi(0)}^{\xi(t)} \frac{|u'(\theta)|}{\sqrt{2W(u(\theta))}} d\theta$$

$$= \int_{\xi(0)}^0 \frac{|u'(\theta)|}{\sqrt{2W(u(\theta))}} d\theta + \int_0^{\xi(t)} \frac{|u'(\theta)|}{\sqrt{2W(u(\theta))}} d\theta. \tag{2.26}$$

By inverting $t \mapsto \hat{x}(t)$, $t = \hat{\phi}(x)$, we obtain $\hat{v}$ in (2.25). We now compose this with the map we obtain by inverting

$$\hat{x}_0(t) = \int_0^{\xi(t)} \frac{|u'(\theta)|}{\sqrt{2W(u(\theta))}} d\theta = \hat{x}(t) + k. \tag{2.27}$$

Hence if $\hat{\phi}_0$ is the inverse of $t \mapsto \hat{x}_0(t)$, we have

$$\hat{\phi}_0(y) = \hat{\phi}(y - k).$$

Thus

$$\hat{v}_0(x) := (\hat{u} \circ \hat{\phi}_0)(x) = \hat{v}(x - k).$$

Next, consider $\hat{v}_0(x) = (\hat{u} \circ \hat{\phi}_0)(x)$. Since $\hat{x}_0 = (x \circ \xi)$, we can see that

$$\hat{\phi}_0 = \hat{x}_0^{-1} = \xi^{-1} \circ x^{-1} = \xi^{-1} \circ \phi,$$

and

$$\hat{v}_0 = \hat{u} \circ \hat{\phi}_0 = (u \circ \xi) \circ \hat{\phi}_0 = u \circ \xi \circ \xi^{-1} \circ \phi = v.$$

Remark 2.2 There is a close analogy with the classical Plateau problem [7]. We recall that in that set-up the roles of the *length* functional and the *action* functional are played by the *area* functional A and the *energy* functional J. There as well $A(u) \leq J(u)$. The equipartition parametrization corresponds to the isothermal parametrization, Proposition 2.1 corresponds to the uniformization theorem, the group of diffeomorphisms leaves both the length functional and the area functional invariant. Finally, the group of translations in our set-up corresponds to the Möbius group in the Plateau problem. In the Plateau problem, one deals with the minimization of J, mainly because the group of diffeomorphisms is much harder to handle. Similarly, in the construction of the connections we utilize Hamilton's principle, because the group of translations (one-dimensional) is easier to handle.

2.3   The Heteroclinic Connection Problem

Let $W : \mathbb{R}^m \to \mathbb{R}$ a smooth nonnegative function that vanishes on a finite set A, with $\#A \geq 2$, Given two distinct points $a^-, a^+ \in A$ we can ask about the existence of a solution $\bar{u} : \mathbb{R} \to \mathbb{R}^m$ of the equation

$$u'' - W_u(u) = 0, \quad x \in \mathbb{R}, \tag{2.28}$$

where $W_u(u) := (\partial W / \partial u_1, \ldots, \partial W / \partial u_m)^\top$, with the conditions

$$\lim_{x \to \pm\infty} u(x) = a^\pm. \tag{2.29}$$

If a solution $\bar{u}$ of (2.28) (2.29) does exist we say that there is a *heteroclinic connection* between a^- and a^+ and, in the mathematical theory of phase transitions, the map $\bar{u}$ describes the behavior of the order parameter across the interface separating the two phases corresponding to a^- and a^+. If x is interpreted as time, (2.28) can be seen as the Newton equation of a particle of unit mass moving in m-dimensional space under a conservative field of force of potential W. Then problem (2.28) is the same as to show that one can choose the position and velocity of the particle at time 0 in such a way that the asymptotic fate of the particle in the future and in the past are a^+ and a^- respectively. In the scalar case ($m = 1$) this observation and the analysis of orbits on the phase plane leads to the existence of connections between neighboring zeros of W. In the vector case $(m > 1)$ this shooting approach becomes significantly harder because the dimension of the phase space is $2m$. On the other hand the variational approach is still tractable, since solutions of (2.28) are, in each bounded interval (x_1, x_2), stationary points of the energy

$$J_{(x_1,x_2)}(u) = \int_{x_1}^{x_2} \left(\frac{1}{2}|u'|^2 + W(u)\right)dx. \tag{2.30}$$

While for a classical solution of Eq. (2.28) we need W to be a C^1 function, the variational problem can be formulated under the assumption that W is merely continuous. As we shall see, with W continuous it is not guaranteed that the time interval required for a minimizer to travel from a^- to a^+ will be infinite, and therefore the function space where we minimize J has to include maps defined on bounded or unbounded intervals. We plan to show that each $a^- \in A$ is connected to some other $a^+ \in A$ by minimizing J on the set of maps $u : (l_-^u, l_+^u) \to \mathbb{R}^m$ defined by

$$\mathscr{A} = \{u \in W_{\text{loc}}^{1,2}((l_-^u, l_+^u); \mathbb{R}^m) : -\infty \leq l_-^u < l_+^u \leq +\infty,$$

$$\lim_{x \to l_-^u} u(x) = a^-, \ \lim_{x \to l_+^u} u(x) \in A \setminus \{a^-\}, \ u((l_-^u, l_+^u)) \subset \mathbb{R}^m \setminus A\}. \tag{2.31}$$

Note that in (2.31) the interval (l_-^u, l_+^u) associated to u is not fixed but is free to change with u.

Without some condition on the behavior of W at infinity, a minimizer of J on $\mathscr{A}$ may not exist. The problem is that J may not be coercive on $\mathscr{A}$ in the sense that there exists a minimizing sequence $\{u_j\} \subset \mathscr{A}$ such that $\|u_j\|_{L^\infty} \to +\infty$ as $j \to +\infty$, while $J(u_j)$ remains bounded. A sufficient condition for coerciveness is

$$\liminf_{|u| \to +\infty} W(u) > 0, \tag{2.32}$$

but, as observed in [18], it is possible to allow potentials W that decay to 0 at infinity provided they satisfy the condition

h : $\sqrt{W(u)} \geq \gamma(|u|)$, for some nonnegative function $\gamma : (0, +\infty) \to \mathbb{R}$ such that $\int_0^{+\infty} \gamma(r)dr = +\infty$.

We have

Theorem 2.1 *Assume that $W : \mathbb{R}^m \to [0, \infty)$ is a continuous function that satisfies* **h**. *Then, given $a^- \in A$, there exist $a^+ \in A \setminus \{a^-\}$ and a Lipschitz continuous map $u : (l_-, l_+) \to \mathbb{R}^m$, $-\infty \leq l_- < l_+ \leq +\infty$, which minimizes $J : \mathscr{A} \to [0, +\infty]$ and satisfies*

$$\frac{1}{2}|u'|^2 - W(u) = 0, \quad \text{a.e. in } (l_-, l_+). \tag{2.33}$$

In particular

$$\text{(i)} \quad \lim_{x \to l_\pm} u(x) = a^\pm, \tag{2.34}$$

$$\text{(ii)} \quad W(u(x)) > 0, \quad x \in (l_-, l_+). \tag{2.35}$$

If W is continuously differentiable in $\mathbb{R}^m \setminus A$, then u is a classical solution of (2.28).

Before giving the proof of Theorem 2.1 we make some observations and present some related results.

Remark 2.3 From Theorem 2.1 we have that, under the assumption that $W \in C^1(\mathbb{R}^m \setminus A; \mathbb{R})$, for each $a^- \in A$ there is an orbit of (2.28) that starts in a^- and terminates in some $a^+ \in A \setminus \{a^-\}$ without any other intersection with A. It follows that there are at least $\frac{\#A}{2}$ such orbits if $\#A$ is even, and $\frac{\#A+1}{2}$ if $\#A$ is odd.

Given $a_i \neq a_j \in A$, one can show (see Proposition 2.6 below) that a sufficient condition for the existence of an orbit that connects a_i to a_j and satisfies (2.35) is $\sigma_{ij} < \sigma_{ih} + \sigma_{hj}$, for all $a_h \in A \setminus \{a_i, a_j\}$, where

$$\sigma_{ij} = \inf_{u \in \mathscr{A}_{ij}} J(u), \tag{2.36}$$

$$\mathscr{A}_{ij} = \{u \in W^{1,2}_{\text{loc}}((l^u_-, l^u_+); \mathbb{R}^m) : -\infty \leq l^u_- < l^u_+ \leq +\infty, \lim_{x \to l^u_-} u(x) = a_i, \lim_{x \to l^u_+} u(x) = a_j\}.$$

In Proposition 2.12 we establish, for a particular choice of W, that the condition above is also necessary.

In the scalar case $m = 1$ it follows from (2.33) and (2.35) that the minimizer u given by Theorem 2.1 is a solution of

$$u' = \sqrt{2W(u)} > 0, \quad x \in (l_-, l_+). \tag{2.37}$$

If a^- and a^+ are two neighboring zeros of $W \in C^1(\mathbb{R} \setminus A; \mathbb{R})$ this equation has a unique solution u that satisfies (2.34) and $u(0) = \frac{a^- + a^+}{2}$, therefore u is the minimizer in Theorem 2.1. For instance, if $W(u) = \frac{1}{2}(1 - u^2)^2$ this solution is given by $u(x) = \tanh x$, $x \in \mathbb{R}$ and satisfies $\lim_{x \to \pm\infty} u(x) = \pm 1$. Note that, if W vanishes at a point a between a^- and a^+, there is no minimizer. Indeed, any continuous function u that travels from a^- to a^+ has to assume the value a, violating (2.35).

A simple criterion for showing that the minimizer given in Theorem 2.1 satisfies $l_\pm = \pm\infty$ is given in

Proposition 2.3 *Assume there exist $c > 0$ and $r_0 > 0$ such that*

$$W(u) \leq c|u - a^+|^2, \quad \text{for } |u - a^+| \leq r_0.$$

Then $l_+ = +\infty$ and an analogous statement applies to l_-.

Proof By (2.34), there is an $x_0 \in (l_-, l_+)$ such that $|u - a^+| \le r_0$ for $x \in [x_0, l_+)$. This and the assumption on W imply

$$\frac{d}{dx}|u - a^+| \ge -|u'| = -\sqrt{2W(u)} \ge -\sqrt{2c}|u - a^+|, \quad \text{for } x \in [x_0, l_+),$$

which yields $|u - a^+| \ge |u(x_0) - a^+|e^{-\sqrt{2c}(x-x_0)}$, for $x \in [x_0, l_+)$. This is compatible with (2.34) only if $l_+ = +\infty$. $\qquad\qquad\square$

Proposition 2.4 *Assume that* $W \in C^2(\mathbb{R}^m; \mathbb{R})$ *and that the Hessian matrix* $W_{uu}(a)$ *of* W *at* a *is positive definite for* $a \in A$. *Let* u *as in Theorem* 2.1. *Then* $l_\pm = \pm\infty$ *and there are positive constants* k, K *such that*

$$|u(x) - a^+| \le Ke^{-kx} \text{ and } |u(x) - a^-| \le Ke^{+kx}, \; \forall x \in \mathbb{R}. \tag{2.38}$$

Proof That $l_\pm = \pm\infty$ follows from Proposition 2.3. To prove the exponential estimates (2.38), let $\phi(x) := |u - a^+|^2$. We recall that $\lim_{x \to +\infty} \phi(x) = 0$ and that $\phi > 0$. Using (2.33) and the assumption on $W_{uu}(a)$, we obtain

$$\phi''(x) = 2|u'(x)|^2 + 2(u(x) - a^+) \cdot W_u(u(x))$$

$$\ge 4W(u(x)) \ge 4c^2\phi(x), \text{ for } x \ge l, \tag{2.39}$$

where l, c are positive constants. It follows that, in each interval of the form (l, L), ϕ is a subsolution of $\varphi'' = 4c^2\varphi$ with Dirichlet conditions $\varphi(l) = \phi(l)$, $\varphi(L) = \phi(L)$ and therefore

$$\phi(x) \le \phi(l)\frac{\sinh 2c(L - x)}{\sinh 2c(L - l)} + \phi(L)\frac{\sinh 2c(x - l)}{\sinh 2c(L - l)}.$$

Since, for fixed $x \ge l$, this is valid for all $L > x$, passing to the limit for $L \to +\infty$ yields $\phi(x) \le \phi(l)e^{-2c(x-l)}$ and (2.38) follows. $\qquad\qquad\square$

Proposition 2.5 *Let* $W \in C^2(\mathbb{R}^m \setminus A; \mathbb{R})$ *be such that* $W_u(u) \cdot (u - a) \ge c^2|u - a|^\gamma$ *in a neighborhood of each* $a \in A$, *for some constants* $c > 0$ *and* $0 < \gamma < 2$. *Then the minimizer* u *of Theorem* 2.1 *satisfies* $l_\pm \in \mathbb{R}$.

Proof Proceeding as in the proof of Proposition 2.4, we find that $\phi(x) := |u - a^+|^2$ satisfies

$$\phi'' \ge 2c^2\phi^{\frac{\gamma}{2}}, \quad x \in [l, l_+),$$

for some $l \in (l_-, l_+)$. It follows that ϕ is convex in $[l, l_+)$. Then, $\phi \to 0$ as $x \to l_+$, and $\phi' \to 0$ as $x \to l_+$ by (2.33), imply that $\phi' < 0$ in $[l, l_+)$. By integrating the inequality $-\phi'\phi'' \geq -2c^2\phi^{\frac{\gamma}{2}}\phi'$ over the interval $[x, l^+)$, with $x \geq l$, we obtain

$$\frac{1}{2}(\phi'(x))^2 \geq \frac{4c^2}{\gamma+2}\phi^{\frac{\gamma+2}{2}}(x) \implies \frac{\phi'(x)}{\phi^{\frac{\gamma+2}{4}}(x)} \leq -\sqrt{\frac{8c^2}{\gamma+2}}.$$

Then, since $\int_l^{l^+} \dfrac{\phi'}{\phi^{\frac{\gamma+2}{4}}} = -\dfrac{4}{2-\gamma}\phi^{\frac{2-\gamma}{4}}(l)$, we deduce that $l_+ \in \mathbb{R}$. The same argument applies to l_-. $\square$

Proof (Theorem 2.1) The first observation is that J is translation invariant on $\mathscr{A}$, in the sense that

$$J(u^\lambda) = J(u), \quad \text{for } u \in \mathscr{A}, \ \lambda \in \mathbb{R},$$

where $u^\lambda = u(\cdot - \lambda) \in \mathscr{A}$. This results in a loss of compactness that manifests itself in the existence of minimizing sequences $\{u_j\} \in \mathscr{A}$ that converges in C^1_{loc} to a map u which fails to satisfy (2.34) in Theorem 2.1. For example, this happens for $m = 1$ and $W = \frac{1}{2}(1 - u^2)^2$: in this case $u = \tanh x$ is a minimizer and $\{\tanh(\cdot - j)\}$ a minimizing sequence that converges to -1. We remove this pathology by an elementary observation. Since $a \in A$ is an isolated zero of W, for small fixed $r_0 > 0$ we have

$$\min_{a \in A, |u-a|=r_0} W(u) = W_0 > 0,$$

and any map $u \in \mathscr{A}$ has to satisfy $W(u(x_0)) = W_0$ for some $x_0 \in (l^u_-, l^u_+)$. Taking $x_0 = 0$ restricts the possible translations to a compact set and removes the obstruction of noncompactness. It follows that we can assume

$$W(u(0)) = W_0, \tag{2.40}$$

and restrict J to the subset of $\mathscr{A}$ where (2.40) holds.

Given $a^- \in A$, let $\bar{a} \in A$ be such that $|a^- - \bar{a}| = \min_{a \in A \setminus \{a^-\}} |a^- - a|$, and set $\tilde{u}(x) = (1 - (x + x_0))a^- + (x + x_0)\bar{a}$, $x \in (-x_0, 1 - x_0)$, where $x_0 \in (0, 1)$ is chosen so that $W(\tilde{u}(0)) = W_0$. Then $\tilde{u} \in \mathscr{A}$, $l^{\tilde{u}}_- = -x_0$, $l^{\tilde{u}}_+ = 1 - x_0$ and

$$J(\tilde{u}) = \sigma < +\infty.$$

In the following, whenever we wish to specify that the energy is relative to some interval (x_1, x_2), we will write $J_{(x_1, x_2)}(u)$.

Next we show that there are constants $M > 0$ and $l_0 > 0$ such that each $u \in \mathscr{A}$ with

$$J(u) \leq \sigma, \tag{2.41}$$

satisfies

$$\|u\|_{L^\infty((l^u_-,l^u_+);\mathbb{R}^m)} \le M,$$
$$l^u_- \le -l_0 < l_0 \le l^u_+. \tag{2.42}$$

The L^∞ bound on u follows from **h**. Indeed, if $|u(\bar{x})| = M$ for some $\bar{x} \in (l^u_-, l^u_+)$, we have

$$\sigma \ge J_{(l^u_-,\bar{x})}(u) \ge \int_{l^u_-}^{\bar{x}} \sqrt{2W(u(x))}|u'(x)|dx \ge \sqrt{2}\int_{|a^-|}^{M} \gamma(s)ds.$$

If $l^u_+, -l^u_- = +\infty$ the existence of l_0 is obvious. If instead $l^u_- > -\infty$ and/or $l^u_+ < +\infty$, we set $d_0 = d(A, \{u : W(u) > W_0\}) > 0$ and observe that from

$$d_0 \le \int_{l^u_-}^{0} |u'(x)|dx \le |l^u_-|^{\frac{1}{2}}(\int_{l^u_-}^{0} |u'(x)|^2 dx)^{\frac{1}{2}} \le |l^u_-|^{\frac{1}{2}}(2\sigma)^{\frac{1}{2}}$$

and from the analogous inequality for $l^u_+ < +\infty$ it follows that we can take $l_0 = \frac{d_0^2}{2\sigma}$.

Let $\{u_j\} \subset \mathscr{A}$ be a minimizing sequence:

$$\lim_{j\to+\infty} J(u_j) = \inf_{u\in\mathscr{A}} J(u) := \sigma_0 \le \sigma. \tag{2.43}$$

We can assume that each u_j satisfies (2.41) and (2.42). By considering a subsequence, still denoted by $\{u_j\}$, we can also assume that there exist l^∞_-, l^∞_+ with $-\infty \le l^\infty_- \le -l_0 < l_0 \le l^\infty_+ \le +\infty$, and a continuous map $u^* : (l^\infty_-, l^\infty_+) \to \mathbb{R}^m$ such that

$$\lim_{j\to+\infty} l^{u_j}_\pm = l^\infty_\pm,$$
$$\lim_{j\to+\infty} u_j(x) = u^*(x), \quad x \in (l^\infty_-, l^\infty_+), \tag{2.44}$$

and in the last limit the convergence is uniform on bounded intervals. This follows from the Ascoli-Arzelá theorem and from (2.42) which implies that the sequence $\{u_j\}$ is equibounded and from (2.41) which yields

$$|u_j(x_1) - u_j(x_2)| \le \left|\int_{x_1}^{x_2} |u'_j(x)|dx\right| \le \sqrt{2\sigma}|x_1 - x_2|^{\frac{1}{2}}, \tag{2.45}$$

so that the sequence is also equicontinuous.

By passing to a further subsequence we can also assume that $u_j \rightharpoonup u^*$ in $W^{1,2}((l_1, l_2); \mathbb{R}^m)$ for each l_1, l_2 with $l_-^\infty < l_1 < l_2 < l_+^\infty$. This follows from (2.41), which implies

$$\frac{1}{2} \int_{l_-^{u_j}}^{l_+^{u_j}} |u_j'|^2 dx \le J(u_j) \le \sigma,$$

and from the fact that each u_j satisfies (2.42) and thus is bounded in $L^2((l_1, l_2); \mathbb{R}^m)$.

We also have

$$J(u^*, (l_-^\infty, l_+^\infty)) \le \sigma_0. \tag{2.46}$$

Indeed, from the lower semicontinuity of the norm, it follows that for each l_1, l_2 with $l_-^\infty < l_1 < l_2 < l_+^\infty$,

$$\int_{l_1}^{l_2} |u^{*\prime}|^2 dx \le \liminf_{j \to +\infty} \int_{l_1}^{l_2} |u_j'|^2 dx.$$

This and the fact that u_j converges to u^* uniformly in $[l_1, l_2]$ imply

$$J_{(l_1, l_2)}(u^*) \le \liminf_{j \to +\infty} J_{(l_1, l_2)}(u_j) \le \liminf_{j \to +\infty} J_{(l_-^{u_j}, l_+^{u_j})}(u_j) = \sigma_0.$$

Since this is valid for each $l_-^\infty < l_1 < l_2 < l_+^\infty$ the claim (2.46) follows.

Lemma 2.2 *Define* $l_-^\infty \le l_- \le -l_0 < l_0 \le l_+ \le l_+^\infty$ *by setting*

$$l_- = \inf\{x \in (l_-^\infty, 0] : u^*((x, 0]) \subset \mathbb{R}^m \setminus A\}$$
$$l_+ = \sup\{x \in [0, l_+^\infty) : u^*([0, x)) \subset \mathbb{R}^m \setminus A\}.$$

Then u^* *with* $l_\pm^{u^*} = l_\pm$ *belongs to* $\mathscr{A}$ *and is a minimizer. That is*

$$J(u^*) = \sigma_0. \tag{2.47}$$

Proof If $l_+ < +\infty$ the existence of

$$a^+ = \lim_{x \to l_+} u^*(x) \tag{2.48}$$

follows from (2.45) which implies that u^* is a $C^{0, \frac{1}{2}}$ map. The limit a^+ belongs to A. Indeed, $a^+ \notin A$ would imply the existence of $\lambda > 0$ such that, for j large enough,

$$d(u_j([l_+, l_+ + \lambda], A) \ge \frac{1}{2} d(a^+, A),$$

in contradiction with the definition of l_+. If $l_+ = +\infty$ and (2.48) does not hold, there is $\delta > 0$ and a diverging sequence $\{x_j\}$, such that

$$d(u^*(x_j), A) \geq \delta.$$

Set $W_\delta = \min_{d(u,A)=\delta} W(u) > 0$. From the uniform continuity of W in $\{|u| \leq M\}$ (M as in (2.42)) it follows that there is $l > 0$ such that

$$|W(u_1) - W(u_2)| \leq \frac{1}{2} W_\delta, \quad \text{for } |u_1 - u_2| \leq l, \ u_1, u_2 \in \{|u| \leq M\}.$$

This and (2.45) imply

$$W(u^*(x)) \geq \frac{1}{2} W_\delta, \quad x \in I_j = \left(x_j - \frac{l^2}{2\sigma}, x_j + \frac{l^2}{2\sigma}\right),$$

and, by passing to a subsequence, we can assume that the intervals I_j are disjoint. Therefore for each $L > 0$ we have

$$\sum_{x_j \leq L} \frac{l^2 W_\delta}{2\sigma} \leq \int_0^{L + \frac{l^2}{2\sigma}} W(u^*(x)) dx \leq \sigma_0,$$

which is impossible for L large. This proves that, also when $l_+ = +\infty$, there exists the limit $a^+ = \lim_{x \to +\infty} u^*(x) \in A$. To show that $a^+ \neq a^-$ we observe that $a^+ = a^-$ implies the existence of a sequence $\{x_j\} \subset [l_0, l_+]$ that satisfies

$$\lim_{j \to +\infty} x_j = l_+, \tag{2.49a}$$

$$\lim_{j \to +\infty} u_j(x_j) = a^-. \tag{2.49b}$$

Since $W(u_j(0)) = W_0$ from the uniform continuity of W in $\{|u| \leq M\}$ and (2.45) it follows that

$$W(u_j(x)) \geq \frac{1}{2} W_0 \quad \text{for } x \in (-\delta, \delta),$$

for some $\delta > 0$. Therefore, for j large,

$$J_{(l_-^{u_j}, x_j)}(u_j) \geq 2\delta W_0.$$

On the other hand from (2.49b) we have

$$J_{(x_j, l_+^{u_j})}(u_j) \geq \sigma_0 - \epsilon_j,$$

where $\epsilon_j \to 0$ as $j \to +\infty$. These inequalities contradict the minimizing character of the sequence $\{u_j\}$ and prove $a^+ \neq a^-$. We have seen that u^* with $l_\pm^{u^*} = l_\pm$

satisfies all the properties required for membership in $\mathscr{A}$. This and (2.46) show that $u^* \in \mathscr{A}$ is indeed a minimizer. The proof of the lemma is complete. $\qquad\square$

Remark 2.4 It is actually possible that $l_+ < l_+^\infty$ and/or $l_- > l_-^\infty$. Assume $W = \frac{\pi^2}{8}(1 - u^2)$ for $u \in (-1, 1)$. Then the solution of (2.37) that satifies $u(0) = 0$ is $u = \sin(\frac{\pi}{2}x)$, $x \in (-1, 1)$, and $J(u) = \frac{\pi^2}{4}$. Consider the sequence $\{u_j\}$ defined by

$$
u_j(x) = \begin{cases}
\sin(\frac{\pi}{2}x) & \text{for } x \in (-1, 1 - \epsilon_j), \\
\sin(\frac{\pi}{2}(1 - \epsilon_j)) & \text{for } x \in (1 - \epsilon_j, x_j), \\
\sin(\frac{\pi}{2}(1 - \epsilon_j + x - x_j)) & \text{for } x \in (x_j, x_j + \epsilon_j),
\end{cases}
$$

where $\epsilon_j \to 0^+$ and $x_j \to +\infty$. We have $J(u_j) = \frac{\pi^2}{4} + \frac{\pi^2}{8}(x_j - 1 + \epsilon_j)\cos^2(\frac{\pi}{2}(1 - \epsilon_j))$ and we can choose the sequence $\{x_j\}$ in such a way that $J(u_j) \to \frac{\pi^2}{4}$. Then $\{u_j\}$ is a minimizing sequence and consequently $1 = l_+ < l_+^\infty = +\infty$.

Lemma 2.3 *The map u^* satisfies (2.33) in (l_-, l_+).*

Proof Given x_0, x_1 with $l_- < x_0 < x_1 < l_+$, let $\phi : [x_0, x_1 + \xi] \to [x_0, x_1]$ be linear, with $|\xi|$ small, and let $\psi : [x_0, x_1] \to [x_0, x_1 + \xi]$ be the inverse of ϕ. Define $u_\xi : [l_-, l_+ + \xi] \to \mathbb{R}^n$ by

$$
u_\xi(x) = \begin{cases}
u^*(x) & \text{for } x \in (l_-, x_0], \\
u^*(\phi(x)) & \text{for } x \in [x_0, x_1 + \xi], \\
u^*(x - \xi) & \text{for } x \in [x_1 + \xi, l_+ + \xi).
\end{cases}
\tag{2.50}
$$

Note that $u_\xi \in \mathscr{A}$ with $l_-^{u_\xi} = l_-$ and $l_+^{u_\xi} = l_+$ if $l_+ = +\infty$, and $l_+^{u_\xi} = l_+ + \xi$ if $l_+ < +\infty$. Since u^* is a minimizer, we have

$$
\frac{d}{d\xi} J_{(l_-^{u_\xi}, l_+^{u_\xi})}(u_\xi)\Big|_{\xi=0} = 0.
\tag{2.51}
$$

From (2.50), using also the change of variables $x = \psi(s)$, it follows

$$
J_{(l_-^{u_\xi}, l_+^{u_\xi})}(u_\xi) - J_{(l_-, l_+)}(u^*)
$$

$$
= \int_{x_0}^{x_1 + \xi} \left(\frac{\phi'^2(x)}{2}|u^{*\prime}(\phi(x))|^2 + W(u^*(\phi(x))) \right)dx - \int_{x_0}^{x_1} \left(\frac{1}{2}|u^{*\prime}(x)|^2 + W(u^*(x)) \right)dx
$$

$$
= \int_{x_0}^{x_1} \left(\frac{1 - \psi'(x)}{2\psi'(x)}|u^{*\prime}(x)|^2 + (\psi'(x) - 1)W(u^*(x)) \right)dx
$$

$$
= \int_{x_0}^{x_1} \left(\frac{-\frac{\xi}{x_1 - x_0}}{2(1 + \frac{\xi}{x_1 - x_0})}|u^{*\prime}|^2 + \frac{\xi}{x_1 - x_0}W(u^*(x)) \right)dx
$$

$$
= -\frac{\xi}{x_1 - x_0} \int_{x_0}^{x_1} \left(\frac{|u^{*\prime}(x)|^2}{2(1 + \frac{\xi}{x_1 - x_0})} - W(u^*(x)) \right)dx.
$$

This and (2.51) imply

$$\int_{x_0}^{x_1} \left(\frac{1}{2} |u^{*\prime}(x)|^2 - W(u^*(x)) \right) dx = 0. \tag{2.52}$$

Since this holds for all x_0, x_1, with $l_- < x_0 < x_1 < l_+$, (2.33) follows. $\square$

Remark 2.5 Since u^* is a minimizer in the class $\mathscr{A}$, we have for each l_1, l_2 with $l_- < l_1 < l_2 < l_+$:

$$J_{(l_1,l_2)}(u^*) = \inf_{v \in W_0^{1,2}([l_1,l_2];\mathbb{R}^m)} J_{(l_1,l_2)}(u^* + v), \tag{2.53}$$

that is, u^* is *minimal* with respect to perturbations with compact support. Indeed, these perturbations coincide with u^* on $(l_-, l_+) \setminus (l_1, l_2)$, and thus belong to $\mathscr{A}$. We also point out that if v^* is a minimizer of J with prescribed boundary condition on a compact interval $[l_1, l_2]$, then, by a slight variation of the argument for the proof of (2.33), it follows that there is a constant $C \in \mathbb{R}$ such that

$$\frac{1}{2} |v^{*\prime}|^2 - W(v^*) = C, \quad \text{a.e. in } [l_1, l_2]. \tag{2.54}$$

On the basis of Lemmas 2.2 and 2.3, $u^* : (l_-, l_+) \to \mathbb{R}^m$ can be identified with the map u in Theorem 2.1. To complete the proof of Theorem 2.1, it remains to show that if W is of class C^1 in $\mathbb{R}^m \setminus A$, then u^* is a classical solution of (2.28). Since u^* satisfies (2.53), if $(l_1, l_2) \subset (l_-, l_+)$ and $w : (l_1, l_2) \to \mathbb{R}^m$ is a smooth map that satisfies $w(l_i) = 0$, $i = 1, 2$, then we have

$$0 = \frac{\mathrm{d}}{\mathrm{d}\lambda} J(u^* + \lambda w) \Big|_{\lambda=0} = \int_{l_1}^{l_2} (u^{*\prime} \cdot w' + W_u(u^*) \cdot w) \mathrm{d}x$$

$$= \int_{l_1}^{l_2} \left(u^{*\prime} - \int_{l_1}^{x} W_u(u^*(s)) \mathrm{d}s \right) \cdot w' \, \mathrm{d}x. \tag{2.55}$$

Since this is valid for all $l_- < l_1 < l_2 < l_+$ and $w' : (l_1, l_2) \to \mathbb{R}^m$ is an arbitrary map with zero average, (2.55) implies

$$u^{*\prime} = \int_{l_1}^{x} W_u(u^*(s)) \mathrm{d}s + \text{Const.}$$

The continuity of u^* and of W_u implies that the right-hand side of this equation is a map of class C^1. It follows that we can differentiate and obtain

$$u^{*\prime\prime} = W_u(u^*), \quad x \in (l_-, l_+).$$

The proof of Theorem 2.1 is complete. $\square$

With σ_{ij} defined in (2.36) we have

Proposition 2.6 *Given* $a_i \neq a_j \in A$, *a sufficient condition for the existence of an orbit that connects* a_i *to* a_j *and satisfies (2.35) is*

$$\sigma_{ij} < \sigma_{ih} + \sigma_{hj}, \quad \forall a_h \in A \setminus \{a_i, a_j\}. \tag{2.56}$$

Proof From (2.56) it follows that there is $\eta > 0$ such that

$$\sigma_{ij} < \min_{h \notin \{i,j\}} \{\sigma_{ih} + \sigma_{hj}\} - \eta. \tag{2.57}$$

Given a small number $r > 0$ and $a \in A$ there is a map $u^r : [0, \lambda_r] \to \mathbb{R}^m$ that satisfies

$$|u^r(0) - a| = r, \quad u^r(\lambda_r) = a,$$
$$J(u^r) \leq r\sqrt{2W_r} := \eta_r, \tag{2.58}$$

where $W_r = \max\{W(z) : z \in \bigcup_{a \in A} \overline{B}_r(a)\}$. Indeed, given $z_0 \in \partial B_r(a)$, the map $u^r(x) = \frac{x}{\lambda_r}(a - z_0) + z_0$, with $\lambda_r = \frac{r}{\sqrt{2W_r}}$, satisfies (2.58). Set

$$\mathscr{A}_{ih}^r = \{v \in W^{1,2}_{\mathrm{loc}}((l_-^v, l^v); \mathbb{R}^m) : -\infty \leq l_-^v < l^v < +\infty, \lim_{x \to l_-^v} v(x) = a_i, |v(l^v) - a_h| = r\} \tag{2.59}$$

and let $\sigma_{ih}^r = \inf_{v \in \mathscr{A}_{ih}^r} J(v)$. Since $\eta_r \to 0$ as $r \to 0$, from (2.58) it follows that we can fix $r > 0$ so that

$$\sigma_{ih}^r \geq \sigma_{ih} - \eta_r \geq \sigma_{ih} - \frac{\eta}{4}. \tag{2.60}$$

Consider now a minimizing sequence $\{u_k\} \subset \mathscr{A}_{ij}$, $\lim_{k \to +\infty} J(u_k) = \sigma_{ij}$. We can assume that

$$J(u_k) \leq \min_{h \notin \{i,j\}} \{\sigma_{ih} + \sigma_{hj}\} - \eta, \quad k = 1, \ldots. \tag{2.61}$$

We claim that we can also assume that

$$u_k((l_-^{u_k}, l_+^{u_k})) \cap B_r(a_h) = \emptyset, \quad \text{for } a_h \notin \{a_i, a_j\}. \tag{2.62}$$

Indeed, if there is $x_k \in (l_-^{u_k}, l_+^{u_k})$ such that $|u_k(x_k) - a_h| = r$, then (2.60) implies

$$J(u_k) = J_{(l_-^{u_k}, x_k)}(u_k) + J_{(x_k, l_+^{u_k})}(u_k) \geq \sigma_{ih}^r + \sigma_{jh}^r \geq \sigma_{ih} + \sigma_{jh} - \frac{\eta}{2},$$

in contradiction with (2.61). This establishes (2.62) and therefore we can proceed as though a_i and a_j were the only zeros of W. The proof is complete. $\square$

2.4 Constrained Minimization, the Standing Wave Revisited

In Theorem 2.1 we gave a general proof of the existence of a minimizing heteroclinic connection under very mild assumptions on W. The main purpose of this section is to present a method together with some tools that contain the basic ingredients of several proofs in this monograph. Minimizing under a constraint is a natural way to construct a solution with required properties. We give an alternative proof of the existence of a standing wave (cf. Theorem 2.2), which is instructive since it can be extended to the traveling wave problem (cf. Sect. 2.6), and contains ideas applicable to the P.D.E. system $\Delta u - W_u(u) = 0$. In the proof of Theorem 2.2 we introduce the polar decomposition of a map (cf. (2.69)). It is a basic tool to produce competitors of a map with less energy, under the assumption that the potential locally is a monotone function on the rays emanating from its minima (cf. hypothesis **H** below). This hypothesis and the polar form are essential for proving later the maximum principle in Chap. 4, and for deriving the density estimates in Chap. 5. We also point out that Lemmas 2.4 and 2.5 in the proof of Theorem 2.2 can be seen as the one-dimensional analogs of Cases 1 and 2 in the proof of the maximum principle for the P.D.E. system $\Delta u - W_u(u) = 0$ (cf. Sect. 4.2).

We assume now that W is a double-well potential, satisfying a monotonicity condition in a neighborhood of its minima, and the coerciveness property already encountered in (2.32). We will establish the following

Theorem 2.2 *Let* $W : \mathbb{R}^m \to \mathbb{R}$ *a function of class* C^2 *that satisfies*

$$W(a^{\pm}) = 0, \quad W > 0 \ \text{on} \ \mathbb{R}^m \setminus \{a^-, a^+\},$$

for some $a^- \neq a^+ \in \mathbb{R}^m$. *Assume that* W *satisfies* (2.32) *and the monotonicity condition*

H :   *There exists* $r_0 > 0$ *such that* $r \to W(a^{\pm} + r\xi)$ *is strictly increasing for each* $r \in (0, r_0)$ *and each unit vector* $\xi \in \mathbb{R}^m$.

Then there exists a classical solution $u : \mathbb{R} \to \mathbb{R}^m$ *to*

$$u'' - W_u(u) = 0, \tag{2.63}$$

with

$$\lim_{x \to \pm\infty} u(x) = a^{\pm}. \tag{2.64}$$

Moreover, u minimizes the action $J_{\mathbb{R}}(v) = \int_{\mathbb{R}} \left(\frac{1}{2}|v'|^2 + W(v) \right) dx$ in the class

$$\mathscr{A} = \{ v \in W^{1,2}_{\mathrm{loc}}(\mathbb{R}; \mathbb{R}^m) : \exists\, x_v^- < x_v^+ \text{ (depending on } v\text{) such that}$$

$$x \leq x_v^- \implies |u(x) - a^-| \leq r_0/2 \,, \quad x \geq x_v^+ \implies |v(x) - a^+| \leq r_0/2 \}, \tag{2.65}$$

$$J(u) = \min_{\mathscr{A}} J(v).$$

Note that condition **H** allows potentials W with C^∞ contact with 0 at $a^\pm$. We divide the proof of Theorem 2.2 in three steps. We first introduce a constrained class of variations $\mathscr{A}_L \subset \mathscr{A}$, depending on a parameter $L > 0$, which eliminates the problem of translations, and thus restores compactness, and also incorporates essentially the boundary conditions (2.64). The direct method applies in a straightforward manner and provides a solution to the problem

$$\min_{\mathscr{A}_L} J. \tag{2.66}$$

However, due to the constraint, the minimizer u_L may not solve the Euler-Lagrange equation. Then we show that for L sufficiently large the constraint is not realized, and hence u_L satisfies (2.63). The translation invariance of (2.63) is crucial here. Finally we show that the minimizer u of (2.66) is actually a minimizer of the action on $\mathscr{A}$.

The Constrained Minimization Problem
For $L > 0$, we set (Fig. 2.3)

$$\mathscr{A}_L = \{ u \in \mathscr{A} : x_u^- \geq -L, \ x_u^+ \leq L \}.$$

Observe that membership in $\mathscr{A}_L$ implies that $u(x)$ is constrained in $\overline{B}_{r_0/2}(a^-)$ for $x \leq -L$ and in $\overline{B}_{r_0/2}(a^+)$ for $x \geq L$. Note that the map $\tilde{u}$ defined by

$$\tilde{u}(x) = \begin{cases} a^- & \text{for } x \leq -1, \\ \frac{1-x}{2}a^- + \frac{1+x}{2}a^+ & \text{for } -1 \leq x \leq 1, \\ a^+ & \text{for } x \geq 1, \end{cases} \tag{2.67}$$

belongs to $\mathscr{A}_L$ for all $L > 1$ and has finite energy

$$J(\tilde{u}) = \sigma < +\infty.$$

Note also that while $u \in \mathscr{A}$ implies that $\mathscr{A}$ contains the whole manifold of the translates of u for $u \in \mathscr{A}_L$, $\tau \leq 2L$ is an upper bound for the translate $u(\cdot - \tau)$ to remain in $\mathscr{A}_L$. This together with the fact that in (2.66) we can restrict to the maps that satisfy $J(u) \leq \sigma$ implies the existence of a minimizing sequence $\{u_j\} \subset \mathscr{A}_L$

Fig. 2.3 A map in $\mathscr{A}_L$

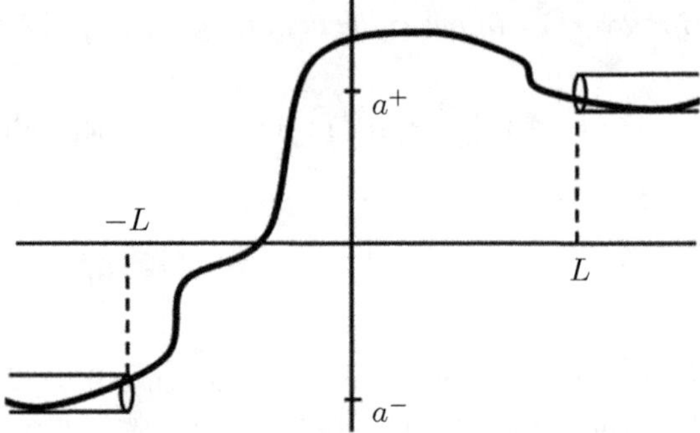

which is equibounded and equicontinuous and standard arguments as in the proof
of Theorem 2.1 show that (2.66) has a solution, and we can state.

Proposition 2.7 *Let $L > 1$, and arbitrary otherwise. Then the variational problem*

$$\min_{\mathscr{A}_L} \int_{\mathbb{R}} \left(\frac{1}{2}|v'|^2 + W(v) \right) dx, \tag{2.68}$$

has a minimizer u_L.

Removing the Constraint
We begin by introducing the *polar form*. For $u : \mathbb{R} \to \mathbb{R}^m$, provided that $|u(x) - a^\pm| > 0$, we can write

$$u(x) - a^\pm = \rho^\pm(x)\boldsymbol{n}^\pm(x), \quad \left(\rho^\pm(x) = |u(x) - a^\pm|, \quad \boldsymbol{n}^\pm(x) = \frac{u(x) - a^\pm}{|u(x) - a^\pm|} \right). \tag{2.69}$$

First, note that formally, for $\rho = \rho^\pm$ and $\boldsymbol{n} = \boldsymbol{n}^\pm$,

$$|u'(x)|^2 = u'(x) \cdot u'(x), \text{ (where } \cdot \text{ denotes the Euclidean inner product)} \tag{2.70}$$

$$= (\rho'(x)\boldsymbol{n}(x) + \rho(x)\boldsymbol{n}'(x)) \cdot (\rho'(x)\boldsymbol{n}(x) + \rho(x)\boldsymbol{n}'(x))$$

$$= |\rho'(x)|^2 + \rho^2(x)|\boldsymbol{n}'(x)|^2,$$

where we utilized that $\boldsymbol{n}'(x) \cdot \boldsymbol{n}(x) = 0$. Given $u \in W^{1,2}_{\text{loc}}(\mathbb{R}; \mathbb{R}^m)$ we have that u is
continuous by the embedding $W^{1,2} \subset C$. Consequently, the set $\Omega^+ := \{x : \rho(x) > 0\}$ is open. We notice that in Ω^+, $\boldsymbol{n}(x)$ is well defined and (2.69) holds. Also ρ is
in $W^{1,2}$, so is absolutely continuous, and therefore so is $\boldsymbol{n}(x)$ in Ω^+. Thus, (2.70)
holds in Ω^+. Now for any $u \in W^{1,2}_{\text{loc}}(\mathbb{R}; \mathbb{R}^m)$ we have that

$$|u'| = 0 \quad \text{a.e. on } \{x : u(x) = 0\}.$$

Consequently, on any measurable set S we can write

$$\int_S |u'(x)|^2 dx = \int_{S \cap \{\rho > 0\}} |u'(x)|^2 dx \tag{2.71}$$

$$= \int_{S \cap \{\rho > 0\}} \left(|\rho'(x)|^2 + \rho^2(x)|\boldsymbol{n}'(x)|^2\right) dx.$$

Next, we introduce a class of variations of u obtained by modifying $\rho(\cdot)$, but keeping $\boldsymbol{n}(\cdot)$ the same. Let $\alpha < \beta$, $r \in \mathbb{R}$, and $u \in W^{1,2}(\mathbb{R}; \mathbb{R}^m)$, and suppose that

$$\begin{cases} 0 < \rho(\alpha) = \rho(\beta) = r \le r_0, \\ r \le \rho(x) \le r_0, \ \forall x \in (\alpha, \beta). \end{cases} \tag{2.72}$$

Let also $\phi : [0, 1] \to \mathbb{R}$ be a C^2 function with $\phi(0) = \phi(1) = 0$, $\phi(s) > 0$ on $(0, 1)$. For $\epsilon \ge 0$ define

$$u^\epsilon(x) = \begin{cases} a + \left(1 - \epsilon\phi\left(\frac{x-\alpha}{\beta-\alpha}\right)\right) r\boldsymbol{n}(x), & \text{for } x \in [\alpha, \beta], \\ u(x), & \text{for } x \in \mathbb{R} \setminus [\alpha, \beta]. \end{cases} \tag{2.73}$$

First note that

$$u^\epsilon \in W^{1,2}(\mathbb{R}; \mathbb{R}^m). \tag{2.74}$$

Indeed, in (α, β), $\rho(x) \ge r > 0$, so $\boldsymbol{n}(x)$ is well defined and in $W^{1,2}([\alpha, \beta]; \mathbb{R}^m)$. Hence the restriction of u^ϵ in $[\alpha, \beta]$ is in $W^{1,2}([\alpha, \beta]; \mathbb{R}^m)$, and since $u^\epsilon(\alpha) = u(\alpha)$, $u^\epsilon(\beta) = u(\beta)$, (2.74) follows, and actually

$$\int_{\mathbb{R}} |u^{\epsilon'}|^2 dx = \int_\alpha^\beta |u^{\epsilon'}|^2 dx + \int_{\mathbb{R} \setminus [\alpha, \beta]} |u'|^2 dx. \tag{2.75}$$

These variations will be utilized in the proof of the following

Lemma 2.4 *Let a be one of the zeros of W and assume that $\mathbf{H}$ is satisfied. Let $u \in W^{1,2}([\alpha, \beta]; \mathbb{R}^m)$ be a map that satisfies (2.72). Then there exists $\tilde{u} \in W^{1,2}([\alpha, \beta]; \mathbb{R}^m)$ with the following properties:*

$$\begin{cases} \tilde{u}(\alpha) = u(\alpha), \quad \tilde{u}(\beta) = u(\beta), \\ \tilde{\rho}(x) < r, \quad \forall x \in (\alpha, \beta), \\ J_{(\alpha,\beta)}(\tilde{u}) < J_{(\alpha,\beta)}(u). \end{cases}$$

Proof We show that, provided $\epsilon > 0$ is sufficiently small, the map $u^\epsilon : [\alpha, \beta] \to \mathbb{R}^m$ defined in (2.73) can be identified with the sought $\tilde{u}$. Thanks to (2.71), we have

$$J_{(\alpha,\beta)}(u^\epsilon) = J_{(\alpha,\beta)}(u^0) - \epsilon r^2 \int_\alpha^\beta \phi |\boldsymbol{n}'|^2 dx + \frac{\epsilon^2}{2} r^2 \int_\alpha^\beta \phi^2 |\boldsymbol{n}'|^2 dx$$

$$- \int_\alpha^\beta \big(W(a + r\boldsymbol{n}) - W(a + (1 - \epsilon\phi)r\boldsymbol{n}) \big) dx + \frac{\epsilon^2}{2(\beta - \alpha)^2} r^2 \int_\alpha^\beta \phi'^2 dx.$$

$$(2.76)$$

Note that $r = \rho^0(x) \le \rho(x)$ by (2.72), and so via **H**, it follows that

$$J_{(\alpha,\beta)}(u^0) \le J_{(\alpha,\beta)}(u).$$

$$(2.77)$$

Now, we will show that for $\epsilon > 0$ small enough, the sum of the four remaining terms on the right-hand side of (2.76) is negative. Indeed, from **H** it follows that

$$- \int_\alpha^\beta \big(W(a + r\boldsymbol{n}) - W(a + (1 - \epsilon\phi)r\boldsymbol{n}) \big) dx + \frac{\epsilon^2}{2(\beta - \alpha)^2} r^2 \int_\alpha^\beta \phi'^2 dx$$

$$= -\epsilon r \int_\alpha^\beta \left(\int_0^1 (\nabla W(a + r\boldsymbol{n} - \tau\epsilon\phi r\boldsymbol{n})) \cdot \phi\boldsymbol{n} d\tau \right) dx + \frac{\epsilon^2}{2(\beta - \alpha)^2} r^2 \int_\alpha^\beta \phi'^2 dx$$

$$< -\epsilon r C + \frac{\epsilon^2}{2(\beta - \alpha)^2} r^2 \int_\alpha^\beta \phi'^2 dx < 0,$$

for some constant $C > 0$, and for $0 < \epsilon \ll 1$. In addition, for $0 < \epsilon \ll 1$,

$$- \epsilon r^2 \int_\alpha^\beta \phi |\boldsymbol{n}'|^2 dx + \frac{\epsilon^2}{2} r^2 \int_\alpha^\beta \phi^2 |\boldsymbol{n}'|^2 dx \le 0.$$

Thus, for $0 < \epsilon \ll 1$ we have:

$$J_{(\alpha,\beta)}(u^\epsilon) < J_{(\alpha,\beta)}(u^0) \le J_{(\alpha,\beta)}(u),$$

and taking $\tilde{u} = u^\epsilon$ with ϵ small, we deduce the lemma.   □

We remark that Lemma 2.4 is of a local type and gives a function that takes the same values as u at the end points of an interval, but in the interior is more efficient in the sense of having a smaller value of the action than u.

Lemma 2.5 *Let a be one of the zeros of W and assume that* **H** *is satisfied. Let $\alpha < \beta \in \mathbb{R}$, $r \in \mathbb{R}$ and $u \in W^{1,2}([\alpha, \beta]; \mathbb{R}^m)$ be such that*

$$0 < \rho(\alpha) = \rho(\beta) = r \le r_0/2,$$

$$\rho(x_0) \ge r, \quad \text{for some } x_0 \in (\alpha, \beta).$$

Then there exists $\tilde{u} \in W^{1,2}([\alpha, \beta]; \mathbb{R}^m)$ with the following properties:

$$\begin{cases} \tilde{u}(\alpha) = u(\alpha), \quad \tilde{u}(\beta) = u(\beta), \\ \tilde{\rho}(x) < r, \quad \forall x \in (\alpha, \beta), \\ J_{(\alpha,\beta)}(\tilde{u}) < J_{(\alpha,\beta)}(u). \end{cases}$$

We remark that in contrast to Lemma 2.4, in Lemma 2.5 no a priori bound is imposed on $\rho_M := \max_{x \in (\alpha,\beta)} \rho(x)$. The intuition behind Lemma 2.5 is that from the point of view of minimizing the action for a curve that starts and comes back well inside the convexity region of W (cf. the assumption $r \leq r_0/2$), it is more efficient to remain in the region rather than making an excursion outside.

Proof (Lemma 2.5) Without loss of generality we can assume that $\rho(x_0) = \rho_M = \max \rho$. We begin with the special case $\rho(x_0) = \rho_M = r$. We can assume that $\rho(x) < r$ for some $x \in (\alpha, x_0)$ (or $x \in (x_0, \beta)$), since otherwise we can apply Lemma 2.4 and thus produce a $\tilde{u}$ with strictly smaller action than u. Utilizing the continuity of u, we conclude that there are $\hat{\alpha} \in (\alpha, x_0)$, $\hat{\beta} \in (x_0, \beta)$, $\hat{r} < r$, such that $\rho(\hat{\alpha}) = \rho(\hat{\beta}) = \hat{r}$, and $\hat{r} < \rho(x) \leq r$, for all $x \in (\hat{\alpha}, \hat{\beta})$. Lemma 2.4 applies and gives a $\tilde{u}$ with strictly smaller action than u. Consequently, we can assume that $\rho_M > r$. If $r < \rho_M \leq r_0$, then again Lemma 2.4 can be applied on the connected component I_0 of the set $\{x \in (\alpha, \beta) : \rho(x) > r\}$ that contains x_0, and in this way we reach the same conclusion.

Therefore, the only case that remains is $\rho_M > r_0$. Suppose I_0 is the connected component of $\{x \in (\alpha, \beta) : \rho(x) > r\}$ that contains x_0. Without loss of generality we may assume that I_0 coincides with (α, β). We define

$$h : [r, r_0] \to \mathbb{R}, \quad v \in W^{1,2}([\alpha, \beta]; \mathbb{R}^m),$$

as follows:

$$h(s) := \frac{r_0 - s}{r_0 - r}, \quad \text{for } r \leq s \leq r_0,$$

$$v(x) = \begin{cases} a + rh(\rho(x))\boldsymbol{n}(x), & \forall x \in [\alpha, \beta], \ \rho(x) \in [r, r_0], \\ a, & \forall x \in [\alpha, \beta], \ \rho(x) \geq r_0. \end{cases}$$

In Fig. 2.4 below we show the deformation of u.

First, we note that the assumption $r \leq r_0/2$ implies that $0 \leq h(s) \leq 1$. In what follows we will compare separately the *kinetic energy* and the *potential energy* of u and v. We begin with the potential part, for $x \in (\alpha, \beta)$:

$$W(v(x)) = W(a + rh(\rho(x))\boldsymbol{n}(x)),$$

$$W(u(x)) = W(a + \rho(x)\boldsymbol{n}(x)).$$

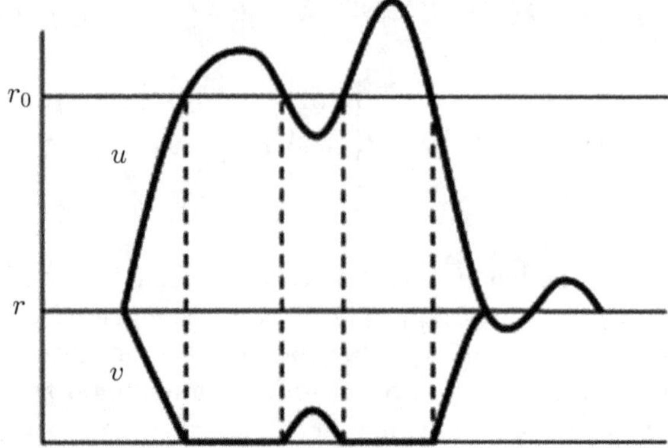

Fig. 2.4 The deformation v

For $\rho(x) \in [r, r_0]$, Hypothesis **H** gives $W(v(x)) \leq W(u(x))$. For $\rho(x) \geq r_0$ we have $W(v(x)) = W(a) = 0$, and so $W(v(x)) < W(u(x))$ for a nonempty set of x's, by assumption **H**

$$\int_\alpha^\beta W(v(x))\mathrm{d}x < \int_\alpha^\beta W(u(x))\mathrm{d}x.$$

Next, we will compare the kinetic parts, for $x \in (\alpha, \beta)$:

$$|h'(s)| = \frac{1}{r_0 - r} \leq \frac{1}{r} \ (\Longleftrightarrow r \leq r_0/2),$$

$$|v'(x)|^2 = (rh'\rho')^2 + r^2h^2|\mathbf{n}'|^2 \leq \rho'^2 + r^2|\mathbf{n}'|^2 \leq \rho'^2 + \rho^2|\mathbf{n}'|^2 = |u'(x)|^2.$$

Consequently

$$\int_\alpha^\beta |v'(x)|^2\mathrm{d}x \leq \int_\alpha^\beta |u'(x)|^2\mathrm{d}x,$$

and so $J_{(\alpha,\beta)}(v) < J_{(\alpha,\beta)}(u)$, and the proof of the lemma is complete. $\qquad\square$

Remark 2.6 Lemmas 2.4 and 2.5 apply to more general functionals than J. For every smooth positive function $\phi : [\alpha, \beta] \to (0, \infty)$, we may also consider the functional $\widetilde{J}(u) := \int_\alpha^\beta \left(\frac{1}{2}|u'(x)|^2 + W(u(x))\right)\phi(x)\mathrm{d}x$. Then, we can check that Lemmas 2.4 and 2.5 still hold for $\widetilde{J}$, since all the arguments in the proofs are based on pointwise deformations. In particular, in Lemma 2.7 below, we will take $\phi(x) = e^{cx}$, with $c > 0$.

Lemma 2.6 *Let u_L be a minimizer of the constrained problem* (2.66). *Then,*

$$\rho_L^+(x_0) = r \leq r_0/2 \Longrightarrow \rho_L^+(x) < r, \quad \forall x > x_0, \tag{2.78}$$

with an analogous result for ρ_L^-.

Proof $J(u_L) < \infty$ implies that $\int_{\mathbb{R}} W(u_L(x))dx < \infty$. By (2.32), it then follows that there is a sequence $x_n \to +\infty$ as $n \to \infty$ such that $\rho_L^+(x_n) \to 0$. If (2.78) does not hold, then $\rho_L^+(\hat{x}) = r$ for some $\hat{x} > x_0$. Then, by Lemma 2.5, $\rho_L^+(x) < r$, for all $x \in (x_0, \hat{x})$. Thus, given $\alpha \in (x_0, \hat{x})$, there is $\beta > \hat{x}$ such that $\rho_L^+(\alpha) = \rho_L^+(\beta) < \rho_L^+(\hat{x}) = r$, in contradiction to Lemma 2.5. $\square$

An obvious consequence of Lemmas 2.5 and 2.6 is that the constraint can be realized, if at all, only at $x = \pm L$. A more important implication of Lemma 2.5 is that the minimizer of (2.66), $x \mapsto u_L(x)$, viewed as a curve in $\mathbb{R}^m$, once it exits the ball $B_r(a^-)$ it cannot reenter it, and similarly, once it enters $B_r(a^+)$, it cannot exit it.

Thus, there exist $x_L^- \in [-L, L)$ and $x_L^+ \in (-L, L]$ such that

$$\{\rho_L^-(x) \le r_0/2\} = (-\infty, x_L^-],$$

$$\{\rho_L^+(x) \le r_0/2\} = [x_L^+, +\infty),$$

and ρ_L^- (ρ_L^+) restricted to $(-\infty, x_L^-]$ (resp., $[x_L^+, +\infty)$) is strictly monotone and converges to zero as $|x| \to \infty$. Indeed, $\lim_{\pm\infty} \rho_L^\pm > 0$ is excluded by the boundedness of $J(u_L)$. It follows that

$$x < x_L^- \iff \rho_L^-(x) < r_0/2,$$

$$x > x_L^+ \iff \rho_L^+(x) < r_0/2.$$

Since (2.32) and $W > 0$ on $\mathbb{R}^m \setminus \{a^-, a^+\}$ imply

$$W(u) > c_0, \quad \forall u \in \mathbb{R}^m \setminus (B_{r_0/2}(a^-) \cup B_{r_0/2}(a^+)), \tag{2.79}$$

for some $c_0 > 0$, we have

$$c_0(x_L^+ - x_L^-) \le \int_{\mathbb{R}} W(u_L)dx \le \sigma \Rightarrow |x_L^+ - x_L^-| \le \frac{\sigma}{c_0} := 2L^*. \tag{2.80}$$

Therefore, if $L > L^* = \frac{\sigma}{2c_0}$, the two conditions

$$\rho_L^-(-L) = r_0/2 \quad \text{and} \quad \rho_L^+(L) = r_0/2$$

are incompatible and

$$\begin{aligned}
\rho_L^-(-L) = r_0/2 &\implies \rho_L^+(x) < r_0/2, \quad \text{for } x \ge L, \\
\rho_L^+(L) = r_0/2 &\implies \rho_L^-(x) < r_0/2, \quad \text{for } x \le -L.
\end{aligned} \tag{2.81}$$

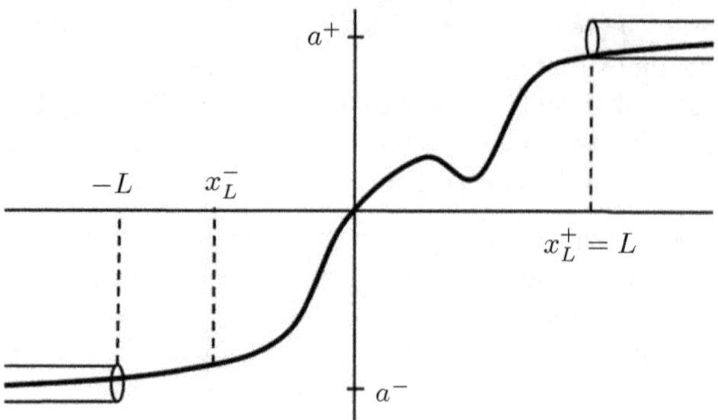

Fig. 2.5 u_L cannot realize the constraint on both rims

(cf. Fig. 2.5). Suppose that $\rho_L^-(-L) = r_0/2$. Since $x \mapsto \rho_L^-(x)$ is strictly increasing in $(-\infty, -L]$ and $x \to \rho_L^+(x)$ is strictly decreasing in $[L, +\infty)$, upon translating $u_L(x)$ to the right by a small amount $0 < \delta \ll 1$, we obtain

$$\rho_L^-(x - \delta) < r_0/2, \quad \text{for } x \in (-\infty, -L],$$

$$\rho_L^+(x - \delta) < r_0/2, \quad \text{for } x \in [L, +\infty)$$

and therefore, since the action is translation invariant, $u_L(\cdot - \delta)$ is still a minimizer in $\mathscr{A}_L$ that does not realize the constraints imposed in $\mathscr{A}_L$ and thus solves the Euler-Lagrange equation. The case $\rho_L^+(L) = r_0/2$ is discussed in a similar way.

The Variational Characterization

We have seen that, if $L > L^* = \sigma/c_0$, the minimizer $u_L \in \mathscr{A}_L$ provided by Proposition 2.7 satisfies (2.63) and (2.64). To complete the proof of Theorem 2.2, we show that u_L is also a minimizer of $J_{\mathbb{R}}$ on $\mathscr{A}$, the set defined in (2.65). To prove this we show that to each map in $\mathscr{A}$ we can associate a map in $\mathscr{A}_L$ without increasing the action. Let $\widetilde{I}_v^{\pm}$ the complement of the set $I_v^{\pm} = \{x \in \mathbb{R} : |v(x) - a^{\pm}| \leq r_0/2\}$. From Lemma 2.5 it follows that, if $\widetilde{I}_v^+$ has a bounded component we can eliminate it and reduce the action, and the same is true for $\widetilde{I}_v^-$. This implies that minimizing on $\mathscr{A}$ is the same as minimizing on the subset of the maps that satisfy

$$I_v^- = (-\infty, x_v^-], \quad I_v^+ = [x_v^+, +\infty).$$

This together with $J_{\mathbb{R}}(v) \leq \sigma$ implies

$$c_0(x_v^+ - x_v^-) \leq \int_{x_v^-}^{x_v^+} W(v(x))dx \leq \sigma,$$

which shows that we can further restrict to the subset of the maps that satisfy $(x_v^+ - x_v^-) \leq 2L^*$ and, by a translation that does not change the action, reduce to the case

$$-L < -L^+ \leq x_v^- < x_v^+ \leq L^* < L.$$

This completes the proof of Theorem 2.2. $\qquad\qquad\qquad\qquad\qquad\qquad\square$

Remark 2.7 Clearly, the u constructed above is not a global minimizer of the action. The only global minimizers are $u^\pm(x) \equiv a^\pm$, and had to be excluded from our admissible class $\mathscr{A}_L$.

Remark 2.8 The monotonicity of the minimizers u_L inside the cylinders (cf. Lemma 2.6) is a relatively easy local fact that could be established directly via the linearized equation at the minima.

2.5 Characterization of Minimizers

Now, we are going to show, in the more general set-up of W's with several global minima, that nonconstant *minimizers* $u : \mathbb{R} \to \mathbb{R}^m$ of the action are heteroclinic connections. We recall that by a *minimizer* of the action we mean a map $u \in W^{1,2}_{\mathrm{loc}}(\mathbb{R}; \mathbb{R}^m) \cap L^\infty(\mathbb{R}; \mathbb{R}^m)$ such that $J_{[\alpha,\beta]}(u) \leq J_{[\alpha,\beta]}(u + v)$, for every $v \in W^{1,2}_0([\alpha, \beta]; \mathbb{R}^m)$ and every $[\alpha, \beta] \subset \mathbb{R}$. It is obvious that minimizers are solutions of (2.63).

Theorem 2.3 *Assume $W : \mathbb{R}^m \to \mathbb{R}$ is C^2 and that there are $N \geq 2$ distinct points $a_1, \ldots, a_N \in \mathbb{R}^m$ such that*

$$0 = W(a_j) < W(u), \ \text{for } u \notin \{a_1, \ldots, a_N\}, \ j = 1, \ldots, N.$$

Then, if $u : \mathbb{R} \to \mathbb{R}^m$ is a minimizer, either $u \equiv a$ for some $a \in \{a_1, \ldots, a_N\}$, or there are $a^- \neq a^+ \in \{a_1, \ldots, a_N\}$ such that

$$\lim_{x \to \pm\infty} u(x) = a^\pm.$$

Moreover

$$J(u) < +\infty.$$

Proof Given $l > 1$ define $v_l : \mathbb{R} \to \mathbb{R}^m$ by

$$v_l(x) = \begin{cases} u(x) & \text{for } x \in (-\infty, -l] \cup [l, +\infty), \\ a + (1 - x - l)(u(-l) - a) & \text{for } x \in [-l, -l + 1], \\ a & \text{for } x \in [-l + 1, +l - 1], \\ a + (1 + x - l)(u(l) - a), & \text{for } x \in [l - 1, l]. \end{cases}$$

Since u is bounded, there is a constant $\bar{J} > 0$ such that

$$J_{(-l,l)}(v_l) < \bar{J}, \quad \text{for all } l > 1. \tag{2.82}$$

Furthermore, the minimality of u implies that $J_{(-l,l)}(u) \leq J_{(-l,l)}(v_l) < \bar{J}$, since $u(-l) = v_l(-l)$ and $u(l) = v_l(l)$. As a consequence, $J(u) < \infty$, and u is uniformly continuous. Our next claim is that $W(u(x)) \to 0$ as $|x| \to \infty$. Indeed, suppose by contradiction that there exists a sequence $|x_n| \to \infty$ such that $W(u(x_n)) > 2\epsilon > 0$. Then, by the uniform continuity of $W(u)$, we can find intervals $[x_n - \delta, x_n + \delta]$ of length 2δ independent of n, such that $W(u(x)) > \epsilon$, for all $x \in [x_n - \delta, x_n + \delta]$, $\forall n$. This contradicts the boundedness of $J(u)$ and proves our claim. It follows that $\lim_{x \to \pm\infty} u(x) = a^{\pm}$, for some $a^{\pm} \in \{a_1, \ldots, a_N\}$. If $a^{-} = a^{+} = a$, one can see that $\lim_{l \to +\infty} J_{(-l,l)}(v_l) = 0$, and thus

$$J(u) = \lim_{l \to +\infty} J_{(-l,l)}(u) = 0 \Rightarrow u \equiv a.$$

The proof is concluded. $\square$

Remark 2.9 The converse of Theorem 2.3 is not true. We give a counterexample. Assume $W = \frac{1}{4}(1 - |u|^2)^2 + \epsilon|u_1|^2$, $u = (u_1, u_2) \in \mathbb{R}^2$. Since $W((-u_1, u_2)) = W((u_1, u_2))$ we can apply Theorem 2.2 to the restriction of W to the line $u_1 = 0$ and deduce the existence of a heteroclinic orbit of the form $\bar{u} = (0, \bar{u}_2)$ that connects the two minima $(0, -1)$ and $(0, 1)$ of W. We have $J(\bar{u}) > 0$. Let $\tilde{u}(x) = (\cos \sqrt{\epsilon}x, \sin \sqrt{\epsilon}x)$, $x \in (-\frac{\pi}{2\sqrt{\epsilon}}, \frac{\pi}{2\sqrt{\epsilon}})$. Since $J(\tilde{u}) = \pi \sqrt{\epsilon}$ for small $\epsilon > 0$, we have $J(\tilde{u}) < J(\bar{u})$, which shows that the connection given by Theorem 2.2 has energy strictly less then $J(\bar{u})$. That is $\bar{u}$ is a connection which is not minimal. Note that this example also shows that in general the minimal connection given by Theorem 2.2 is not unique.

We also refer to [6] to see that only potentials W that are bounded below and attain their minimum allow for the existence of minimizers. Finally, let us mention two situations in the vector case that cannot occur for scalar potentials.

(i) The existence of nonconstant solutions u to (2.63) that may connect at $\pm\infty$ the same minimum, that is, $\lim_{x \to \pm\infty} u(x) = a$, with $a \in \{a_1, \ldots, a_N\}$ (see [6]) .

(ii) The existence of periodic connections, that is, solutions to (2.63) such that $u(x + T) = u(x)$, for all $x \in \mathbb{R}$ and for some constant $T > 0$, satisfying in addition $u(0) = a_i$ and $u(T/2) = a_j$, with $a_i \neq a_j \in \{a_1, \ldots, a_N\}$ (see [22]). In the scalar case, this situation is excluded, since all bounded solutions u have a nonpositive Hamiltonian (cf. Modica's inequality in [17]). Thus, if $W(u(x)) = 0$ for some $x \in \mathbb{R}$, u must be constant. For vector potentials this property does not hold in general.

Clearly, the solutions in (i) and (ii) above are not minimal.

2.6   Heteroclinic Connections for Double-Well Unbalanced Potentials; the Traveling Wave

In this section we consider potentials W as in Fig. 2.6 below, having a global minimum at a^-, and a local minimum at a^+: $W(a^-) < 0 = W(a^+)$. The traveling wave problem is

$$\begin{cases} u'' - W_u(u) = -cu', & u : \mathbb{R} \to \mathbb{R}^m, \\ \lim_{x \to \pm\infty} u(x) = a^{\pm}, \end{cases} \tag{2.83}$$

where the unknown now is the pair (c, u), $c \in \mathbb{R}$.

We recall that solutions of $U_t = U_{xx} - W_u(U)$ of the form $U(x, t) = u(x - ct)$ are called, naturally, traveling waves, and this explains the equation in (2.83). In (2.7), (2.8), we see that $c > 0$ can be interpreted as a friction coefficient, and so it is useful to have in mind the mechanical analog of a ball rolling on a potential landscape given by $-W$, from the global maximum $-W(a^-)$ down to the local maximum $-W(a^+)$. So we can see the boundary conditions in (2.83).

It was noted by Fife and McLeod in [9] that (2.83) is variational with respect to the weighted action

$$J_c(u) = \int_{\mathbb{R}} \left(\frac{1}{2} |u'|^2 + W(u) \right) e^{cx} dx. \tag{2.84}$$

In this section we will adopt the following hypotheses:

$\mathbf{H_1}$:   The potential $W : \mathbb{R}^m \to \mathbb{R}$ is of class C^2, with two minima a^-, a^+, $W(a^-) < W(a^+) = 0$, $W(u) > W(a^-)$ for $u \neq a^-$, and $\liminf_{|u| \to \infty} W(u) > 0$.

$\mathbf{H_2}$: $\{u : W(u) \leq 0\} = C_0^- \cup \{a^+\}$, $\mathrm{dist}(C_0^-, a^+) > 0$, where C_0^- is a strictly convex set with C^2 boundary ∂C_0^-.

$\mathbf{H_3}$:

(i) $\nabla W \cdot \nu > 0$ on ∂C_0^-, ν the outward normal on ∂C_0^-.

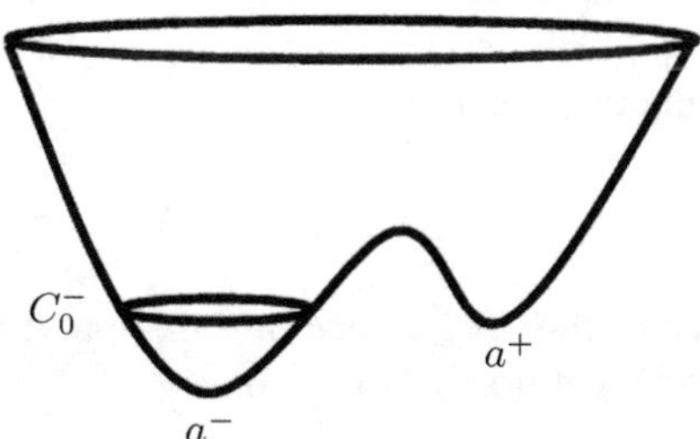

Fig. 2.6 A representative W and the convex set C_0^-

(ii) $W_{uu} \geq \epsilon_0 I$ on ∂C_0^-, for some constant $\epsilon_0 > 0$, where W_{uu} and I denote respectively the Hessian matrix of W and the identity.

H$_4$:

(i) There is $r_0 > 0$ such that the map $r \mapsto W(a^+ + r\xi)$ has a strictly positive derivative for $r \in (0, r_0]$, $|\xi| = 1$.
(ii) The map $r \mapsto W(a^- + r\xi)$ has a strictly positive derivative as long as $a^- + r\xi \in C_0^-$, $|\xi| = 1$, $r > 0$. We assume that $B(a^-, r_0) \subset\subset C_0^-$.

Theorem 2.4 *Under Hypotheses* **H$_1$**–**H$_4$**, *there exist* $c^* > 0$ *and* u *satisfying* (2.83) *above. In addition, we have the variational characterization*

$$c^* = \sup_{c>0}\{\inf_{\mathscr{A}} J_c(v) < 0\}, \tag{2.85}$$

where

$$\mathscr{A} = \{v \in W_{\mathrm{loc}}^{1,2}(\mathbb{R}; \mathbb{R}^m) : \exists\, x_v^- < x_v^+ \text{ (depending on v) such that}$$

$$x \leq x_v^- \implies |v(x) - a^-| \leq r_0/2\,,\; x \geq x_v^+ \implies |v(x) - a^+| \leq r_0/2\},$$

and $\inf_{\mathscr{A}} J_{c^*}(v) = \min_{\mathscr{A}} J_{c^*}(v) = J_{c^*}(u)$. *Moreover* c^*, u *satisfy the two conditions*

$$J_{c^*}(u) = 0, \quad c^* = \frac{-W(a^-)}{\int_{\mathbb{R}} |u'|^2 dx}.$$

Finally, the speed c^* *is unique in the class* $\mathscr{A}$ *of wave profiles, while the corresponding profile* u *in general is not.*

The first part of the proof proceeds along lines similar to Theorem 2.2: $c > 0$ is taken as an arbitrary parameter and a constrained minimization problem produces a minimizer u_L that depends on c. Then, the replacement lemmas are adjusted to control the time when u_L crosses the spheres $\partial B(a^\pm, r_0/2)$.

The second part deals with the special choice of c so that the constraint can be removed. The special choice of c in (2.85) can be motivated by the following argument due to Muratov [19]. Notice that there is an incompatibility between the Eq. (2.83) and the functional (2.84), since the first is translation invariant, while the functional is not:

$$J_c(u(\cdot - \delta)) = e^{c\delta} J_c(u(\cdot)). \tag{2.86}$$

Thus, the only choice of c that preserves the minimizing property of u does satisfy the condition $J_{c^*}(u) = 0$.

Proof (Theorem 2.4)

Part I: $c > 0$ as a parameter

Step 1: The Constrained Problem
For $L > 0$, we define

$$\mathscr{A}_L = \{v \in W^{1,2}_{\text{loc}}(\mathbb{R}; \mathbb{R}^m) : |v(x) - a^+| \le r_0/2, \forall x \ge L,$$

$$\text{and } |v(x) - a^-| \le r_0/2, \forall x \le -L\}, \qquad (2.87)$$

Proposition 2.8 *Let $L > 1$, and arbitrary otherwise. Then the variational problem*

$$\min_{\mathscr{A}_L} \int_{\mathbb{R}} \left(\frac{1}{2}|v'|^2 + W(v)\right)e^{cx}\mathrm{d}x,$$

has a minimizer u_L depending on $c > 0$.

Proof First, we check that $-\infty < \inf_{\mathscr{A}_L} J_c$. Indeed, since $v \in \mathscr{A}_L$ implies $W(v) > 0$ for $x \ge L$, we have

$$J_c(v) \ge \int_{-\infty}^{L} \left(\frac{1}{2}|v'|^2 + W(v)\right)e^{cx}\mathrm{d}x \ge \frac{e^{cL}W(a^-)}{c}. \qquad (2.88)$$

Then, setting $W^+ := \max(W, 0)$, $W^- := \max(-W, 0)$, we note that the u_{aff} defined in (2.67) provides an easy bound as follows:

$$\sup_{L \ge 1} \inf_{\mathscr{A}_L}\{J_c(u)\} \le \sup_{L \ge 1} J_c(u_{\text{aff}}) = J_c(u_{\text{aff}}),$$

$$J_c(u_{\text{aff}}) = \int_{-\infty}^{-1} W(a^-)e^{cx}\mathrm{d}x + \int_{1}^{\infty} W(a^+)e^{cx}\mathrm{d}x$$

$$+ \int_{-1}^{1} \left\{\frac{1}{2}\left|\frac{a^+ - a^-}{2}\right|^2 + W\left(\frac{1-x}{2}a^- + \frac{1+x}{2}a^+\right)\right\} e^{cx}\mathrm{d}x$$

$$J_c(u_{\text{aff}}) \le \int_{-1}^{1} \left\{\frac{1}{2}\left|\frac{a^+ - a^-}{2}\right|^2 + W^+\left(\frac{1-x}{2}a^- + \frac{1+x}{2}a^+\right)\right\} e^{cx}\mathrm{d}x + \frac{1}{c}e^{-c}W(a^-).$$

Hence if we set

$$J_c^+(u) := \int_{\mathbb{R}} \left(\frac{1}{2}|u'|^2 + W^+(u)\right)e^{cx}\mathrm{d}x,$$

we obtain

$$J_c(u_{\text{aff}}) \leq e^{-c}\frac{W(a^-)}{c} + e^c J_0^+(u_{\text{aff}}). \tag{2.89}$$

since $W(a^+) = 0$ and u_{aff} is a bounded Lipschitz map, there is a constant $C_{\text{aff}} > 0$ such that

$$J_c(u_{\text{aff}}) = \int_{-\infty}^{-1} W(a^-)e^{cx}dx + \int_{-1}^{1}\left(\frac{1}{2}|u'_{\text{aff}}|^2 + W(u_{\text{aff}})\right)e^{cx}dx$$

$$\leq \frac{e^{-c}W(a^-)}{c} + e^c C_{\text{aff}}.$$

It follows that $\{u \in \mathscr{A}_L : J_c(u) \leq J_c(u_{\text{aff}})\}$ is an equibounded and equicontinuous set. To see this we observe that, since $W(u) \geq 0$ for $x \geq L$, we have

$$\frac{1}{2}\int_{\mathbb{R}}|u'|^2 e^{cx}dx \leq J_c(u_{\text{aff}}) - \int_{\mathbb{R}}W(u)e^{cx}dx$$

$$\leq J_c(u_{\text{aff}}) - \int_{-\infty}^{L}W(u)e^{cx}dx \leq J_c(u_{\text{aff}}) - \frac{e^{cL}W(a^-)}{c}. \tag{2.90}$$

This implies the equicontinuity. Indeed, for each bounded interval (x_1, x_2), we have

$$|u(x_2) - u(x_1)| \leq |x_2 - x_1|^{\frac{1}{2}}(\int_{x_1}^{x_2}|u'|^2 dx)^{\frac{1}{2}}$$

$$\leq |x_2 - x_1|^{\frac{1}{2}}e^{\frac{c}{2}x_2}\left(\int_{x_1}^{x_2}|u'|^2 e^{cx}dx\right)^{\frac{1}{2}}. \tag{2.91}$$

In particular, for $x \in (-L, L)$ we obtain

$$|u(x) - u(-L)| \leq |2Le^{cL}|^{\frac{1}{2}}(\int_{-L}^{L}|u'|^2 e^{cx}dx)^{\frac{1}{2}}. \tag{2.92}$$

This and the boundedness of $u \in \mathscr{A}_L$ in $(-\infty, -L] \cup [L, +\infty)$ imply

$$\|u\|_{L^\infty(\mathbb{R};\mathbb{R}^m)} \leq \max\{|a^-|, |a^+|\} + \frac{r_0}{2} + 2|2Le^{cL}|^{\frac{1}{2}}(J_c(u_{\text{aff}}) - \frac{e^{cL}W(a^-)}{c}). \tag{2.93}$$

Let $\{u_n\}$ be a minimizing sequence in $\mathscr{A}_L$,

$$J_c(u_n) \Longrightarrow \inf J_c. \tag{2.94}$$

Thus, we see, thanks to (2.93), that the sequence $\{u_n\}$ is uniformly bounded, and equicontinuous on every compact interval. Utilizing the Ascoli-Arzelà theorem and a diagonal argument, we obtain a subsequence, still called $\{u_n\}$, that converges uniformly on compact intervals to $u_L \in C(\mathbb{R}; \mathbb{R}^m)$. Next, we work in the space $L^2_\mu(\mathbb{R}; \mathbb{R}^m)$ with weight $x \mapsto e^{cx}$, the standard Lebesgue measure $\mathrm{d}x$ being replaced by the absolutely continuous one $\mathrm{d}\mu(x) = e^{cx}\mathrm{d}x$. For a subsequence still called $\{u_n\}$, we have

$$u'_n \rightharpoonup v \text{ in } L^2_\mu(\mathbb{R}; \mathbb{R}^m), \tag{2.95}$$

and weak lower semicontinuity gives

$$\liminf \int_{\mathbb{R}} |u'_n|^2 e^{cx}\mathrm{d}x \geq \int_{\mathbb{R}} |v|^2 e^{cx}\mathrm{d}x. \tag{2.96}$$

By a standard argument, we check that $v = u'_L$, and $u_L \in \mathscr{A}_L$. Finally, we apply the Fatou lemma to the sequence $\{W(u_n) - W(a^-)\chi_{(-\infty,L]}\}$ (where $\chi_{(-\infty,L]}$ denotes the characteristic function of the interval $(-\infty, L]$):

$$\int_{\mathbb{R}} (W(u_L) - W(a^-)\chi_{(-\infty,L]})e^{cx}\mathrm{d}x \leq \liminf \int_{\mathbb{R}} (W(u_n) - W(a^-)\chi_{(-\infty,L]})e^{cx}\mathrm{d}x. \tag{2.97}$$

Collecting the previous results, we deduce that $J_c(u_L) \leq \liminf J_c(u_n) = \inf_{\mathscr{A}_L} J_c$. The proof of Proposition 2.8 is complete. $\qquad\square$

Step 2: Removing the Constraint Inside the Cylinders
Next we show that the constraint cannot be realized in the interior of the cylinders, thus the only potentially dangerous points are the 'rims' $x = \pm L$.

Proposition 2.9 *All u_L's, $L > 1$, satisfy*

$$u'' - W_u(u) = -cu'$$

in $C^2(\mathbb{R} \setminus \{\pm L\})$. Moreover $u_L(x) \to a^\pm$ as $x \to \pm\infty$.

Proof The proof of the proposition follows from the two lemmas below which make use of the projection P on a smooth convex set. We recall that if $u \in W^{1,2}_{\mathrm{loc}}(\mathbb{R}; \mathbb{R}^m)$, then $v := P(u) \in W^{1,2}_{\mathrm{loc}}(\mathbb{R}; \mathbb{R}^m)$, and $|v'| \leq |u'|$ since P is a contraction. $\qquad\square$

Lemma 2.7 *Let $\rho^-_L(x) := |u_L(x) - a^-|$. Then, the equation $\rho^-_L(x) = r_0/2$ has a unique solution $\lambda^-_L \geq -L$. In addition, the function ρ^-_L is strictly increasing in the interval $(-\infty, \lambda^-_L]$ and $\lim_{x \to -\infty} \rho^-_L(x) = 0$.*

Proof In Remark 2.6 we mentioned that Lemma 2.5 is based on a pointwise deformation and thus it also holds for the functional J_c. Let us first prove that for every $r \in (0, r_0/2]$, the equation $\rho^-_L(x) = r$ has a unique solution. Thanks to Lemma 2.5, it is clear that this equation has at most two solutions. Suppose by contradiction that $\rho_L^-(x_1) = \rho_L^-(x_3) = r$ for $x_1 < x_3$. Then, according to Lemma 2.5, there exists $x_2 \in (x_1, x_3)$ such that $\min_{[x_1,x_3]} \rho_L^- = \rho_L^-(x_2)$. Moreover $\min_{(-\infty,x_1]} \rho_L^- = r$, because otherwise the equation $\rho_L^-(x) = r - \epsilon$ will have more than two solutions for $\epsilon > 0$ small. Writing $u_L(x) = a^- + \rho_L^-(x)\boldsymbol{n}_L^-(x)$, we define

$$v(x) := \begin{cases} u_L(x) & \text{for } x \geq x_2, \\ a^- + \rho_L^-(x_2)\boldsymbol{n}_L^-(x) & \text{for } x \leq x_2, \end{cases} \tag{2.98}$$

and obtain $J_c(v) < J_c(u_L)$. This proves our claim, which yields, due to the constraint, the existence of a unique $\lambda_L^- \geq -L$ such that

$$\rho_L^-(\lambda_L^-) = r_0/2 \text{ and } x < \lambda_L^- \iff \rho_L^-(x) < r_0/2.$$

It also follows that ρ_L^- is strictly increasing in the interval $(-\infty, \lambda_L^-]$, and that u_L solves the equation $u'' - W_u(u) = -cu'$ in the interval $(-\infty, \lambda_L^-)$. To conclude, suppose by contradiction that $\lim_{x \to -\infty} \rho_L^-(x) = \epsilon > 0$. Setting $f(x) := (\rho_L^-(x))^2$, we obtain, thanks to Hypothesis $\mathbf{H}_4$, that

$$f'' + cf' = 2|u'|^2 + 2W_u(u_L) \cdot (u_L - a^-) \geq \delta > 0$$

in the interval $(-\infty, \lambda_L^-)$. Integrating this inequality, we deduce that for $s < t < x < \lambda_L^-$:

$$f'(t) \geq \delta(t - s) - c\frac{r_0^2}{2} + f'(s) \geq \delta(t - s) - c\frac{r_0^2}{2},$$

and

$$f(x) - f(s) = \int_s^x f'(t)\mathrm{d}t \geq \frac{\delta}{2}(x - s)^2 - c\frac{r_0^2}{2}(x - s),$$

which contradicts that f is bounded. Thus, $\lim_{x \to -\infty} \rho_L^-(x) = 0$. $\square$

Note From the proof of Lemma 2.7 we can also see that the function ρ_L^- is strictly increasing as long as $\rho_L^-(x) \leq r$ and $B(a^-, 2r) \subset C_0^-$.

Lemma 2.8 *Let $\rho_L^+(x) := |u_L(x) - a^+|$. Then, the equation $\rho_L^+(x) = r_0/2$ has a unique solution $\lambda_L^+ \leq L$. In addition, the function ρ_L^+ is strictly decreasing in the interval $[\lambda_L^+, \infty)$ and $\lim_{x \to \infty} \rho_L^+(x) = 0$.*

Proof As in the proof of Lemma 2.6, we first note that $J_c(u_L) < \infty$ implies the existence of a sequence $x_n \to +\infty$ as $n \to \infty$ such that $\rho_L^+(x_n) \to 0$. Next, we show that for every $r \in (0, r_0/2]$, the equation $\rho_L^+(x) = r$ has a unique solution.

Suppose by contradiction that $\rho_L^+(\alpha) = \rho_L^+(\beta) = r$ for $\alpha < \beta$. If $u_L(y_2) \in \partial C_0^-$ for some $y_2 \in (\alpha, \beta)$, there also exists $y_1 < \alpha$ such that $u_L(y_1) \in \partial C_0^-$. Defining

$$v(x) := \begin{cases} P(u_L(x)) & \text{for } x \in [y_1, y_2], \\ u_L(x) & \text{for } x \in (-\infty, y_1] \cup [y_2, \infty), \end{cases} \tag{2.99}$$

where P is the projection onto the convex set C_0^-, we see that $J_c(v) < J_c(u_L)$. Thus, the curve $x \mapsto u_L(x)$ does not intersect C_0^- when $x \in [\alpha, \beta]$, and Lemma 2.5 applies. It follows that

$$\rho_L^+(\alpha) = \rho_L^+(\beta) = r \implies \rho_L^+(x) < r, \ \forall x \in (\alpha, \beta),$$

and utilizing the fact that $\rho_L^+(x_n) \to 0$ as $n \to \infty$, we reach a contradiction. This proves our claim, which yields, due to the constraint, the existence of a unique $\lambda_L^+ \leq L$ such that

$$\rho_L^+(\lambda_L^+) = r_0/2 \text{ and } x > \lambda_L^+ \Leftrightarrow \rho_L^+(x) < r_0/2.$$

The monotonicity of ρ_L^+ is another straightforward consequence. $\qquad\square$

Step 3: A Replacement Lemma for the Convex Set C_α^-
Hypotheses $\mathbf{H}_1$–$\mathbf{H}_3$ imply that the set $\{u : W(u) \leq \alpha\}$ for $\alpha \in (0, \alpha_0]$ $(0 < \alpha_0 \ll 1)$, is made up of two components, which we denote by C_α^- and C_α^+, with C_α^- strictly convex and enclosing a^-. Moreover, decreasing α_0 if necessary, we may assume that $C_\alpha^+ \subset B_{r_0/2}(a^+)$, $B_{r_0}(a^+)$ is disjoint from C_α^-, and $W(u) > \alpha$ for $x \notin C_\alpha^- \cup B_{r_0/2}(a^+)$.

Lemma 2.9 *For every $\alpha \in (0, \alpha_0)$, there exists a unique $\lambda_L^{\alpha-} \in (\lambda_L^-, \lambda_L^+)$ such that $u_L(\lambda_L^{\alpha-}) \in \partial C_\alpha^-$ and $u_L(x) \in C_\alpha^- \Longleftrightarrow x \leq \lambda_L^{\alpha-}$,*

Proof Suppose by contradiction that $u_L(x_2) \in \partial C_\alpha^-$ and $u_L(x_4) \in \partial C_\alpha^-$ for $x_2 < x_4$. If $u_L(x) \notin C_\alpha^-$ for some $x \in (x_2, x_4)$, we know thanks to Lemma 2.8 that $u_L(x) \notin B(a^+, r_0/2)$ for $x \in [x_2, x_4]$. Setting

$$v(x) := \begin{cases} u_L(x) & \text{for } x \in (-\infty, x_2] \cup [x_4, \infty), \\ P(u_L(x)) & \text{for } x \in [x_2, x_4], \end{cases} \tag{2.100}$$

where P is the projection onto the convex set C_α^-, we obtain $J_c(v) < J_c(u_L)$, which is a contradiction. On the other hand, if $0 < W(u_L(x_3)) < \alpha$ for some $x_3 \in (x_2, x_4)$, then there exists $x_1 < x_2$ such that $W(u_L(x_1)) = W(u_L(x_3))$, and $W(u_L(x_3)) < W(u_L(x_2))$, in contradiction with what precedes. So far, we have proved that $u_L(x) \in \partial C_\alpha^-$ for $x \in [x_2, x_4]$. We are going to show that this situation is impossible due to the strict convexity of C_α^-. Indeed, u_L solves the equation $u'' - W_u(u) = -cu'$ in the interval $[x_2, x_4]$, and it is not constant by Hypothesis $\mathbf{H}_3$, so there exists $y \in (x_2, x_4)$ such that $u'_L(y) \neq 0$. Setting for $\delta > 0$ small enough

$$v_\delta(x) := \begin{cases} u_L(x) & \text{for } x \in (-\infty, y] \cup [y + \delta, \infty), \\ P_\delta(u_L(x)) & \text{for } x \in [y, y + \delta], \end{cases} \tag{2.101}$$

where P_δ is the projection onto the line going through $u_L(y)$ and $u_L(y+\delta)$. In view of the strict convexity of C_α^-, $v_\delta((y, y + \delta))$ is included in the interior of C_α^-. Thus, we obtain $J_c(v_\delta) < J_c(u_L)$, which is a contradiction. This completes the proof of the lemma. $\square$

Recapitulating the previous results, we have

1. u_L exits C_α^- precisely once at $x = \lambda_L^{\alpha-}$, $u_L^{-1}(C_\alpha^-) = (-\infty, \lambda_L^{\alpha-}]$;
2. $u_L^{-1}(B_{r_0/2}(a^+)) = (\lambda_L^+, \infty)$;
3. $x \in [\lambda_L^{\alpha-}, \lambda_L^+] \implies W(u_L(x)) \geq \alpha$;
4. $\lambda_L^\pm$ are well defined as the unique x-values at which the curve u_L crosses the spheres $\partial B_{r_0/2}(a^\pm)$.

Part II: Determination of c

So far $c > 0$ is a free parameter. The first time in the proof where c has to be chosen in a special way appears in Lemma 2.10. The notation $u_{L,c}$ means what we used to call u_L so far, but now emphasizes also the dependence on c. In all situations where the value of c is not self-evident, we will always write explicitly the dependence on L and c.

Step 4: Implications of the Replacement Lemmas—The Bound on $|\lambda_L^+ - \lambda_L^-|$
In the following lemma, we establish an L-independent bound on $|\lambda_L^+ - \lambda_L^{\alpha-}|$. We note that

$$-L \leq \lambda_L^{0-} := \sup\{x \in \mathbb{R} : u_L(x) \in \partial C_0^-\} \leq \lambda_L^{\alpha-} \leq \lambda_L^+,$$

and set $\mathrm{dist}(C_\alpha^-, B(a^+, r_0/2)) =: d_\alpha$.

Lemma 2.10 *For all $\alpha \in (0, \alpha_0]$, $L \geq 1$ and $c > 0$ such that $J_c(u_L) \leq 0$, we have the estimate*

$$|\lambda_L^+ - \lambda_L^{\alpha-}| \leq \frac{1}{c} \ln\left(1 + \frac{W^-(a^-)}{\alpha}\right) := \Lambda_{\alpha,+}. \tag{2.102}$$

Proof Consider the identity

$$J_c(u_L) = -\int_{-\infty}^{\lambda_L^{0-}} W^-(u_L)e^{cx}dx + \int_{\lambda_L^{0-}}^{\infty} W^+(u_L)e^{cx}dx + \frac{1}{2}\int_{\mathbb{R}} |u'_L|^2 e^{cx}dx. \tag{2.103}$$

Recall that $W(u_L) \geq \alpha$ on $[\lambda_L^{\alpha-}, \lambda_L^+]$, $W^-(u_L) \leq W^-(a^-)$. We estimate separately

$$\int_{-\infty}^{\lambda_L^{0-}} W^-(u_L)e^{cx}dx \leq \frac{W^-(a^-)}{c}e^{c\lambda_L^{0-}},$$

$$\int_{\lambda_L^{0-}}^{\infty} W^+(u_L)e^{cx}dx \geq \int_{\lambda_L^{\alpha-}}^{\lambda_L^+} W^+(u_L)e^{cx}dx \geq \frac{\alpha}{c}[e^{c\lambda_L^+} - e^{c\lambda_L^{\alpha-}}],$$

$$d_\alpha \leq |u_L(\lambda_L^{\alpha-}) - u_L(\lambda_L^+)| \leq \int_{\lambda_L^{\alpha-}}^{\lambda_L^+} |u_L'|dx$$

$$\leq \left(\int_{\lambda_L^{\alpha-}}^{\lambda_L^+} e^{-cx}dx\right)^{1/2} \left(\int_{\lambda_L^{\alpha-}}^{\lambda_L^+} |u_L'|^2 e^{cx}dx\right)^{1/2}.$$

Thus in view of the assumption $J_c(u_L) \leq 0$, we have

$$0 \geq J_c(u_L) \geq -\frac{W^-(a^-)}{c}e^{c\lambda_L^{0-}} + \frac{\alpha}{c}[e^{c\lambda_L^+} - e^{c\lambda_L^{\alpha-}}] + \frac{cd_\alpha^2}{2(e^{-c\lambda_L^{\alpha-}} - e^{-c\lambda_L^+})} \tag{2.104}$$

$$\geq e^{c\lambda_L^{\alpha-}}\left\{-\frac{W^-(a^-)}{c} + \frac{\alpha}{c}(e^{c(\lambda_L^+ - \lambda_L^{\alpha-})} - 1) + \frac{cd_\alpha^2}{2(1 - e^{-c(\lambda_L^+ - \lambda_L^{\alpha-})})}\right\}$$

$$\geq \frac{e^{c\lambda_L^{\alpha-}}\alpha}{c}\left\{-\left(\frac{W^-(a^-)}{\alpha} + 1\right) + e^{c(\lambda_L^+ - \lambda_L^{\alpha-})}\right\},$$

from which (2.102) follows. $\qquad\square$

Next, we are going to establish an L-independent bound on $|\lambda_L^{\alpha-} - \lambda_L^-|$. Set

$$R_{\max}^\alpha := \max_{u \in \partial C_\alpha^-} |u - a^-|,$$

$$w^* := \min\left\{\frac{d}{dt}\Big|_{t=r} W(a^- + t\xi) \ : \ \frac{r_0}{2} \leq r \leq R_{\max}^\alpha, \ |\xi| = 1, \ a^- + r\xi \in C_\alpha^-\right\}.$$

Lemma 2.11 *We have the estimate*

$$|\lambda_L^{\alpha-} - \lambda_L^-| \leq \frac{1}{w^*}\left\{cR_{\max}^\alpha + [(cR_{\max}^\alpha)^2 + 2w^*(R_{\max}^\alpha - r_0/2)]^{1/2}\right\} =: \Lambda_{\alpha,-}. \tag{2.105}$$

Proof Consider the polar form $u_L = a^- + \rho_L^- n_L^-$, and define $\rho(x) := \rho_L^-(x)$ and observe that, in $(\lambda_L^-, \lambda_L^{\alpha-})$, u_L satisfies the equation $u'' - W_u(u) = -cu'$. Then, by scalar multiplication of $u_L'' - W_u(u_L) = -cu_L'$ by n_L^- and recalling that $(n_L^-)' \cdot n_L^- = 0$, $(n_L^-)'' \cdot n_L^- = -|(n_L^-)'|^2$, we have

$$\rho'' + c\rho' = \rho|(n_L^-)'|^2 + \nabla W(a^- + \rho n_L^-) \cdot n_L^- \geq w^* > 0, \qquad (2.106)$$

where $w^* > 0$ on $(\lambda_L^-, \lambda_L^{\alpha-}] \subseteq (-L, L]$ follows from Hypothesis $\mathbf{H_4}$. Integrating this once, and utilizing that $\rho'(x) \geq 0$ when $x \to \lambda_L^-$, we obtain

$$\rho'(x) + c\rho(x) \geq w^*(x - \lambda_L^-),$$

for $x \in (\lambda_L^-, \lambda_L^{\alpha-}]$. Integrating once more, and utilizing that $0 \leq \rho \leq R_{\max}^\alpha$, we deduce that

$$(R_{\max}^\alpha - r_0/2) + (\lambda_L^{\alpha-} - \lambda_L^-)cR_{\max}^\alpha \geq \frac{w^*}{2}(\lambda_L^{\alpha-} - \lambda_L^-)^2.$$

From this (2.105) follows. $\square$

Note $\rho_L^-(x) = |u_L(x) - a^-|$ is strictly increasing on $(-\infty, \lambda_L^{\alpha-}]$. This follows from (2.106).

Combining now Lemma 2.10 and 2.11, we obtain the desired L-independent bound on $\lambda_L^+ - \lambda_L^- = |\lambda_L^+ - \lambda_L^-|$:

Corollary 2.2 *For all* $\alpha \in (0, \alpha_0]$, $L \geq 1$, *and* $c > 0$ *such that* $J_c(u_L) \leq 0$, *we have*

$$|\lambda_L^+ - \lambda_L^-| \leq \Lambda, \quad \Lambda \ L\text{-independent}.$$

Proof $\lambda_L^+ - \lambda_L^- = (\lambda_L^+ - \lambda_L^{\alpha-}) + (\lambda_L^{\alpha-} - \lambda_L^-) \leq \Lambda_{\alpha,+} + \Lambda_{\alpha,-} =: \Lambda$. $\square$

Step 5: Uniform Bounds
At this stage we consider that $L > 1$ is fixed, and compute bounds for the minimizers $u_{L,c}$, when c belongs to a compact subset of $(0, \infty)$.

Proposition 2.10 *Let* $L > 1$ *and* $c_0 > 0$ *be fixed. Then, there exists a constant* $k > 0$ *such that for every* $c \in [c_0/2, 2c_0]$ *we have:*

(i) $|u_{L,c}(x)| \leq k, \ \forall x \in \mathbb{R}$;

(ii) $|u'_{L,c}(x)| \leq k, \ \forall x \neq \pm L$;

(iii) $|u_{L,c}(x) - a^+| \leq ke^{-cx}$, $|W(u_{L,c}(x))| \leq ke^{-2cx}$, *and* $|u'_{L,c}(x)| \leq ke^{-cx}$, $\forall x > L$;

(iv) *the mininizers* $u_{L,c}$ *are equicontinuous on bounded intervals for* $c \in [c_0/2, 2c_0]$.

Proof (i) and (iv) follow respectively from (2.93) and (2.91). To prove (ii) we first establish the uniform bound in the intervals $(-\infty, -L - 1)$, $(-L + 1, L - 1)$ and $(L + 1, \infty)$. Setting $v_{L,c}(x) := e^{cx/2}u_{L,c}(x)$, we have

$$v''_{L,c}(x) = e^{cx/2}\Big(W_u(u_{L,c}(x)) + \frac{c^2}{4}u_{L,c}(x)\Big).$$

From this expression and property (i), we deduce that in the three aforementioned intervals:

$$|v'_{L,c}(x)| \leq Me^{cx/2} \iff |u'_{L,c}(x)| \leq M, \text{ for some constant } M > 0.$$

To extend this uniform bound on all $\mathbb{R}$, we note that

$$\frac{\mathrm{d}}{\mathrm{d}x}\Big(W(u_{L,c}(x)) - \frac{1}{2}|u'_{L,c}(x)|^2\Big) = c|u'_{L,c}(x)|^2 \geq 0, \ \forall x \neq \pm L, \tag{2.107}$$

and thus, by monotonicity, the one-sided limits of $|u'_{L,c}(x)|$ exist at $\pm L$. We claim that these limits are uniformly bounded. Indeed, if we integrate (2.107), we obtain

$$|u'_{L,c}(-L^+)|^2 \leq 2ce^{cL}\int_{-L}^{0}|u'_{L,c}(x)|^2 e^{cx}\mathrm{d}x - 2W(u_{L,c}(0))$$

$$+ 2W(u_{L,c}(-L)) + |u'_{L,c}(0)|^2,$$

$$|u'_{L,c}(L^+)|^2 \leq 2ce^{-cL}\int_{L}^{\infty}|u'_{L,c}(x)|^2 e^{cx}\mathrm{d}x + 2W(u_{L,c}(L)) + M^2,$$

and we can see as before that these quantities are uniformly bounded. Then, utilizing again the monotonicity of the function $x \mapsto W(u_{L,c}(x)) - \frac{1}{2}|u'_{L,c}(x)|^2$, we deduce the desired uniform bound for $u'_{L,c}$ in $\mathbb{R} \setminus \{\pm L\}$. Finally, to show (iii), we consider the function $\rho_{L,c}(x) := |u_{L,c}(x) - a^+|$, which satisfies for $x > L$ the inequality $\rho'' + c\rho' \geq 0$ (cf. (2.106)). Utilizing (ii), we find that $-\rho'_{L,c}(x) \leq Me^{-cx}$ for $x > L$, with $M > 0$ independent of c. Then, by an integration, we obtain the first inequality in (iii). The second inequality follows immediately. To prove the third, we introduce $s_{L,c}(x) := e^{cx/2}(u_{L,c}(x) - a^+)$, and deduce from the previous inequality that $s'_{L,c} \leq Me^{-cx/2}$, for all $x > L$, and for a constant $M > 0$. Since $u'_{L,c} = e^{-cx/2}s'_{L,c} - \frac{c}{2}(u_{L,c} - a^+)$, we obtain the desired inequality for $u'_{L,c}$. $\quad\square$

Corollary 2.3 *Let $L > 1$ and $c_0 > 0$ be fixed. Then, the function $c \mapsto J_c(u_{L,c_0})$ is continuous in the interval $(c_0/2, 2c_0)$.*

Proof According to Proposition 2.10, we can find constants $k_1, k_2 > 0$ such that

$$\Big|\frac{1}{2}|u'_{L,c_0}(x)|^2 + W(u_{L,c_0}(x))\Big| \leq k_1, \ \forall x \leq L,$$

while

$$\left| \frac{1}{2}|u'_{L,c_0}(x)|^2 + W(u_{L,c_0}(x)) \right| \le k_2 e^{-2c_0 x}, \quad \forall x \ge L.$$

Thus we have for $x \le L$ and $c \in (c_0/2, 2c_0)$ that

$$e^{cx}\left| \frac{1}{2}|u'_{L,c_0}(x)|^2 + W(u_{L,c_0}(x)) \right| \le k_1 e^{\frac{3c_0 L}{2}} e^{\frac{c_0}{2}x},$$

while for $x \ge L$ and $c \in (c_0/2, 2c_0)$,

$$e^{cx}\left| \frac{1}{2}|u'_{L,c_0}(x)|^2 + W(u_{L,c_0}(x)) \right| \le k_2 e^{(c-2c_0)x}.$$

Finally, we apply the dominated convergence theorem to conclude. $\qquad\square$

Corollary 2.4 *Let $L > 1$ be fixed, and let $c_n > 0$ be a sequence converging to $c^* > 0$. If $J_{c_n}(u_{L,c_n}) \le 0$ for every n, then also $J_{c^*}(u_{L,c^*}) \le 0$.*

Proof According to Proposition 2.10, the sequence u_{L,c_n} is uniformly bounded and equicontinuous on compact intervals. Thus, applying the theorem of Ascoli-Arzelá via a diagonal argument, we can find a subsequence, still called u_{L,c_n}, which converges uniformly on compact intervals to a continuous function u^*. According to Proposition 2.10, the sequence $u'_{L,c_n} e^{c_n x/2}$ is also uniformly bounded in $L^2(\mathbb{R}; \mathbb{R}^m)$. As a consequence, there exists $v^* \in L^2_{\mathrm{loc}}(\mathbb{R}; \mathbb{R}^m)$ such that for a subsequence $u'_{L,c_n} e^{c_n x/2} \rightharpoonup v^* e^{c^* x/2}$ in L^2, and

$$\int_{\mathbb{R}} |v^*(x)|^2 e^{c^* x} dx \le \liminf \int_{\mathbb{R}} |u'_n(x)|^2 e^{c_n x} dx. \tag{2.108}$$

Furthermore, we can prove that $v^* = (u^*)'$, and thus $u^* \in \mathscr{A}_L$. On the other hand, thanks to (iii) in Proposition 2.10, we have by dominated convergence

$$\int_{\mathbb{R}} W(u^*(x)) e^{c^* x} dx = \lim \int_{\mathbb{R}} W(u_n(x)) e^{c_n x} dx. \tag{2.109}$$

Gathering the previous results, we deduce that

$$J_{c^*}(u_{L,c^*}) = \int_{\mathbb{R}} \left\{ \frac{1}{2}|v^*(x)|^2 + W(u^*(x)) \right\} e^{c^* x} dx \le 0.$$

$$\square$$

Step 6: The Continuity Argument
We will show the existence of a unique c^* such that

$$J_{c^*}(u_L) = \inf_{\mathscr{A}_L} J_{c^*} = 0, \quad L \ge \Lambda. \tag{2.110}$$

For this purpose, following an idea due to S. Heinze, we introduce the set

$$C := \{c > 0 : \exists L \geq 1 : \ J_c(u_{L,c}) < 0\}.$$

Lemma 2.12 *C is nonempty, open, and*

$$\sup C \leq \sqrt{2W^-(a^-)} \, (d_0)^{-1},$$

where $d_0 = \mathrm{dist}(C_0^-, B_{r_0/2}(a^+))$.

Proof It is convenient to work with the following equivalent definition of C:

$$C = \{c > 0 : \exists L \geq 1 \text{ and } \exists v \in \mathscr{A}_L : \ J_c(v) < 0\}.$$

Let us first prove that C is open. Indeed, if $c_0 \in C$, that is if $J_{c_0}(u_{L,c_0}) < 0$, then by Corollary 2.3 $J_c(u_{L,c_0}) < 0$ for c close to c_0. Now, recall that $u_{\mathrm{aff}} \in \bigcap_{L \geq 1} \mathscr{A}_L$ and note that, by the estimate (2.89),

$$f(c) \geq J_c(u_{\mathrm{aff}}),$$

where

$$f(c) := e^{-c} \left(-\frac{1}{c} W^-(a^-) + e^{2c} J_0^+(u_{\mathrm{aff}}) \right).$$

Since $f(0) = -\infty$, $f' > 0$ on $(0, \infty)$, $f(+\infty) = +\infty$, there is a unique c_0 such that $f(c_0) = 0$. Thus $(0, c_0) \subseteq C$, hence $C \neq \emptyset$. Moreover, for $c \in C$ fixed,

$$0 > J_c(v) \geq J_c(u_L) \geq e^{c\lambda_L^{\alpha-}} \left\{ -\frac{W^-(a^-)}{c} + \frac{cd_\alpha^2}{2(1 - e^{-c(\lambda_L^+ - \lambda_L^{\alpha-})})} \right\}$$

by (2.104) above, which implies that

$$0 \geq c^2 d_\alpha^2 - 2W^-(a^-).$$

Letting $\alpha \to 0$ we complete the proof. $\qquad\square$

We define now

$$c^* := \sup \ C.$$

We will show (2.110), that is,

$$J_{c*}(u_{L,c*}) = 0, \quad \text{for all } L \geq \Lambda,$$

where Λ was introduced in Corollary 2.2. Let $\{c_m\} \subset C$ be a sequence such that $c_m \to c^*$ as $m \to +\infty$. By the definition of C, there exists another sequence $\{L_m\}$ such that

$$J_{c_m}(u_{L_m,c_m}) < 0.$$

Thus, by the uniform bound in Corollary 2.2,

$$\lambda_{L_m}^+ - \lambda_{L_m}^- \leq \Lambda.$$

Moreover, since $J_{c_m}(u_{L_m,c_m}) < 0$, we necessarily have $\lambda_{L_m}^+ = L_m$, since otherwise a translation to the right would contradict the minimality of u_{L_m,c_m}. Now for $L_m \geq \Lambda$ we see that the translate to left $u_{L_m,c_m}(\cdot + L_m)$ is in $\mathscr{A}_\Lambda$, and

$$J_{c_m}(u_{\Lambda,c_m}) \leq J_{c_m}(u_{L_m,c_m}(\cdot + L_m)) = e^{-c_m L_m} J_{c_m}(u_{L_m,c_m}) < 0.$$

Passing to the limit as $m \to +\infty$ we have, according to Corollary 2.4,

$$J_{c^*}(u_{\Lambda,c^*}) \leq 0,$$

and in addition

$$J_{c^*}(u_{L,c^*}) \leq J_{c^*}(u_{\Lambda,c^*}) \leq 0, \quad \text{for all } L \geq \Lambda.$$

Since C is open and $c^* = \sup C < +\infty$, it follows that $c^* \notin C$. Thus we obtain

$$J_{c^*}(u_{L,c^*}) \geq 0, \quad \text{for all } L \geq 1.$$

Combining the previous inequalities, we deduce that $J_{c^*}(u_{L,c^*}) = 0$, for all $L \geq \Lambda$. (2.110) has been established.

Step 7: Existence
Choose $L > \Lambda + \delta$. By (2.110), $J_{c^*}(u_{L,c^*}) = 0$. Since $|\lambda_L^+ - \lambda_L^-| < \Lambda$, u_{L,c^*} cannot touch both rims. We can always translate by δ so as to avoid the dangerous rim. Moreover, $J_{c^*}(u_{L,c^*}(\cdot \pm \delta)) = 0$. Thus $u_{L,c^*}(\cdot \pm \delta)$ is still a minimizer, and so it satisfies the Euler-Lagrange equation $u'' - W_u(u) = -c^* u'$. The asymptotic limits hold by Proposition 2.9. The proof of existence is complete.

Step 8: Uniqueness of the Speed for Minimizers
If (u, c) is a solution to $u'' - W_u(u) = -cu'$, then by multiplying the equation by u' one obtains the identity

$$\frac{|u'|^2}{2} + W(u) = e^{-cx}\left\{\frac{e^{cx}}{c}\left(W(u) - \frac{|u'|^2}{2}\right)\right\}'.$$

Let (u_1, c_1^*), (u_2, c_2^*) be solutions, with u_i minimizing. We will show that $c_1^* = c_2^*$. We proceed by contradiction. So assume that $0 \leq c_1^* < c_2^*$. Considering the identity above for $c = c_2^*$ and $u = u_2^*$, and multiplying by $e^{c_1^* x}$, one can obtain after a few manipulations the identity

$$c_1^* J_{c_1^*}(u_2, (-t, t)) = (c_1^* - c_2^*) \int_{-t}^{t} |u_2'|^2 e^{c_1^* x} dx + \left[e^{c_1^* x} \left(W(u_2) - \frac{|u_2'|^2}{2} \right) \right]_{-t}^{t},$$

where the notation $J_c(u, (-t, t))$ means that the integration is over $(-t, t)$. We can take along a sequence $t_n \to +\infty$ the limit and obtain, utilizing $J_{c_2^*}(u_2) = 0$, $c_1^* < c_2^*$,

$$c_1^* J_{c_1^*}(u_2) = (c_1^* - c_2^*) \int_{\mathbb{R}} |u_2'|^2 e^{c_1^* x} dx < 0,$$

which contradicts the fact that $c_1^* J_{c_1^*}(u_2) \geq 0$.

Note We will see in the following section that heteroclinic connections (standing waves) are in general not unique. Thus we do not expect uniqueness of the profile u in general.

Step 9: The Variational Characterization
First we note that the sets

$$\widehat{C} = \{\hat{c} > 0 : \exists v \in \mathscr{A} \text{ such that } J_{\hat{c}}(v) < 0\}$$

and

$$C = \{c > 0 : \exists L \geq 1, \text{ and } \exists v \in \mathscr{A}_L \text{ such that } J_c(v) < 0\}$$

coincide ($\mathscr{A}$ as defined in the statement of Theorem 2.4), since $\mathscr{A} = \bigcup_{L \geq 1} \mathscr{A}_L$. Therefore, by Lemma 2.12,

$$c^* = \sup C = \sup \widehat{C},$$

and so

$$c^* = \sup_{c > 0} \left\{ \inf_{\mathscr{A}} J_c(v) < 0 \right\}.$$

Next we recall that for the profile u we have

$$u = u_{L^*, c^*}, \quad J_{c^*}(u) = 0,$$

where $L^* > \Lambda + \delta$, and arbitrary otherwise,

$$0 = J_{c^*}(u) = \min_{\mathscr{A}_L^*} J_{c^*}(v) \geq \inf_{\mathscr{A}} J_{c^*}(v) \geq 0.$$

Finally, multiplying the equation by u' and integrating as in (2.107), and utilizing Proposition 2.10, we obtain

$$c^* = \frac{-W(a^-)}{\int_{\mathbb{R}} |u'|^2 dx}.$$

The proof of Theorem 2.4 is complete. $\square$

2.7 Remarks on the Problem of Heteroclinic Connections for Potentials Possessing Three or More Global Minima

In this subsection, $W \in C^2(\mathbb{R}^m; \mathbb{R})$, and its minima are nondegenerate (i.e., the Hessian matrix W_{uu} is positive definite at the minima). Moreover, we assume that

$$W > 0 \text{ on } \mathbb{R}^m \setminus A, \quad A = \{a_1, \ldots, a_N\}, \; N \geq 3. \tag{2.111}$$

We are interested in the existence of a classical solution $u : \mathbb{R} \to \mathbb{R}^m$ to

$$u'' - W_u(u) = 0, \tag{2.112}$$

with

$$\lim_{x \to -\infty} u(x) = a_i, \quad \lim_{x \to +\infty} u(x) = a_j, \quad a_i \neq a_j \in A. \tag{2.113}$$

The new phenomenon for $N \geq 3$ is that it may happen that there is a pair (a_i, a_j) that cannot be connected, hence (2.113) may not have a solution. On the other hand, if $N \geq 3$ and $m = 1$ there is always a pair that is not connected (Fig. 2.7), and so for coexistence of three or more phases a vector order parameter is needed, that is, m should be larger than or equal to 2.

The problem of developing computable criteria for deciding existence is open for general m. However, for $m = 2$ a lot can be said for geometric reasons. As we have seen in Sect. 2.2, connections can be obtained also as geodesics for $\mathbb{R}^m$ equipped with the metric $W(u) du_i du_j$. For $m = 2$ we identify $z = u_1 + i u_2$ and write the metric as $W(z) dz d\bar{z}$. We note that it is reasonable to consider potentials of the form

$$W(z) = |f(z)|^2, \tag{2.114}$$

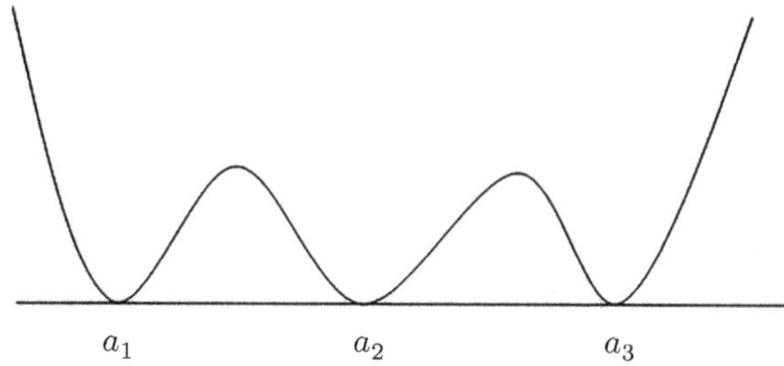

Fig. 2.7 a_1 is not connected to a_3

where f a is holomorphic or even a meromorphic function, since this class includes products of squares of distances from $a_1, \ldots, a_N$, i.e.

$$f(z) = (z - a_1)(z - a_2) \cdots (z - a_N).$$

For example, the potential $W(z) = |z^3 - 1|^2$ belongs to this class. The next observation is that the transformation

$$z \longmapsto \omega, \ \omega = g(z), \ g' = f, \tag{2.115}$$

is an isometry from $(\mathbb{C}, W(z)\mathrm{d}z\mathrm{d}\overline{z})$ to the Euclidean plane. We will see this later. Of course, care is needed because $g'(a_i) = 0$, and so g is not a bijection. The problem of existence then reduces to showing that the preimage of the line segment $[\omega_i, \omega_j]$, $\omega_i = g(a_i)$, $\omega_j = g(a_j)$, is contained in the same branch leaf of the Riemann surface $g^{-1}(\omega)$. Moreover, the fact that g maps geodesics to straight lines allows explicit formulas, that is, we obtain solutions in closed form.

The following calculation is illuminating. Consider parametrizations $x \mapsto u(x) \in C^1$, $|u'(x)| \neq 0$, with $\lim_{x \to -\infty} u(x) = a_i$, $\lim_{x \to +\infty} u(x) = a_j$. By the nondegeneracy of the minima, the integrals below are finite,

$$L(u) = \sqrt{2} \int_{\mathbb{R}} \sqrt{W(u(x))}\, |u'(x)|\mathrm{d}x = \sqrt{2} \int_{\mathbb{R}} |f(z(x))||z'(x)|\mathrm{d}x$$

$$= \sqrt{2} \int_{\mathbb{R}} \left| \frac{\mathrm{d}}{\mathrm{d}x} g(z(x)) \right| \mathrm{d}x, \tag{2.116}$$

where $z(x) := u_1(x) + iu_2(x)$. Hence

$$d_W(a_i, a_j) := \inf L(u), \ (u(\cdot) \text{ as above}) \tag{2.117}$$

$$= \inf \left\{ \sqrt{2} \int_{\mathbb{R}} |\omega'(x)|\mathrm{d}x \ : \ \omega(x) = g(z(x)) \right\}$$

$$\geq \min \left\{ \sqrt{2} \int_{\mathbb{R}} |\omega'(x)|\mathrm{d}x \ : \ x \to \omega(x)\, C^1, \right.$$

$$\left. \lim_{x \to -\infty} \omega(x) = g(a_i), \ \lim_{x \to +\infty} \omega(x) = g(a_j) \right\}$$

$$= \sqrt{2}\, |g(a_i) - g(a_j)|.$$

Notice that the min above is taken over all paths $\omega(\cdot)$ connecting $g(a_i)$ and $g(a_j)$, and of course it is achieved by the line segment. On the other hand, the inf is taken over paths that can be realized as images under g of curves connecting a_i and a_j. Hence in general the inequality above is strict, and when it is there is no connection. That is the case when the line segment $[g(a_i), g(a_j)]$ is not the image under g of a curve $z(x)$ connecting a_i and a_j. We will show this later. We begin by showing that solutions to $u'' - W_u(u) = 0$ satisfying the equipartition relation are mapped under g to line segments.

Theorem 2.5 *Identify the point (u_1, u_2) with the complex number $z = u_1 + iu_2$, and write $W(u_1, u_2) = |f(z)|^2$. Assume $f = g'$ is holomorphic in D, an open set in $\mathbb{R}^2$. Let $\gamma = \{u(x) : x \in (a, b)\}$ be a smooth curve in D, where x is an equipartition parameter, i.e., $\frac{1}{2}|u'|^2 = W(u)$. Set $\alpha = u(a)$, $\beta = u(b)$. Then u is a solution to $u'' - W_u(u) = 0$ on (a, b) if and only if*

$$\mathrm{Im}\left(\frac{g(z) - g(\alpha)}{g(\beta) - g(\alpha)}\right) = 0, \quad \text{for all } z \in \gamma. \tag{2.118}$$

In addition, when u is a solution, the set $g(\gamma) := \{g(z) : z \in \gamma\}$ is a line segment with end points $g(\alpha)$, $g(\beta)$, and the (partial) action is given by

$$\int_a^y \left(\frac{1}{2}|u'|^2 + W(u)\right) dx = \sqrt{2}\int_a^y \left|\frac{d}{dx}\, g(u)\right| dx = \sqrt{2}\,|g(u(y)) - g(\alpha)|, \ \forall y \in (a, b]. \tag{2.119}$$

Remark 2.10 We are particularly interested in $\alpha = a_i$, $\beta = a_j$; then $(a, b) = (-\infty, +\infty)$, and

$$a_i = \lim_{x \to -\infty} u(x), \quad a_j = \lim_{x \to +\infty} u(x),$$

$$0 = u' \cdot (u'' - W_u(u)) = \left(\frac{1}{2}|u'|^2 - W(u)\right)',$$

hence $\frac{1}{2}|u'|^2 - W(u) = C$, and utilizing the limits we conclude that $C = 0$. Thus u is an equipartition parametrization.

Proof (Theorem 2.5) ($\Longrightarrow$) $u = (u_1, u_2) : (a, b) \to D$ is a solution to

$$u'' - W_u(u) = 0, \quad \frac{1}{2}|u'|^2 = W(u). \tag{2.120}$$

Set $z(x) = u_1(x) + iu_2(x)$. Then (2.120) is equivalent to $z'' = 2f(z)\overline{f'(z)}$, and $\frac{1}{2}|z'|^2 = f(z)\overline{f(z)}$.

Let L be the total arclength and l be the arclength parameter defined by

$$L = \int_a^b \sqrt{W(u)}\,|u'|\,dx, \quad l = \int_a^x \sqrt{W(u(\hat{x}))}\,|u'(\hat{x})|\,d\hat{x}.$$

Then

$$\frac{d}{dl} = \left(\sqrt{W(u)}\,|u'|\right)^{-1}\frac{d}{dx} = \left(\sqrt{W(u)}\sqrt{2W(u)}\right)^{-1}\frac{d}{dx} = \frac{1}{\sqrt{2}}(\overline{f}f)^{-1}\frac{d}{dx},$$

so that

$$\frac{dg(z)}{dl} = \frac{1}{\sqrt{2}}\frac{g'(z)z'}{f(z)\overline{f(z)}} = \frac{1}{\sqrt{2}}\frac{f(z)z'}{f(z)\overline{f(z)}} = \frac{1}{\sqrt{2}}\frac{z'}{\overline{f}},$$

$$\frac{d^2 g(z)}{dl^2} = \frac{1}{\sqrt{2}}\frac{\overline{f}\,z'' - z'\overline{f'z'}}{\overline{f}^2 f\overline{f}} = \frac{z'' - 2f\overline{f'}}{\sqrt{2}\,f\overline{f}^2} = 0.$$

Thus $\frac{d}{dl}g(z) = m$, a constant.[2] Integrating this equation and evaluating it at $l = L$ gives respectively $g(z(l)) = g(\alpha) + ml$, $mL = g(\beta) - g(\alpha)$. Upon noting that

$$m = \frac{d}{dl}\,g(z) = \frac{1}{\sqrt{2}}\frac{z'}{\overline{f}},$$

$$|m|^2 = \frac{1}{2}\frac{|z'|^2}{|\overline{f}|^2} = 1, \quad L = |g(\beta) - g(\alpha)|,$$

$$m = \frac{g(\beta) - g(\alpha)}{|g(\beta) - g(\alpha)|}, \quad g(z(l)) = \frac{L-l}{L}g(\alpha) + \frac{l}{L}g(\beta).$$

These equations imply (2.118) and (2.119).

($\Longleftarrow$) Next assume that $\gamma = u(a, b)$ satisfies (2.118) and x, the parameter for u, is an equipartition parameter, namely $\frac{1}{2}|u'|^2 = W(u)$. Equation (2.118) can be written as

$$g(z(x)) - g(\alpha) = s(x)(g(\beta) - g(\alpha)),$$

where $s(x)$ is a real valued function. Upon differentiation, we obtain

$$s'(g(\beta) - g(\alpha)) = g'(z)z' = f(z)z'.$$

---

[2] If $m = 0$, it follows that $g(z(t)) = g(\alpha)$, and since g is analytic $z(t) = \alpha$. Thus, if $u(x)$ is a heteroclinic, it follows that $m \neq 0$ and so $g(z_i) \neq g(z_j)$.

This equation implies

$$|s'| = \frac{|f(z)||z'|}{|g(\beta) - g(\alpha)|} = \frac{|z'|^2}{\sqrt{2}\,|g(\beta) - g(\alpha)|} > 0.$$

As $s(x)$ is real valued and $s(\alpha) = 0$, $s(\beta) = 1$, we must have $s'(x) > 0$. Hence,

$$s'(x) = \frac{\sqrt{2}\,|f(z)|^2}{|g(\beta) - g(\alpha)|},$$

and consequently

$$z' = \frac{s'(g(\beta) - g(\alpha))}{f(z)} = \frac{\sqrt{2}\,|f(z)|^2(g(\beta) - g(\alpha))}{f(z)|g(\beta) - g(\alpha)|} = \sqrt{2}\,m\overline{f(z)},$$

$$m = \frac{g(\beta) - g(\alpha)}{|g(\beta) - g(\alpha)|}.$$

Thus

$$z'' = \sqrt{2}\,m\overline{f'(z)z'} = \sqrt{2}\,m\overline{f'(z)}\sqrt{2}\,\overline{m}\,f(z) = 2|m|^2 f\overline{f'} = 2f\overline{f'}.$$

This equation is equivalent to u being a solution to $u'' - W_u(u) = 0$. The proof of Theorem 2.5 is complete. $\qquad\square$

As an important application of the last theorem we can show the following.

Theorem 2.6 *There exists at most one trajectory connecting any two minima of a potential $W(z) = |f(z)|^2$, with $f(z)$ holomorphic.*

Proof Let g be an antiderivative of f and suppose that γ_1 and γ_2 are two trajectories to $u'' - W_u(u) = 0$ with the same end points α, β. Since $\int_\alpha^\beta \left(\frac{1}{2}|u'|^2 + W(u)\right) dx > 0$, it follows that $|g(\beta) - g(\alpha)| > 0$, hence $g(\alpha) \neq g(\beta)$, and so we can define an entire function

$$\tilde{g}(z) = \frac{|g(\beta) - g(\alpha)|}{g(\beta) - g(\alpha)}(g(z) - g(\alpha)), \quad z \in \mathbb{C}.$$

Then $\tilde{g}$ is real-analytic on $\gamma_1 \cup \gamma_2$. Now if $\gamma_1 \neq \gamma_2$, then γ_1 and γ_2 enclose an open domain D in $\mathbb{C}$. As the imaginary part of $\tilde{g}$ on $\partial D = \gamma_1 \cup \gamma_2 \cup \{\alpha, \beta\}$ is zero, it has to be identically zero in D, by the uniqueness in the Dirichlet problem for harmonic functions. But then $\tilde{g}$ is constant by the open mapping theorem, which is a contradiction. The proof of Theorem 2.6 is complete. $\qquad\square$

We now analyze in detail (2.117) above.

Proposition 2.11 *Let* $W(u) = |f(z)|^2$, $f(z) = (z - z_1)(z - z_2) \cdots (z - z_n)$, *with* $z_j \neq z_k$ *for* $j \neq k$, *and suppose that* $\{g(z_1), g(z_2), g(z_k)\}$ *is a non-degenerate triangle for every* $k \neq 1, 2, \ldots$. *Then there is a connection between* z_1 *and* z_2 *if and only if*

$$d_W(z_1, z_2) = \sqrt{2}\,|g(z_1) - g(z_2)|.$$

Proof

1. Suppose there is a connection $u(x)$ between z_1 and z_2. Then by Remark 2.10, x is an equipartition parameter. Thus by (2.119), $d_W(z_1, z_2) = \sqrt{2}\,|g(z_1) - g(z_2)|$. We note that by (2.117), u is a minimizer of $L(u)$, and by Proposition 2.2, u is a minimizer of $J(u)$ (see (2.10), (2.11)).
2. Conversely, suppose that $d_W(z_1, z_2) = \sqrt{2}\,|g(z_1) - g(z_2)| = \inf L$ (by (2.117)). Also by (2.117),

$$d_W(z_1, z_k) \geq \sqrt{2}\,|g(z_1) - g(z_k)|, \quad d(z_2, z_k) \geq \sqrt{2}\,|g(z_2) - g(z_k)|, \quad \forall k = 3, \ldots, n.$$
$$\tag{2.121}$$

By the assumption of nondegeneracy of the triangle it follows from (2.121) that

$$
\begin{aligned}
d_W(z_1, z_2) &= \sqrt{2}\,|g(z_1) - g(z_2)| \\
&< \sqrt{2}\,|g(z_1) - g(z_k)| + \sqrt{2}\,|g(z_2) - g(z_k)| \\
&\leq d_W(z_1, z_k) + d_W(z_2, z_k),
\end{aligned}
$$

that is,

$$d_W(z_1, z_2) < \min_{k=3,\ldots,n} [d_W(z_1, z_k) + d_W(z_2, z_k)]. \tag{2.122}$$

From (2.122) it follows that there is a curve in $X_{z_1 z_2}$ realizing $L_{z_1 z_2} = d(z_1, z_2)$ (see Sect. 2.2 for notation). We sketch the argument. By (2.122), there is a $\delta > 0$ such that

$$d_W(z_1, z_2) < d_W(z_1, z_k) + d_W(z_2, z_k) - \delta, \quad \forall k = 3, \ldots, n. \tag{2.123}$$

Let $\{\gamma_j\}$ be a minimizing sequence in $X_{z_1 z_2}$ with $L(\gamma_j) \to d_W(z_1, z_2)$. It follows from (2.123) that γ_j is bounded away from z_k, $k = 3, \ldots, n$, a fixed distance. Consequently, the potential W can be modified near z_k so that the modification has only two minima precisely at z_1 and z_2. The hypotheses (10), (11), (12), p. 801 in [26] hold and the lemma applies and ensures compactness for the sequence $\{\gamma_j\}$, and therefore the existence of a γ realizing $d_W(z_1, z_2)$. Moreover, since the curves $\{\gamma_j\}$ are bounded away from the z_k, γ minimizes $d_W(z_1, z_2)$ for the original potential W. By Proposition 2.1, we can reparametrize $\gamma = \{u(t) : t \in \mathbb{R}\}$ via an equipartition

Fig. 2.8 The triple-well case

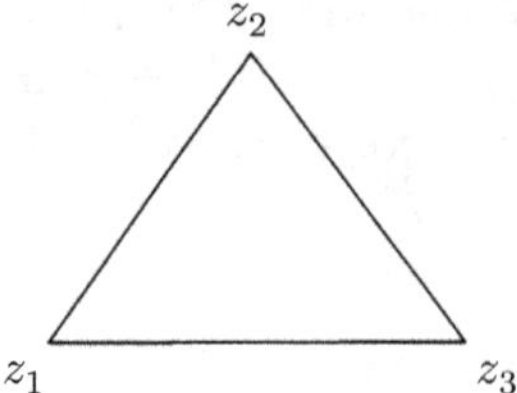

parametrization, $t = \phi(x)$, so that $u \circ \phi$ is a connection. We note that an easy additional argument is needed to handle the possibility that $|u'(t)|$ may be zero at certain points. $\qquad\square$

We now apply the Proposition above to the triple-well case (cf. Fig. 2.8).

Proposition 2.12 *Let $W(u) = |(z - z_1)(z - z_2)(z - z_3)|^2$, with z_1, z_2, z_3 distinct. Then, there is a connection between a pair if and only if the strict triangle inequality holds. For example, there is a connection between z_1 and z_2 if and only if*

$$d_W(z_1, z_2) < d_W(z_1, z_3) + d_W(z_3, z_2). \tag{2.124}$$

Proof

1. We already established in Proposition 2.6 that condition (2.124) is sufficient. Let us give an alternative proof based on the material of the present section. First we note that two of the three possible connections always exist (cf. also the comments after Theorem 2.1). Indeed, we may assume the ordering

$$d_W(z_2, z_3) \le d_W(z_1, z_3) \le d_W(z_1, z_2).$$

Hence, we have the strict inequalities

$$d_W(z_2, z_3) < d_W(z_2, z_1) + d_W(z_1, z_3),$$

and

$$d_W(z_1, z_3) < d_W(z_1, z_2) + d_W(z_2, z_3).$$

Thus, by the argument in Proposition 2.11 above, based on (2.122), we conclude that there exist connections between z_2 and z_3, and between z_1 and z_3. Thus by the first part of Proposition 2.11 we have

$$d_W(z_1, z_3) = \sqrt{2}\,|g(z_1) - g(z_3)|, \quad d(z_2, z_3) = \sqrt{2}\,|g(z_2) - g(z_3)|. \tag{2.125}$$

2. All we need to show is that (2.124) is necessary for the existence of a connection between z_1 and z_2. So suppose that

$$d_W(z_1, z_2) = d_W(z_1, z_3) + d_W(z_3, z_2). \tag{2.126}$$

Case 1: $g(z_1), g(z_2), g(z_3)$ form a nondegenerate triangle.
   In this case

$$\sqrt{2}\,|g(z_1) - g(z_2)| < \sqrt{2}\,|g(z_1) - g(z_3)| + \sqrt{2}\,|g(z_3) - g(z_2)|$$
$$= d_W(z_1, z_2), \ (\text{cf. } (2.125), (2.126)),$$

hence nonexistence follows by Proposition 2.11.

Case 2: $g(z_1), g(z_2), g(z_3)$ lie on a straight line. Then exactly one connection does not exist.

To establish this we need a lemma.

Lemma 2.13 *Let $f(z)$ be holomorphic, and suppose that z_i, z_j, z_l are three geometrically distinct roots of $f(z)$, and that there exist connections $z_{ij}(x)$ and $z_{jl}(x)$ with trajectories γ_{ij} and γ_{jl} exist. Then*

$$\overline{\gamma_{ij}} \cap \overline{\gamma_{il}} = \{z_i\}. \tag{2.127}$$

Proof We proceed by contradiction. Assume that the trajectories of the connections intersect at some other point z^* (cf. Fig. 2.9). By Theorem 2.5, the images of γ_{ij} and γ_{il} under g lie in the line segments $[g(z_i), g(z_j)]$ and $[g(z_i), g(z_l)]$ respectively. Since $\{g(z_i), g(z^*)\} \subset g(\gamma_{ij}) \cap g(\gamma_{il})$, the line segments would have to lie on the same straight line unless $g(z^*) = g(z_i)$. This possibility is excluded since it would imply reversal of the course of the image, which is not allowed ($|s'| > 0$, see proof of Theorem 2.5). Thus the triangle has to be degenerate. The plan next is to reach a contradiction by showing that the open map g sends the interior of the Jordan curve $z_i z^* z_i$ into the line segment $[g(z_i), g(z^*)]$. In the remainder of the proof, we pay particular attention to the definition of z^*.

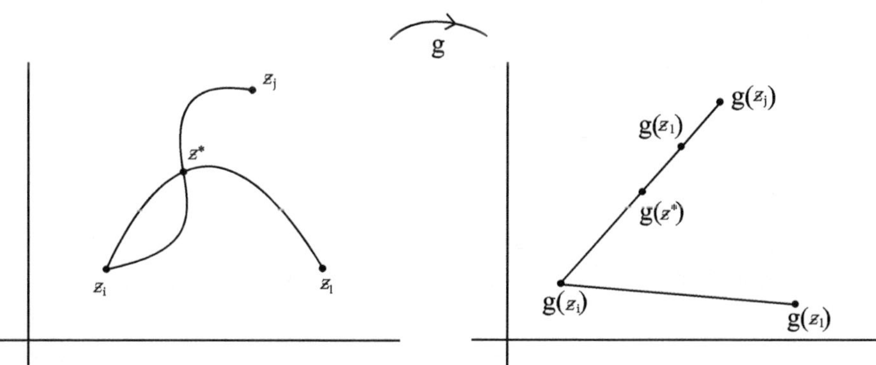

Fig. 2.9 The images of γ_{il}, γ_{ij} and z^* under g

(i) First, we note that $z_l \notin \overline{\gamma_{ij}}$. Indeed, otherwise $z_{ij}(\hat{x}) = z_l$ for some $\hat{x}$. From the equipartition relation, $\frac{1}{2}|z'_{ij}(\hat{x})|^2 = W(z_{ij}(\hat{x})) = 0$. Thus by uniqueness $z_{ij}(x) \equiv z_l$, a contradiction.

(ii) Since $\overline{\gamma_{ij}}$ is a compact set, from $z_l \notin \overline{\gamma_{ij}}$ we conclude that $z_{il}(x) \notin \gamma_{ij}$ for $x > M$, M appropriately large. Consider now the set $X = \{x \in \mathbb{R} : z_{il}(x) \in \gamma_{ij}\}$. By assumption, $X \neq \varnothing$. Also, X is bounded above. Let σ be its supremum. By compactness, $z_{il}(\sigma) \in \gamma_{ij}$ and so $z^* = z_{il}(\sigma)$ is the maximal intersection of $\gamma_{ij} \cap \gamma_{il}$.

(iii) Without loss of generality we may assume that $z_{ij}(\sigma) = z_{il}(\sigma)$. By uniqueness, $z'_{ij}(\sigma) \neq z'_{il}(\sigma)$, and locally for x near σ, γ_{ij} and γ_{il} are distinct, say for $[\sigma - \epsilon, \sigma), \epsilon > 0$ appropriate.

Now we are ready to finish. Consider the distance $d(x) = d(z_{il}(x), \gamma_{ij})$, which is well defined by the compactness of γ_{ij}. We know that $d(\sigma - \epsilon) > 0$. Let $D = \{x \leq \sigma - \epsilon : d(x) = 0\}$, and let m be the supremum of D. Either $m = -\infty$, or m is finite. In either case consider the Jordan curve $\{\gamma_{il}(x) : m \leq x \leq \sigma\} \cup \{\gamma_{ij}(x) : m \leq x \leq \sigma\}$, and notice that its interior is mapped under g into the line segment $[g(z_i), g(z^*)]$. This is in contradiction with the open mapping theorem. The proof of the lemma is complete. $\qquad\square$

Now we return to the proof of Case 2, and proceed by contradiction. So assume that all three connections exist, and that $g(z_l) \in (g(z_i), g(z_j))$, the open line segment. Denote by $z_{il}(x)$, $z_{ij}(x)$, $z_{lj}(x)$ representatives of the three connections corresponding to γ_{il}, γ_{ij}, γ_{lj}. By Theorem 2.6, the images of the trajectories under g lie in $[g(z_i), g(z_l)]$, $[g(z_i), g(z_j)]$, $[g(z_l), g(z_j)]$ respectively. Thus by the assumption above, all three lie in $[g(z_i), g(z_j)]$. By Lemma 2.13, $\gamma_{il} \cup \gamma_{lj} \cup \gamma_{ji}$ is a Jordan curve whose interior is mapped under g into $[g(z_i), g(z_j)]$, in contradiction to the open mapping theorem, as before. The proof of Proposition 2.12 is complete. $\qquad\square$

Two Examples

Example 1
$W(z) = |z^n - 1|^2$, where $n \geq 2$ is an integer. The set of minima is

$$A = \{e^{2ik\pi/n} : k = 0, \ldots, n - 1\}.$$

In this we take

$$f(z) = 1 - z^n, \quad g(z) = \int_0^z f(z)\mathrm{d}z = z\left(1 - \frac{z^n}{n+1}\right).$$

Given two different wells $e^{2ik\pi/n}$, $e^{2il\pi/n}$, a trajectory to (2.112) is determined by the preimage under g of the line segment connecting $g(e^{2ik\pi/n})$ and $g(e^{2il\pi/n})$. This amounts to finding $z(t)$, for each t, from the equation

$$z\left(1 - \frac{z^n}{n+1}\right) = \frac{n}{n+1}(te^{2ik\pi/n} + (1-t)e^{2il\pi/n}), \quad t \in (0,1). \tag{2.128}$$

We observe the following:

1. Restricted to the closed disk $D = \{z \in \mathbb{C} : |z| \leq 1\}$, the map g is one-to-one.
2. $\min_{|z|=1} g(z) = \frac{n}{n+1}$; hence $g(D)$ contains the disk $\{\omega \in \mathbb{C} : |\omega| \leq \frac{n}{n+1}\}$.
3. The right-hand side of (2.128) is contained in the disk $\{\omega \in \mathbb{C} : |\omega| \leq \frac{n}{n+1}\}$.

Thus (2.128) is uniquely solvable in D. By the uniqueness, we know that the solution in D of (2.128) provides the needed trajectory. In terms of the polar coordinates $z = re^{i\theta}$, (2.128) with $|z| \leq 1$ can be written in the non-parametric form

$$\begin{cases} (n+1)r\cos\left(\theta - \frac{k+l}{n}\pi\right) = r^{n+1}\cos\left((n+1)\theta - \frac{k+l}{n}\pi\right) + n\cos\left(\frac{k+l}{n}\pi\right), \\ \frac{k\pi}{n} \leq \theta \leq \frac{l\pi}{n}, \quad 0 < r < 1. \end{cases}$$

$$\tag{2.129}$$

In conclusion, there exists exactly one trajectory of (2.112), (2.113) that connects any two roots (minima). The action of the connection between $e^{2ik\pi/n}$ and $e^{2il\pi/n}$, k, l integers, is given by

$$\frac{2n}{n+1}\left|\sin\left(\frac{k-l}{n}\pi\right)\right|.$$

Moreover, the trajectory is given in non-parametric closed form by (2.129), alternatively in parametric form by the solutions to (2.128) in the unit disk.

Example 2
Let $W(u) = |(1 - z^2)(z - i\epsilon)|^2$, $u = (u^1, u^2)$, $z = u^1 + iu^2$. We will show that there is a connection between 1 and -1 if and only if $|\epsilon| > \sqrt{2\sqrt{3} - 3} =: \epsilon^*$.

Proof We have

$$f(z) = i(1 - z^2)(z - i\epsilon),$$

$$z_1 = i\epsilon, \quad z_2 = -1, \quad z_3 = 1,$$

$$g(z) = \epsilon z\left(1 - \frac{1}{3}z^2\right) - \frac{i}{4}(z^2 - 1)^2, \quad g(i\epsilon) = \frac{i}{12}(\epsilon^4 + 6\epsilon^2 - 3), \quad g(\pm 1) = \pm\frac{2}{3}\epsilon.$$

Let ϵ^* be the positive root of $\epsilon^4 + 6\epsilon^2 - 3 = 0$. Without loss of generality we will consider $0 \leq \epsilon$. Figure 2.10 below displays the image under g of the typical curve

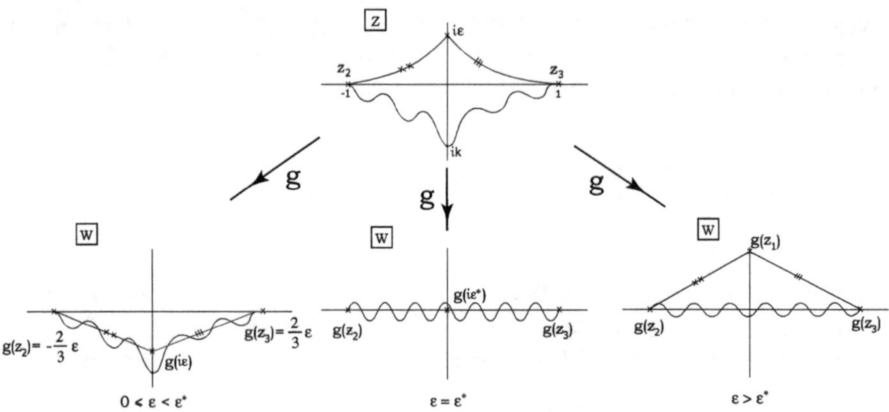

Fig. 2.10 Images under g of the triangle $\{-1, i\epsilon, 1\}$ for $\epsilon < \epsilon^*, \epsilon = \epsilon^*, \epsilon > \epsilon^*$; images under g of a typical $z(t)$ joining -1 and 1, for $\epsilon < \epsilon^*$, and $\epsilon > \epsilon^*$

$z(t)$ with end points at -1 and 1 depending on whether $0 \le \epsilon < \epsilon^*, \epsilon = \epsilon^*$, and $\epsilon > \epsilon^*$. Note that the triangle $\{g(z_1), g(z_2), g(z_3)\}$ is in the lower half-plane in the first case, in the upper half-plane in the third case, and degenerate at $\epsilon = \epsilon^*$.

1. We will show that in the range $0 \le \epsilon \le \epsilon^*$ we have

$$d_W(z_2, z_3) := \inf_{z(0)=z_2,\, z(1)=z_3} E_1(z) = \sqrt{2}\,|g(z_2)-g(z_1)| + \sqrt{2}\,|g(z_1)-g(z_3)|$$

$$= d_W(z_1, z_2) + d_W(z_1, z_3), \qquad (2.130)$$

from which it follows via Proposition 2.12 that there is no connection between $z_2 = -1$ and $z_3 = 1$.

Proof (of (2.130)) Let $z(t)$ be a smooth curve in the z-plane with endpoints at z_2, z_3. Necessarily then $z(t)$ intersects the imaginary axis at a point ik, $k \in \mathbb{R}$. We consider the image of such a point under g:

$$g(ik) = i\left(\epsilon k\left(1 + \frac{1}{3}k^2\right) - \frac{1}{4}(1 + k^2)^2\right) =: i\phi(k).$$

Thus the image $g(ik)$ lies on the imaginary axis in the w-plane. We will now argue that it lies below $g(i\epsilon)$. Indeed, from $\phi'(k) = (\epsilon - k)(1 + k^2)$ and the fact $\phi(k) \to -\infty$, as $|k| \to \infty$, it follows that $\max \phi = \phi(\epsilon)$. Hence $\phi(k) \le \phi(\epsilon) \le 0$, $(0 \le \epsilon \le \epsilon^*)$. It is therefore geometrically evident (Fig. 2.11(i)) that

$$\inf_{z(0)=z_2,\, z(1)=z_3} E_1(z) = \inf\left\{\sqrt{2}\int |w'(t)|dt : w(t) = g(z(t)),\ z(0) = z_2,\ z(1) = z_3\right\}$$

$$\ge \sqrt{2}\,|g(z_2) - g(z_1)| + \sqrt{2}\,|g(z_1) - g(z_3)|. \qquad (2.131)$$

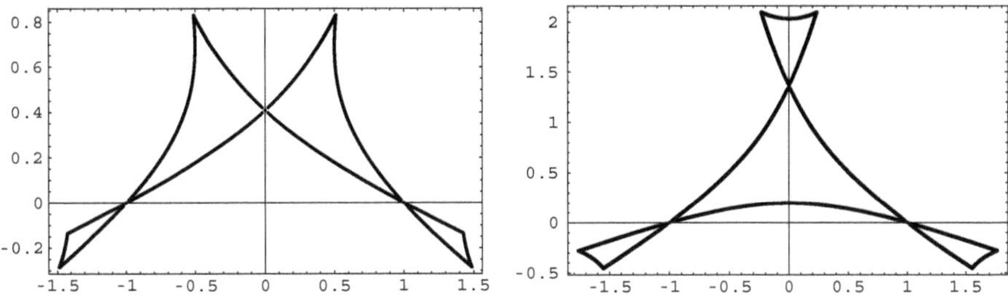

Fig. 2.11 Computation of the inverse image of the triangle $\{g(-1), g(i\epsilon), g(1)\}$ under g^{-1} for specific values of $\epsilon < \epsilon^*$, and $\epsilon > \epsilon^*$

We note that the other two connections exist and so by Proposition 4.1 2.12,

$$d_W(z_1, z_2) = \sqrt{2}\,|g(z_1) - g(z_2)|, \quad d_W(z_1, z_3) = \sqrt{2}\,|g(z_1) - g(z_3)|. \qquad (2.132)$$

Relations (2.132), (2.131) imply, via $d_W(z_2, z_3) \leq d_W(z_1, z_2) + d_W(z_1, z_3)$, relations (2.130). In Fig. 2.11(i) below we show an actual numerical result, the inverse image of the triangle $\{g(z_1), g(z_2), g(z_3)\}$ under g^{-1} calculated for a specific ϵ in the range $0 \leq \epsilon < \epsilon^*$. $\qquad\qquad\qquad\qquad\qquad\qquad\qquad\qquad\qquad\qquad\qquad\quad\square$

2. We will show that in the range $\epsilon > \epsilon^*$ there is a connection between $z_2 = -1$ and $z_3 = 1$, which is unique by Theorem 2.6. By Theorem 2.5 above, if there is such a connection, then it has to be mapped under g onto the line segment $[g(z_2), g(z_3)] = [-\frac{2}{3}\epsilon, \frac{2}{3}\epsilon]$. Thus we are led to investigating the following equation for $z(t)$:

$$g(z(t)) = (1 - t)g(-1) + tg(1) = (1 - t)\left(-\frac{2}{3}\epsilon\right) + t\left(\frac{2}{3}\epsilon\right). \qquad (2.133)$$

Differentiating we obtain that necessarily such a $z(t)$ satisfies

$$z'(t) = \frac{4\epsilon}{3}\frac{1}{f(z(t))}, \qquad (2.134)$$

which we consider together with the initial condition

$$z\left(\frac{1}{2}\right) = i\tau_1, \qquad (2.135)$$

where $0 < \tau_1(\epsilon) < \epsilon < \tau_2(\epsilon)$ are the two real roots of $g(i\tau) = 0$, which exists if and only if $\epsilon > \epsilon^*$. The plan is to establish the existence of the solution curve to (2.133) via the initial value problem (2.134), (2.135). We first consider $t \geq \frac{1}{2}$. By the continuation Lemma 4.1 in [2], the local solution of (2.134), (2.135) can be continued up to $t = 1$. Note that $z(1) \in g^{-1}\{g(1)\} = \{1, c_1^1, c_1^2\}$, where we have used that g is of 4th order and 1 is a double root of $g(z) = g(1)$. We will show that the other preimages c_1^1, c_1^2 are not options for $z(1)$. We proceed by contradiction, so assume that $z(1) \in \{c_1^1, c_1^2\}$. We can calculate c_1^1, c_1^2 which are roots to $g(z) = \frac{2}{3}\epsilon$ by factoring the equation

$$g(z) - \frac{2}{3}\epsilon = -\frac{i}{4}(z-1)^2\left[z^2 + 2z\left(1 - \frac{2i}{3}\epsilon\right) + \left(1 - \frac{8i}{3}\epsilon\right)\right] = 0,$$

$$c_1^{1,2} = -1 + \frac{2i}{3}\epsilon \pm \sqrt{-\frac{4}{9}\epsilon^2 + \frac{4i}{3}\epsilon},$$

$$\mathrm{Re}\, c_1^{1,2} = -1 \pm \frac{\sqrt{2}}{3}\sqrt{\epsilon}[\sqrt{\epsilon^2 + 9} - \epsilon]^{1/2} \le -1 + \frac{\sqrt{2}}{3}\sqrt{\epsilon}[\sqrt{\epsilon^2 + 9} - \epsilon]^{1/2} < 0,$$

for $\epsilon \ge 0$. We now note that on the imaginary axis the vector field (2.134) is parallel to the real axis, and that it reverses its orientation at $i\epsilon$. Hence $z(t)$ must intersect the imaginary axis for a second time at $t = \tau < 1$, and this has to occur at $z = i\tau_2$. We now utilize the symmetry of the vector field and show that the part of the solution on the right half-plane can be reflected in x to produce a closed trajectory. This will lead to contradiction. Let $Sz := -x + iy$ be the reflection in x. It can be checked that $f(Sz) = -Sf(z)$, $S(1/z) = 1/(Sz)$. Let $z(t)$, $t \in [\frac{1}{2}, \tau]$, $\tau < 1$, and define $\hat{z}(t) = Sz(1-t)$ for $t \in [1 - \tau, \frac{1}{2}]$. We see that $\hat{z}$ satifies (2.134), and so extends $z(t)$ on $[1 - \tau, \frac{1}{2}] \cup [\frac{1}{2}, \tau]$. The extended z is a rectifiable closed Jordan curve, that is mapped under g into a line segment $[-\frac{2}{3}\epsilon, \frac{2}{3}\epsilon]$, contradicting the open mapping theorem for analytic functions. Thus $z(1) = 1$ is the only option left. By the symmetry of the vector field $z(0) = -1$. In Fig. 2.11 above we show a numerical result for the image of the triangle $\{g(z_1), g(z_2), g(z_3)\}$ under g^{-1} for a specific ϵ in the range $\epsilon > \epsilon^*$. $\square$

2.8 Scholia on Chap. 2

The first existence proofs of a heteroclinic connection in the vector case for a double-well potential where given by Rabinowitz [20] by minimization of the action functional and by Sternberg [25, 26] who utilized the Jacobi principle under somewhat restrictive hypotheses on the behavior of W at the minima. In [20] aside from smoothness no other assumption is required on the behavior of W at the minima.

Recently Zuniga and Sternberg [27] reexamined the problem via the Jacobi method and established existence under general conditions on W. At about the same time, Monteil and Santambrogio [18], utilizing very similar methods, obtained existence under comparable hypotheses on W (see **H** in Sect. 2.3). Alberti [1] gave a variational existence proof in the scalar case for a double-well potential. His proof is based on the relationship between the action J and the length functional L (cf Lemma 2.1), and on the equipartition relationship (2.14), which in the scalar case reduces the problem to a first order O.D.E.

Theorem 2.1 in Sect. 2.3 is from Fusco et al. [10, 11]. Our presentation in Sect. 2.4 follows [2]. The method is based on a constraint that is removed afterwards via a replacement result, the so called 'cut-off lemma'. The idea of minimizing under constraints that are eventually removed via a comparison argument was already used by Rabinowitz and Coti Zelati in [8]. This procedure is extended to higher dimensions and is followed also in Chap. 4 for constructing P.D.E. connections on periodic domains unbounded in one direction. The cut-off lemma excludes oscillations and localizes the transition, and thus restores compactness for a minimizing sequence via translations. We refer to Sect. 4.3.4 for more precise explanations.

Stefanopoulos [23] allows an anisotropic gradient and extends [2]. Katzourakis [14] establishes existence by minimizing the action and restoring compactness along the lines of the concentration compactness principle of Lions. He requires stronger assumptions on W than [2].

Recently, Antonopoulos and Smyrnelis [6] established the following very general result with a rather short proof, utilizing a lemma by Sourdis [24], and a variant of the arguments presented in Sect. 2.4.

H$_1$: Let $W \in C^2(\mathbb{R}^m, \mathbb{R})$, and let Ω be a connected component of the set $\{u \in \mathbb{R}^m : W(u) > 0\}$, such that $\partial\Omega$ is partitioned into two disjoint compact subsets A^- and A^+. Thus, $W(u) = 0$, for all $u \in A^\pm$.

H$_2$: $\liminf_{u \in \Omega, \, |u| \to +\infty} W(u) > 0$, if Ω is not bounded.

Theorem 2.7 *Assume* $W : \mathbb{R}^m \to \mathbb{R}$ *satisfies conditions* **H$_1$** *and* **H$_2$**. *Then* $J_\mathbb{R}(u) = \int_\mathbb{R} \left\{ \frac{1}{2} |u'(x)|^2 + W(u(x)) \right\} dx$ *admits a minimizer* $\bar{u} \in \mathscr{A}$,

$$\mathscr{A} = \left\{ u \in W^{1,2}_{\text{loc}}(\mathbb{R}; \overline{\Omega}) : \begin{array}{l} d(u(x), A^-) \leq \bar{q}, \text{ for } x \leq x_u^-, \\ d(u(x), A^+) \leq \bar{q}, \text{ for } x \geq x_u^+, \end{array} \text{ for some } x_u^- < x_u^+ \right\}.$$

where $\bar{q} \in \left(0, \frac{d(A^-, A^+)}{2} \right)$, *and* d *is the Euclidean distance,*

$$J_\mathbb{R}(\bar{u}) = \min_{u \in \mathscr{A}} J_\mathbb{R}(u) < +\infty,$$

and moreover

$$\lim_{x \to \pm\infty} d(\bar{u}(x), A^\pm) = 0.$$

From this general result by particularizing the sets $A^\pm$, they obtain heteroclinic, homoclinic and periodic orbits. To explain this, let

He: $\nabla W(u) = 0$, for all $u \in A^*$.

Ho: $\nabla W(u) \neq 0$, for all $u \in A^*$.

Corollary 2.5 (Heteroclinic)   *If **He** holds on A^- and A^+, then there exists a heteroclinic connection v:*

$$\lim_{x \to \pm\infty} d(v(x), A^\pm) = 0,$$

and $v(x) \in \Omega$, $\forall x \in \mathbb{R}$. In particular, if $W \geq 0$ and $\{u \in \mathbb{R}^m :\ W(u) = 0\} = \{a_1, \ldots, a_N\}$, then for every a_i, there exists a heteroclinic connection to some a_j, $j \in \{1, \ldots, N\}$, $j \neq i$.

Corollary 2.6 (Homoclinic)   *If **He** holds on A^- and **Ho** holds on A^+, then there exists an even homoclinic connection v,*

$$\lim_{x \to \pm\infty} d(v(x), A^-) = 0,$$

$$v(x) \in A^+ \iff x = 0, \ v(x) \in \Omega, \ \forall x \neq 0.$$

Corollary 2.7 (Periodic)   *If **Ho** holds on A^- and A^+, then there exists a periodic solution v of period T,*

$$v\left(x + T\right) = v(x),$$

$$v\left(x + \frac{T}{2}\right) = v\left(-x + \frac{T}{2}\right)$$

$$v(x) \in A^- \iff x \in T\mathbb{Z},$$

$$v(x) \in A^+ \iff x + \frac{T}{2} \in T\mathbb{Z}.$$

Very recently, Fusco et al. [10] have extended some of the results in [6], by allowing critical points (cf. **Ho** above).

Connections between zero sets of W containing nontrivial connected components have also been studied by André and Shafrir [5], and by Lin et al. [15].

Our presentation in Sect. 2.6 is based on a simplified and improved version of Alikakos and Katzourakis [4]. Here the traveling wave problem is handled by means of a variational approach similar to Sect. 2.4. Risler [21] at about the same time had established a similar result by studying the general problem of connections within the associated parabolic gradient flow. Independently, also Lucia et al. [16], utilizing variational methods treated a P.D.E. generalization which when particularized gives a traveling wave in the present setting, under different hypotheses on the potential W. Heinze [12] before had developed appropriate variational settings for related problems and introduced a basic topological argument (see Lemma 2.12). Analogous characterizations have been obtained in Heinze et al. [13], and Heinze [12]. Terman [28] already in 1987 addressed the traveling wave problem in the vector case and established existence of infinitely many traveling waves (in the sense of infinitely many speeds), connecting two local minima of W. His methods are topological, utilizing the Conley index.

Among the advantages of the variational method is its simplicity. Also, it provides a variational characterization of the wave profile and also of the speed, and thus an estimate on the speed, which in the class of minimizers is unique.

The material in Sect. 2.7 is mainly from Alikakos et al. [3], part of which was further elaborated in Alikakos and Fusco [2]. Stefanopoulos [23] gives a sufficient condition for existence of connections for a triple-well potential in $\mathbb{R}^m$.

References

1. Alberti, G.: Variational models for phase transitions, an approach via Gamma convergence. In: Ambrosio, L., Dancer, N. (eds.) Calculus of Variations and Partial Differential Equations, pp. 95–114. Springer, Berlin (2000)
2. Alikakos, N.D., Fusco, G.: On the connection problem for potentials with several global minima. Indiana Univ. Math. J. **57**, 1871–1906 (2008)
3. Alikakos, N.D., Betelú, S.I., Chen, X.: Explicit stationary solutions in multiple well dynamics and non-uniqueness of interfacial energies. Eur. J. Appl. Math. **17**, 525–556 (2006)
4. Alikakos, N.D., Katzourakis, N.: Heteroclinic travelling waves of gradient diffusion systems. Trans. Am. Math. Soc. **363**, 1362–1397 (2011)
5. André, N., Shafrir, I.: On a vector-valued singular perturbation problem on the sphere. In: Proceedings of the International Conference on Nonlinear Analysis, Recent Advances in Nonlinear Analysis, pp. 11–42. World Scientific Publishing, Singapore (2008)
6. Antonopoulos, P., Smyrnelis, P.: On minimizers of the Hamiltonian system $u'' = \nabla W(u)$, and on the existence of heteroclinic, homoclinic and periodic orbits. Indiana Univ. Math. J. **65**(5), 1503–1524 (2016)
7. Colding, T.H., Minicozzi, W.P.: A Course in Minimal Surfaces. Graduate Studies in Mathematics, vol. 121. American Mathematical Society, Providence (2011)
8. Coti Zelati, V., Rabinowitz, P.H.: Homoclinic orbits for second order Hamiltonian systems possessing superquadratic potentials. J. Am. Math. Soc. **4**, 693–727 (1991)
9. Fife, P.C., McLeod, J.B.: The approach of solutions of nonlinear diffusion equations to travelling front solutions. Arch. Ration. Mech. Anal. **65**(4), 335–361 (1977)
10. Fusco, G., Gronchi, G.F., Novaga, M.: On the existence of connecting orbits for critical values of the energy. J. Differ. Equ. **263**, 8848–8872 (2017)
11. Fusco, G., Gronchi, G.F., Novaga, M.: On the existence of heteroclinic connections. Sao Paulo J. Math. Sci. **12**, 1–14 (2017)
12. Heinze, S.: Travelling waves for semilinear parabolic partial differential equations in cylindrical domains. PhD thesis, Heidelberg University (1988)
13. Heinze, S., Papanicolaou, G., Stevens, A.: Variational principles for propagation speeds in inhomogeneous media. SIAM J. Appl. Math. **63**(1), 129–148 (2001)
14. Katzourakis, N.: On the loss of compactness in the vectorial heteroclinic connection problem. Proc. Roy. Soc. Edinb. Sect. A **146**(3), 595–608 (2016)
15. Lin, F., Pan, X.B., Wang, C.: Phase transition for potentials of high-dimensional wells. Commun. Pure Appl. Math. **65**(6), 833–888 (2012)
16. Lucia, M., Muratov, C., Novaga, M.: Existence of traveling wave solutions for Ginzburg-Landau-type problems in infinite cylinders. Arch. Ration. Mech. Anal. **188**(3), 475–508 (2008)
17. Modica, L.: A Gradient bound and a Liouville Theorem for nonlinear Poisson equations. Commun. Pure. Appl. Math. **38**(5), 679–684 (1985)

18. Monteil, A., Santambrogio, F.: Metric methods for heteroclinic connections. Math. Methods Appl. Sci. **41**(3), 1019–1024 (2018)
19. Muratov, C.B.: A global variational structure and propagation of disturbances in reacting-diffusion systems of gradient type. Discrete Contin. Dyn. Syst. Ser. B **4**, 867–892 (2004)
20. Rabinowitz, P.H.: Periodic and heteroclinic orbits for a periodic hamiltonian system. Ann. Inst. Henri Poincaré **6**(5), 331–346 (1989)
21. Risler, R.E.: Global convergence towards travelling fronts in nonlinear parabolic systems with a gradient structure. Ann. Inst. Henri Poincaré (C) Non Linear Anal. **25**(2), 381–424 (2008)
22. Smyrnelis, P.: Gradient estimates for semilinear elliptic systems and other related results. Proc. Roy. Soc. Edinb. Sect. A **145**(6), 1313–1330 (2015)
23. Stefanopoulos, V.: Heteroclinic connections for multiple well potentials: the anisotropic case. Proc. Roy. Soc. Edinb. Sect. A **138**, 1313–1330 (2008)
24. Sourdis, C.: The heteroclinic connection problem for general double-well potentials. Mediterr. J. Math. **13**, 4693–4710 (2016)
25. Sternberg, P.: The effect of a singular perturbation on nonconvex variational problems. Arch. Ration. Mech. Anal. **101**(3), 209–260 (1988)
26. Sternberg, P.: Vector-valued local minimizers of nonconvex variational problems. Rocky Mountain J. Math. **21**, 799–807 (1991)
27. Sternberg, P., Zuniga, A.: On the heteroclinic problem for multi-well gradient systems. J. Differ. Equ. **261**, 3987–4007 (2016)
28. Terman, D.: Infinitely many traveling wave solutions of a gradient system. Trans. Am. Math. Soc. **301**(2), 537–556 (1987)

Chapter 3
Basics for P.D.E. Systems

Abstract The main object in this chapter is the stress-energy tensor, which is an algebraic fact implying several useful identities like the (weak) monotonicity formula, Gui's Hamiltonian identities, and Pohozaev' identities, for all solutions and all potentials $W \geq 0$. Modica's inequality holds in the scalar case and implies a strong monotonicity formula, but is not generally valid in the vector case. The triple junction on the plane is also introduced.

3.1 The Stress-Energy Tensor

We consider the system

$$\Delta u - W_u(u) = 0, \quad u : \mathbb{R}^n \to \mathbb{R}^m, \tag{3.1}$$

with $W \in C^2(\mathbb{R}^m, \mathbb{R})$, $W \geq 0$, where, $W_u(u) := (\partial W / \partial u_1, \ldots, \partial W / \partial u_m)^\top$.

Distinguished examples are:

(a) The phase-transition model, or vector Allen-Cahn equation, where W has a finite number of global minima $a_1, \ldots, a_N$.

(b) The Ginzburg-Landau system $\Delta u - (|u|^2 - 1)u = 0$, where $W(u) = \frac{1}{4}(|u|^2 - 1)^2$ vanishes on $\mathbb{S}^{n-1}$.

(c) The segregation models in population dynamics, $\Delta u_i - \sum_{j \neq i} u_i u_j^2 = 0$, where $W(u) = \frac{1}{2} \sum_{i < j} u_i^2 u_j^2$.

Solutions to systems (a) have distinctly different behavior from those of systems (b), (c). The key difference lies in the connectedness of $\{W = 0\}$. This is in sharp contrast with the scalar counterpart of (3.1), where for the class of results we are interested in, the form of W does not play a major role, and the analysis proceeds essentially by eliminating W via differentiation (cf. [17]).

© Springer Nature Switzerland AG 2018

N. D. Alikakos et al., *Elliptic Systems of Phase Transition Type*,
Progress in Nonlinear Differential Equations and Their Applications 91,
https://doi.org/10.1007/978-3-319-90572-3_3

System (3.1) is the Euler-Lagrange equation for the *free energy functional*

$$J_{\mathbb{R}^n}(u) := \int_{\mathbb{R}^n} \left(\frac{1}{2} |\nabla u|^2 + W(u) \right) dx, \tag{3.2}$$

where $\nabla u = (\partial u_i / \partial x_j)$, $i = 1, \ldots, m$; $j = 1, \ldots, n$ and $|\cdot|$ is the Euclidean norm of the matrix.

Equation (3.1) can be written as a divergence-free condition, that is,

$$\operatorname{div} T = (\nabla u)^\top \left(\Delta u - W_u(u) \right) = 0 \tag{3.3}$$

for the stress-energy tensor

$$T_{ij}(u, \nabla u) := u_{x_i} \cdot u_{x_j} - \delta_{ij} \left(\frac{1}{2} |\nabla u|^2 + W(u) \right), \tag{3.4}$$

where $\cdot$ stands for the Euclidean inner product. Relation (3.3) is an algebraic fact and goes back to E. Noether; (3.3) is easy to check. Writing $T = (T_1, T_2, \ldots, T_n)^\top$ and $\operatorname{div} T = (\operatorname{div} T_1, \operatorname{div} T_2, \ldots, \operatorname{div} T_n)^\top$, we calculate

$$\operatorname{div} T_j = \sum_{k \neq j} \left(u_{x_j} \cdot u_{x_k} \right)_{x_k} + \frac{1}{2} \left(|u_{x_j}|^2 - \sum_{i \neq j} |u_{x_i}|^2 - 2W(u) \right)_{x_j}$$

$$= \sum_{k \neq j} \left(u_{x_j x_k} \cdot u_{x_k} + u_{x_j} \cdot u_{x_k x_k} \right) + u_{x_j} \cdot u_{x_j x_j} - \sum_{i \neq j} u_{x_i} \cdot u_{x_i x_j} - W_u \cdot u_{x_j}$$

$$= u_{x_j} \cdot \left(\Delta u - W_u(u) \right),$$

and so (3.3) follows. We also note that T is invariant under rotations of the coordinate system, that is, it transforms as a tensorial quantity. To see this, consider an orthogonal transformation Q, and a new coordinate system $x' = Qx$. Letting u' be the map acting in the new coordinates, with $u'(x') = u(x)$, the chain rule gives that its gradient is transformed via $\nabla u' = (\nabla u)Q^\top$, where the prime indicates that the derivatives are taken with respect to the new coordinate system. Then, by (3.11) below, the transformed tensor T' is given by $T' = QTQ^\top$, and there holds

$$T'_{ij} = u'_{x_i} \cdot u'_{x_j} - \delta_{ij} \left(\frac{1}{2} |\nabla' u'|^2 + W(u') \right),$$

where again the prime indicates that the tensor is calculated in the new coordinate system. That is, the transformed tensor has exactly the same expression as the original one, except for the fact that it acts in the transformed coordinates.

Next, we calculate the trace of T:

$$\operatorname{tr} T = \sum_{i=1}^{n} T_{ii} = \sum_{i=1}^{n} |u_{x_i}|^2 - \frac{n}{2} \sum_{i=1}^{n} |u_{x_i}|^2 - n W(u)$$

$$= \frac{2-n}{2} \sum_{i=1}^{n} |u_{x_i}|^2 - n W(u). \tag{3.5}$$

Finally, we introduce the *interface-energy density*:

$$g(u) := \frac{1}{2} |\nabla u|^2 + W(u). \tag{3.6}$$

Notice that

$$\operatorname{tr} T = -n\, g(u) + |\nabla u|^2, \tag{3.7}$$

$$\operatorname{tr} T \leq -(n-2)\, g(u). \tag{3.8}$$

In this chapter we will derive properties of general solutions to (3.1) for arbitrary $W \geq 0$ as a consequence of the divergence-free condition (3.3).

3.2 The Monotonicity Formula

Theorem 3.1 *Assume $W \geq 0$ and let u be a $W_{\mathrm{loc}}^{1,2}(\mathbb{R}^n; \mathbb{R}^m) \cap L_{\mathrm{loc}}^{\infty}(\mathbb{R}^n; \mathbb{R}^m)$ solution to (3.1). Then, we have*

$$\frac{\mathrm{d}}{\mathrm{d}r} \left(r^{-(n-2)} J_{B_r}(u) \right) \geq 0, \ \text{for } r > 0, \tag{3.9}$$

where

$$J_{B_r}(u) = \int_{B_r} \left(\frac{1}{2} |\nabla u|^2 + W(u) \right) \mathrm{d}x, \tag{3.10}$$

with $x_0 \in \mathbb{R}^n$ arbitrary and $B_r := B_r(x_0)$ the r-ball in $\mathbb{R}^n$ centered at x_0.

Proof We begin by noting the simple fact that

$$T + g(u)\mathrm{Id} = (\nabla u)^{\top}(\nabla u) \geq 0 \quad \text{(semidefiniteness)}, \tag{3.11}$$

where Id stands for the identity matrix on $\mathbb{R}^n$. Take $x_0 = 0$ for convenience. We have

$$\sum_{i,j} \int_{B_r} (x^i T_{ij})_{x_j} = \sum_{i,j} \int_{B_r} (\delta_{ij} T_{ij} + x^i T_{ij,x_j})$$

$$= \sum_i \int_{B_r} T_{ii} \quad \text{(by (3.3))}$$

$$= -\int_{B_r} \left(\frac{n-2}{2} |\nabla u|^2 + n W(u) \right) \quad \text{(by (3.5))} \tag{3.12}$$

$$\leq -(n-2) \int_{B_r} g(u) \quad \text{(by (3.6))}. \tag{3.13}$$

On the other hand, by the divergence theorem and for $\nu = x/r$,

$$\sum_{i,j} \int_{B_r} (x^i T_{ij})_{x_j} = \sum_{i,j} \int_{\partial B_r} x^i T_{ij} \nu_j$$

$$= r \sum_{i,j} \int_{\partial B_r} T_{ij} \nu_i \nu_j$$

$$= r \int_{\partial B_r} (T\nu) \cdot \nu$$

$$= -r \int_{\partial B_r} \left(g(u) - \left| \frac{\partial u}{\partial \nu} \right|^2 \right) \quad \text{(by (3.11))} \tag{3.14}$$

$$\geq -r \int_{\partial B_r} g(u) = -r \frac{dJ_{B_r}(u)}{dr} \quad \text{(by the co-area formula)}. \tag{3.15}$$

Combining (3.13) and (3.15) we obtain

$$-(n-2) J_{B_r}(u) \geq -r \frac{dJ_{B_r}(u)}{dr},$$

or, equivalently,

$$\frac{d}{dr} \left(r^{-(n-2)} J_{B_r}(u) \right) \geq 0. \tag{3.16}$$

The proof of the theorem is complete. $\square$

An immediate consequence of (3.9) is the lower bound

$$J_{B_r}(u) \geq c r^{n-2} \tag{3.17}$$

for nonconstant solutions. Actually, as we will see later, the following is true:

$$\begin{cases} J_{B_r}(u) = o(r^{n-2}), \ \text{as } r \to \infty, \ n \geq 3 \Longrightarrow u \equiv \text{Const.}, \\ J_{B_r}(u) = o(\ln r), \ \text{as } r \to \infty, \ n = 2 \Longrightarrow u \equiv \text{Const.} \end{cases} \tag{3.18}$$

Remark 3.1 Combining (3.12) and (3.14), one also obtains the Pohozaev identity:

$$\int_{B_r} \left(\frac{n-2}{2} |\nabla u|^2 + n W(u) \right) = r \int_{\partial B_r} \left(\frac{1}{2} |\nabla u|^2 + W(u) - \left| \frac{\partial u}{\partial v} \right|^2 \right), \tag{3.19}$$

which holds for any solution to (3.1), and any ball B_r of radius r contained in the domain of u. A similar identity can be derived for smooth domains $\Omega \subset \mathbb{R}^n$, provided the solution u satisfies appropriate boundary conditions (cf. (3.54)).

From (3.19), it follows that

$$-(n-2)J_{B_r} + r\frac{dJ_{B_r}}{dr} = 2\int_{B_r} W(u) + r\int_{\partial B_r} \left| \frac{\partial u}{\partial v} \right|^2, \tag{3.20}$$

and integrating we obtain the identity

$$\frac{d}{dr}(r^{-(n-2)}J_{B_r}) = \frac{2}{r^{n-1}} \int_{B_r} W(u) + \frac{1}{r^{n-2}} \int_{\partial B_r} \left| \frac{\partial u}{\partial v} \right|^2. \tag{3.21}$$

From (3.20) we see that, more generally,

$$\frac{d}{dr}(r^{-\mu}J_{B_r}) = \frac{1}{r^{\mu+1}} \int_{B_r} \left((n-\mu)W(u) + \frac{(n-2-\mu)}{2}|\nabla u|^2 \right) + \frac{1}{r^{\mu}} \int_{\partial B_r} \left| \frac{\partial u}{\partial v} \right|^2. \tag{3.22}$$

Of particular importance is the case $\mu = n - 1$ which gives

$$\frac{d}{dr}(r^{-(n-1)}J_{B_r}) = \frac{1}{r^n} \int_{B_r} \left(W(u) - \frac{1}{2}|\nabla u|^2 \right) + \frac{1}{r^{n-1}} \int_{\partial B_r} \left| \frac{\partial u}{\partial v} \right|^2. \tag{3.23}$$

3.3 The Validity of the Modica Inequality

Modica established in [28] that given a non-negative potential $W \in C^2(\mathbb{R}, \mathbb{R})$, every bounded entire solution $u \in C^2(\mathbb{R}^n, \mathbb{R})$ of the scalar equation

$$\Delta u - W'(u) = 0 \tag{3.24}$$

satisfies the gradient bound

$$\frac{1}{2}|\nabla u(x)|^2 \le W(u(x)), \quad \text{for all } x \in \mathbb{R}^n, \tag{3.25}$$

referred to now as the Modica estimate. The proof of (3.25) is based on the use of the so-called P-functions (cf. [33]): to every solution $u : \mathbb{R}^n \to \mathbb{R}$ of (3.24), one associates the P-function $P(u; x) := \frac{1}{2}|\nabla u(x)|^2 - W(u(x))$. With this choice of P, we obtain the inequality

$$|\nabla u|^2 \Delta P \ge \frac{1}{2}|\nabla P|^2 + 2W'(u)\nabla u \cdot \nabla P. \tag{3.26}$$

Then, the maximum principle is applied to P to show that $P(u; x) \le 0$, for every bounded solution u and every $x \in \mathbb{R}^n$. Assuming that the solutions are entire is an essential hypothesis in proving the Modica estimate. We note that other gradient bounds can be obtained for solutions of (3.24) defined in proper domains of $\mathbb{R}^n$ (cf. [18]).

In the case of bounded solutions $u : \mathbb{R}^n \to \mathbb{R}^m$ to the system (3.1) with $W \ge 0$, the Modica estimate (3.25) does not hold in general. This is due to the fact that in the vector case one cannot obtain for P an appropriate inequality like (3.26) to which the maximum principle can be applied. However, assuming that a solution to the system (3.1) does satisfy the Modica estimate, the two following corollaries (cf. [28] and [10]) still hold for any dimension $m \ge 1$ of the range:

Corollary 3.1 (Liouville Type Theorem) *Let $W \in C^2(\mathbb{R}^m, \mathbb{R})$ be a non-negative potential, and let $u : \mathbb{R}^n \to \mathbb{R}^m$ be a solution to the system (3.1) satisfying $|\nabla u|^2 = O(W(u))$ (in particular u may be any bounded solution of (3.24)). Then the condition $W(u(x_0)) = 0$ for some $x_0 \in \mathbb{R}^n$, implies that u is a constant.*

Proof Let $a = u(x_0)$ and $A = \{x \in \mathbb{R}^n : u(x) = a\}$. As A is nonempty and closed, it suffices to prove that A is open. Let $x_1 \in A$. Since $W \ge 0$ and $W(a) = 0$, there exists $k \ge 0$ such that $W(u) \le k|u - a|^2$, $\forall u \in B_\delta(a)$, provided that δ is small enough. Now, if $n \in \mathbb{R}^n$ and $|n| = 1$, and if we define $\phi(t) = u(x_1 + tn) - u(x_1)$, $\psi(t) = |\phi(t)|^2$, for $|t|$ small, we have by assumption $|\phi'(t)|^2 \le O(|\phi(t)|^2)$, and $|\psi'(t)|^2 \le O(|\psi(t)|^2)$. Finally, since $\psi(0) = 0$ it follows that $\phi \equiv 0$. Hence u is constant in the ball $B_\delta(x_1)$. $\square$

Corollary 3.2 (Strong Monotonicity) *Let $W \in C^2(\mathbb{R}^m, \mathbb{R})$ be a non-negative potential, and let $u : \mathbb{R}^n \to \mathbb{R}^m$ be a solution to the system (3.1) satisfying (3.25) (in particular u may be any bounded solution of (3.24)). Then for every $x_0 \in \mathbb{R}^n$, the quotient $\frac{J_{B_r}(u)}{r^{n-1}}$ is an increasing function of $r > 0$ (where $B_r := B_r(x_0)$ as in Theorem 3.1).*

Proof In view of (3.25) and (3.23), we obtain $\frac{d}{dr}(r^{-(n-1)}J_{B_r}) \geq 0$. $\qquad\square$

In the vector case, every solution $u : \mathbb{R} \to \mathbb{R}^m$ of (2.28) satisfying the boundary conditions (2.29) is equipartitioned:

$$\frac{1}{2}|u'(x)|^2 = W(u(x)). \tag{3.27}$$

To see this, we recall the inequality

$$|f'(x)| \leq \frac{1}{2}(|f(x+1)| - |f(x-1)|) + \max_{y \in [x-1, x+1]}|f''(y)|, \tag{3.28}$$

which holds for any function $f \in C^2(\mathbb{R}, \mathbb{R})$. Applying (3.28) to the components of u and utilizing (2.29), we obtain that $\lim_{x \to \pm\infty} \frac{1}{2}|u'(x)|^2 = 0$. On the other hand, since $\lim_{x \to \pm\infty} W(u(x)) = 0$ and the Hamiltonian $H := \frac{1}{2}|u'(x)|^2 - W(u(x))$ is constant along solutions, we deduce (3.27).

However, solutions $u : \mathbb{R} \to \mathbb{R}^m$ of (2.28) violating the Modica estimate can easily been constructed when $m = 2$. Indeed, in the case of the O.D.E. $u'' - (|u|^2 - 1)u = 0$ corresponding to the Ginzburg-Landau potential $W(u) = \frac{1}{4}(1 - |u|^2)^2$, there exists, for every $r \in (0, 1)$, a periodic solution $u_r : \mathbb{R} \to \mathbb{R}^2 \simeq \mathbb{C}$, $u_r(x) = re^{i\sqrt{1-r^2}x}$, whose Hamiltonian $H_r = \frac{-3r^4 + 4r^2 - 1}{4}$ is positive if and only if $\sqrt{1/3} < r < 1$. Similarly, let us consider a phase transition potential $W : \mathbb{R}^2 \to \mathbb{R}$ satisfying for every $u \in \mathbb{R}^2$ such that $|u| = r > 0$:

$$W(u) = \lambda \text{ and } \nabla W(u) = -\mu u, \text{ with } \lambda, \mu > 0, \text{ constants.}$$

Then, $u : \mathbb{R} \to \mathbb{R}^2 \simeq \mathbb{C}$, $u(x) = re^{i\sqrt{\mu}x}$ solves the O.D.E. $u'' - W_u(u) = 0$, and its Hamiltonian $H = \frac{1}{2}|u'|^2 - W(u) = \frac{r^2\mu}{2} - \lambda$ may become positive and arbitrarily big.

Finally, let us mention the construction in [32] of a solution $u : \mathbb{R} \to \mathbb{R}^2$ to $u'' - W_u(u) = 0$ such that

(i) $W : \mathbb{R}^2 \to [0, \infty)$ is a double-well potential with two nondegenerate minima a^+ and a^-;

(ii) u is T-periodic and attains $a^\pm$ at finite times: $u(0) = a^+$, $u(T/2) = a^-$;

(iii) $u'(0) \neq 0$ and $u'(T/2) \neq 0$.

The existence of this new kind of orbit (cf. Remark 2.9) shows that no Modica type estimate holds in general for system (3.1), since $W(u(0)) = 0$, while $u'(0) \neq 0$. Clearly, the aforementioned Liouville theorem also fails, since $W(u(0)) = 0$, while u is not constant.

Next we examine counterexamples in space dimensions $n \geq 2$. For the Ginzburg-Landau potential $W(u) = \frac{1}{4}(|u|^2 - 1)^2$, Hagan [23] constructed nontrivial solutions $u : \mathbb{R}^2 \to \mathbb{R}^2$ for which $\int_{\mathbb{R}^2} W(u)dx < \infty$. Since necessarily for nonconstant solutions $\int_{\mathbb{R}^n} \left(\frac{1}{2}|\nabla u|^2 + W(u) \right) dx = \infty$ (by (3.18)), it follows that the Modica type estimate is not possible for such potentials. Furthermore, Farina [15] established the existence of solutions $u : \mathbb{R}^3 \to \mathbb{R}^3$ for the Ginzburg-Landau potential which satisfy the estimate

$$\lim_{r \to \infty} \frac{1}{r^{n-2}} \int_{B_r} \left(\frac{n-2}{2}|\nabla u|^2 + nW(u) \right) dx$$

$$= \lim_{r \to \infty} \left(\frac{(n-2)}{r^{n-2}} J_{B_r}(u) + \frac{2}{r^{n-2}} \int_{B_r} W(u)dx \right) = l > 0, \tag{3.29}$$

and therefore $J_{B_r}(u) \leq Cr^{n-2}$. This shows that the lower bound (3.17) is in general sharp and confirms that the Modica estimate and the strong monotonicity formula $\frac{d}{dr}(r^{-(n-1)}J_{B_r}) \geq 0$ which implies the lower bound $J_{B_r}(u) \geq cr^{n-1}$ do not hold for arbitrary solutions for the Ginzburg-Landau potential.

Even minimal solutions (cf. Definition 4.1) may not satisfy the Modica esti-mate. We show below that the radial solution $u : \mathbb{R}^2 \to \mathbb{R}^2$, $u(x) = \eta(|x|)\frac{x}{|x|}$ to the Ginzburg-Landau system $\Delta u - (|u|^2 - 1)u = 0$, or equivalently

$$\eta''(r) + \frac{\eta'(r)}{r} - \frac{\eta(r)}{r^2} = \eta^3(r) - \eta(r) \ \text{ on } (0, \infty), r = |x|, \tag{3.30}$$

with $\eta : \mathbb{R} \to \mathbb{R}$ a smooth odd function, is such that

$$|\nabla u(x)|^2 = |\eta'(r)|^2 + \frac{|\eta(r)|^2}{r^2} > 2W(u(x)), \quad \text{ for all } x \in \mathbb{R}^2.$$

The minimality of this solution is established in [26]. Multiplying (3.30) by η' and integrating from $r = |x|$ to $+\infty$ we obtain that since $\lim_{+\infty} \eta = 1$ and $\lim_{+\infty} \eta' = 0$,

$$- W(u(x)) + \frac{(\eta'(r))^2}{2} = \int_r^\infty \left(\frac{(\eta'(s))^2}{s} - \frac{\eta(s)\eta'(s)}{s^2} \right) ds. \tag{3.31}$$

Next, it follows from

$$\frac{(\eta(r))^2}{2r^2} = \int_r^\infty \left(\frac{(\eta(s))^2}{s^3} - \frac{\eta(s)\eta'(s)}{s^2} \right) ds,$$

that for every $x \in \mathbb{R}^2$, we have

$$\frac{1}{2}|\nabla u(x)|^2 - W(u(x)) = \int_{|x|}^\infty \frac{(s\eta'(s) - \eta(s))^2}{s^3} ds > 0.$$

3.4 Hamiltonian Identities

The conservation of the mechanical energy for the Hamiltonian system (2.5) is expressed by

$$\frac{1}{2}|u'(x)|^2 - 2W(u(x)) = \text{Const.} \tag{3.32}$$

For heteroclinic solutions the constant is zero and (3.32) becomes the equipartition relation. Gui [21] discovered analogs of (3.32) for the P.D.E. system (3.1) which rather appropriately he called *Hamiltonian identities*. We begin with a sample from [21].

Theorem 3.2 *Let $u : \mathbb{R}^2 \to \mathbb{R}^m$ be a solution of system* (3.1), $x = (x_1, x_2)$. *If u is bounded and $u(x_1, x_2)$ converges to $a(x_2)$, $b(x_2)$ as x_1 tends to $\pm\infty$, respectively, then the following identity holds for u:*

$$\int_{-\infty}^{\infty} \left[\frac{1}{2}\left(|u_{x_1}|^2 - |u_{x_2}|^2\right) + W(u(x))\right] dx_1 = \text{Const.,} \quad \text{for all } x_2 \in \mathbb{R}, \tag{3.33}$$

provided that the integral is finite at least for some value of x_2.

Proof Note that the stress-energy tensor in this case takes the form

$$T = \begin{bmatrix} \frac{1}{2}\left(|u_{x_1}|^2 - |u_{x_2}|^2\right) - W(u), & u_{x_1} \cdot u_{x_2} \\ u_{x_1} \cdot u_{x_2}, & \frac{1}{2}\left(|u_{x_2}|^2 - |u_{x_1}|^2\right) - W(u) \end{bmatrix} = \begin{bmatrix} T_{11} & T_{12} \\ T_{21} & T_{22} \end{bmatrix}. \tag{3.34}$$

Applying the divergence theorem over the rectangle $\mathscr{R} = PQRS$ (cf. Fig. 3.1) we have, by (3.3),

$$0 = \int_{\mathscr{R}} \text{div}(T_{21}, T_{22}) dx = \int_{\partial\mathscr{R}} (T_{21}, T_{22}) \cdot \nu dS,$$

thus

$$\int_{SR} \left[\frac{1}{2}(|u_{x_1}|^2 - |u_{x_2}|^2) + W(u(x))\right] dx_1 - \int_{PQ} \left[\frac{1}{2}(|u_{x_1}|^2 - |u_{x_2}|^2) + W(u(x))\right] dx_1$$

$$= \int_{SP} (u_{x_1} \cdot u_{x_2}) dx_2 - \int_{RQ} (u_{x_1} \cdot u_{x_2}) dx_2. \tag{3.35}$$

Fig. 3.1 The rectangle $\mathscr{R}$

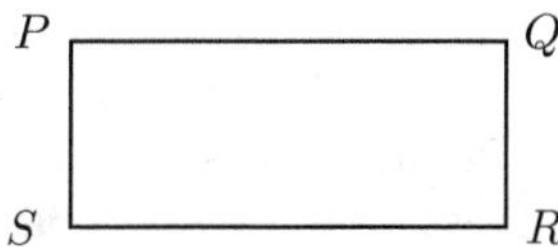

Let $Q = (Q_1, Q_2)$. We will show that $\lim_{Q_1 \to +\infty} v(Q_1, Q_2) = 0$ uniformly for Q_2 in bounded sets. Notice that $u(x_1 + Q_1, x_2) \to v(x_1, x_2)$ along a subsequence, as $Q_1 \to +\infty$ in $C^1_{\text{loc}}(\mathbb{R}^2; \mathbb{R}^m)$ by linear elliptic estimates, and $v(x_1, x_2)$ solves (3.1). By hypothesis, $u(x_1, x_2) \to a(x_2)$ as $x_1 \to +\infty$, hence $v(x_1, x_2) = a(x_2)$ and the limit as $Q_1 \to +\infty$ exists, and moreover

$$u_{x_1}(x_1 + Q_1, x_2) \to 0 \quad \text{as } Q_1 \to +\infty. \tag{3.36}$$

By a similar argument,

$$u_{x_1}(x_1 + P_1, x_2) \to 0 \quad \text{as } P_1 \to -\infty. \tag{3.37}$$

Passing to the limit in (3.35) completes the proof. $\qquad\square$

Note: We note that $u_{x_2}(x_1, x_2) \to a'(x_2)$ as $x_1 \to +\infty$. Similarly, we have

Theorem 3.3 *Let $u : \mathbb{R}^2 \to \mathbb{R}^m$ be a solution of system (3.1), $x = (x_1, x_2)$. If u is bounded and $u(x_1, x_2)$ converges to $a^\pm$ as x_1 tends to $\pm\infty$, with $W(a^\pm) = 0$, then the following identity holds for u:*

$$\int_{-\infty}^{\infty} u_{x_1} \cdot u_{x_2} dx_1 = \text{Const.}, \quad \text{for all } x_2 \in \mathbb{R}, \tag{3.38}$$

provided that the integral is finite at least for some value of x_2.

Proof Here we apply the divergence theorem over the rectangle $\mathscr{R} = PQRS$ (cf. Fig. 3.1) to the first row of the stress-energy tensor:

$$0 = \int_{\mathscr{R}} \text{div}(T_{11}, T_{12}) dx = \int_{\partial\mathscr{R}} (T_{11}, T_{12}) \cdot \nu dS,$$

thus

$$\int_{QR} \left[\frac{1}{2}(|u_{x_1}|^2 - |u_{x_2}|^2) + W(u(x))\right] dx_2 - \int_{PS} \left[\frac{1}{2}(|u_{x_1}|^2 - |u_{x_2}|^2) + W(u(x))\right] dx_2$$

$$= -\int_{PQ} (u_{x_1} \cdot u_{x_2}) dx_1 + \int_{SR} (u_{x_1} \cdot u_{x_2}) dx_1. \tag{3.39}$$

Proceeding as in Theorem 3.3 we can see that $\lim_{Q_1 \to +\infty} u(Q_1, Q_2) = a^+$, and $\lim_{Q_1 \to +\infty} \nabla u(Q_1, Q_2) = 0$ uniformly for Q_2 in bounded sets. Passing to the limit in (3.39) we complete the proof. $\qquad\square$

The identities (3.33) and (3.38) play an essential role for the proof of Theorems 9.1, 9.2 and 9.3 in Chap. 9.

Before discussing the next application, again from Gui [21], we make a digression to minimal partitions, a topic that is of independent interest. We refer to Appendix and the references therein. Consider an open set $U \subset \mathbb{R}^n$ occupied by N immiscible fluids, or phases. Associated to each pair of phases i and j, there is a surface energy density σ_{ij}, with $\sigma_{ij} > 0$, for $i \neq j$, and $\sigma_{ij} = \sigma_{ji}$, with $\sigma_{ii} = 0$. Hence if D_i denotes the subset of U occupied by phase i, then U is the disjoint union

$$U = D_1 \cup D_2 \cup \cdots \cup D_N,$$

and the energy of the partition $P = \{D_i\}_{i=1}^N$ is

$$E(P) = \sum_{0<i<j\leq N} \sigma_{ij}\mathscr{H}^{n-1}(\partial D_i \cap \partial D_j),$$

where H^k the k Hausdorff measure. For $n = 3$, it is simply the area of $\partial D_i \cap \partial D_j$. If U is unbounded, for example $U = \mathbb{R}^n$, the quantity above in general will be infinite. Thus, for each $V \subset U$ open with $V \subsetneq U$, we consider the energy

$$E(P; V) = \sum_{0<i<j\leq N} \sigma_{ij}\mathscr{H}^{n-1}(I_{ij} \cap V), \quad \text{where } I_{ij} := \partial D_i \cap \partial D_j.$$

Definition 3.1 The partition P is a *minimizing N partition*, if given any $V \subsetneq U$, and any N-partition P' of U with

$$\bigcup_{i=1}^{N}(D_i \bigtriangleup D_i') \subsetneq V,$$

we have

$$E(P; V) \leq E(P'; V),$$

where $D_i \bigtriangleup D_j'$ is the symmetric difference of the sets D_i, D_j'.

It is well known that if P is a minimizing partition in $U = \mathbb{R}^2$ with $N = 3$, and with the surface tension coefficients σ_{ij} satisfying

$$\sigma_{ik} < \sigma_{ij} + \sigma_{jk}, \quad \text{for } j \neq k, i, j, k \in \{1, 2, 3\},$$

then ∂P is a triod. Moreover, it is well known that the triod generates a minimizing partition and it is the unique singular minimal cone in $\mathbb{R}^2$. See for example the expository article [2].

Now we turn to the next application of the stress-energy tensor. If the asymptotic behavior of the solution at infinity is known, then the divergence theorem applied on an expanding sphere renders a balancing relationship (zero flux) at infinity which

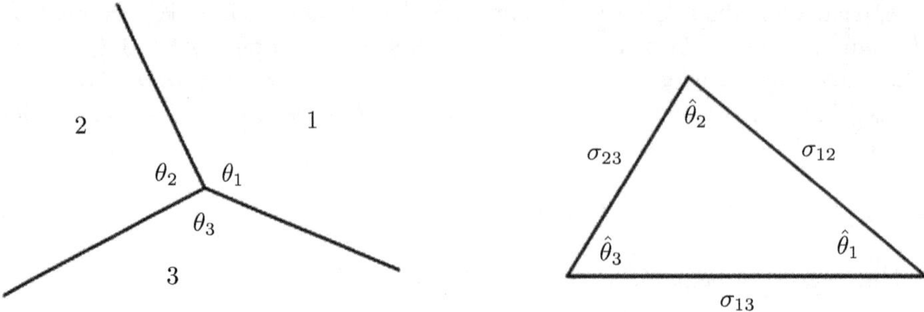

Fig. 3.2 The triod

implies a certain geometric rigidity, at least asymptotically. A good example of such a solution is the *triple junction*, which we now define.

Let $W > 0$ on $\mathbb{R}^2 \setminus \{a_1, a_2, a_3\}$, W a C^3 potential, with nondegenerate minima a_i, $i = 1, 2, 3$, and satisfying

$$\liminf_{|u| \to \infty} W(u) > 0.$$

An example of such a potential is

$$W(u) = |f(z)|^2, \quad z = u_1 + iu_2, \quad f(z) = (z - a_1)(z - a_2)(z - a_3). \quad (3.40)$$

Consider the partition $P = \{D_1, D_2, D_3\}$ of $\mathbb{R}^2$ as in the Fig. 3.2 together with the minimal cone C with vertex at the origin.

Definition 3.2 A *triple-junction* solution u to (3.1) is a solution which satisfies the following estimates

(1) $\forall x \in D_i$, $\forall i = 1, 2, 3$: $|u(x) - a_i| \leq Ce^{-cd(x, \partial D_i)}$, for positive constants c and C. From this it follows that $\lim_{\lambda \to +\infty} u(\lambda \xi) = a_i$, for $\xi \in D_i$.
(2) Given any line parallel to the wall $\partial D_i \cap \partial D_j$, $d(x, \partial D_i \cap \partial D_j) = \mu$ with d the signed distance, it holds that $\lim_{|x| \to \infty, \, d(x, \partial D_i \cap \partial D_j) = \mu} u(x) = e_{ij}(\mu)$, where e_{ij} is a heteroclinic connection connecting a_i to a_j, $i \neq j$, $i, j = 1, 2, 3$ (cf. Sect. 2.3).

Remark 3.2 We note that Bronsard et al. [8] established the existence of such solutions in the case $\theta_1 = \theta_2 = \theta_3 = \frac{2\pi}{3}$, for W with the symmetries of the equilateral triangle (cf. [8]). These solutions are minimal in the equivariance class. The construction of triple-junction solutions that are minimal with respect to general perturbations is a major open problem.

We recall from Proposition 2.6 that if the strict triangle inequality

$$d_W(a_i, a_j) < d_W(a_i, a_k) + d_W(a_k, a_j) \quad (3.41)$$

is satisfied for i, j, k distinct, then all three connections exist, and by Theorem 2.6 for potentials as is (3.40), the connections are also unique.

Theorem 3.4 *Let u be a triple-junction solution and assume (3.41) holds. Then, Young's relation*

$$\frac{\sin \hat{\theta}_1}{\sigma_{23}} = \frac{\sin \hat{\theta}_2}{\sigma_{13}} = \frac{\sin \hat{\theta}_3}{\sigma_{12}} \tag{3.42}$$

holds, where $\hat{\theta}_i = \pi - \theta_i$, $\sigma_{ij} = \int_{\mathbb{R}} \left(\frac{1}{2} |e'_{ij}(\eta)|^2 + W(e_{ij}(\eta)) \right) d\eta$ (see Fig. 3.2). Moreover, the cone C is minimal with respect to the group of the surface tension coefficients σ_{ij}, $i, j = 1, 2, 3$.

Note: Recall that $\sigma_{ij} = d_W(a_i, a_j)$.

Proof (Theorem 3.4) It is sufficient to prove the balance of forces relation

$$\sigma_{12} \nu_{12} + \sigma_{23} \nu_{23} + \sigma_{31} \nu_{31} = 0, \tag{3.43}$$

where ν_{ij} is the unit vector along the interface between D_i and D_j, as it is evident by choosing for example ν_{12} along the y-axis and projecting (3.43) along the x-axis. We take a disk B_r centered at the origin and apply the divergence theorem utilizing (3.3). This gives

$$0 = \int_{B_r} \mathrm{div} T \, dx = \int_{\partial B_r} T \cdot \nu \, dS, \tag{3.44}$$

where ν is the outer unit normal to the boundary ∂B_r. We need to study the limit

$$\lim_{r \to \infty} \int_{\partial B_r} T \cdot \nu \, dS,$$

in order to utilize the hypotheses on the solution at infinity. We align the x_2-axis with the $\partial D_1 \cap \partial D_2$ interface, and define the sector S_{12} with angle $\psi_1(r)$ about this interface. We choose ψ_1 so that

$$r \sin \psi_1(r) \to +\infty, \quad \psi_1(r) \to 0, \quad \text{as } r \to +\infty. \tag{3.45}$$

Analogously, we define sectors about the other two interfaces. We now split the integration $\int_{\partial B_r} T \cdot \nu \, dS$ into that over $S \cap \partial B_r$, $S = S_{12} \cup S_{23} \cup S_{31}$, and that over its complement. From (1) of Definition 3.2, via linear elliptic estimates, we obtain

$$|\nabla u(x)| \leq \mathrm{Const.} \, e^{-cd(x, \partial D_i)}, \tag{3.46}$$

and thus utilizing (3.45), we see that

$$\left|\int_{\partial B_r \cap S^c} T \cdot v \, dS\right| \leq \int_{\partial B_r \cap S^c} |T||v| \, dS \leq \int_{\partial B_r \cap S^c} e^{-cr \sin \psi_1(r)} \, dS \to 0 \quad \text{as } r \to \infty.$$
$$(3.47)$$

Next, we consider

$$\lim_{r \to \infty} \int_{\partial B_r \cap S} T \cdot v \, dS.$$

For example, consider

$$\int_{\partial B_r \cap S_{12}} T \cdot v \, dS = \int_{-r \sin \psi_1(r)}^{r \sin \psi_1(r)} T \cdot v \frac{dy_1}{\cos \theta} \quad (y_1 = r \sin \theta, \ y_2 = r \cos \theta). \quad (3.48)$$

To apply the Lebesgue dominated convergence theorem, we notice that by (3.46)

$$|T_{ij}| \leq C e^{-|y_1|}.$$

Moreover we notice that on $\partial B_r \cap S_{12}$, for fixed y_1,

$$\begin{cases} \lim_{r \to \infty} u(y_1, y_2) = \lim_{y_2 \to +\infty} u(y_1, y_2) = e_{12}(y_1), \\ \lim_{r \to \infty} u_{y_1}(y_1, y_2) = \lim_{y_2 \to +\infty} u_{y_1}(y_1, y_2) = e'_{12}(y_1), \\ \lim_{r \to \infty} u_{y_2}(y_1, y_2) = \lim_{y_2 \to +\infty} u_{y_2}(y_1, y_2) = 0. \end{cases} \quad (3.49)$$

Also by (3.45) on $\partial B_r \cap S_{12}$

$$\lim_{r \to \infty} v = \lim_{r \to \infty} (\sin \theta, \cos \theta) = (0, 1) = v_{12}. \quad (3.50)$$

Therefore, from (3.49) and (3.50) we obtain

$$\lim_{r \to \infty} \int_{\partial B_r \cap S_{12}} T \cdot v \, dS = -\int_{\mathbb{R}} \left(\frac{1}{2}|e'_{12}(x_1)|^2 + W(e_{12}(x_1))\right) dx_1 \, v_{12} = -\sigma_{12} v_{12}.$$

Since T is a tensor and invariant under rotations, we can apply the same procedure for the other two interfaces for appropriately rotated coordinate systems. This concludes the proof of (3.43), and thus that of the theorem. $\square$

3.5 A Liouville Theorem

Theorem 3.5 *Let u be a $W^{1,2}_{loc}(\mathbb{R}^n; \mathbb{R}^m) \cap L^\infty_{loc}(\mathbb{R}^n; \mathbb{R}^m)$ solution to the system (3.1), under the hypothesis that $W \in C^2(\mathbb{R}^m; \mathbb{R})$ and $W \geq 0$. Then, the following estimate* *holds*

$$\begin{cases} J_{B_r}(u) = \mathrm{o}(r^{n-2}), & as\ r \to \infty,\ n \geq 3 \quad \Longrightarrow \quad u \equiv \mathrm{Const.}, \\ J_{B_r}(u) = \mathrm{o}(\log r), & as\ r \to \infty,\ n = 2 \quad \Longrightarrow \quad u \equiv \mathrm{Const.} \end{cases} \tag{3.51}$$

In particular, for $n \geq 2$ we have that

$$\int_{\mathbb{R}^n} \left(\frac{1}{2} |\nabla u|^2 + W(u) \right) dx < \infty \quad \Longrightarrow \quad u \equiv \mathrm{Const.} \tag{3.52}$$

Proof For $n \geq 3$, by the monotonicity formula (3.9), for $r \geq r_0$, we have

$$\frac{1}{r_0^{n-2}} \int_{|x|<r_0} \left(\frac{1}{2} |\nabla u|^2 + W(u) \right) dx \leq \frac{1}{r^{n-2}} \int_{|x|<r} \left(\frac{1}{2} |\nabla u|^2 + W(u) \right) dx.$$

By taking the limit $r \to \infty$ we are set.

For $n = 2$, (3.5) gives

$$\mathrm{tr}\, T = -2W(u).$$

Following the proof of the monotonicity formula, we have

$$\sum_{i,j} \int_{B_r} (x^i T_{ij})_{x_j} dx = - \int_{B_r} 2W(u) dx$$

and thus we obtain, as in (3.15),

$$r \frac{dJ_{B_r}(u)}{dr} \geq 2 \int_{B_r} W(u)\, dx \geq 2 \int_{B_{r_0}} W(u)\, dx, \quad \text{for all } r \geq r_0,$$

for $r \geq r_0$. Hence, integrating we obtain

$$J_{B_r}(u) \geq J_{B_{r_0}}(u) + 2 \log \frac{r}{r_0} \int_{B_{r_0}} W(u)\, dx.$$

The hypothesis (3.51) for $n = 2$ implies that $W(u) \equiv 0$. But then the components of u are harmonic:

$$\Delta u_i = 0, \quad i = 1, \ldots, m,$$

and differentiating,

$$\Delta \left(\frac{\partial u_i}{\partial x_k} \right) = 0.$$

By the hypothesis, $\int_{B_r} |\nabla u|^2 dx = o(\ln r)$. On the other hand, by the mean value theorem,

$$\frac{\partial u_i}{\partial x_k}(0) = \fint_{B_r} \frac{\partial u_i}{\partial x_k} dx \leq \frac{c}{r} \left(\int_{B_r} \left| \frac{\partial u_i}{\partial x_k} \right|^2 dx \right)^{1/2} \leq \frac{c}{r} (\ln r)^{1/2}.$$

Hence, taking $r \to \infty$ we conclude that $\frac{\partial u_i}{\partial x_k}(0) = 0$, and since the equation is invariant under translations, we conclude that $\frac{\partial u_i}{\partial x_k} \equiv 0$, thus $u_i \equiv$ Const. $\qquad\square$

3.6   Pohozaev Identities

In this final section, we recall the Derrick-Pohozaev identity (cf. [12, p. 554]) that played a role in the development of the theory of nonlinear elliptic equations and also served as the basis for the Hamiltonian and other identities. We present a derivation based on the stress-energy tensor [3]. Consider the more general system

$$\text{div}(\varphi'(|\nabla u|^2)\nabla u) - W_u(u) = 0, \tag{3.53}$$

where $\varphi \in C^2(\mathbb{R}^+)$ such that $\varphi(0) = 0$ and $\varphi'(s) \geq 0, \ \forall s \geq 0$.

Proposition 3.1 *Let Ω be a smooth open domain in $\mathbb{R}^n$, and $u \in C^1(\overline{\Omega}) \cap C^2(\Omega)$ a solution of (3.53), such that $u : \Omega \to \mathbb{R}^m$, and $u = a$ on $\partial\Omega$. Let W be a C^2 potential satisfying $W(a) = 0$, and let $x_0 \in \Omega$. Then, the following identity holds true:*

$$0 = \frac{n}{2} \int_\Omega \psi(|\nabla u|^2) dx + n \int_\Omega W(u) dx + \frac{1}{2} \int_{\partial\Omega} (x - x_0) \cdot \nu \widetilde{\psi}\left(\left| \frac{\partial u}{\partial \nu} \right|^2 \right) dS, \tag{3.54}$$

where

$$\begin{aligned} \psi(s) &= \varphi(s) - \frac{2}{n} s\varphi'(s), \\ \widetilde{\psi}(s) &= 2s\varphi'(s) - \varphi(s). \end{aligned} \tag{3.55}$$

Proof First, we introduce the stress-energy tensor

$$T_{ij} = \varphi'(|\nabla u|^2) u_{x_i} \cdot u_{x_j} - \delta_{ij} \left(\frac{1}{2}\varphi(|\nabla u|^2) + W(u) \right) \tag{3.56}$$

and note that (3.53) yields

$$T_{ij,x_j} = 0, \tag{3.57}$$

where the summation convention is adopted. To prove now (3.54), we integrate $(x_i T_{ij})_{x_j} = T_{ii}$ over Ω and apply the divergence theorem to obtain

$$\int_\Omega T_{ii}\, dx = \int_{\partial\Omega} x_i T_{ij} \nu_j\, dS. \tag{3.58}$$

From (3.56) we obtain

$$T_{ii} = \varphi'(|\nabla u|^2)|\nabla u|^2 - n\left(\frac{1}{2}\varphi(|\nabla u|^2) + W(u)\right) \tag{3.59}$$

$$= -\frac{n}{2}\psi(|\nabla u|^2) - nW(u).$$

For $x \in \partial\Omega$ we have

$$x = (x\cdot\nu)\nu + (x\cdot h)h, \quad h \in T_x(\partial\Omega), \quad |h| = 1\ (h\cdot\nu = 0),$$

and moreover

$$x_i T_{ij}\nu_j = (x\cdot\nu)\nu_i T_{ij}\nu_j + (x\cdot h)h_i T_{ij}\nu_j$$

$$= (x\cdot\nu)\varphi'(|\nabla u|^2)\left|\frac{\partial u}{\partial\nu}\right|^2 - (x\cdot\nu)L \tag{3.60}$$

$$+ (x\cdot h)\varphi'(|\nabla u|^2)u_{k,x_i}h_i u_{k,x_j}\nu_j - (x\cdot h)(h\cdot\nu)L,$$

where $L(u, p) = \frac{1}{2}\varphi(|p|^2) + W(u)$, $p = \nabla u$. Since the boundary condition $u = a$ implies $u_{x_i}h_i = 0$ on $\partial\Omega$, we see that

$$x_i T_{ij}\nu_j = (x\cdot\nu)\left(\varphi'(|\nabla u|^2)\left|\frac{\partial u}{\partial\nu}\right|^2 - \frac{1}{2}\varphi(|\nabla u|^2) - W(a)\right). \tag{3.61}$$

Recalling that

$$W(a) = 0, \quad \left|\frac{\partial u}{\partial\nu}\right| = |\nabla u| \text{ on } \partial\Omega$$

we obtain from (3.61)

$$x_i T_{ij}\nu_j = (x\cdot\nu)\frac{1}{2}\tilde\psi\left(\left|\frac{\partial u}{\partial\nu}\right|^2\right), \quad x \in \partial\Omega. \tag{3.62}$$

Combining (3.58), (3.59) and (3.62) yields (3.54). The proof is complete.   □

Remark 3.3 We notice that for certain choices of φ the functions ψ, $\widetilde{\psi}$ are nonnegative. For example, for the minimal surface choice

$$\varphi(s) = 2(\sqrt{1+s} - 1)$$

we have

$$\psi(s) = 2\left(\sqrt{1+s} - 1 - \frac{1}{n}\frac{s}{\sqrt{1+s}}\right) \geq 0$$

and

$$\widetilde{\psi}(s) = \frac{2s}{\sqrt{1+s}} - 2(\sqrt{1+s} - 1) = 2\frac{\sqrt{1+s} - 1}{\sqrt{1+s}} \geq 0.$$

Remark 3.4 Recalling that $u|_{\partial\Omega} = a$, from (3.53), after taking the inner product with $u - a$ and integrating over Ω, we obtain via the divergence theorem

$$-\int_\Omega \varphi'(|\nabla u|^2)|\nabla u|^2 \mathrm{d}x = \int_\Omega W_u(u) \cdot (u - a)\mathrm{d}x,$$

equivalently

$$-\int_\Omega \widetilde{\varphi}(|\nabla u|^2)\mathrm{d}x = \int_\Omega W_u(u) \cdot (u - a)\mathrm{d}x, \tag{3.63}$$

where

$$\widetilde{\varphi}(s) = s\varphi'(s). \tag{3.64}$$

From Pohozaev's identity (3.54), for $\Omega = \mathbb{R}^n$ and provided $\left|\dfrac{\partial u}{\partial \nu}\right| \to 0$ as $|x| \to \infty$, sufficiently fast, we obtain

$$-\frac{1}{2}\int_{\mathbb{R}^n} \psi(|\nabla u|^2)\mathrm{d}x = \int_{\mathbb{R}^n} W(u)\mathrm{d}x. \tag{3.65}$$

Before closing this section, we return to system (3.1), and observe that, for $\phi(s) = s$, (3.54) reduces to

$$0 = \int_\Omega \left(\frac{n-2}{2}|\nabla u|^2 + nW(u)\right)\mathrm{d}x + \frac{1}{2}\int_{\partial\Omega} \left|\frac{\partial u}{\partial \nu}\right|^2 (x - x_0) \cdot \nu \mathrm{d}S. \tag{3.66}$$

Note that, if $W \geq 0$ and Ω is star shaped, for $n \geq 2$ all three terms in this identity are nonnegative. It follows that the only solution of (3.1) is the trivial solution $u \equiv a$.

Taking $\Omega = \mathbb{R}^n$ and assuming that $u(x) \to a$ sufficiently fast, we can derive from (3.66) by dropping the boundary term, the identity

$$\frac{2n}{n-2} \int_{\mathbb{R}^n} W(u)\mathrm{d}x = \int_{\mathbb{R}^n} W_u(u) \cdot (u-a)\mathrm{d}x, \tag{3.67}$$

where we have eliminated the gradient term by utilizing (3.1). Typically from this, one obtains nonexistence results. For example, for $m = 1$ and $W(u) = -\frac{u^2}{2} + \frac{1}{\sigma+2}|u|^\sigma u^2$, $a = 0$, (3.67) gives

$$\left(\frac{2n}{n-2} - 1\right) \int_{\mathbb{R}^n} u^2 \mathrm{d}x = \left(\frac{2n}{(n-2)(\sigma+2)} - 1\right) \int_{\mathbb{R}^n} |u|^\sigma u^2 \mathrm{d}x,$$

and thus for $u \not\equiv 0$, $\sigma < \frac{4}{n-2}$, a rather well-known fact.

In the particular case where Ω is a ball, no boundary conditions are needed to derive Pohozaev's identity (cf. (3.19)). When Ω is a planar domain, we shall also state the most general version of these identities without assuming any boundary conditions.

Proposition 3.2 *Let $\Omega \subset \mathbb{R}^2$ be a smooth open domain, let $x_0 \in \mathbb{R}^2$ be an arbitrary point, and let $u \in C^1(\overline{\Omega}; \mathbb{R}^m) \cap C^2(\Omega; \mathbb{R}^m)$ be a solution to (3.1). On $\partial\Omega$ we consider the positively oriented orthonormal basis (v, τ), where v is the outer normal and τ the tangent vector. Then,*

$$2\int_\Omega W(u) = \int_{\partial\Omega} \left(\left(\frac{1}{2}|\nabla u|^2 + W(u) - |u_v|^2\right)(x-x_0)\cdot v - (u_v \cdot u_\tau)(x-x_0)\cdot \tau\right), \tag{3.68}$$

and

$$0 = \int_{\partial\Omega} \left(\left(\frac{1}{2}|\nabla u|^2 + W(u) - |u_v|^2\right)(x-x_0)\cdot \tau + (u_v \cdot u_\tau)(x-x_0)\cdot v\right). \tag{3.69}$$

Proof Without loss of generality we take $x_0 = 0$. We derive (3.68) and (3.69) by applying the divergence theorem to the vector fields $X = Tx = x_1 T_1 + x_2 T_2$ and $Y = T\Pi x = -x_2 T_1 + x_1 T_2$, respectively ($x = \begin{pmatrix} x_1 \\ x_2 \end{pmatrix}$, $T = [T_1, T_2]$, Π the rotation by $\pi/2$). We have

$$\mathrm{div}\, X = (x_1 T_{i1} + x_2 T_{i2})_{x_i} = \delta_{1i} T_{i1} + x_1 T_{i1,x_i} + \delta_{2i} T_{i2} + x_2 T_{i2,x_i}$$

$$= \mathrm{tr}\, T \quad \text{(since div } T = 0\text{)}$$

$$= -2W(u) \quad \text{(by (3.5), } n = 2\text{)}.$$

Fix $x \in \partial\Omega$. To compute $X \cdot v$, we use the basis $\{v, \tau\}$ at x and let x' the coordinates of x in $\{v, \tau\}$. We have $x' = Ox$ with $O \in SO_2$, and

$$X \cdot v = Tx \cdot v = OTx \cdot Ov = OTO^\top x' \cdot Ov = T'x' \cdot Ov,$$

where T' is the expression of T with respect to $\{v, \tau\}$. Since $Ov = \begin{pmatrix} 1 \\ 0 \end{pmatrix}$ and

$$T'x' = \begin{pmatrix} |u_v|^2 - g(u) & u_v \cdot u_\tau, \\ u_\tau \cdot u_v & |u_v|^2 - g(u) \end{pmatrix} \begin{pmatrix} x \cdot v \\ x \cdot \tau \end{pmatrix}$$

we obtain

$$X \cdot v = (|u_v|^2 - g(u))x \cdot v + (u_v \cdot u_\tau)x \cdot \tau$$

and (3.68) follows.

Similarly we have

$$\mathrm{div}\, Y = (-x_2 T_{i1} + x_1 T_{i2})_{x_i} = -T_{21} + T_{12} = 0$$

and

$$Y \cdot v = OT\Pi x \cdot Ov = OTO^\top (\Pi x)' \cdot Ov = T'(\Pi x)' \cdot Ov = T'\Pi x' \cdot Ov.$$

Since $\Pi x' = \begin{pmatrix} -x \cdot \tau \\ x \cdot v \end{pmatrix}$ we obtain

$$Y \cdot v = -(|u_v|^2 - g(u))x \cdot \tau + (u_v \cdot u_\tau)x \cdot v.$$

This completes the proof. $\square$

Remark 3.5 If $\Omega = B_R(x_0)$, then the identities of Proposition 3.2 reduce to (3.19) and

$$0 = \int_{\partial B_R(x_0)} u_v \cdot u_\tau. \tag{3.70}$$

Clearly, there is an analogy with the Hamiltonian identities in Theorems 3.2 and 3.3.

Remark 3.6 Identities (3.68) and (3.69) can be utilized to determine the constants appearing in (3.33) and (3.38). If we consider for instance that the assumptions of Theorem 3.3 hold, we can show that $C := \int_{-\infty}^{\infty} (u_{x_1} \cdot u_{x_2}) dx_1$ vanishes, provided $u(x_1, x_2)$ converges to $a^{\pm}$ sufficiently fast as $x_1 \to \pm\infty$. Indeed, by applying (3.69) in the rectangle $\mathscr{R}$ (cf. Fig. 3.1) and proceeding as in the proof of Theorem 3.3, we obtain:

$$\int_{-\infty}^{\infty} x_1\left[\frac{1}{2}\left(|u_{x_1}|^2 - |u_{x_2}|^2\right) + W(u(x))\right]dx_1 + Cx_2 = \text{Const.}, \quad \text{for all } x_2 \in \mathbb{R},$$

(3.71)

provided $x_1|\nabla u(x_1, x_2)|^2 \to 0$ as $|x_1| \to \infty$, uniformly when x_2 is bounded. Then, if the integrals

$$\int_{-\infty}^{\infty} x_1\left[\frac{1}{2}\left(|u_{x_1}|^2 - |u_{x_2}|^2\right) + W(u(x))\right]dx_1$$

are uniformly bounded for $x_2 \in \mathbb{R}$, it follows that $C = 0$, and in addition the following weighted Hamiltonian identity holds (cf. [22]):

$$\int_{-\infty}^{\infty} x_1\left[\frac{1}{2}\left(|u_{x_1}|^2 - |u_{x_2}|^2\right) + W(u(x))\right]dx_1 = \text{Const.}, \quad \text{for all } x_2 \in \mathbb{R}. \quad (3.72)$$

This situation occurs in particular in the context of Lemma 9.12.

3.7 Scholia on Chap. 3

The Ginzburg-Landau system was first studied by Bethuel et al. [6]. Segregation models and related systems are studied in Caffarelli et al. [11], Caffarelli and Lin [9], Berestycki et al. [5], and Farina [16] among others.

Stress-energy tensors have been known in physics for a wide class of Lagrangians [25, 34], including the one introduced here, which was rediscovered in joint work with S. Betelu, motivated by calculations in Bronsard and Reitich [7]. Ilmanen [24] in his work on the linking of motion by curvature with the scalar Allen-Cahn equation had already employed a predecessor of this tensor. Later the tensor was used in the Ginzburg-Landau context by Sandier and Serfaty [29]. The mathematical foundation of all this is classical and goes back to Noether's work, where the divergence-free condition is derived by considering variations with respect to the domain. We refer to Giaquinta and Hildebrandt [20]. Faliagas [13] notes the equivalence of (3.1) and (3.3).

Our derivation of the monotonicity formula (see [1]) is based on the divergence-free formulation of the stress-energy tensor and was inspired by Schoen [30]. The limit of the rescaled tensor

$$T_{ij}^{\epsilon} = \epsilon u_{x_i} \cdot u_{x_j} - \delta_{ij}\left(\frac{\epsilon}{2}|\nabla u|^2 + \frac{W(u)}{\epsilon}\right)$$

is $T^0 = \nabla d \otimes \nabla d - I$, that can be identified with the orthogonal projection to the interface $\{d(x) = 0\}$, d the distance function. L. Simon in his book [31] had already used the relationship of T^0 to the minimal surface equation to derive the monotonicity formula for minimal surfaces. Modica [27] had derived the (strong) monotonicity formula for solutions to the scalar Allen-Cahn in a different, more complicated way. Farina has a sharper version of the Modica inequality in [19]. Caffarelli et al. [10] have derived the inequality for a broader class of equations. Farina was aware of the (weak) monotonicity formula and he utilized it in [15]. He had also obtained Liouville type theorems for the Ginzburg-Landau system [14].

Gui [21] derived certain identities for the solutions to the system (3.1) which extend the equipartition relation (3.27). The use of these identities requires asymptotic information on the solution. Gui's derivation is based on Pohozaev identity type arguments. We instead derive these identities by utilizing the stress-energy tensor. These identities had been already employed by Gui to derive Young's law for triple junction solutions on the plane (Theorem 3.4). We refer to [4] for the analogous result for triple-junction solutions in 3-space.

For minimal partitions, minimal cones and general surface tension coefficients we refer to White [35]. The contents in Sects. 3.1, 3.2, 3.5 are taken from [1]. The argument in the proof of Theorem 3.4 is taken from [4]. The contents of Sects. 3.3 and 3.6 are from [32] and [3] respectively.

References

1. Alikakos, N.D.: Some basic facts on the system $\Delta u - \nabla W(u) = 0$. Proc. Am. Math. Soc. **139**, 153–162 (2011)
2. Alikakos, N.D.: On the structure of phase transition maps for three or more coexisting phases. In: Chambolle, A., Novaga, M., Valdinoci, E. (eds.) Geometry and Partial Differential Equations, Proceedings, CRM series. Edizioni della Normale, Pisa (2013)
3. Alikakos, N.D., Faliagas, A.: The stress-energy tensor and Pohozaev's identity for systems. Acta Math. Sci. **32**(1), 433–439 (2012)
4. Alikakos, N.D., Antonopoulos, P., Damialis, A.: Plateau angle conditions for the vector-valued Allen–Cahn equation. SIAM J. Math. Anal. **45**(6), 3823–3837 (2013)
5. Berestycki, H., Terracini, S., Wang, K., Wei, J.: On entire solutions of an elliptic system modeling phase transitiona. arxiv: 1204.1038
6. Bethuel, F., Brezis, H., Helein, F.: Ginzburg-Landau vortices. In: Progress in Nonlinear Differential Equations and Their Applications, vol. 13. Birkhäuser, Basel (1994)
7. Bronsard, L., Reitich, F.: On three-phase boundary motion and the singular limit of a vector-valued Ginzburg–Landau equation. Arch. Ration. Mech. Anal. **124**(4), 355–379 (1993)
8. Bronsard, L., Gui, C., Schatzman, M.: A three-layered minimizer in $\mathbb{R}^2$ for a variational problem with a symmetric three-well potential. Commun. Pure Appl. Math. **49**(7), 677–715 (1996)
9. Caffarelli, L., Lin, F.H.: Singularly perturbed elliptic systems and multi-valued harmonic functions with free boundaries. J. Am. Math. Soc. **21**(3), 847–862 (2008)
10. Caffarelli, L., Garofalo, N., Segala, F.: A Gradient bound for entire solutions of quasi-linear equations and its consequences. Commun. Pure Appl. Math. **47**(11), 1457–1473 (1994)
11. Caffarelli, L., Karakhanyan, A.L., Lin, F.H.: The geometry of solutions to a segregation problem for nondivergence systems. J. Fixed Point Theory Appl. **5**, 319–351 (2009)

12. Evans, L.C.: Partial Differential Equations. Graduate Studies in Mathematics, vol. 19, 2nd edn. American Mathematical Society, Providence (2010)
13. Faliagas, A.C.: On the equivalence of Euler-Lagrange and Noether equations. Math. Phys. Anal. Geom. **19**, 1 (2016). https://doi.org/10.1007/s11040-016-9203-3
14. Farina, A.: Finite-energy solutions, quantization effects and Liouville-type results for a variant of the Ginzburg–Landau systems in $\mathbb{R}^K$. C. R. Acad. Sci. Paris Ser. I Math. **325**(5), 487–491 (1997)
15. Farina, A.: Two results on entire solutions of Ginzburg–Landau system in higher dimensions. J. Funct. Anal. **214**(2), 386–395 (2004)
16. Farina, A.: Some symmetry results for entire solutions of an elliptic system arising in phase separation. Discrete Contin. Dynam. Syst. **34**(6), 2505–2511 (2014)
17. Farina, A., Valdinoci, E.: The state of art for a conjecture of De Giorgi and related questions. In: Reaction-Diffusion Systems and Viscosity Solutions. World Scientific, Singapore (2008)
18. Farina, A., Valdinoci, E.: A pointwise gradient estimate in possibly unbounded domains with nonnegative mean curvature. Adv. Math. **225**, 2808–2827 (2010)
19. Farina, A., Sciunzi, B., Valdinoci, E.: Bernstein and De Giorgi type problems: new results via a geometric approach. Ann. Scuola Norm. Sup. Pisa Cl. Sci. **5**(7), 741–791 (2008)
20. Giaquinta, M., Hildebrandt, S.: Calculus of Variations I. The Lagrangian Formalism. Grundlehren der mathematischen Wissenschaften, vol. 310. Springer, Berlin (2010)
21. Gui, C.: Hamiltonian identities for partial differential equations. J. Funct. Anal. **254**(4), 904–933 (2008)
22. Gui, C., Malchiodi, A., Xu, H.: Axial symmetry of some steady state solutions to nonlinear Schrodinger equations. Proc. Am. Math. Soc. **139**, 1023–1032 (2011)
23. Hagan, P.S.: Spiral waves in reaction-diffusion equations. SIAM J. Appl. Math. **42**(4), 762–786 (1982)
24. Ilmanen, T.: Convergence of the Allen–Cahn equation to Brakke's motion by mean curvature. J. Differ. Geom. **38**(2), 417–461 (1993)
25. Landau, L.D., Lifschitz, E.M.: Course of Theoretical Physics. Classical Field Theory, vol. 2, 4th edn. Butterworth-Heinemann, Oxford (1980)
26. Mironescu, P.: Local minimizers for the Ginzburg-Landau equation are radially symmetric. C. R. Acad. Sci. Paris Sér. I Math. **323**, 593–598 (1996)
27. Modica, L.: Monotonicity of the energy for entire solutions of semilinear elliptic equations. In: Colombini, F., Marino, A., Modica, L. (eds.) Partial differential equations and the calculus of variations. Essays in honor of Ennio De Giorgi, vol. 2, pp. 843–850. Birkhäuser, Boston (1989)
28. Modica, L.: A Gradient bound and a Liouville Theorem for nonlinear Poisson equations. Commun. Pure Appl. Math. **38**(5), 679–684 (1985)
29. Sandier, E., Serfaty, S.: Vortices in the Magnetic Ginzburg-Landau Model. Progress in Nonlinear Differential Equations and Their Applications, vol. 70. Birkhäuser, Boston (2007)
30. Schoen, R.: Lecture Notes on General Relativity. Stanford University, Stanford (2009)
31. Simon, L.: Lectures on Geometric Measure Theory. In: Proceedings of the Centre for Mathematics and Its Applications, vol. 3. Australian National University, Canberra (1984)
32. Smyrnelis, P.: Gradient estimates for semilinear elliptic systems and other related results. Proc. R. Soc. Edinb. Sect. A **145**(6), 1313–1330 (2015)
33. Sperb, R.: Maximum Principles and Their Applications. Mathematics in Science and Engineering, vol. 157. Academic Press, New York (1981)
34. Struwe, M.: Variational Methods, Applications to Nonlinear Partial Differential Equations and Hamiltonian Systems. A Series of Modern Surveys in Mathematics, vol. 34, 4th edn. Springer, Berlin (2008)
35. White, B.: Existence of least energy configurations of immiscible fluids. J. Geom. Anal. **6**, 151–161 (1996)

Chapter 4
The Cut-Off Lemma and a Maximum Principle

Abstract In this chapter we establish a maximum principle type result that provides pointwise control on minimal solutions. In contrast to the usual maximum principle, it does not hold for solutions in general, not even for local minimizers in the scalar case. We obtain it as a corollary of a replacement lemma modeled after Lemmas 2.4 and 2.5.

4.1 Introduction and Statements

Assume that $W : \mathbb{R}^m \to \mathbb{R}$ is non-negative and for $\Omega \subset \mathbb{R}^n$ open and bounded define $J_\Omega : W^{1,2}(\Omega; \mathbb{R}^m) \to [0, +\infty]$ by

$$J_\Omega(v) = \int_\Omega \left(\frac{1}{2}|\nabla u|^2 + W(u)\right)\mathrm{d}x. \tag{4.1}$$

In this chapter we will deal with bounded solutions of

$$\Delta u - W_u(u) = 0, \tag{4.2}$$

which are defined in an open set $\mathcal{O} \subset \mathbb{R}^n$, generally unbounded, and which are *minimal* (alternatively, *minimizers*) in the sense that they minimize, for each $\Omega \subset \mathcal{O}$, the energy J_Ω, subject to their Dirichlet values. More precisely,

Definition 4.1 (Minimality) Let $\mathcal{O} \subset \mathbb{R}^n$ open; a map $u \in W^{1,2}_{\mathrm{loc}}(\mathcal{O}; \mathbb{R}^m) \cap L^\infty(\mathcal{O}; \mathbb{R}^m)$ is called a *minimizer* or a *minimal map* if

$$J_\Omega(u) \le J_\Omega(u + v), \quad \text{for } v \in W^{1,2}_0(\Omega; \mathbb{R}^m) \cap L^\infty(\Omega; \mathbb{R}^m), \tag{4.3}$$

for every open bounded Lipschitz set $\Omega \subset \mathcal{O}$.

Clearly, if $\mathcal{O}$ is bounded, this reduces to the standard definition of minimizer with Dirichlet conditions. Otherwise, it is a natural extension that accounts for the fact

© Springer Nature Switzerland AG 2018

N. D. Alikakos et al., *Elliptic Systems of Phase Transition Type*,
Progress in Nonlinear Differential Equations and Their Applications 91,
https://doi.org/10.1007/978-3-319-90572-3_4

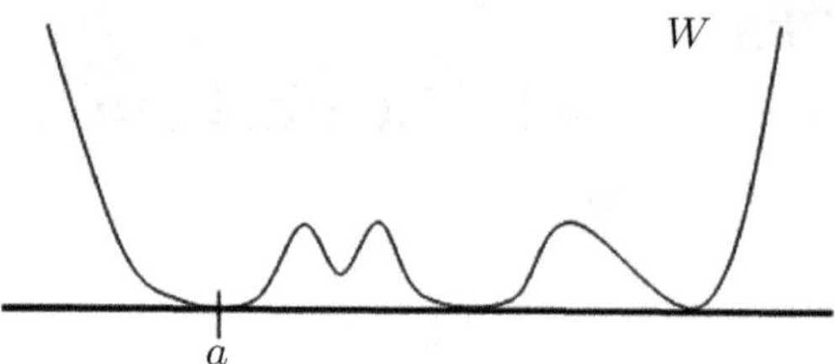

Fig. 4.1 The potential W

that for nontrivial solutions of (4.2) we may have $J_{\mathscr{O}}(u) = +\infty$ if $\mathscr{O}$ is unbounded (cf. (3.52)).

Note that, under sufficient smoothness of W, from the L^∞ bound on u and elliptic regularity it follows that a minimal map $u : \mathscr{O} \to \mathbb{R}^m$ is a classical solution of (5.1) which is the Euler-Lagrange equation associated to J_Ω.

We remark that the terminology in not uniform and some authors refer to minimizers in the sense of Definition 4.1 as global minimizers, other as local minimizers. We simply call them minimizers or minimal maps, in particular to avoid confusion with a local minimizer u where local means in a neighborhood of u.

A. *Hypothesis on W*:

$W \in C(\mathbb{R}^m; \mathbb{R})$ is nonnegative, and has a zero $a \in \mathbb{R}^m$, $W(a) = 0$, satisfying the hypothesis:

$\widehat{\mathbf{H}}$: There exists $r_0 > 0$ such that for every $\xi \in \mathbb{R}^m$ with $|\xi| = 1$:
     In Fig. 4.1 we sketch a W that satisfies $\widehat{H}$. $(0, r_0] \ni r \mapsto W(a + r\xi)$ is nondecreasing, and moreover $W(a + r_0\xi) > 0$.[1]

We will begin by stating a maximum principle for a class of solutions of (4.2), in the simplest possible set-up.

B. *Maximum principle*

Let $A \subset \mathbb{R}^n$, open and bounded, and with Lipschitz boundary.

Theorem 4.1 *Let $v(\cdot) \in W^{1,2}(A; \mathbb{R}^m) \cap L^\infty(A; \mathbb{R}^m)$ be a minimizer of J_A. Assume*

$$|v(x) - a| \le r \quad \text{on } \partial A, \quad 0 < 2r \le r_0. \tag{4.4}$$

Then,

$$|v(x) - a| \le r \quad \text{on } A. \tag{4.5}$$

Moreover, if $u \mapsto W_u(u)$ is Lipschitz, then the attainment of equality in (4.5) at an interior point of A,

$$|v(\hat{x}) - a| = r, \quad \text{for some } \hat{x} \in A, \tag{4.6}$$

---

[1] In particular, $\widehat{\mathbf{H}}$ allows for potentials that vanish in a ball centered in a.

implies that

$$v(x) \equiv \text{Const.} \quad \text{in the connected component of } \hat{x} \text{ in } A. \tag{4.7}$$

The following extension is also true:

Theorem 4.2 *Let $v(\cdot) \in W^{1,2}(A; \mathbb{R}^m) \cap L^\infty(A; \mathbb{R}^m)$ and let $\partial A = \partial_D A \cup \partial_N A$, $\partial_D A$ the subset of ∂A where $u = v$, $\partial_N A = \partial A \setminus \partial_D A$, and assume that ∂A and $\partial_D A$ are Lipschitz. Let*

$$J_A(v) = \min\{J_A(u), \ u = v \text{ on } \partial_D A\}.$$

If $\partial_D A \cap \partial A_i \neq \varnothing$ for every one of the finitely many connected components A_i of A, then the condition

$$|v(x) - a| \leq r \quad \text{on } \partial_D A, \ 0 < 2r \leq r_0, \tag{4.8}$$

implies the same conclusions as in Theorem 4.1 *above.*

Remark 4.1 Theorem 4.1 follows from Theorem 4.2 since for every connected component A_i we have $\partial A_i \subset \partial A = \partial_D A$ ($\partial_N A = \varnothing$).

C. *Comparison with the usual maximum principle*

The theorems above are different from the usual maximum principle in the following respects:

(a) W is not convex, hence the usual maximum principle is not valid even in the scalar case $m = 1$.
(b) Condition $\widehat{\mathbf{H}}$ on W at $u = a$ is extremely mild, and allows applicability in situations where degeneracy is natural [3].
(c) The usual maximum principle is a calculus fact and applies to *all* solutions. This is not true for the result above. Consider the O.D.E.

$$u'' - W'(u) = 0, \quad u : \mathbb{R} \to \mathbb{R}, \tag{4.9}$$

with $W : \mathbb{R} \to \mathbb{R}$ as in Fig. 4.2 (e.g., $W(u) = \frac{1}{4}(u^2 - 1)^2$).
   Notice that (4.9) has periodic solutions satisfying $u(-L/2) = u(L/2) = -1 + r$, for $r > 0$ as small as desired, and the period L chosen accordingly. By choosing $A = [-L/2, L/2]$ (Fig. 4.3) we see that Theorem 4.1 does not apply.

Fig. 4.2 The potential $W(u) = \frac{1}{4}(u^2 - 1)^2$

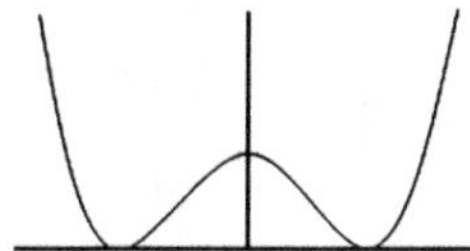

Fig. 4.3 Periodic solutions violate Theorem 4.1

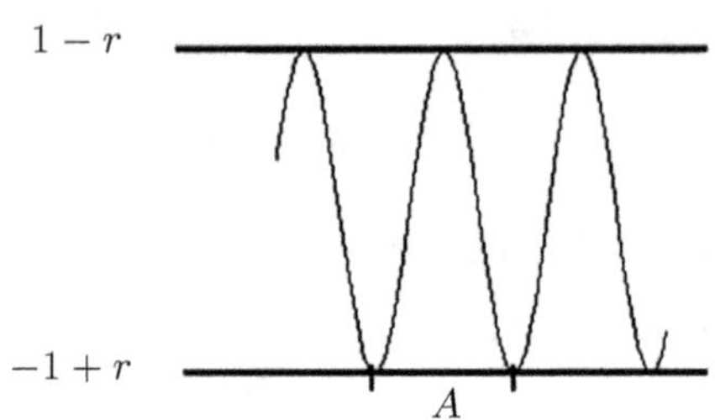

Fig. 4.4 $\kappa(p_i) < 0$, $i = 1, 2$, with the sign convention that $\kappa = 1$ for the unit circle

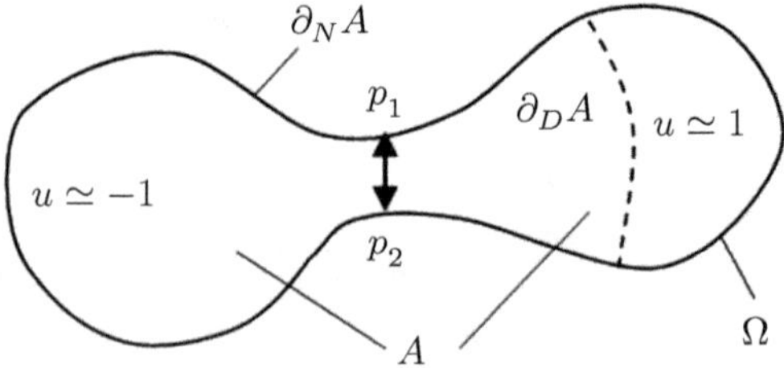

The following example shows that Theorem 4.1 does not apply even to local minimizers (stable solutions of (4.2), defined in terms of the definiteness of the sign of the second variation). Consider the scalar P.D.E.

$$\begin{cases} \epsilon^2 \Delta u - W'(u) = 0 \text{ in } \Omega, \\ \dfrac{\partial u}{\partial n}\Big|_{\partial \Omega} = 0, \end{cases} \qquad (4.10)$$

where W is as in Fig. 4.2 and Ω is a dumbbell domain as in Fig. 4.4. It is well known from Kohn and Sternberg [12][2] that for $0 < \epsilon$ sufficiently small, (4.10) has a stable solution which on the left and right of the neck is as close to -1 and respectively $+1$ as desired (by taking $\epsilon > 0$ sufficiently small). By choosing the set A as in Fig. 4.4 we can secure that $|u(x) - 1| \leq r$ on $\partial_D A$, and therefore we see that Theorem 4.2 does not apply.

The notion of minimal (minimizing) solution is useful for unbounded domains A ($A = \mathbb{R}^n$ in particular), where the energy $J_A(u)$ is infinite (cf. (3.52)). This is a reflection of a familiar property of minimal surfaces: it is not altogether surprising since the functional in (4.1) is linked to the perimeter functional if scaled appropriately [15, see Chapter 4, Section 5.1].

D. *The cut-off lemma*

Theorem 4.1 is a corollary of the following replacement result.

---

[2]cf. related work of Casten-Holland [6], and independently Matano [14].

Lemma 4.1 *Let W as in part* A. *above, and let $A \subset \mathbb{R}^n$, open, bounded, with Lipschitz boundary. Suppose that $u(\cdot) \in W^{1,2}(A; \mathbb{R}^m) \cap L^\infty(A; \mathbb{R}^m)$. If the following two conditions hold,*

(I) $|u(x) - a| \le r$ *on* ∂A, $0 < 2r \le r_0$,
(II) $\mathscr{L}^n(A \cap \{|u(x) - a| > r\}) > 0$ *($\mathscr{L}^n(E)$, the n-dimensional Lebesgue measure),*

then, there is $\tilde{u}(\cdot) \in W^{1,2}(A; \mathbb{R}^m) \cap L^\infty(A; \mathbb{R}^m)$ such that

$$\begin{cases} \tilde{u} = u, & \text{on } \partial A, \\ |\tilde{u}(x) - a| \le r, & \text{on } A, \\ J_A(\tilde{u}) < J_A(u). \end{cases} \qquad (4.11)$$

The following variant of the cut-off lemma is more flexible in applying it to specific situations where usually the difficulty is in choosing the set A so that condition (II) is satisfied. For this purpose, we introduce a pair of sets, A and Ω. Such an example is in Fig. 4.4.

Lemma 4.2 *Let $\Omega \subset \mathbb{R}^n$, be open, bounded and connected, and let A be an open Lipschitz subset of Ω with $\partial A \cap \Omega \ne \varnothing$. Assume W as in part* A. *above, and let $u(\cdot) \in W^{1,2}(\Omega; \mathbb{R}^m) \cap L^\infty(\Omega; \mathbb{R}^m)$. If*

(I) $|u(x) - a| \le r$ *on* $\partial A \cap \Omega$, $0 < 2r \le r_0$, *and*
(II) $\mathscr{L}^n(A \cap \{|u(x) - a| > r\}) > 0$,

then, there is $\tilde{u}(\cdot) \in W^{1,2}(\Omega; \mathbb{R}^m) \cap L^\infty(\Omega; \mathbb{R}^m)$ such that

$$\begin{cases} \tilde{u} = u, & \text{on } \Omega \setminus A, \\ |\tilde{u}(x) - a| \le r, & \text{on } A, \\ J_\Omega(\tilde{u}) < J_\Omega(u). \end{cases} \qquad (4.12)$$

Remark 4.2 The reason Theorems 4.1 and 4.2 do not apply to local minimizers (in the sense of the second variation being positive definite) is that the variations (replacements) $\tilde{u}$ in Lemmas (4.1), (4.2) are generally large.

E. *The polar form*

For a map $u(x)$, consider the *polar representation*

$$u(x) = a + |u(x) - a| \frac{u(x) - a}{|u(x) - a|} =: a + \rho(x)\boldsymbol{n}(x), \qquad (4.13)$$

where

$$\rho(x) := |u(x) - a|, \quad n(x) := \begin{cases} \frac{u(x)-a}{|u(x)-a|}, & \text{if } u(x) \neq a, \\ 0, & \text{if } u(x) = a. \end{cases} \qquad (4.14)$$

If u is smooth and ρ does not vanish, we see that

$$|\nabla u(x)|^2 = |\nabla \rho(x)|^2 + \rho^2(x)|\nabla n(x)|^2, \qquad (4.15)$$

and consequently, we have the *polar form* of the free energy

$$J_A(u) = \int_A \left\{ \frac{1}{2}(|\nabla \rho|^2 + \rho^2|\nabla n|^2) + W(a + \rho n) \right\} dx. \qquad (4.16)$$

The following calculus facts for Sobolev functions hold true:

(a) If $u(\cdot) \in W^{1,2}(A; \mathbb{R}^m) \cap L^\infty(A; \mathbb{R}^m)$, then

$$\int_A |\nabla u|^2 dx = \int_A (|\nabla \rho|^2 + \rho^2|\nabla n|^2) dx. \qquad (4.17)$$

(b) Let $u(\cdot) \in W^{1,2}(A; \mathbb{R}^m) \cap L^\infty(A; \mathbb{R}^m)$, and let $f : \mathbb{R} \to \mathbb{R}$ be a locally Lipschitz function, with $f(0) = 0$, and consider

$$\tilde{u}(x) = a + f(\rho(x))n(x), \qquad (4.18)$$

where ρ and n as in (4.14) above. Then $\tilde{u}(\cdot) \in W^{1,2}(A; \mathbb{R}^m) \cap L^\infty(A; \mathbb{R}^m)$ and

$$\int_A |\nabla \tilde{u}|^2 dx = \int_A (|f'(\rho)\nabla \rho|^2 + f^2(\rho)|\nabla n|^2) dx. \qquad (4.19)$$

Functions $\tilde{u}$ as in (4.18), are the type of variations that we utilize in the proof of the cut-off lemma, for appropriate cut-off functions f. The point is that we keep the direction $n(x)$ of the vector $u(x)-a$ and modify only its modulus $\rho(x) = |u(x)-a|$. Actually, the variations we will be using have the more special form

$$f(s) = sg(s), \qquad (4.20)$$

where $g : \mathbb{R} \to \mathbb{R}$ is a locally Lipschitz function. For the form (4.20), the calculus facts above become completely elementary since one can by-pass $n(x)$, and justify (4.19) directly by establishing the pointwise equality of the corresponding integrands. In fact, the following identity holds rigorously for $u(\cdot) \in W^{1,2}(A; \mathbb{R}^m) \cap L^\infty(A; \mathbb{R}^m)$, and f as in (4.20):

$$|\nabla \tilde{u}(x)|^2 = (f'(\rho))^2|\nabla \rho|^2 + (|\nabla u|^2 - |\nabla \rho|^2)\left(\frac{f(\rho)}{\rho}\right)^2. \qquad (4.21)$$

Indeed, by (4.18), (4.20) and (4.13),

$$\tilde{u}(x) = a + f(\rho(x))\mathbf{n}(x) = a + g(\rho(x))(u(x) - a), \tag{4.22}$$

which is in $W^{1,2}(A; \mathbb{R}^m) \cap L^\infty(A; \mathbb{R}^m)$. Hence

$$\tilde{u}_{x_i} = g'(\rho)\rho_{x_i}(u - a) + g(\rho)u_{x_i} \quad (g \text{ Lipschitz}), \tag{4.23}$$

and with the summation convention we have

$$\begin{aligned}
\tilde{u}_{x_i}\tilde{u}_{x_i} &= (g'(\rho))^2\rho^2|\nabla\rho|^2 + g^2(\rho)|\nabla u|^2 + 2g(\rho)g'(\rho)\rho|\nabla\rho|^2 \\
&= (f'(\rho))^2|\nabla\rho|^2 + g^2(\rho)(|\nabla u|^2 - |\nabla\rho|^2),
\end{aligned} \tag{4.24}$$

where it has been noticed that

$$\frac{1}{2}(\rho^2)_{x_i} = \rho_{x_i}\rho = u_{x_i}\cdot(u - a).$$

Remark 4.3 If $|f'| \le 1$ and $|g| \le 1$, then since $|\nabla u|^2 \ge |\nabla\rho|^2$ we see from (4.24) that

$$|\nabla\tilde{u}(x)|^2 \le |\nabla u(x)|^2. \tag{4.25}$$

Thus the gradient term of the free energy is not increased.

Remark 4.4 The cut-off lemma has been established for $n = 1$ in Chap. 2 (cf. Lemmas 2.4 and 2.5), with a proof similar in broad terms to the one we are about to give, but naturally much simpler, and again based on the polar form (2.69).

4.2 Proofs

Proof (Lemma 4.1) Without loss of generality we may assume, by Remark 4.1, that A is connected.

Case 1: We will first establish the lemma under the additional hypothesis that

$$\rho(x) \le r_0. \tag{4.26}$$

The argument here is easy, since u stays in the monotonicity region of W about a.

Let (cf. Fig. 4.5)

$$f(s) = \frac{\min\{s, r\}}{s}s = g(s)s.$$

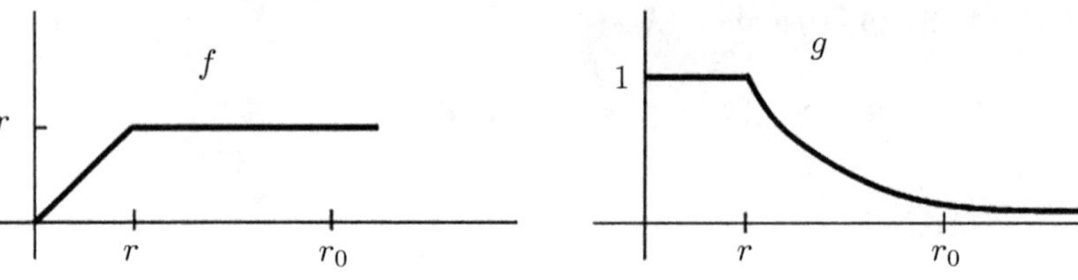

Fig. 4.5 The cut-off functions f and g

By Remark 4.3 we obtain

$$\int_A |\nabla \tilde{u}(x)|^2 \mathrm{d}x \le \int_A |\nabla u(x)|^2 \mathrm{d}x. \tag{4.27}$$

A closer look shows that in case of equality in (4.27), that is,

$$0 = \int_A |\nabla \tilde{u}|^2 \mathrm{d}x - \int_A |\nabla u|^2 \mathrm{d}x \tag{4.28}$$

$$= \int_A |\nabla \rho|^2 ((f'(\rho))^2 - 1)\mathrm{d}x + \int_A (|\nabla u|^2 - |\nabla \rho|^2)(g^2(\rho) - 1)\mathrm{d}x \text{ (via (4.21))}$$

$$\le -\int_{A \cap \{\rho \ge r\}} |\nabla \rho|^2 \mathrm{d}x,$$

from which it follows that

$$\nabla \rho = 0 \text{ a.e. on } A \cap \{\rho \ge r\},$$

and therefore

$$\nabla(\tilde{\rho} - \rho) = 0 \text{ a.e. on } A,$$

where $\tilde{\rho}(x) = f(\rho(x))$ via (4.22). Since $\tilde{\rho} - \rho \in W^{1,2}(A)$, we have by connectedness

$$\tilde{\rho}(x) - \rho(x) = \text{Const. a.e. in } A,$$

and since $\tilde{\rho} - \rho = 0$ on ∂A in the sense of trace, we obtain

$$\tilde{\rho}(x) - \rho(x) = 0 \text{ a.e. in } A,$$

in contradiction to assumption (II) in Lemma 4.1. Therefore we have strict inequality in (4.27).

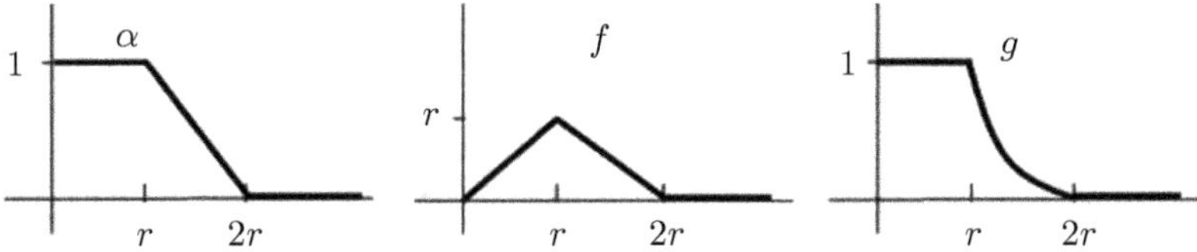

Fig. 4.6 The cut-off functions α, f and g

On the other hand, since $a + f(\rho(x))\boldsymbol{n}(x)$ is measurable and W is continuous, the composition is measurable, and we have

$$\int_A W(\tilde{u}(x))\mathrm{d}x = \int_A W(a + f(\rho(x))\boldsymbol{n}(x))\mathrm{d}x \tag{4.29}$$

$$\leq \int_A W(a + \rho(x)\boldsymbol{n}(x))\mathrm{d}x \quad (\text{by } \mathbf{H})$$

$$= \int_A W(u(x))\mathrm{d}x,$$

hence Case 1 is settled. Notice that in this case the strictness in (4.11) (c) was obtained via the gradient term.

Case 2: Assume

$$\mathscr{L}^n(A \cap \{\rho > r_0\}) > 0. \tag{4.30}$$

Consider the following cut-off functions (Fig. 4.6):

$$\alpha(s) := \begin{cases} 1, & \text{for } s \leq r, \\ \frac{2r-s}{r}, & \text{for } r \leq s \leq 2r, \\ 0, & \text{for } s \geq 2r, \end{cases} \tag{4.31}$$

$$f(s) = \min\{s, r\}\alpha(s), \quad g(s) = \frac{f(s)}{s}. \tag{4.32}$$

Define

$$\tilde{u}(x) = a + g(\rho(x))(u(x) - a) \tag{4.33}$$

(cf. (4.20)). By Remark 4.3,

$$|\nabla \tilde{u}(x)|^2 \leq |\nabla u(x)|^2. \tag{4.34}$$

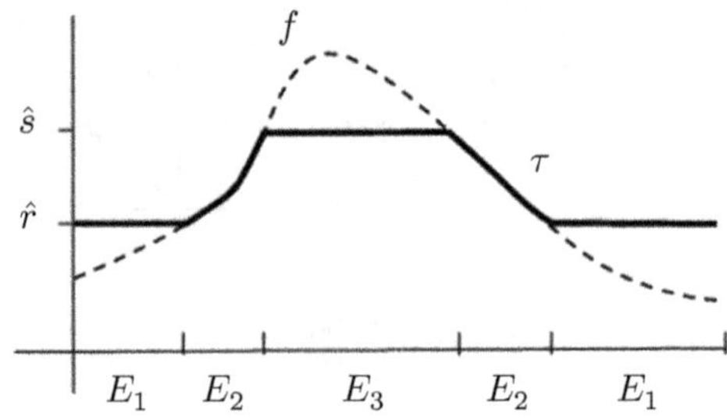

Fig. 4.7 The dashed and thick curves stand for f and τ respectively.
$E_1 = A \cap \{f \le \hat{r}\}$,
$E_2 = A \cap \{\hat{r} < f \le \hat{s}\}$,
$E_3 = A \cap \{\hat{s} < f\}$

We note in passing that $\tilde{u}$ is a reflection of u along $|u - a| = r$, and thus (4.34) is expected. Unlike Case 1, the strictness of the inequality in (4.11) (c) will follow from the potential term. We will need the following

Lemma 4.3 ('Continuity' of Sobolev Functions) *Let $A \subset \mathbb{R}^n$, open, bounded and connected, with Lipschitz boundary, and assume that $f \in W^{1,2}(A; \mathbb{R})$ satisfies*

$$\begin{cases} f \le \hat{r} \text{ on } \partial A \text{ (in the sense of trace)}, \\ \mathcal{L}^n(A \cap \{\hat{s} < f\}) > 0 \text{ for some } \hat{r} < \hat{s}. \end{cases} \tag{4.35}$$

Then, $\mathcal{L}^n(A \cap \{\hat{r} < f \le \hat{s}\}) > 0$.

Proof Let $\sigma, \tau : A \to \mathbb{R}$ be defined by

$$\sigma(x) = \min\{f(x), \hat{s}\} = \begin{cases} f(x), & \text{for } , x \in E_1 := A \cap \{f \le \hat{r}\}, \\ \hat{s}, & \text{for } x \in E_3 := A \cap \{\hat{s} < f\}, \end{cases} \tag{4.36}$$

and (Fig. 4.7)

$$\tau(x) = \max\{\sigma(x), \hat{r}\} = \begin{cases} \hat{r}, & \text{for } x \in E_1, \\ \hat{s}, & \text{for } x \in E_3. \end{cases} \tag{4.37}$$

Suppose, by contradiction, that $\mathcal{L}^n(A \cap \{\hat{r} < f \le \hat{s}\}) = 0$. Therefore, τ is a step function, and so $\nabla \tau = 0$ a.e. in A. On the other hand, σ, τ are in $W^{1,2}(A; \mathbb{R})$ (cf. [11, p. 130]). This and the connectedness of A imply that $\tau \equiv$ Const. (cf. [9, p. 307]). Hence $\tau \equiv \hat{s}$, since $\mathcal{L}^n(E_3) > 0$. It follows that $\mathcal{L}^n(E_1) = 0$ and $f > \hat{s}$ a.e. in A. Thus, $f \ge \hat{s}$ on ∂A in the sense of trace, which contradicts (4.35). The proof is complete. $\square$

Conclusion Let $\epsilon > 0$ such that $W(u) > 0$ on $r_0 \le |u - a| \le r_0 + \epsilon$. We define the sets

$$E_1 := A \cap \{\rho \le r_0\}, \quad E_2 := A \cap \{r_0 < \rho \le r_0 + \epsilon\}, \quad E_3 := A \cap \{\rho > r_0 + \epsilon\}.$$

From (4.30), we obtain that if $\mathscr{L}^n(E_2) = 0$, then necessarily $\mathscr{L}^n(E_3) > 0$. But $\rho \leq r < r_0 + \epsilon$ on ∂A, hence by Lemma 4.3

$$\mathscr{L}^n(E_2) > 0. \tag{4.38}$$

Therefore, (4.38) holds under any circumstances. On $A \cap \{r \leq \rho \leq 2r\}$ we have:

$$W(\tilde{u}(x)) = W(a + r\alpha(\rho(x))\boldsymbol{n}(x)) \tag{4.39}$$
$$\leq W(a + r\boldsymbol{n}(x))$$
$$\leq W(a + \rho(x)\boldsymbol{n}(x)) = W(u(x)).$$

On the other hand, on $A \cap \{\rho > 2r\}$

$$0 = W(\tilde{u}(x)) \leq W(u(x)), \tag{4.40}$$

while on E_2

$$0 = W(\tilde{u}(x)) < W(u(x)). \tag{4.41}$$

Therefore, $J_A(\tilde{u}) < J_A(u)$ and the proof of Lemma 4.1 is complete. $\qquad\square$

Proof (Theorem 4.1) It is sufficient to establish the theorem for A connected, since if A_i is a connected component of A, we have $\partial A_i \subset \partial A$. We proceed by contradiction. So suppose that (4.5) does not hold, hence $\mathscr{L}^n(A \cap \{|v(x) - a| > r\}) > 0$. But this contradicts the minimality of v by Lemma 4.1. Thus (4.5) holds. Next, suppose (4.6) holds and notice that $|v - a|^2 \in W^{1,2}(A; \mathbb{R}) \cap L^\infty(A; \mathbb{R})$ satisfies

$$\Delta|v - a|^2 = 2|\nabla v|^2 + 2(v - a) \cdot W_u(v) \geq 0,$$

thanks to Hypothesis $\widehat{\mathbf{H}}$ on W. By the strong maximum principle it follows that $|v - a|^2$ is constant in A. As a consequence $\nabla v \equiv 0$ in A, and v is constant. The proof of Theorem 4.1 is complete. $\qquad\square$

Proof (Theorem 4.2) First we note that if in Lemma 4.1, specifically in condition (I), one replaces ∂A with $\partial_\mathrm{D} A$, then the same conclusion (4.11) holds, where ∂A is now replaced with $\partial_\mathrm{D} A$. The argument is completely unaltered. Similarly, in Lemma 4.3, ∂A is replaced with $\partial_\mathrm{D} A$, without change in the proof. The proof of Theorem 4.2 is complete. $\qquad\square$

Proof (Lemma 4.2) The proof here is identical to the proof of the modification of Lemma 4.1 implemented in the proof of Theorem 4.2 above, with the identification of $\partial_\mathrm{D} A$ with $\partial A \cap \Omega$. The connectedness of Ω guarantees that if A_i is a connected component of A, then $\partial A_i \cap \Omega \neq \emptyset$. The proof of Lemma 4.2 is complete. $\qquad\square$

4.3 Applications

4.3.1 First Application

From Lemma 4.1 we learn that, from the point of view of minimizing the energy, for a map $u : A \to \mathbb{R}^m$ whose boundary values lie in a small neighborhood of a where W is convex it is more convenient to remain in the convexity region throughout A rather then make an excursion away from a. This suggests uniqueness for minimization problems with Dirichlet conditions near a.

Theorem 4.3 *Let $W : \mathbb{R}^m \to \mathbb{R}$ be a nonnegative function of class C^2. Let $a \in \mathbb{R}^m$ be a zero of W that satisfies $\xi^\perp W_{uu}(u)\xi \geq c^2|\xi|^2$, for a constant $c > 0$, and for every $\xi \in \mathbb{R}^m$, and $|u - a| \leq r_0$. Let $A \subset \mathbb{R}^n$ be open and bounded with Lipschitz boundary and suppose that $u \in W^{1,2}(A; \mathbb{R}^m)$ satisfies the boundary condition* (I) *in the cut-off Lemma 4.1, and*

$$J_A(u) = \min_{w \in u + W_0^{1,2}(A;\mathbb{R}^m)} J_A(w).$$

Then, if $v = u$ on ∂A, and $J_A(v) = J_A(u)$, it follows that $v \equiv u$.

Proof Lemma 4.1 implies

$$|u - a| \leq \frac{r_0}{2} \quad \text{on} \quad A,$$

$$|v - a| \leq \frac{r_0}{2} \quad \text{on} \quad A. \tag{4.42}$$

Therefore,

$$W(v) - W(u) - W_u(u) \cdot (v - u) = \int_0^1 (W_u(s(v - u) + u) - W_u(u)) \cdot (v - u)\mathrm{d}s$$

$$= \int_0^1 s \int_0^1 W_{uu}(ts(v - u) + u)(v - u) \cdot (v - u)\mathrm{d}t\mathrm{d}s \geq \frac{1}{2}c^2|v - u|^2, \tag{4.43}$$

where we have observed that (4.42) implies $|ts(v - u) + u - a| \leq r_0$. Now note that from the Euler-Lagrange equation for u

$$\int_A (\nabla u \cdot \nabla \eta + W_u(u)\eta)\mathrm{d}x = 0, \quad \text{for} \quad \eta \in W_0^{1,2}(A; \mathbb{R}^m),$$

for $\eta = v - u$, it follows

$$\int_A (\nabla u \cdot \nabla(v - u) + W_u(u)(v - u))\mathrm{d}x = 0.$$

This and (4.43) yield

$$0 = J_A(v) - J_A(u) = \int_A \left(\frac{1}{2}(|\nabla v|^2 - |\nabla u|^2) + W(v) - W(u) \right) dx$$

$$= \int_A \left(\frac{1}{2}|\nabla(v - u)|^2 + \nabla u \cdot (\nabla v - \nabla u) + W(v) - W(u) \right) dx$$

$$\geq \int_A \left(\nabla u \cdot (\nabla v - \nabla u) + W(v) - W(u) \right) dx$$

$$= \int_A \left(W(v) - W(u) - W_u(u)(v - u) \right) dx \geq \frac{1}{2}c^2 \int_A |v - u|^2 dx$$

which concludes the proof. $\qquad\qquad\qquad\qquad\qquad\qquad\qquad\qquad\qquad\qquad\qquad\square$

Another consequence of Lemma 4.1 is a useful comparison lemma for minimal solutions of (4.2).

Lemma 4.4 *Assume W, a, A and u as in Theorem 4.3. Then*

$$|u(x) - a|^2 \leq \varphi(x)r^2, \quad \text{on } A, \tag{4.44}$$

where φ is the solution to

$$\begin{cases} \Delta\varphi = c^2\varphi, & \text{on } A, \\ \varphi = 1, & \text{on } \partial A. \end{cases} \tag{4.45}$$

In particular, $|u(x) - a| < r$ in A.

Proof As in the proof of Theorem 4.3, Lemma 4.1 implies

$$|u(x) - a| \leq r, \quad \text{on } A.$$

On the other hand, by $W_u(a) = 0$ and the non-degeneracy assumption on a,

$$W_u(u) \cdot (u - a) = (W_u(u) - W_u(a)) \cdot (u - a) \geq c^2|u - a|^2. \tag{4.46}$$

This and the fact that u is a solution of (4.2) imply

$$\Delta|u - a|^2 \geq 2\Delta u \cdot (u - a) = W_u(u) \cdot (u - a) \geq c^2|u - a|^2.$$

Therefore, by the classical maximum principle, we obtain (4.44). The proof of the lemma is complete. $\qquad\qquad\qquad\qquad\qquad\qquad\qquad\qquad\qquad\qquad\square$

This comparison result can be applied to *minimal* solutions $u \in W^{1,2}_{\text{loc}}(\mathscr{O}; \mathbb{R}^m) \cap L^\infty(\mathscr{O}; \mathbb{R}^m)$ defined on unbounded domains $\mathscr{O} \subset \mathbb{R}^n$, that is, solutions which minimize $J_A(\cdot)$ subject to their Dirichlet values on ∂A, for any open bounded set $A \subset \mathcal{O}$.

Lemma 4.5 *Assume W and a as in Theorem 4.3 and assume that $u \in W^{1,2}_{\mathrm{loc}}(\mathcal{O}; \mathbb{R}^m) \cap L^\infty(\mathcal{O}; \mathbb{R}^m)$ is a solution of (4.2) that satisfies the condition*

$$|u - a| \leq r, \quad \text{on } \mathcal{O}, \tag{4.47}$$

for some $r \in (0, r_0]$. Then there exist $k > 0$, $K > 0$ such that

$$|u - a| \leq K e^{-kd(x, \partial \mathcal{O})}, \quad \text{for } x \in \mathcal{O}.$$

Proof Given $\rho_0 > 0$, for any $\rho \geq \rho_0$ and $A = B_\rho(x)$, the ball of center x and radius ρ, the solution φ of (4.45) satisfies (cf. Lemma A.1)

$$\varphi(x) \leq e^{-k_0 \rho}, \quad \rho \geq \rho_0, \tag{4.48}$$

for some $k_0 > 0$ independent of $\rho \geq \rho_0$. Therefore, (4.47) and Lemma 4.4 imply

$$|u(x) - a| \leq r e^{-\frac{k_0}{2} d(x, \partial \mathcal{O})}$$

for each $x \in \mathcal{O}$ with $d(x, \partial \mathcal{O}) \geq \rho_0$. On the other hand, from (4.47) we have

$$|u(x) - a| \leq K e^{-\frac{k_0}{2} d(x, \partial \mathcal{O})},$$

with $K = r e^{\frac{k_0}{2} \rho_0}$ if $d(x, \partial \mathcal{O}) < \rho_0$. The proof is complete. $\qquad\square$

4.3.2   Second Application: A Liouville Type Theorem

Theorem 4.4 *Let a be a minimum of $W \geq 0$ satisfying Hypothesis $\widehat{\mathbf{H}}$, and let $u : \mathbb{R}^n \to \mathbb{R}^m$ be a minimal solution to (4.2) such that $u(x) \to a$ as $|x| \to \infty$. Then, $u \equiv a$.*

Proof It is a direct application of Theorem 4.1. For every $\epsilon > 0$, there is $R_\epsilon > 0$ such that $|u(x) - a| < \epsilon$ when $|x| \geq R_\epsilon$. Thus, applying Theorem 4.1 in the ball $B_R := \{|x| < R\}$ of radius $R \geq R_\epsilon$ we obtain: $|u(x) - a| < \epsilon$, for $|x| \leq R$, which implies immediately that $u \equiv a$. $\qquad\square$

Remark 4.5 In the scalar case $m = 1$, Theorem 4.4 holds for any potential $W \geq 0$ and any zero a of W (cf. [17]). To see this, one can utilize the strong monotonicity formula (cf. Corollary 3.2), together with the comparison argument in Lemma 5.1, which gives the estimate $J_{B_R}(u) = o(R^{n-1})$, in view of the hypothesis that $u(x) \to a$ as $|x| \to \infty$. Consequently $J_{B_R}(u) \equiv 0$, and $u \equiv a$.

4.3.3 Third Application: A General Property of Minimizers

Assume that $W : \mathbb{R}^m \to \mathbb{R}$, $n \geq 2$, is nonnegative with a finite number of zeros $a_1, \ldots, a_N$ and let $u : \mathbb{R}^n \to \mathbb{R}^m$ be a minimizer. Aside from the scalar case $m = 1$ for which, in low dimension, there is a good understanding of the structure of u, the only general statement that, under the above assumption on W, is valid for all n and all m is the upper bound

$$J_{B_r(x)}(u) \leq Cr^{n-1}, \quad r > 0,$$

where $C > 0$ is independent of $x \in \mathbb{R}^n$, see Lemma 5.1. From this upper bound it follows that given a small number $q > 0$, the set where u is near to one of the zeros of W has full measure, and one is naturally led to focus on the set

$$I_q := \{x \in \mathbb{R}^n : \min_{1 \leq j \leq N} |u(x) - a_j| > q\} \tag{4.49}$$

which separates the regions where $|u(x) - a_j| \leq q$ and can be regarded as a *diffuse interface*. Clearly understanding the structure of I_q for small $q > 0$ is basic and essentially equivalent to understanding the structure of u. Lemma 4.1 implies a property of I_q which is valid for all $n \geq 2$ and all $m \geq 1$.

Theorem 4.5 *Assume $W : \mathbb{R}^m \to \mathbb{R}$ is such that $W > 0$ on $\mathbb{R}^m \setminus \{a_1, \ldots, a_N\}$. Assume that W satisfies $\widehat{\mathbf{H}}$ for $a = a_j$, $j = 1, \ldots, N$. Let $u : \mathbb{R}^n \to \mathbb{R}^m$, $n \geq 2$, be a nonconstant minimizer, regular enough so that the Morse-Sard theorem applies. Then, if $I_q \neq \emptyset$ and $q \in (0, \frac{r_0}{2}]$, all the connected components of I_q are unbounded.*

Proof Suppose instead that $D_q \subset I_q$ is a bounded nonempty connected component of I_q. By the Morse-Sard theorem, there is a sequence $q_k \to q^+$ such that the boundary of $D_k := I_{q_k} \cap D_q$ is a C^1 manifold. Let $D'_k \subset D_k$ a connected component of D_k. Since D_q is bounded so are D'_k and $\partial D'_k$. It follows that $\mathbb{R}^n \setminus \partial D'_k$ has a unique unbounded connected component O_k. The complement $\widetilde{O}_k$ of O_k is a bounded set that contains D'_k and is connected. This is because D'_k is connected, and because if D''_k is another bounded connected component of $\mathbb{R}^n \setminus \partial D'_k$, then $\partial D''_k \subset \partial D'_k$ and therefore $D'_k \cup \overline{D''_k}$ is connected. Since O_k and $\widetilde{O}_k$ are connected, we deduce that $\partial O_k = \partial \widetilde{O}_k$ is connected [7]. By definition we have

$$\min_j |u - a_j| = q_k, \quad \text{on} \quad \partial O_k.$$

Actually, since ∂O_k is connected and u is continuous, we can conclude that there is $a \in \{a_1, \ldots, a_N\}$ such that

$$|u - a| = q_k, \quad \text{on} \quad \partial O_k. \tag{4.50}$$

Equation (4.50) implies that the assumptions of Lemma 4.1 are verified for $A = \mathbb{R}^n \setminus \overline{O_k}$. It follows

$$|u - a| \leq q_k \quad \text{on} \quad \tilde{O}_k.$$

Since ∂D_k is a C^1 manifold, D_k has a finite number of connected components. Therefore, after applying the above argument a finite number of times we obtain

$$\min_j |u - a_j| \leq q_k, \quad \text{on} \quad D_k$$

and since this is true for all k and $D_k \subset D_{k+1}$, $D_q = \bigcup_k D_k$, we conclude

$$\min_j |u - a_j| \leq q, \quad \text{on} \quad D_q,$$

in contradiction with the definition of I_q and D_q. The proof is concluded. $\square$

Other properties of the set I_q will be derived in Chap. 5, see Lemma 5.5.

Remark 4.6 The Morse-Sard theorem has been proved under the assumption of $C^{n-m,1}$, $n \geq m$, regularity ($C^{0,1}$, if $m \geq n$) in [4], and more conveniently for our purposes in the Sobolev setting $W_{\text{loc}}^{n-m+1,p}$, $p > n$ in [8]. For example, if $m = n$, then for Theorem 4.5 to apply, it is sufficient that W is locally Lipschitz.

4.3.4   Fourth Application: Standing Waves on Periodic Domains

The cut-off lemma can be utilized for establish the existence and study the qualitative behavior of solutions to the system

$$\Delta u - W_u(u) = 0, \quad u : \Omega \to \mathbb{R}^m, \ \Omega \subset \mathbb{R}^n. \tag{4.51}$$

The construction of the set A, and particularly the verification of condition (I) on ∂A, are key steps. We now sketch a typical situation, referring to [1] for more details. Assume

(h1) $W(u) > 0$ on $\mathbb{R}^m \setminus \{a_+, a_-\}$ with $W \in C^2$, $\left(\frac{\partial^2 W}{\partial u_i \partial u_j}\right)$ diagonal with $\frac{\partial^2 W}{\partial u_i^2} \geq c^2$, $i = 1, \ldots, m$.

(h2) $\Omega = \{(s, y) : s \in \mathbb{R}, y \in \Omega_s \subset \mathbb{R}^{n-1}\}$, Ω_s is a bounded Lipschitz cross-section, $\Omega_s \neq \varnothing$ (hence $\min_s \mathcal{L}^{n-1}(\Omega_s) \geq \mu_0 > 0$). Ω is T-periodic in s; equivalently, Ω is invariant under the translation $(s, y) \to (s + T, y)$, (Fig. 4.8).

Fig. 4.8 A typical Ω

Problem Existence of a solution to (4.51) connecting $a_\pm$,

$$u(s, y) \to a_\pm \text{ as } s \to \pm\infty. \tag{4.52}$$

The major difficulty is the T-translation invariance of the free energy functional

$$J_\Omega(u) = \int_\Omega \left(\frac{1}{2}|\nabla u|^2 + W(u)\right)dx, \tag{4.53}$$

which is a source of noncompactness. We elaborate this point a little further. Consider for example the minimization of (4.53) in the special case $n = 1$, in the class

$$\mathscr{A} = \{u \in W^{1,2}_{\text{loc}}(\mathbb{R}; \mathbb{R}^m) : \lim_{x \to \pm\infty} u(x) = a_\pm\}.$$

First, note that $\mathscr{A}$ is not closed with respect to weak convergence in $W^{1,2}_{\text{loc}}(\mathbb{R}; \mathbb{R}^m)$. For example, a sequence $\{u_k\} \subset \mathscr{A}$ may converge to a map that does not satisfy the boundary conditions at $\pm\infty$ required for membership in $\mathscr{A}$. Indeed, if

$$u_k = \begin{cases} -1, & \text{for } x \in (-\infty, -k), \\ \sin\left(\frac{\pi x}{2k}\right), & \text{for } x \in [-k, k], \\ 1, & \text{for } x \in (k, \infty), \end{cases}$$

then $\lim_{x \to \pm\infty} u_k(x) = \pm 1$ for all k, but u_k converges in C^1 over compacts to $u_\infty \equiv 0$. Note that in this case the measure $\mu_k = \left(\frac{1}{2}|\nabla u_k|^2 + W(u_k)\right)dx$ is not concentrated in a bounded subset of $\mathbb{R}$, but is dispersed on the whole of $\mathbb{R}$. Note that $J_\mathbb{R}(u_k) < \infty$, while $J_\mathbb{R}(u_\infty) = \infty = \lim_{k \to \infty} J_\mathbb{R}(u_k)$. The cut-off lemma, in the case of minimizing sequences $\{u_k\}$ can be used for

(a) exclusion of *oscillations*,
(b) localisation of the *interface*.

Returning to the problem above, let $\{u_k\}$ be a minimizing sequence with $u_k(s, y) \to a_\pm$, as $s \to \pm\infty$. We may assume for the minimizing sequence the bounds $|u_k(s, y)| \le M$, $|\nabla u_k(s, y)| \le M$, which hold for the solution. This by general facts following from the Ekeland ϵ-variational principle (see [5, Lemma 2.2]), and in our particular case can be obtained by regularization with a higher-order elliptic operator. Thus we derive equicontinuity for $\{u_k\}$. The second easy bound for domains that are unbounded only in one direction is

$$J_\Omega(u_k) \leq M. \tag{4.54}$$

From these two bounds it follows that $u_k(s, y)$ is close to one of the minima of W along a large set of cross-sections of Ω. Indeed, for $\epsilon_0 > 0$ consider the open set

$$\bar{S}_k = \{\bar{s} \in (-L, L) : \exists\, y(\bar{s}) \in \Omega_{\bar{s}} \text{ such that } W(u_k(\bar{s}, y(\bar{s}))) > \epsilon_0\}, \tag{4.55}$$

where $L > 0$ will be selected later. By equicontinuity, there is $\delta_0 = \delta_0(\epsilon_0, \mu_0) > 0$, uniform over k and $\bar{s}$, such that

$$\int_{\Omega_{\bar{s}}} W(u_k(s, y))\mathrm{d}y \geq \delta_0 \frac{\epsilon_0}{2}. \tag{4.56}$$

Hence, by (4.54) and (4.56),

$$M \geq \int_{\bar{S}_k} \int_{\Omega_s} W(u_k(s, y))\, \mathrm{d}y\, \mathrm{d}s$$
$$= \mathscr{L}^1(\bar{S}_k)\delta_0 \frac{\epsilon_0}{2},$$

and therefore we obtain the bound

$$\mathscr{L}^1(\bar{S}_k) \leq \frac{2M}{\epsilon_0\delta_0}, \qquad \text{uniformly in } k. \tag{4.57}$$

Hence for $s \in [-L, L] \setminus \bar{S}_k$ $(2L > 2L_0 = \frac{2M}{\epsilon_0\delta_0})$ we have:

$$W(u_k(s, y)) \leq \epsilon_0, \quad \text{for all } y \in \Omega_s. \tag{4.58}$$

Suppose Ω_s is connected. Then, by taking $\epsilon_0 > 0$ small, we conclude that $u_k(s, y)$ is $r(\epsilon_0)$-close to one of the minima on Ω_s. To see how oscillations are eliminated, assume that (Fig. 4.9)

$$|u_k(s_1, y) - a_+| \leq r, \quad |u_k(s_2, y) - a_+| \leq r, \tag{4.59}$$

for $y \in \Omega_{s_i}, i = 1, 2$.
  By Theorem 4.2

$$|u_k(s, y)) - a_+| \leq r, \quad \text{for all } s \in [s_1, s_2], \tag{4.60}$$

that is, u_k cannot get close to a_- for $s \in [s_1, s_2]$. As far as the localization of the interface is concerned, that is the speed of the transition region between a_- and a_+,

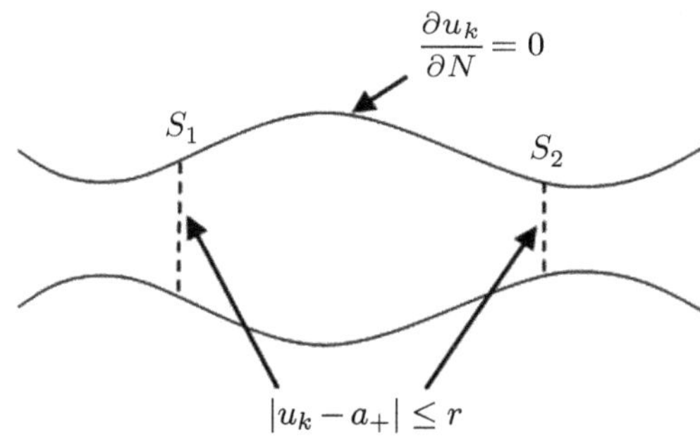

Fig. 4.9 Eliminating oscillations

we note that from

$$\min_{-L \le s \le L,\; y \in \Omega_s} \{|u_k(s, y)) - a_+|, |u_k(s, y)) - a_-|\} > \epsilon_0, \tag{4.61}$$

we obtain the uniform in k lower bound

$$\int_{-L}^{L} \int_{\Omega_s} W(u_k(s, y))\, dy\, ds \ge 2L\mu_0 w(\epsilon_0), \tag{4.62}$$

for some $w(\epsilon_0) > 0$, which in conjunction with the upper bound (4.54) gives the estimate

$$L \le \frac{M}{2\mu_0 w(\epsilon_0)}. \tag{4.63}$$

This is a uniform bound on the width of the transition zone. Finally, since the transition set may depend on k, we can utilize the translation invariance of J_Ω to obtain from $\{u_k\}$ a family of minimizers $\{\hat{u}_k\}$ with a common transition zone, thus restoring compactness.

Remark 4.7 An application of Lemma 4.1 similar to the one described above is given in the proof of Lemma 9.17.

4.4 Scholia on Chap. 4

The main results of this chapter are taken from [1]. The considerations in Sect. 4.3.4 above fall into the set-up of the Concentration-Compactness (trichotomy) Lemma of Lions [13, p. 115], as stated for instance in [16, p. 39]. Specifically, the first option in that lemma is the compactness of the measures $\mu_k = \left(\frac{1}{2}|\nabla u_k|^2 + W(u_k)\right)dx$

which is characterized by the property

$$\int_{B_R(x_k)} \mathrm{d}\mu_k \geq 1 - \epsilon \quad \left(\int \mathrm{d}\mu_k = 1 \right),$$

uniformly in k. The cut-off lemma provides such a uniform estimate for an appropriate sequence of translates.

Recently, Antonopoulos and Smyrnelis [2] established generalizations of Theorems 4.1, 4.2, and Lemma 4.1 for solutions of (4.2). They assume that W vanishes at the boundary of a convex set $C_0 \subset \mathbb{R}^m$, which is either C^2 smooth or reduces to a point $\{a\}$. The point case corresponds to the theorems presented in this chapter. For $u(\cdot) \in W^{1,2}(A; \mathbb{R}^m) \cap L^\infty(A; \mathbb{R}^m)$, $A \subset \mathbb{R}^n$, open, bounded, and Lipschitz, consider the decomposition

$$u(x) = p(x)a + (u(x) - p(x)a), \tag{4.64}$$

where $p(x) = p(u(x))$ is the projection of $u(x)$ on C_0. Define $d(x) = d(u(x), C_0) = |u(x) - p(x)|$, and consider the deformation

$$\tilde{u}(x) = p(x) + g(d(x))(u(x) - p(x)), \tag{4.65}$$

in analogy to (4.22). Under the hypotheses $|f'| \leq 1$, $g \leq 1$, f, g as in (4.20), one can verify that $|\nabla \tilde{u}(x)|^2 \leq |\nabla u(x)|^2$ as in (4.25). Assume

(i) $W \in C(\mathbb{R}^m; \mathbb{R})$, with $W\big|_{\partial C_0} = 0$, $W \geq 0$ on $\mathbb{R}^m \setminus C_0$.
(ii) There exists $r_0 > 0$ such that for every $\xi \perp \partial C_0$ at the point p, with $|\xi| = 1$, $(0, r_0) \ni r \mapsto W(p + r\xi)$ is increasing, and moreover $W(a + r_0\xi) > 0$.

Lemma 4.6 *Let W satisfy hypotheses* (i), (ii) *above, let C_0 be either C^2 convex or $C_0 = \{a\}$, and let $A \subset \mathbb{R}^n$, open, bounded, and with Lipschitz boundary. Suppose that $u(\cdot) \in W^{1,2}(A; \mathbb{R}^m) \cap L^\infty(A; \mathbb{R}^m)$. Suppose the following two conditions hold:*

(I) $d(x) \leq r$ *on* ∂A, $0 < 2r \leq r_0$,
(II) $\mathscr{L}^n(A \cap \{d(x) > r\}) > 0$

Then, there is $\tilde{u}(\cdot) \in W^{1,2}(A; \mathbb{R}^m) \cap L^\infty(A; \mathbb{R}^m)$ such that

$$\begin{cases} \tilde{u} = u, & \text{on } \partial A, \\ d(x) \leq r, & \text{on } A, \\ J_A(\tilde{u}) < J_A(u). \end{cases} \tag{4.66}$$

An immediate corollary is the following maximum principle under the hypotheses above:

Theorem 4.6 *Let* $v(\cdot) \in W^{1,2}(A; \mathbb{R}^m) \cap L^\infty(A; \mathbb{R}^m)$ *be a minimizer of* $J_A(u)$ *subject to its boundary conditions on* ∂A,

$$J_A(v) = \min\{J_A(u), \ u = v \ on \ \partial A\}.$$

Assume

$$d(x) := d(v(x), C_0) \le r \quad on \ \partial A, \ 0 < 2r \le r_0. \tag{4.67}$$

Then,

$$d(x) \le r \quad on \ A. \tag{4.68}$$

Moreover, if $u \mapsto W_u(u)$ *is Lipschitz, then the attainment of equality in (4.68) at an interior point of the set* A, *assumed now to be connected,*

$$d(\hat{x}) = r, \quad for \ some \ \hat{x} \in A, \tag{4.69}$$

implies that

$$d(x) = r, \ \forall x \in A, \ and \ in \ addition \ if \ C_0 \ is \ strictly \ convex, \ v(x) \equiv Const. \ in \ A. \tag{4.70}$$

The strong maximum principle part of the theorem above follows from the usual maximum principle as developed in Weinberger [18] and Evans [10]. Its proof is based on the (strict) convexity of the function $u \mapsto d^2(u, C_0)$, in the complement of C_0. The strong version was suggested to us by Andreas Savas-Halilaj.

In our use of the polar form, we were able to justify (4.19) by circumventing the difficulties involved in the definition of $\boldsymbol{n}$, simply because it was sufficient for our purposes to consider cut-off functions f with the property that $\frac{f(s)}{s}$ is Lipschitz. However, the formula holds for Lipschitz f with $f(0) = 0$ (cf. statements (a), (b), Sect. 4.1(E)). We sketch very briefly the arguments. Put

$$A_+ := \{x \in A : \rho > 0\}, \quad A_0 := \{x \in A : \rho = 0\}.$$

Given a sequence $\{h\} \to 0$, there exists a subsequence $\{h'\} \subset \{h\}$ such that

(i) $\boldsymbol{n}_{x_i}(x) := \lim_{h' \to 0} \frac{n(x+h'e_i)-n(x)}{h'}$ for $i = 1, \dots, n$ exists a.e. as an extended measurable function on A, where $e_i = (0, \dots, 0, 1, 0, \dots, 0)$ is the standard unit vector, and the following product rule holds

(ii) $\rho(x)\boldsymbol{n}_{x_i}(x) = u_{x_i}(x) - \rho_{x_i}(x)\boldsymbol{n}(x)$ a.e.,

where $u_{x_i}(x), \rho_{x_i}(x)$ are strong L^2 derivatives which coincide with the weak $W^{1,2}$ derivatives of u and ρ. In particular, by (ii) above, $\boldsymbol{n}_{x_i}(x)$ is independent of the subsequence on A_+ and so is well defined. In addition, the following identity holds:

(iii)

$$\int_A |\nabla u|^2 \mathrm{d}x = \int_A |\nabla \rho|^2 \mathrm{d}x + \int_{A_+} \rho^2(x)|\nabla n(x)|^2 \mathrm{d}x.$$

We note that although $n_{x_i}(x)$ is not necessarily independent of the subsequence on A_0, nevertheless by (ii) the product $\rho(x)n_{x_i}(x)$ is independent and well defined, and

$$\int_{A_0} \rho^2(x)|\nabla n(x)|^2 \mathrm{d}x = 0$$

since $|\nabla u| = 0$, $|\nabla \rho| = 0$ a.e. on A_0, and $n = 0$ on A_0. Hence,

$$\int_A \rho^2(x)|\nabla n(x)|^2 \mathrm{d}x$$

is independent of the subsequence and unambiguous.

The proof of (i), (ii) proceeds by considering the difference quotient version of (ii) and taking the limit. The proof of (iii) proceeds via mollification, $u^\epsilon := u * \eta_\epsilon$, and uses the formula

$$|\nabla u^\epsilon(x)|^2 = |\nabla \rho^\epsilon(x)|^2 + (\rho^\epsilon(x))^2|\nabla n^\epsilon(x)|^2 \quad \text{on } A_+^\epsilon = \{x \in A : \rho^\epsilon > 0\}.$$

Thus (4.17) is justified.

To justify (4.19), it is sufficient to establish that for $f : \mathbb{R} \to \mathbb{R}$, locally Lipschitz, and $f(0) = 0$, $\tilde{u}$ as defined in (4.18), lies in $W^{1,2}(A; \mathbb{R}^m) \cap L^\infty(A; \mathbb{R}^m)$. Set $\tilde{u}(x) = a + \tilde{\rho}(x)n(x)$, $\tilde{\rho}(x) = f(\rho(x))$. Then one shows that $\tilde{u}_{x_i}(x) = \tilde{\rho}\, n_{x_i} + \tilde{\rho}_{x_i}\, n$ in the sense of distributions $\mathscr{D}'$. This can be done by using difference quotients. Finally, one shows that $\tilde{u}_{x_i}(x) = \tilde{\rho}\, n_{x_i} + \tilde{\rho}_{x_i}\, n \in L^2(A)$ by noting that

$$\tilde{\rho}\, n_{x_i} + \tilde{\rho}_{x_i}\, n = f(\rho)\, n_{x_i} + f'(\rho)\rho_{x_i}\, n = \frac{f(\rho)}{\rho}\rho\, n_{x_i} + f'(\rho)\rho_{x_i}\, n.$$

References

1. Alikakos, N.D., Fusco, G.: A maximum principle for systems with variational structure and an application to standing waves. J. Eur. Math. Soc. **17**(7), 1547–1567 (2015)
2. Antonopoulos, P., Smyrnelis, P.: A maximum principle for the system $\Delta u - \nabla W(u) = 0$. C. R. Acad. Sci. Paris Ser. I **354**, 595–600 (2016)
3. Ball, J.M., Crooks, E.C.M.: Local minimizers and planar interfaces in a phase-transition model with interfacial energy. Calc. Var. Partial Differ. Equ. **40**(3), 501–538 (2011)
4. Bates, S.M.: Toward a precise smoothness hypothesis in Sard's theorem. Proc. Am. Math. Soc. **117**(1), 279–283 (1993)

5. Boccardo, L., Ferone, V., Fusco, N., Orsina, L.: Regularity of minimizing sequences for functionals of the Calculus of Variations via the Ekeland principle. Differ. Integral Equ. **12**(1), 119–135 (1999)
6. Casten, R.G., Holland, C.J.: Instability results for reaction-diffusion equations with Neumann boundary conditions. J. Differ. Equ. **27**, 266–273 (1978)
7. Czarnecki, A., Kulczychi, M., Lubawski, W.: On the connectedness of boundary and complement for domains. Ann. Pol. Math. **103**, 189–191 (2011)
8. de Pascale, L.: The Morse-Sard theorem in Sobolev spaces. Indiana Univ. Math. J. **50**(3), 1371–1386 (2001)
9. Evans, L.C.: Partial Differential Equations. Graduate Studies in Mathematics, vol. 19, 2nd edn. American Mathematical Society, Providence (2010)
10. Evans, L.C.: A strong maximum principle for parabolic systems in a convex set with arbitrary boundary. Proc. Am. Math. Soc. **138**(9), 3179–3185 (2010)
11. Evans, L.C., Gariepy, R.F.: Measure Theory and Fine Properties of Functions. CRC Press, Boca Raton (1992)
12. Kohn, R.V., Sternberg, P.: Local minimisers and singular perturbations. Proc. R. Soc. Edinb. Sect. A **111**(1–2), 69–84 (1989)
13. Lions, P.L.: The concentration-compactness principle in the Calculus of Variations. The locally compact case, Part 1. Ann. I. H. Poincaré Anal. Nonlinear **1**, 109–145 (1984)
14. Matano, H.: Asymptotic behavior and stability of solutions of semilinear diffusion equations. Publ. Res. Inst. Math. Sci. **15**(2), 401–454 (1979)
15. Modica, L., Mortola, S.: Un esempio di Γ-convergenza. Boll. Unione Mat. Ital. Sez B **14**, 285–299 (1977)
16. Struwe, M.: Variational Methods, Applications to Nonlinear Partial Differential Equations and Hamiltonian Systems. A Series of Modern Surveys in Mathematics, vol. 34, 4th edn. Springer, Berlin (2008)
17. Villegas, S.: Nonexistence of nonconstant global minimizers with limit at ∞ of semilinear elliptic equations in all of $\mathbb{R}^N$. Commun. Pure Appl. Anal. **10**(6), 1817–1821 (2011)
18. Weinberger, H.: Invariant sets for weakly coupled parabolic and elliptic systems. Rend. di Matem. Ser. VI **8**, 295–310 (1975)

Chapter 5
Estimates

Abstract This chapter together with Chap. 4 contain some general tools for obtaining estimates for systems.

5.1 The Basic Estimate

In this chapter we continue the analysis of bounded solutions of

$$\Delta u - W_u(u) = 0, \tag{5.1}$$

which are defined in an open set $\mathscr{O} \subset \mathbb{R}^n$, generally unbounded, and which are *minimal* (alternatively *minimizers*) in the sense that minimize the energy J subject to their Dirichlet values. See Definition 4.1. We assume a gradient bound on u besides the L^∞ bound required in Definition 4.1

$$|u(x) - a| < M, \ |\nabla u(x)| < M \ \text{on} \ \mathscr{O}, \tag{5.2}$$

and set

$$W_M = \max_{|u-a| \leq M} W(u). \tag{5.3}$$

We note that the gradient bound follows from the L^∞ estimate on u under sufficient regularity on W. We refer the reader to Remark 5.2 for nonsmooth W's, in Sect. 5.2.2 below. The basic estimate for minimal maps is given in

Lemma 5.1 *Let $W : \mathbb{R}^m \to \mathbb{R}$ be continuous, $W \geq 0$, and assume that $\{W = 0\} \neq \emptyset$. Let u be minimal, satisfying the estimates (5.2). Then there is a constant $\widehat{C}_0 > 0$, $\widehat{C}_0 = \widehat{C}_0(W, M)$, independent of x_0 and such that*

$$B_r(x_0) \subset \mathscr{O} \quad \Rightarrow \quad J_{B_r(x_0)}(u) \leq \widehat{C}_0 r^{n-1}, \ \text{for} \ r > 0. \tag{5.4}$$

© Springer Nature Switzerland AG 2018

N. D. Alikakos et al., *Elliptic Systems of Phase Transition Type*,
Progress in Nonlinear Differential Equations and Their Applications 91,
https://doi.org/10.1007/978-3-319-90572-3_5

Proof From (5.2), $g(u) := \frac{1}{2}|\nabla u|^2 + W(u)$ is bounded on $\mathcal{O}$ and it follows

$$J_{B_r(x_0)}(u) \leq C_1 r^n \leq C_1 r^{n-1}, \quad \text{for } r \leq 1 \tag{5.5}$$

for some $C_1 > 0$ independent of x_0. For $r > 1$ define $v : \mathcal{O} \to \mathbb{R}^m$ by

$$v(x) = \begin{cases} a, & \text{for } |x - x_0| \leq r - 1, \\ (r - |x - x_0|)a + (|x - x_0| - r + 1)u(x), & \text{for } |x - x_0| \in (r - 1, r], \\ u(x), & \text{for } |x - x_0| > r. \end{cases}$$

This definition and the minimality of u over balls imply (Fig. 5.1)

$$J_{B_r(x_0)}(u) \leq J_{B_r(x_0)}(v) = J_{B_r(x_0) \setminus B_{r-1}(x_0)}(v) \leq C_2 r^{n-1}, \tag{5.6}$$

where we have also used that (5.6) and (5.2) imply that $g(v)$ is bounded on $\mathcal{O}$. The lemma follows from (5.5) and (5.6) with $\widehat{C}_0 = \max\{C_1, C_2\}$; $\widehat{C}_0$ is clearly independent of x_0 and depends on u only through the bound M. $\qquad\square$

An elementary consequence of the lemma above is that periodic maps cannot be minimal.

Corollary 5.1 *Assume* $u : \mathbb{R}^n \to \mathbb{R}^m$ *is a nonconstant solution to* (5.1) *that satisfies*

$$u(x + h\eta_i) = u(x) \quad \text{for } x \in \mathbb{R}^n, \ h \in \mathbb{Z}, \ i = 1, \dots, n,$$

for some linearly independent vectors $\eta_1, \dots, \eta_n \in \mathbb{R}^n$. *Then* u *is not minimal.*

Proof $\mathbb{R}^n$ is the union of nonoverlapping copies of the unit cell

$$K = \{x = \sum_{i=1}^{n} t_i \eta_i, \ 0 \leq t_i \leq 1\}.$$

The number of such cells contained in B_R is bounded below by $C R^n$, for some constant $C > 0$. From the assumption that u is nonconstant, we obtain that

Fig. 5.1 The measure of the annulus of fixed width grows like r^{n-1} for $r \to \infty$, $W(a) = 0$

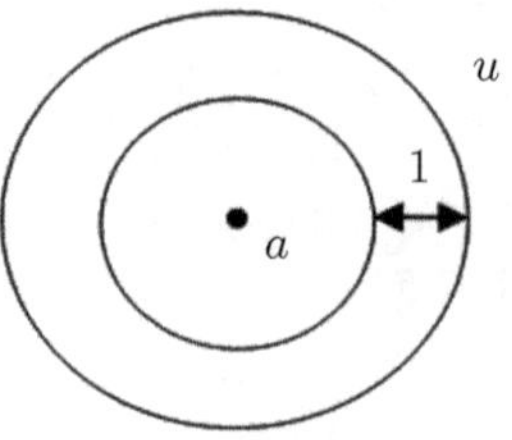

$J(u; K) \geq \bar{J} > 0$. Consequently, $J(u; B_R) \geq \sum_{\sharp \text{ cells } \subset B_R} J(u; K) \geq (CR^n)\bar{J}$. Thus, by the basic estimate, u cannot be minimal. $\square$

5.2 Density Estimates

5.2.1 Introduction

Following [5], we begin with the sharp interface analog that is behind this type of estimate. Consider a minimal surface $\Sigma^{n-1} = \partial D$, as in Fig. 5.2.

Let $x \in \Sigma^{n-1}$. The surface Σ^{n-1} partitions the ball $B_r(x)$ into two parts, D_r and D_r^c. Let $V(r) = \mathcal{L}^n(D_r)$, $A(r) = \mathcal{H}^{n-1}(\Sigma^{n-1} \cap B_r)$, $\mathcal{H}^n$ the n-dimensional Hausdorff measure and S_r the spherical cup bounding D_r. Consider the following formal computation:

$$V(r) \leq C[\mathcal{H}^{n-1}(\Sigma^{n-1} \cap B_r) + \mathcal{H}^{n-1}(S_r)]^{\frac{n}{n-1}}, \text{ by the isoperimetric inequality,} \tag{5.7}$$

$$\leq C[2\mathcal{H}^{n-1}(S_r)]^{\frac{n}{n-1}}, \text{ by minimality since } \partial(\Sigma^{n-1} \cap \partial B_r) = \partial S_r,$$

$$\leq C[V'(r)]^{\frac{n}{n-1}}, \text{ by the coarea formula (cf. for instance [10, Appendix C]).}$$

From (5.7), it follows that

$$V(r) \geq Cr^n, \quad C = C(n), \ \forall r > 0. \tag{5.8}$$

The estimate (5.8) expresses the fact that both D and D^c have uniform positive density at each x, all the way from $r = 0$ to $r = \infty$:

$$0 < \lambda_1 \leq \frac{\mathcal{L}^n(D \cap B_r(x))}{\mathcal{L}^n(B_r(x))} \leq \lambda_2 < 1. \tag{5.9}$$

Our interest in (5.9) is at $r = \infty$, which relates to Bernstein type theorems. The estimate at $r = 0$ leads to regularity results. We recall that minimal sets of codimension 1 in $\mathbb{R}^n$ can be conveniently viewed as boundaries of minimizing partitions of open sets in $\mathbb{R}^n$. The point is that the partition P of a set U can

Fig. 5.2 Σ^{n-1} partitioning surface, $D_r = D \cap B_r$, $D_r^c = D^c \cap B_r$, separating the phases 1 and 2

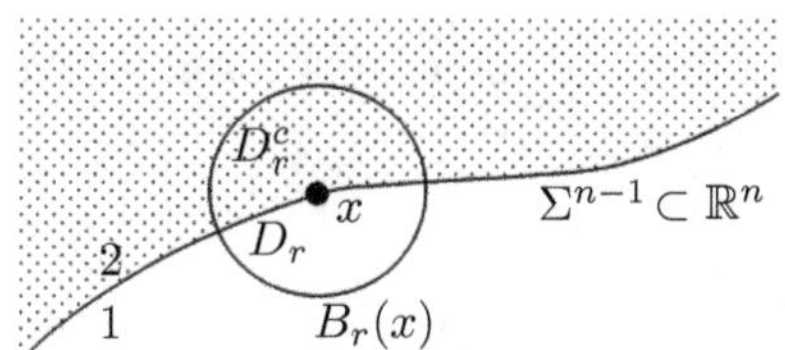

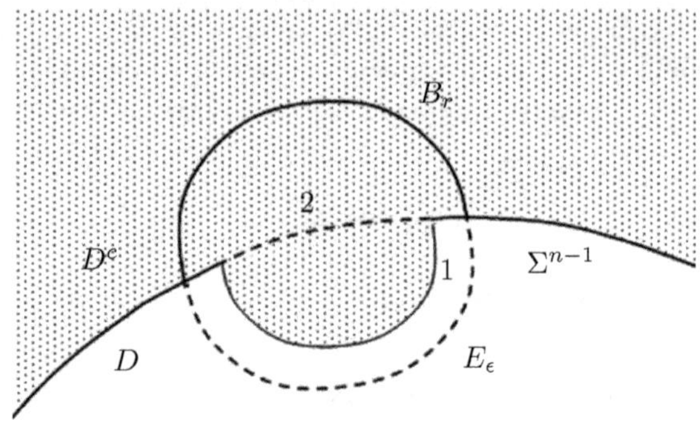

Fig. 5.3 Minimality in balls is enough for (5.4). The partition above (ϵ-boundary layer) has the same Dirichlet values in B_r, with the partition in Fig. 5.2

be identified with a piecewise constant function g on U, the norm of P equals $\int_U |g(x)|dx$, and the perimeter of P, which we seek to minimize, equals $\|g\|_{BV}$, the BV norm of g, and coincides with $\mathcal{H}^{n-1}(\partial P \cap U)$. The sets of finite perimeter are those for which $\|g\|_{BV} < \infty$. The argument above leading to (5.9) can be made precise in the context of minimizing partitions with Dirichlet values as follows. Fix a point '$0 \in \partial D$', which is meant in the sense that $\mathcal{L}^n(B_\delta \cap D) > 0$, and $\mathcal{L}^n(B_\delta \cap D^c) > 0$, for all small δ, $\delta \le \delta_0$. Fix now $r > 0$. Since $\{D, D^c\}$ is minimal

$$P_{B_r}(D) \le P_{B_r}(E) \tag{5.10}$$

for each set E of finite perimeter which coincides with D outside B_r. Take $E_\epsilon = D \setminus B_{r-\epsilon}$ as in Fig. 5.3. Then

$$P_{B_r}(D) \le P_{B_r}(E_\epsilon), \ \forall \epsilon > 0, \tag{5.11}$$

hence

$$P_{B_r}(D) \le \mathcal{H}^{n-1}(D \cap \partial B_r). \tag{5.12}$$

Set $V(r) = \mathcal{L}^n(D \cap B_r)$. Then

$$c_0 V(r)^{\frac{n-1}{n}} \le \mathcal{H}^{n-1}(\partial(D \cap B_r)) \ \text{(isoperimetric)} \tag{5.13}$$

$$= P_{B_r}(D) + \mathcal{H}^{n-1}(D \cap \partial B_r)$$

$$\le 2\mathcal{H}^{n-1}(D \cap \partial B_r) \ \text{(by (5.12))}.$$

This and the coarea formula $V'(r) \ge \mathcal{H}^{n-1}(D \cap \partial B_r)$ yield the differential inequality

$$V'(r) \ge \frac{c_0}{2}(V(r))^{\frac{n-1}{n}}. \tag{5.14}$$

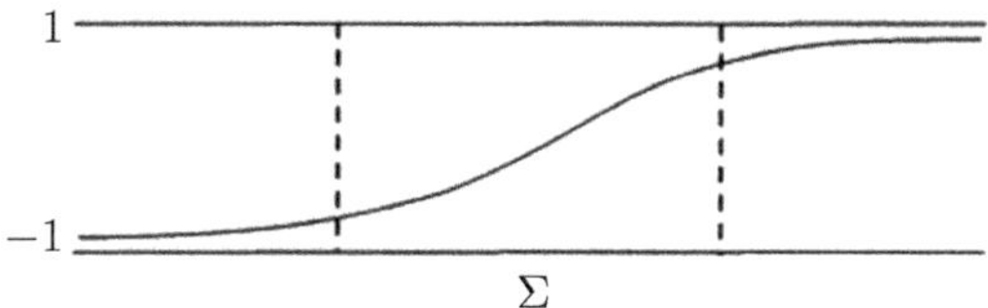

Fig. 5.4 The case of two phases, the profile of u, and the diffuse interface Σ

Thus, for each $\delta \in (0, \delta_0)$, we have

$$V(r) \geq (\frac{c_0}{2n}(r - \delta) + V(\delta)^{\frac{1}{n}})^n \geq \left(\frac{c_0}{2n}(r - \delta)\right)^n, \quad \text{for } r \geq \delta, \tag{5.15}$$

and we conclude that

$$V(r) \geq \left(\frac{c_0}{2n}r\right)^n. \tag{5.16}$$

The analogy with the diffuse interface (Fig. 5.4) problem is via the identification

$$A(r) = \int_{B_r \cap \{|u-a| \leq \lambda\}} W(u)\mathrm{d}x, \quad V(r) = \mathscr{L}^n(B_r \cap \{|u - a| > \lambda\}), \tag{5.17}$$

where a is a phase, $W(a) = 0$, and $\lambda > 0$ is any number such that

$$d_0 = \mathrm{dist}(a, \{W = 0\} \setminus \{a\}) \geq \lambda. \tag{5.18}$$

The interface corresponding to phase a is measured by the set close to a where W does not vanish, while $V(r) = \mathscr{L}^n(B_r \cap \{|u - a| > \lambda\})$ measures the volume of the set where u is close to $\{W = 0\} \setminus \{a\}$. The more singular the potential W, the less diffused the interface, and the easier the derivation of the density estimates, as it gets closer to the argument above. However, in all cases the differential inequality (5.14) is replaced by a difference inequality (cf. (5.51), (5.57)). The basic estimate in Lemma 5.1 above is essential for localizing the (diffuse) interface, and making the specific value of $\lambda \in (0, d_0)$ irrelevant. The original density estimates in [5] are for a class of potential, modeled after $W_\alpha(u) = |1 - u^2|^\alpha$, $0 < \alpha \leq 2$, $W_0(u) = \mathbb{1}_{\{|u|<1\}}$, $\alpha = 0$ (Fig. 5.5). Our extension to the vector case is for $0 < \alpha \leq 2$, and is presented in Sect. 5.2.2 below. The range $\alpha \in [0, 2)$ gives rise to a free boundary problem. We conclude the present section with Savin's derivation of the estimate for $\alpha = 0$, [18]. We stress that this material is not from the technique point of view a prerequisite for understanding the contents of Sect. 5.2.2.

Consider

$$J_\mathcal{O}(u) = \int_\mathcal{O} \left(\frac{1}{2}|\nabla u|^2 + \mathbb{1}_{\{|u|<1\}}\right)\mathrm{d}x, \quad \mathcal{O} \subset \mathbb{R}^n, \text{ open}, \tag{5.19}$$

and assume that u is a minimizer of $J_\mathcal{O}$ in the class of maps with values in $[-1, 1]$.

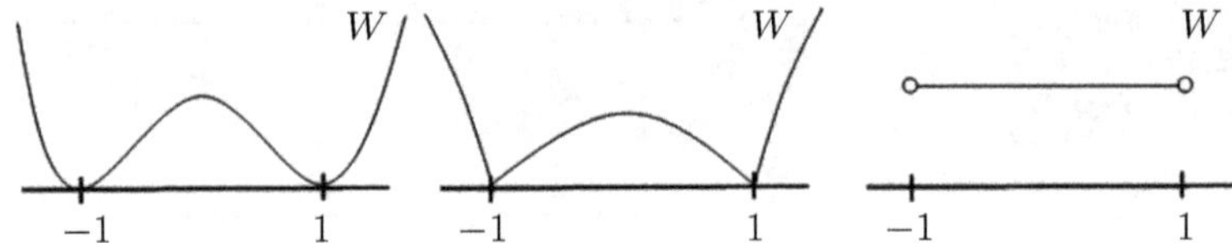

Fig. 5.5 The model potential is $W(u) = |1 - u^2|^\alpha$, with $1 < \alpha \leq 2$ on the left, and $0 < \alpha \leq 1$ in the middle. For $\alpha = 0$ it reduces to $W = \mathbb{1}_{\{|u|<1\}}$, on the right

Theorem 5.1 ([18]) *Assume that $u(0) = 0$. There exist universal positive constants c and C, such that*

$$V(r) = \mathscr{L}^n(\{u > 0\} \cap B_r(0)) \geq Cr^n, \quad \text{for } r \geq c, \ \forall B_r(0) \subset \mathcal{O}.$$

Remark 5.1 Formally we see that the minimizer should satisfy

$$\Delta u = 0 \text{ in } \{|u| < 1\}. \tag{5.20}$$

Assuming that the levels $u = 1$ and $u = -1$, are smooth hypersurfaces, we obtain also the free boundary condition

$$|\nabla u| = \sqrt{2} \text{ on } \partial\{|u| < 1\}. \tag{5.21}$$

To see (5.21) formally one can resort to Hadamard's formula (cf. [10, p.369]) for differentiating $\lambda(\tau) = \frac{1}{2}\int_{U(\tau)} |\nabla u(\tau)|^2 dx$, where $u(\tau) = u(\cdot, \tau)$ is a perturbation of the minimizer $u = u(\cdot, 0)$, $U(\tau)$ a perturbation of a set $U(0)$, $1 < u(\cdot, \tau) < 1$ in $U(\tau)$, $u(\cdot, \tau) = 1$ on $\partial U(\tau)$. More specifically consider an $x \in \partial U(0)$, and then perturb the set $U(0)$ around x. Let $\partial U(\tau) = \{X(s, \tau), s \in \Omega < \mathbb{R}^{n-1}\}$ a local parametrization of ∂U about x. Assuming smoothness we obtain

$$\dot{\lambda}(\tau) = \int_{U(\tau)} \nabla u(\tau) \nabla u_\tau(\tau) dx + \frac{1}{2}\int_{\partial U(\tau)} |\nabla u(\tau)|^2 V \cdot \nu dS,$$

where $V = \frac{\partial X}{\partial \tau}$ is the velocity of the boundary $\partial U(\tau)$, and ν the outward unit normal. By utilizing (5.20) we obtain

$$\dot{\lambda}(\tau) = -\int_{U(\tau)} \Delta u(\tau) u_\tau(\tau) dx + \int_{\partial U(\tau)} \frac{\partial u}{\partial \nu}(\tau) u_\tau(\tau) dS + \frac{1}{2}\int_{\partial U(\tau)} |\nabla u(\tau)|^2 V \cdot \nu dS$$

$$= \int_{\partial U(\tau)} \frac{\partial u}{\partial \nu}(\tau) u_\tau(\tau) dS + \frac{1}{2}\int_{\partial U(\tau)} |\nabla u(\tau)|^2 V \cdot \nu dS.$$

From $u(X(\epsilon, \tau), \tau) = 1$ we obtain

$$O = \frac{\partial}{\partial \tau}[u(X(s, \tau), \tau)] = \frac{\partial u}{\partial \nu}\frac{\partial X}{\partial \tau} \cdot \nu + u_\tau = \frac{\partial u}{\partial \nu}V \cdot \nu + u_\tau,$$

and thus $u_\tau \frac{\partial u}{\partial v} = -|\frac{\partial u}{\partial v}|^2 V \cdot v$. Substituting this above gives $\dot\lambda(\tau) = -\frac{1}{2}\int_{\partial U(\tau)} |\nabla u(\tau)|^2 V \cdot v dS$. On the other hand differentiating $\mu(\tau) = \int_{U(\tau)} dx$ gives $\dot\mu(\tau) = \int_{\partial U(\tau)} V \cdot v dS$. Stationarity $\dot\lambda(0) + \dot\mu(0) = 0$ suggests (5.21). By classical results of Alt and Caffarelli (see [7]), the solution u of (5.20)-(5.21) is globally Lipschitz, and u is a nontrivial minimizer taking all the values in $(-1, 1)$, so we can assume $u(0) = 0$.

Proof (Theorem 5.1) We take $a = -1$, $\lambda = 1$ in (5.17), and

$$A(r) = \mathscr{L}^n(\{|u| < 1\} \cap B_r).$$

For uniformity with our proof of Theorem 5.2, we write u in the polar form (4.13), i.e.,

$$u(x) = -1 + q^u \text{ (in the vector case } q^u \text{ is multiplied by a unit vector)},$$

and define the perturbation

$$\sigma(x) = -1 + \min\{q^u, q^h\} = -1 + q^\sigma(x),$$

where $q^h(x) = (|x| - r)^+$, $h(x) = -1 + (|x| - r)^+$. So $\sigma = \min\{u, h\}$, and since $u(x)$ takes values in $[-1, 1]$, the definition of σ is not affected if we replace h by $\min\{h, 1\}$. We note that the heteroclinic connection between -1 and 1 is a linear function for this potential. This explains the form of $q^h(x)$ (cf. (5.52)). In the diffuse interface context we are considering, r should be replaced by $r + 2$, and ϵ by 2, thus σ corresponds to the set $E_\epsilon = D \setminus B_r$ sketched in Fig. 5.3. To see this, note that to describe the set $E = D \setminus B_r$, instead of the characteristic function we can equally well use a step function σ_0 which equals 1 on D and -1 on D^c. Then, since $\sigma = -1$ in B_r and $\sigma = u$ in $\mathscr{O} \setminus B_{r+2}$, σ is a Lipschitz analog of σ_0. In the same spirit the set $\{u > \sigma\} \subset B_{r+2}$ corresponds to the set $D \cap B_r$. Set $K := \{u > \sigma\} \cap B_{r+2}$. For $s \in (-1, 0)$ we have $\{u > 0\} \cap B_r \subset \{\sigma < s < u\} \cap B_r \subset \{\sigma < s < u\} \cap K$ (since $\sigma = -1$ on B_r). Hence,

$$V(r) = \mathscr{L}^n(\{0 < u\} \cap B_r) \leq \mathscr{L}^n(\{\sigma < s < u\} \cap K), \tag{5.22}$$

and since for each $s \in (-1, 1)$

$$\partial(\{\sigma < s < u\} \cap K) = (\{u = s\} \cup \{\sigma = s\}) \cap K,$$

the isoperimetric inequality yields

$$c(\mathscr{L}^n(\{\sigma < s < u\} \cap K))^{\frac{n-1}{n}} \leq \mathscr{H}^{n-1}(\{u = s\} \cap K) + \mathscr{H}^{n-1}(\{\sigma = s\} \cap K)$$

and by (5.22)

$$c(V(r))^{\frac{n-1}{n}} \le \mathcal{H}^{n-1}(\{u = s\} \cap K) + \mathcal{H}^{n-1}(\{\sigma = s\} \cap K). \tag{5.23}$$

Observe now that the particular form of W we are considering and the coarea formula imply

$$J_K(v) = \int_K (\frac{1}{2}|\nabla v|^2 + \mathbb{1}_{\{|v|<1\}}) dx \ge \sqrt{2} \int_K |\nabla v| dx = \sqrt{2} \int_{-1}^1 \mathcal{H}^{n-1}(\{v = s\} \cap K) ds$$

for each function v with values in $[-1, 1]$. Therefore, integrating (5.23) we get

$$c\int_{-1}^0 (V^{\frac{n-1}{n}}(r)) ds \le \sqrt{2} \int_{-1}^0 \mathcal{H}^{n-1}(\{u = s\} \cap K) ds + \sqrt{2} \int_{-1}^0 \mathcal{H}^{n-1}(\{\sigma = s\} \cap K) ds$$

$$\le J_K(u) + J_K(\sigma) \le 2J_K(\sigma) = 2J_K(h),$$

where we used the minimality of u on K, since on ∂K we have $u = \sigma$. Hence, we obtain the lower bound

$$c(V(r))^{\frac{n-1}{n}} \le J_K(h), \tag{5.24}$$

which corresponds to the first inequality in (5.7). Next, we obtain an upper bound for the right-hand side of (5.24) by using minimality, as in the second inequality in (5.7). From the definition of h and using again the special form of the potential W we have, for $x \in B_{r+2} \setminus B_r$,

$$\frac{1}{2}|\nabla h|^2 + \mathbb{1}_{\{|h|<1\}} = \frac{3}{2}$$

which implies $J_K(h) = \frac{3}{2}\mathcal{L}^n(\{-1 < h\} \cap K)$. It follows that

$$\frac{2}{3}J_K(h) = \mathcal{L}^n(\{-1 < h\} \cap K) = \mathcal{L}^n(\{-1 < h\} \cap K \cap (B_{r+2} \setminus B_r))$$

$$\le \mathcal{L}^n(\{-1 < u\} \cap K \cap (B_{r+2} \setminus B_r)) \le \mathcal{L}^n(\{-1 < u\} \cap (B_{r+2} \setminus B_r))$$

$$= \mathcal{L}^n(\{-1 < u \le 0\} \cap (B_{r+2} \setminus B_r)) + \mathcal{L}^n(\{u > 0\} \cap (B_{r+2} \setminus B_r))$$

$$\le \mathcal{L}^n(\{|u| < 1\} \cap (B_{r+2} \setminus B_r)) + \mathcal{L}^n(\{u > 0\} \cap (B_{r+2} \setminus B_r))$$

$$= A(r + 2) - A(r) + V(r + 2) - V(r). \tag{5.25}$$

We also note that, using the special form of W, we have

$$A(r) = \mathcal{L}^n(\{|u| < 1\} \cap B_r) \le J_{B_r}(u) = J_{B_r \cap K}(u) \le J_K(u) \le J_K(h)$$

$$\le \frac{3}{2}\mathcal{L}^n(B_{r+2} \setminus B_r) \le \frac{3}{2}\frac{\mathcal{L}^n(B_{r+2} \setminus B_r)}{r^{n-1}}r^{n-1} \le Cr^{n-1}, \quad \text{for } r \ge 1,$$

$$\tag{5.26}$$

for some $C > 0$ (cf. Lemma 5.1). From $A(r) \le J_K(h)$ and (5.25) we have

$$A(r) + c(V(r))^{\frac{n-1}{n}} \le 3\Big((V(r+2) + A(r+2)) - (V(r) + A(r))\Big). \qquad (5.27)$$

Since u is Lipschitz, from the assumption $u(0) = 0$ it follows that

$$A(r) \ge A(1) \ge c_0 > 0, \ r \ge 1,$$

and therefore $c_0^{\frac{1}{n}}(A(r))^{\frac{n-1}{n}} \le A(r)$ for $r \ge 1$. This and $(x + y)^p \le x^p + y^p$, $x, y \ge 0$, $p \in [0, 1]$, provided $\tilde{c} \le \{c, c_0^{\frac{1}{n}}\}$, imply

$$\tilde{c}(A(r) + V(r))^{\frac{n-1}{n}} \le A(r) + c(V(r))^{\frac{n-1}{n}}, \quad \text{for } r \ge 1, V(r) \ge 0,$$

which together with (5.27) leads to

$$c_1(A(r) + V(r))^{\frac{n-1}{n}} \le (V(r+2) + A(r+2)) - (V(r) + A(r)), \qquad (5.28)$$

where $c_1 = \tilde{c}/3$. Note that the discretization of (5.14) is instead

$$c_1(V(r))^{\frac{n-1}{n}} \le V(r+2) - V(r),$$

and (5.28) is a modification of this which, as we will see, leads to the lower bound

$$V(r) + A(r) \ge \gamma_1 r^n.$$

Utilizing the upper bound $A(r) \le Cr^{n-1}$ in (5.26), we reach the conclusion of the theorem. We now solve the difference equation (5.28). If we set $y_0 = V(1) + A(1) \ge c_0$ and $y_k = V(2k + 1) + A(2k + 1)$ then, (5.28) becomes

$$y_k - y_{k-1} \ge c_1 y_{k-1}^{\frac{n-1}{n}}, \quad k = 1, \ldots \qquad (5.29)$$

Therefore,

$$y_1 \ge y_0 + c_1 y_0^{\frac{n-1}{n}},$$

$$y_k \ge y_0 + c_1 y_0^{\frac{n-1}{n}} + c_1 \sum_{j=1}^{k-1} y_j^{\frac{n-1}{n}}, \quad \text{for } k \ge 2. \qquad (5.30)$$

Set $\gamma = \min\{(\frac{c_1}{n2^n})^n, y_0 + c_1 y_0^{\frac{n-1}{n}}\}$. We claim that (5.30) implies

$$y_k \ge \gamma k^n, \quad \text{for } k = 1, \ldots \qquad (5.31)$$

This is obviously true for $k = 1$, and by the induction hypothesis we have, for $k \geq 2$

$$y_k \geq c_1 \gamma^{\frac{n-1}{n}} \sum_{j=1}^{k-1} j^{n-1} \geq c_1 \gamma^{\frac{n-1}{n}} \frac{1}{n}(k-1)^n = c_1 \gamma^{\frac{n-1}{n}} \frac{(k-1)^n}{nk^n} k^n \geq \frac{c_1 \gamma^{\frac{-1}{n}}}{n2^n} \gamma k^n \geq \gamma k^n.$$

From (5.31) it follows $V(r) + A(r) \geq \gamma_1 r^n$ for $r \geq 1$, for some constant $\gamma_1 > 0$, so in view of (5.26)

$$V(r) \geq \gamma_1 r^n - C r^{n-1} \geq \frac{\gamma_1}{2} r^n, \quad \text{for } r \geq \frac{2C}{\gamma_1}.$$

The estimate is thus established for r an odd integer. It is extended to $r \geq 2$, possibly with a different constant, by utilizing the monotonicity of V. The proof of the theorem is complete. $\qquad\qquad\square$

Note: Savin's proof is different from the original proof in [5] in two aspects. It utilizing minimality on sets Ω that are not necessarily balls, and employs the isoperimetric inequality as opposed to its Sobolev relative (5.42) below.

5.2.2 The Density Estimate

We consider nonnegative potentials $W \in C(\mathbb{R}^m; [0, \infty))$ with $\{W = 0\} \neq \varnothing$. Let $W(a) = 0$. We model W near a after $|u - a|^\alpha$, and thus introduce the following hypothesis (Fig. 5.6):

$$\mathbf{H} \begin{cases} 0 < \alpha < 2 : W \text{ is differentiable in a deleted neighborhood of } a \text{ and satisfies} \\ \frac{\mathrm{d}}{\mathrm{d}\rho} W(a + \rho\xi) \geq \alpha C^* \rho^{\alpha-1}, \ \forall \rho \in (0, \rho_0], \ \forall \xi \in \mathbb{R}^m : |\xi| = 1, \\ \text{for some constants } \rho_0 > 0, \ C^* > 0 \text{ independent of } \alpha. \\ \alpha = 2 : W \text{ is } C^2 \text{ in a neighborhood of } a \text{ , and} \\ c_0 \leq \xi^\top W_{uu}(u)\xi \leq c_0', \ \forall u : |u - a| \leq q_0, \ \forall \xi : |\xi| = 1, \\ \text{for some constants } q_0 > 0, \ c_0' > c_0 > 0. \end{cases}$$

$$\text{(5.32)}$$

We note that for all $\alpha \in (0, 2]$, $s \mapsto W(a + s\xi)$ is increasing near $s = 0$. In addition, for $\alpha \in [1, 2]$, W is also convex near $u = a$. Also note that $\mathbf{H}$ implies that a is isolated in $\{W = 0\}$, hence

$$d_0 = \min_{\xi}\{|a - z| : z \neq a, \ W(z) = 0\} > 0.$$

Theorem 5.2 (cf. [2]) *Assume W satisfies hypothesis $\mathbf{H}$, $\mathcal{O}$ is open, $n \geq 1$, and $u : \mathcal{O} \subset \mathbb{R}^n \to \mathbb{R}^m$ is minimal as in (4.3). Then for any $\mu_0 > 0$ and any $\lambda \in (0, d_0)$,* *the condition*

$$\mathscr{L}^n(B_{r_0}(x_0) \cap \{|u - a| > \lambda\}) \geq \mu_0, \tag{5.33}$$

implies

$$\mathscr{L}^n(B_r(x_0) \cap \{|u - a| > \lambda\}) \geq Cr^n, \quad \text{for } r \geq r_0, \tag{5.34}$$

as long as $B_r(x_0) \subset \mathscr{O}$, where $C = C(W, \mu_0, \lambda, r_0, M)$.

Remark 5.2 The case $\alpha = 0$ is treated in [3]. For $0 < \alpha < 1$, the L^∞ gradient bound is not an appropriate hypothesis, since $u \in C^\beta_{\mathrm{loc}}(\mathbb{R}^n; \mathbb{R}^m)$ with $\beta \in (0, 1)$ is the available estimate. The only prerequisite for the proof in the case $0 < \alpha < 2$ is Lemma 5.1 which in [3] is established with a different proof that does not require the gradient bound in (5.2), using instead the Hölder continuity. This is also stated, without details, in [6, Lemma 1]. We will take it for granted here.

Remark 5.3 We will assume that $n \geq 2$. At the end of the proof we indicate the necessary modifications for $n = 1$.

Remark 5.4 It is a simple consequence of the basic estimate (5.4) that the validity of the theorem for one value of $\lambda \in (0, d_0)$, implies its validity for all $\lambda' \in (0, d_0)$. Indeed suppose

$$\mathscr{L}^n(B_{r_0'}(x_0) \cap \{|u - a| > \lambda'\}) \geq \mu_0' > 0.$$

It is enough to check for $\lambda' \in (\lambda, d_0)$. Set $w_\lambda^{\lambda'} = \min_{q \in [\lambda, \lambda'], v \in \mathbb{S}^{m-1}} W(a + qv) > 0$. Then,

$$w_\lambda^{\lambda'} \mathscr{L}^n(B_r(x_0) \cap \{\lambda < |u - a| \leq \lambda'\}) \leq J_{B_r(x_0)}(u) \leq C_0 r^{n-1},$$

by Lemma 5.1. Thus

$$\mathscr{L}^n(B_r(x_0) \cap \{|u - a| > \lambda'\})$$
$$= \mathscr{L}^n(B_r(x_0) \cap \{|u - a| > \lambda\}) - \mathscr{L}^n(B_r(x_0) \cap \{\lambda < |u - a| \leq \lambda'\})$$
$$\geq Cr^n - \frac{C_0}{w_\lambda^{\lambda'}} r^{n-1}, \quad \text{for } r \geq r_0,$$

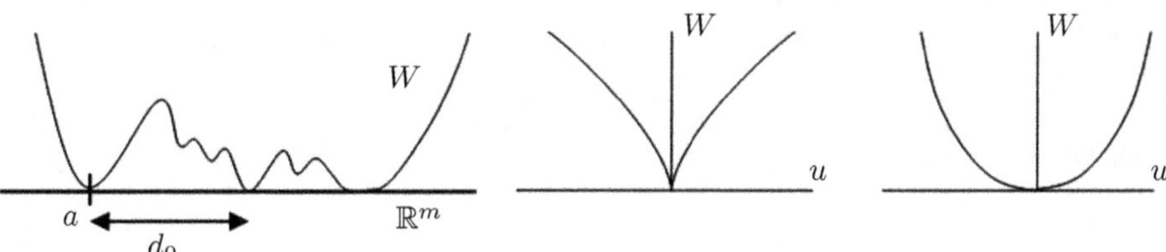

Fig. 5.6 The potential W, and its behavior at a for $0 < \alpha \leq 1$, and $1 < \alpha \leq 2$

and therefore

$$\mathscr{L}^n(B_r(x_0) \cap \{|u - a| > \lambda'\}) \geq \frac{C}{2}r^n, \quad \text{for } r \geq \bar{r} := \max\left\{r_0, \frac{2C}{w_\lambda^{\lambda'} C_0}\right\}.$$

This proves the claim with $C' = \frac{C}{2}$ if $\bar{r} \leq r_0'$. Otherwise we conclude by observing that

$$\mathscr{L}^n(B_{r_0'}(x_0) \cap \{|u - a| > \lambda'\}) \geq \mu_0' \geq \frac{\mu_0'}{\bar{r}^n}r^n, \quad \text{for } r \in [r_0', \bar{r})$$

and by setting $C' = \min\{\frac{C}{2}, \frac{\mu_0'}{\bar{r}^n}\}$. Note that here we take λ in (5.17), (5.18) strictly less than the distance from the rest of the minima of W.

Remark 5.5 Another immediate consequence of Theorem 5.2 and the basic estimate is the following

Proposition 5.1 *Assume there are $a_1 \neq a_2 \in \mathbb{R}^m$ such that $W(a_1) = W(a_2) = 0$, $W > 0$ otherwise, and assume that* **H** *holds at $a = a_j$, $j = 1, 2$. Let $u : \mathcal{O} \subset \mathbb{R}^n \to \mathbb{R}^m$, $n \geq 1$, as in Theorem 5.2. Then given $0 < \theta < |a_1 - a_2|$, the condition $\mathscr{L}^n(B_1(x_0) \cap \{|u - a_1| \leq \theta\}) \geq \mu_0 > 0$, implies the estimate $\mathscr{L}^n(B_r(x_0) \cap \{|u - a_1| \leq \theta\}) \geq Cr^n$ for $r \geq 1$, as long as $B_r(x_0) \subset \mathcal{O}$.*

Remark 5.6 The proof of the case $\alpha \in (0, 2)$ requires the basic estimate (5.4) independently on how small $\lambda > 0$ is taken. On the other hand, the proof for $\alpha = 2$, $\lambda \in (0, q_0)$, does not require (5.4) for small $\lambda > 0$.

Remark 5.7 The possibility of adapting the proof in [5] to the vector case $m \geq 1$ is based on the use of the *polar form* $u(x) = a + q^u(x)\boldsymbol{n}^u(x)$ of a vector map u and of the corresponding energy written in the form

$$J_\Omega(u) = \int_\Omega \left(\frac{1}{2}(|\nabla q^u|^2 + (q^u)^2|\nabla \boldsymbol{n}^u|^2 + W(a + q^u\boldsymbol{n}^u)\right)dx, \tag{5.35}$$

and on the choice of test functions $\sigma : \mathcal{O} \to \mathbb{R}^m$ which are constructed by modifying the scalar quantity q^u without changing the direction vector $\boldsymbol{n}^u$:

$$\sigma = a + q^\sigma \boldsymbol{n}^u. \tag{5.36}$$

The point of departure is that for a map of this form that coincides with u outside $B_r = B_r(x_0)$ ($q^\sigma = q^u$ for $|x - x_0| \geq r$) the minimality of u implies

$$J_{B_r}(u) \leq J_{B_r}(\sigma). \tag{5.37}$$

The polar form has already been implemented in Chap. 2 in the proof of Theorem 2.2, and in Chap. 4 in the proof of Lemma 4.1. We refer the reader to Sect. 4.1 for relevant calculus facts that are used below. Otherwise, we follow quite closely, as far as the essentials are concerned, the argument in [5], including a certain improvement from [23]. In particular, we use minimality only over balls.

5.3 Proof of Theorem 5.2

5.3.1 The Identity

We will use the *polar form* of a vector map $u(\cdot) \in W^{1,2}(B_r; \mathbb{R}^m) \cap L^\infty(B_r; \mathbb{R}^m)$,

$$u(x) = a + q^u(x)\boldsymbol{n}^u(x), \tag{5.38}$$

where

$$q^u(x) = |u(x) - a|, \quad \boldsymbol{n}^u(x) = \frac{u(x) - a}{|u(x) - a|}, \quad \text{if } u(x) \neq a, \tag{5.39}$$

$q^u \in W^{1,2}(B_r) \cap L^\infty(B_r)$, and $\nabla \boldsymbol{n}^u$ is measurable and such that $q^u|\nabla \boldsymbol{n}^u| \in L^2(B_r)$. In addition, we can write $u_{x_i} = q^u_{x_i}\boldsymbol{n}^u + q^u\boldsymbol{n}^u_{x_i}$ and

$$u_{x_i} \cdot u_{x_j} = q^u_{x_i}q^u_{x_j} + (q^u)^2\boldsymbol{n}^u_{x_i} \cdot \boldsymbol{n}^u_{x_j},$$

where we have also used the fact that $\boldsymbol{n}^u \cdot \boldsymbol{n}^u = 1$, which implies $\boldsymbol{n}^u_{x_i} \cdot \boldsymbol{n}^u = 0$. It follows (cf. Sect. 4.1(E) and also Sect. 4.4) that

$$\int_{B_r} |\nabla u|^2 \mathrm{d}x = \int_{B_r} |\nabla q^u|^2 \mathrm{d}x + \int_{B_r} (q^u)^2|\nabla \boldsymbol{n}^u|^2 \mathrm{d}x. \tag{5.40}$$

In the vector framework, instead of the scalar comparison function considered in [5], we introduce the vector maps $h = a + q^h(x)\boldsymbol{n}^u(x)$ and $\sigma = a + q^\sigma\boldsymbol{n}^u(x)$, $q^\sigma = \min\{q^h, q^u\}$, where $q^h \in W^{1,2}(B_r) \cap L^\infty(B_r)$, $q^h \geq 0$ is a suitable radial C^1 map defined later, $q^h \geq q^u$ on ∂B_r. We note that (5.40) with $u = \sigma$ yields

$$\int_{B_r} |\nabla \sigma|^2 \mathrm{d}x = \int_{B_r} |\nabla q^\sigma|^2 \mathrm{d}x + \int_{B_r} (q^\sigma)^2|\nabla \boldsymbol{n}^u|^2 \mathrm{d}x.$$

This point requires attention and is addressed again when the specific q^h is introduced. Thus we derive the identity

$$\frac{1}{2} \int_{B_r} \left(|\nabla q^u|^2 - |\nabla q^\sigma|^2 \right) dx$$

$$= J_{B_r}(u) - J_{B_r}(\sigma) + \frac{1}{2} \int_{B_r} \left((q^\sigma)^2 - (q^u)^2 \right) |\nabla \boldsymbol{n}^u|^2 dx + \int_{B_r} (W(\sigma) - W(u)) dx$$

$$\leq \int_{B_r} (W(\sigma) - W(u)) dx,$$

$$(5.41)$$

where in deriving the last inequality we used the fact that $q^\sigma \leq q^u$ and the minimality of u. We took this difference above in order to generate a difference scheme corresponding to the differential inequality (5.7). The test function σ will be related to the discretization of r, and is equal to u on ∂B_r.

5.3.2   The Isoperimetric Estimate

We will bound the left-hand side of (5.41) from below, and derive the analog of the first inequality in (5.7) (cf. (5.47) below). We recall the Sobolev inequality

$$\left(\int_{\mathbb{R}^n} |f|^{\frac{n}{n-1}} dx \right)^{\frac{n-1}{n}} \leq C(n) \int_{\mathbb{R}^n} |\nabla f| dx, \ \forall f \in W^{1,2}(\mathbb{R}^n), \ \forall n \geq 2, \qquad (5.42)$$

where $C(n)$ is a universal constant depending only on the dimension n, and in particular independent of the support of f. With ρ_0 as in **H**, we define the cut-off function

$$\beta = \min\{q^u - q^\sigma, \lambda\} \text{ in } B_r, \text{ with } \lambda > 0 \text{ small}, \lambda \leq \rho_0, \qquad (5.43)$$

and applying (5.42) to β^2 we obtain, for $n \geq 2$,

$$\left(\int_{B_r} \beta^{\frac{2n}{n-1}} dx \right)^{\frac{n-1}{n}} = \left(\int_{B_r} (\beta^2)^{\frac{n}{n-1}} dx \right)^{\frac{n-1}{n}}$$

$$\leq C(n) \int_{B_r} |\nabla(\beta^2)| dx \leq 2C(n) \int_{B_r \cap \{q^u - q^\sigma \leq \lambda\}} |\nabla \beta| |\beta| dx,$$

$$(5.44)$$

where we have utilized that $\beta = 0$ on ∂B_r, and the fact that $\nabla \beta = 0$ a.e. on $q^u - q^\sigma > \lambda$. By Young's inequality, for $A > 0$ we have

$$
\begin{aligned}
\left(\int_{B_r} \beta^{\frac{2n}{n-1}} \mathrm{d}x \right)^{\frac{n-1}{n}} &\leq 2C(n) \int_{B_r \cap \{q^u - q^\sigma \leq \lambda\}} |\nabla \beta||\beta| \mathrm{d}x \\
&\leq C(n)A \int_{B_r \cap \{q^u - q^\sigma \leq \lambda\}} |\nabla \beta|^2 \mathrm{d}x + \frac{C(n)}{A} \int_{B_r \cap \{q^u - q^\sigma \leq \lambda\}} \beta^2 \mathrm{d}x \\
&\leq C(n)A \int_{B_r} |\nabla(q^u - q^\sigma)|^2 \mathrm{d}x + \frac{C(n)}{A} \int_{B_r \cap \{q^u - q^\sigma \leq \lambda\}} (q^u - q^\sigma)^2 \mathrm{d}x.
\end{aligned}
\tag{5.45}
$$

Noting the identity

$$
|\nabla(q^u - q^\sigma)|^2 = |\nabla q^u|^2 - |\nabla q^\sigma|^2 - 2\nabla q^\sigma \cdot (\nabla q^u - \nabla q^\sigma),
\tag{5.46}
$$

we can bound the right-hand side of (5.45) utilizing (5.41). Thus, we obtain

$$
\begin{aligned}
\left(\int_{B_r} \beta^{\frac{2n}{n-1}} \mathrm{d}x \right)^{\frac{n-1}{n}} &\leq 2C(n)A\left(\int_{B_r} (W(\sigma) - W(u))\mathrm{d}x - \int_{B_r} \nabla q^\sigma \cdot (\nabla q^u - \nabla q^\sigma)\mathrm{d}x \right) \\
&\quad + \frac{C(n)}{A} \int_{B_r \cap \{q^u - q^\sigma \leq \lambda\}} (q^u - q^\sigma)^2 \mathrm{d}x.
\end{aligned}
\tag{5.47}
$$

Assuming that $q^h \in W^{1,2}(B_r) \cap L^\infty(B_r)$ can be chosen so that

$$
q^h = 0 \text{ on } B_{r-T}, \text{ for some fixed } T > 0,
\tag{5.48}
$$

and thus

$$
q^\sigma = 0 \text{ on } B_{r-T} \iff \sigma = a \text{ on } B_{r-T},
$$

we can estimate

$$
\left(\int_{B_r} \beta^{\frac{2n}{n-1}} \mathrm{d}x \right)^{\frac{n-1}{n}} \geq \left(\int_{B_{r-T} \cap \{q^u > \lambda\}} \beta^{\frac{2n}{n-1}} \right)^{\frac{n-1}{n}} \mathrm{d}x \geq \lambda^2 \mathcal{L}^n (B_{r-T} \cap \{q^u > \lambda\})^{\frac{n-1}{n}},
\tag{5.49}
$$

and obtain from (5.47) that

$$
\begin{aligned}
\lambda^2 &\mathcal{L}^n (B_{r-T} \cap \{q^u > \lambda\})^{\frac{n-1}{n}} \\
&\leq 2C(n)A\left(\int_{B_r} (W(\sigma) - W(u))\mathrm{d}x - \int_{B_r} \nabla q^\sigma \cdot (\nabla q^u - \nabla q^\sigma)\mathrm{d}x \right) \\
&\quad + \frac{C(n)}{A} \int_{B_r \cap \{q^u - q^\sigma \leq \lambda\}} (q^u - q^\sigma)^2 \mathrm{d}x.
\end{aligned}
\tag{5.50}
$$

5.3.3 Comments-Preview

Note that so far we have managed to get by only with minimality. The plan now is to bound the right-hand side of (5.50) solely by the first term involving W, and thus generate a perfect analog of (5.7) leading to a similar conclusion. Indeed, if we ignore for the moment the last two terms, then in terms of (5.17)

$$\lambda^2 (V(r-T))^{\frac{n-1}{n}}$$

$$\leq 2CA \int_{B_r} (W(\sigma) - W(u)) dx$$

(by (5.48), positivity of W, and monotonicity near $u = a : q^u \geq q^\sigma \geq 0$)

$$\leq 2CA \int_{(B_r \setminus B_{r-T}) \cap \{q^u \geq \lambda\}} (W(\sigma) - W(u)) dx$$

$$\leq 2CAW_M \mathscr{L}^n ((B_r \setminus B_{r-T}) \cap \{q^u \geq \lambda\}) = 2CAW_M (V(r) - V(r-T)).$$
$$(5.51)$$

Actually, handling the third term on the right-hand side of (5.47) is not difficult in the subquadratic case ($\alpha < 2$), since $W(a+q^u \mathbf{n}) - W(a+q^\sigma \mathbf{n}) \geq C^* ((q^u)^\alpha - (q^\sigma)^\alpha)$, $0 \leq q^\sigma \leq q^u \leq \lambda$. This estimate is certainly valid for $\alpha = 2$ as well. The trouble however is that it is not possible in that case to choose q^σ vanishing in B_{r-T}, which facilitates at the same time the handling of the second term as we are about to show. For the second term, we need detailed information on the solution near $u = a$. Recall that $\mathbf{H}$ is modeled after $W(u) \sim |u - a|^\alpha$ for $u \sim a$, thus the following O.D.E. is relevant:

$$\begin{cases} q' = \frac{2}{2-\alpha} C^{\frac{\alpha}{2}} q^{\frac{\alpha}{2}}, \quad q'' = \frac{2\alpha}{(2-\alpha)^2} C^\alpha q^{\alpha-1} \quad (C > 0), \\ q(0) = 0. \end{cases}$$
$$(5.52)$$

For $\alpha < 2$, the nonlinearity is not Lipschitz and so (5.52) has the family of nontrivial solutions $q(s) = C^{\frac{\alpha}{2-\alpha}} s^{\frac{2}{2-\alpha}}$, $q(s) = 0$ for $s \leq 0$. We will take

$$q^h(x) = q(|x| - (r - T)).$$

Before calculating, we need to check that $\sigma(x) = a + q^\sigma(x) \mathbf{n}(x)$, $q^\sigma = \min\{q^h, q^u\}$ is Sobolev, and that the polar representation rendering (5.41) is valid. For this purpose, consider the difference quotient

$$\frac{\sigma(x + he_j) - \sigma(x)}{h} = \frac{q^\sigma(x + he_j) - q^\sigma(x)}{h} \mathbf{n}(x)$$

$$+ \frac{q^\sigma(x + he_j) - q^\sigma(x)}{h} (\mathbf{n}(x + he_j) - \mathbf{n}(x)) + \frac{q^\sigma(x)}{q^u(x)} q^u(x) \frac{\mathbf{n}(x + he_j) - \mathbf{n}(x)}{h},$$
$$(5.53)$$

and first notice that q^σ is Sobolev by [11, 4.2.2]. It can be seen by appealing to the discussion at the end of Sect. 4.4 that it converges as $h \to 0$ in L^2, and that the formula $\sigma_{,i} = q^\sigma_{,i} n + q^\sigma n_{,i}$ holds. Thus, (5.41) is now justified, and for $x \in B_r \setminus B_{r-T}$ we have

$$\Delta q^h = (q^h)'' + \frac{n-1}{\rho}(q^h)' = \frac{2\alpha}{(2-\alpha)^2} C^\alpha (q^h)^{\alpha-1} + \frac{n-1}{\rho}\frac{2}{2-\alpha} C^{\frac{\alpha}{2}}(q^h)^{\frac{\alpha}{2}}, \ (\rho = |x|)$$

$$\leq \frac{2}{2-\alpha} C^{\frac{\alpha}{2}}\left[\frac{\alpha C^{\frac{\alpha}{2}}}{2-\alpha} + \frac{n-1}{|x|-(r-T)}\right](q^h)^{\alpha-1} \ (q^h < 1). \tag{5.54}$$

We require $q^h = 0$ on B_{r-T}, $q^h = M \geq q^u$ on ∂B_r, which gives $C = M^{\frac{2-\alpha}{2}}T^{-1}$. At the end we will choose $r - T \geq 1$. Focusing on the second term now, we have

$$-\int_{B_r} \nabla q^h (\nabla q^u - \nabla q^\sigma)\mathrm{d}x = -\int_{B_r \setminus B_{r-T}} \nabla q^h (\nabla q^u - \nabla q^\sigma)\mathrm{d}x$$

$$= \int_{B_r \setminus B_{r-T}} \Delta q^h (q^u - q^\sigma)\mathrm{d}x \ \left(\text{since } \left.\frac{\partial q^h}{\partial n}\right|_{\partial B_{r-T}} = 0, \ q^\sigma = q^u \text{ on } \partial B_r\right)$$

$$= \int_{(B_r \setminus B_{r-T}) \cap \{q^h < q^u\}} \Delta q^h (q^u - q^h)\mathrm{d}x. \tag{5.55}$$

Thus for λ small, the part of the integral above near $u = a$ can be estimated as follows:

$$\int_{(B_r \setminus B_{r-T}) \cap \{q^h < q^u\} \cap \{q^u < \lambda\}} \Delta q^h (q^u - q^\sigma)\mathrm{d}x$$

$$\leq \tilde{C} \int_{(B_r \setminus B_{r-T}) \cap \{q^h < q^u\} \cap \{q^u < \lambda\}} (q^h)^{\alpha-1}(q^u - q^h)\mathrm{d}x \quad (\text{cf. (5.54)})$$

$$\leq C \int_{(B_r \setminus B_{r-T}) \cap \{q^h < q^u\} \cap \{q^u < \lambda\}} (q^u)^{\alpha-1}q^u \mathrm{d}x$$

$$\leq C \int_{(B_r \setminus B_{r-T}) \cap \{q^h < q^u\} \cap \{q^u < \lambda\}} W(u)\mathrm{d}x \quad (\text{cf. } \mathbf{H})$$

$$\leq C(A(r) - A(r-T)) \quad (\text{cf. (5.17)}). \tag{5.56}$$

We note that in the third inequality from below we took $\alpha \geq 1$. As we will see in Claim 3, in Sect. 5.3.4 below, this is without loss of generality. Also, C stands for a constant whose value may vary from line to line. This calculation suggests a modification of the difference scheme (5.51) (cf. (5.28))

$$\lambda^2((V(r-T))^{\frac{n-1}{n}} \leq (V(r) - V(r-T))$$

to

$$C(\lambda)(V(r-T)^{\frac{n-1}{n}} + A(r-T)) \leq (V(r) - V(r-T)) + (A(r) - A(r-T)), \tag{5.57}$$

which as we will see later leads to

$$V(r) + A(r) \geq Cr^n.$$

Using now the basic estimate of Lemma 5.1: $A(r) \leq \widehat{C}_0 r^{n-1}$, we can reach the conclusion of the theorem. This is a sketch of the argument for $\alpha < 2$ and we will follow it pretty closely. For $\alpha = 2$, it is not possible to define a comparison function $q^h(x)$ that vanishes in the ball B_{r-T}. Instead, by resorting to the linear equation (5.58) below, we construct a comparison function that is exponentially small with respect to T. Consider

$$\begin{cases} \Delta\varphi = c_0\varphi, & \text{in } B_r, \\ \varphi = 1, & \text{on } \partial B_r, \end{cases} \tag{5.58}$$

with $c_0 > 0$ as in **H**. In Appendix A.1, we provide a detailed account of the radial solution to (5.58), the main feature of which is that $\varphi(x; r) = \Phi(|x|; r)$, with $\Phi' > 0$ for $|x| \in (0, r]$,

$$\Phi(s; r) \leq e^{-c_1(r-s)} \text{ for } s \in (0, r], \; r \geq r_0; \text{ here } c_1 > 0 \text{ depends on } r_0 \text{ (cf. (A.3).} \tag{5.59}$$

Thus (cf. (5.2))

$$q^h(x) = \begin{cases} \varphi M, & \text{on } B_r, \\ M, & \text{on } \mathcal{O} \setminus B_r, \end{cases} \qquad q^\sigma = \min\{q^u, q^h\}, \tag{5.60}$$

is an appropriate test function such that

$$q^h(x) \leq Me^{-c_1 T} \quad \text{on } B_{r-T}, \tag{5.61}$$

thus satisfying only approximately (5.48) for large T. The difference scheme in this case is directly for $V(r)$.

5.3.4 The Case $0 < \alpha < 2$

We estimate the right-hand side of (5.50). We begin with B_{r-T}.

Since $q^\sigma = 0$ on B_{r-T}, the right-hand side reduces to

$$I = -2C(n)A \int_{B_{r-T}} W(u)\mathrm{d}x + \frac{C(n)}{A} \int_{B_{r-T}\cap\{q^u\leq\lambda\}} (q^u)^2\mathrm{d}x \qquad (5.62)$$

$$\leq -2C(n)A \int_{B_{r-T}\cap\{q^u\leq\lambda\}} W(u)\mathrm{d}x + \frac{C(n)}{A} \int_{B_{r-T}\cap\{q^u\leq\lambda\}} (q^u)^2\mathrm{d}x. \qquad (5.63)$$

Claim 1 Assume $\lambda \leq \rho_0$, with ρ_0 the constant in **H**. Then there exists $A_0 > 0$ independent of r such that

$$I \leq -\frac{C(n)}{2}A \int_{B_{r-T}\cap\{q^u\leq\lambda\}} W(u)\mathrm{d}x, \quad \text{for } A > A_0 = \sqrt{2\lambda^{2-\alpha}/3C^*}. \qquad (5.64)$$

Proof To see this, note that from **H** and $q^u \leq \lambda \leq \rho_0$ it follows that

$$W(u) = \int_0^{q^u} W_u(a + s\mathbf{n}^u) \cdot \mathbf{n}^u \mathrm{d}s \geq C^*(q^u)^\alpha,$$

hence

$$-\frac{3A}{2}W(u) + \frac{1}{A}(q^u)^2 \leq (q^u)^\alpha\left(\frac{-3AC^*}{2} + \frac{\lambda^{2-\alpha}}{A}\right),$$

and therefore for $A > \sqrt{2\lambda^{2-\alpha}/3C^*}$ we obtain

$$-\frac{3C(n)}{2}A \int_{B_{r-T}\cap\{q^u\leq\lambda\}} W(u)\mathrm{d}x + \frac{C(n)}{A} \int_{B_{r-T}\cap\{q^u\leq\lambda\}} (q^u)^2\mathrm{d}x \leq 0.$$

(5.64) now follows. $\qquad\square$

Next we consider the right-hand side of (5.50) on $B_r \setminus B_{r-T}$. Set

$$I_1 = 2C(n)A \int_{B_r\setminus B_{r-T}} (W(\sigma) - W(u))\mathrm{d}x + \frac{C(n)}{A} \int_{(B_r\setminus B_{r-T})\cap\{q^u-q^\sigma\leq\lambda\}} (q^u - q^\sigma)^2\mathrm{d}x,$$

$$I_2 = -2C(n)A \int_{B_r\setminus B_{r-T}} \nabla q^\sigma \cdot (\nabla q^u - \nabla q^\sigma)\mathrm{d}x.$$

Claim 2 Assume $\lambda \leq \min\{\rho_0, 1\}$. Then there exists a constant $\widetilde{C} > 0$ independent of r such that

$$I_1 \leq \widetilde{C} A \mathscr{L}^n((B_r \setminus B_{r-T}) \cap \{q^u > \lambda\}) + \frac{\widetilde{C}}{A} \int_{(B_r \setminus B_{r-T}) \cap \{q^u \leq \lambda\}} W(u)\mathrm{d}x, \quad \text{for} \ A > 0.$$

$$(5.65)$$

Proof To establish this, we split the integration over $B_r \setminus B_{r-T}$ into integrations over $\{q^u \leq \lambda\}$ and $\{q^u > \lambda\}$. From $q^\sigma \leq q^u$, $q^u \leq \lambda \leq \rho_0$ we have by the monotonicity of W near $u = a$:

$$\int_{(B_r \setminus B_{r-T}) \cap \{q^u \leq \lambda\}} (W(\sigma) - W(u))\mathrm{d}x \leq 0,$$

and therefore from (5.3) it follows that

$$\int_{B_r \setminus B_{r-T}} (W(\sigma) - W(u))\mathrm{d}x \leq W_M \mathscr{L}^n((B_r \setminus B_{r-T}) \cap \{q^u > \lambda\}). \qquad (5.66)$$

As in the proof of Claim 1, for $q^\sigma \leq q^u \leq \lambda \leq \min\{\rho_0, 1\}$, we get

$$W(u) \geq C^*(q^u)^\alpha \geq C^*(q^u - q^\sigma)^\alpha \geq C^*(q^u - q^\sigma)^2$$

which implies

$$\int_{(B_r \setminus B_{r-T}) \cap \{q^u \leq \lambda\}} (q^u - q^\sigma)^2 \mathrm{d}x \leq \frac{1}{C^*} \int_{(B_r \setminus B_{r-T}) \cap \{q^u \leq \lambda\}} W(u)\mathrm{d}x.$$

Inequality (5.65) now follows with $\widetilde{C} = C(n) \max\{1/C^*, 2W_M\}$. $\qquad \square$

Claim 3 Assume $\lambda < \max\{1, \rho_0\}$. Then there exists $\widehat{C} > 0$ independent of r, but depending on M, T such that

$$I_2 \leq \widehat{C} A \mathscr{L}^n((B_r \setminus B_{r-T}) \cap \{q^u > \lambda\}) + \widehat{C} A \int_{(B_r \setminus B_{r-T}) \cap \{q^u \leq \lambda\}} W(u)\mathrm{d}x, \quad \text{for} \ A > 0.$$

$$(5.67)$$

Proof To see this, we proceed as in the sketch of the proof above, but with a modification (following [5]): we replace α in (5.52) by $\tau = \max\{\alpha, 1\}$, and $q(s)$ by $\bar{q}(s) = Cs^{\frac{2}{2-\tau}}$, $\bar{q}(s) = 0$, for $s \leq 0$, and we set

$$q^h(x) = \bar{q}(|x| - (r - T)).$$

Then, proceeding as in (5.55),

$$I_2 = 2CA \int_{B_r \setminus B_{r-T}} \Delta q^\sigma (q^u - q^\sigma) \mathrm{d}x$$

$$= 2CA \int_{(B_r \setminus B_{r-T}) \cap \{q^h < q^u\}} \Delta q^\sigma (q^u - q^\sigma) \mathrm{d}x. \qquad (5.68)$$

We now split the integral, and write $I_2 = I_2^+ + I_2^-$, where I_2^+ and I_2^- correspond to the integration over $\{q^u > \lambda\}$ and $\{q^u \leq \lambda\}$ respectively. Then we have the estimates:

$$I_2^+ \leq 2CAC_M M \mathscr{L}^n((B_r \setminus B_{r-T}) \cap \{q^u > \lambda\}),$$

$$I_2^- \leq 2CAC_1 \int_{(B_r \setminus B_{r-T}) \cap \{q^h < q^u\} \cap \{q^u \leq \lambda\}} (q^h)^{\tau-1}(q^u - q^h) \mathrm{d}x$$

where $\Delta q^h \leq C_M$ and $\Delta q^h \leq C_1(q^h)^{\tau-1}$, with C_M and C_1 two constants depending on M, τ, T (cf. (5.54)), and C_1 bounded as $\alpha \to 0$. Next, since

$$(q^h)^{\tau-1}(q^u - q^h) \leq (q^u)^{\tau-1}(q^u - q^h) \leq (q^u)^\alpha \leq \frac{1}{C^*} W(u), \qquad (5.69)$$

we obtain

$$I_2^- \leq 2CAC_1 \frac{1}{C^*} \int_{(B_r \setminus B_{r-T}) \cap \{q^h < q^u\} \cap \{q^u \leq \lambda\}} W(u) \mathrm{d}x \qquad (5.70)$$

The claim follows with $\widehat{C} = 2C \max\{C_M M, C_1/C^*\}$. $\qquad\qquad\square$

We are now in the position to complete the proof of Theorem 5.2 for the case $0 < \alpha < 2$. Recalling (5.17), and collecting all the estimates above, we have for fixed $A > A_0$:

$$\lambda^2 (V_{r-T})^{\frac{n-1}{n}} + (CA)A(r - T)$$

$$\leq (\widetilde{C} + \widehat{C})A(V(r) - V(r - T)) + \left(\frac{\widetilde{C}}{A} + \widehat{C}A\right)(A(r) - A(r - T)),$$

and consequently we obtain the same difference scheme as in [5]:

$$C(\lambda)\left((V_{r-T})^{\frac{n-1}{n}} + A(r-T)\right) \leq (V(r) - V(r-T)) + (A(r) - A(r-T)), \qquad (5.71)$$

with

$$C(\lambda) = \frac{\min\{\lambda^2, CA\}}{\max\{(\widetilde{C} + \widehat{C})A, \frac{\widetilde{C}}{A} + \widehat{C}A\}}.$$

Claim 4 There exists $C' = C'(\lambda, \mu_0)$ such that

$$V(kT) + A(kT) \geq C'k^n, \ \text{for } k \geq k_0, \ (k \text{ integer}). \tag{5.72}$$

Proof We will prove (5.72) by induction. We begin the argument and in the process we will choose the appropriate k_0, C'. Assume (5.72) holds for $k = \bar{k}$, and set $r - T = \bar{k}T$. Then, by (5.71),

$$V((\bar{k} + 1)T) + A((\bar{k} + 1)T) \geq V(\bar{k}T) + A(\bar{k}T) + C(\lambda)\big((V(\bar{k}T))^{\frac{n-1}{n}} + A(\bar{k}T)\big). \tag{5.73}$$

Now choose k_0 large enough so that $k_0 T \geq 1$, $k_0 C(\lambda) > 2^{n+1}$. We fix such a k_0. By the hypothesis of the theorem, $V(k_0 T) \geq \mu_0$. Choose now $C' > 0$ small enough so that $\mu_0 \geq C'k_0^n$. Thus (5.72) holds for $k = k_0$. Further, choose C' possibly smaller so that $C(\lambda)(C')^{-1/2} \geq 2^{n+1}$. We need to verify the induction step for $k = \bar{k} + 1$. From (5.72) for $k = \bar{k}$, we have that either $A(\bar{k}T) \geq \frac{1}{2}C'\bar{k}^n$ or $V(\bar{k}T) \geq \frac{1}{2}C'\bar{k}^n$. If $A(\bar{k}T) \geq \frac{1}{2}C'\bar{k}^n$, then

$$C(\lambda)A(\bar{k}T) \geq \frac{C(\lambda)C'}{2}\bar{k}^n = \frac{k_0 C(\lambda)}{2}C'\frac{\bar{k}^n}{k_0} \geq 2^n C'\bar{k}^{n-1}.$$

Otherwise, if $V(\bar{k}T) \geq \frac{1}{2}C'\bar{k}^n$, then

$$C(\lambda)[V(\bar{k}T)]^{\frac{n-1}{n}} = C(\lambda)\left[\frac{1}{2}C'\bar{k}^n\right]^{\frac{n-1}{n}} \geq \frac{C(\lambda)}{2}(C')^{\frac{n-1}{n}}\bar{k}^{n-1} \geq 2^n C'\bar{k}^{n-1}.$$

Therefore, $V((\bar{k} + 1)T) + A((\bar{k} + 1)T) \geq C'\bar{k}^n + 2^n C'\bar{k}^{n-1} \geq C'(\bar{k} + 1)^n$. This completes the induction, and so (5.72) has been established. $\qquad\square$

Finally, we show that

$$V(kT) \geq C'k^n, \ \forall k \geq k_0. \tag{5.74}$$

By the basic estimate (5.4),

$$A(kT) \leq \widehat{C}_0(kT)^{n-1},$$

and thus (5.74) follows. The estimate (5.34) is established for $r = kT$, $k \geq k_0$. It is extended to $r \geq k_0$, possibly with a different constant C, by taking into account the monotonicity of $V(r)$. The proof of the density theorem for $0 < \alpha < 2$ is complete.

Remark 5.8 ($n = 1$) We note that $\int_{B_r} |\nabla(\beta^2)|dx \geq \lambda^2$, thus (5.50) holds for $n = 1$. It follows that (5.71) holds for $n = 1$. Since the basic estimate (Lemma 5.1) is valid for $n = 1$, we conclude that (5.74) holds for $n = 1$, and therefore so does (5.34) is valid for $n = 1$.

5.3.5 The Case $\alpha = 2$

Up to inequality (5.47), the derivation is as before, with one notable exception: in the definition of β, we introduce a new parameter λ':

$$\beta = \min\{q^u - q^\sigma, \lambda'\} \text{ in } B_r. \tag{5.75}$$

We set $\lambda' = \frac{\lambda}{2}$, but any $\lambda' \in (0, \lambda)$ will do. Given $T > 0$, (5.60) and (5.61) imply

$$q^\sigma \leq q^h \leq Me^{-c_1 T}, \quad \text{for } x \in B_{r-T}, \; r > r_0 + T.$$

Therefore, if $T > 0$ is such that

$$Me^{-c_1 T} \leq \lambda - \lambda' \tag{5.76}$$

we have

$$q^u - q^\sigma = q^u - q^h > \lambda', \quad \text{for } x \in B_{r-T} \cap \{q^u > \lambda\}.$$

It follows that

$$B_{r-T} \cap \{q^u > \lambda\} \cap \{q^u - q^h \leq \lambda'\} = \emptyset \tag{5.77}$$

and

$$(\lambda')^2 \left(\mathscr{L}^n(B_{r-T} \cap \{q^u > \lambda\}) \right)^{\frac{n-1}{n}} \leq \left(\int_{B_r} \beta^{\frac{2n}{n-1}} dx \right)^{\frac{n-1}{n}}. \tag{5.78}$$

Thus (5.47) is replaced, after an integration by parts, by

$$\left(\int_{B_r} \beta^{\frac{2n}{n-1}} dx \right)^{\frac{n-1}{n}} \leq 2CA \left(\int_{B_r} (W(\sigma) - W(u))dx + \int_{B_r} \Delta q^\sigma (q^u - q^\sigma)dx \right)$$

$$+ \frac{C}{A} \int_{B_r \cap \{q^u - q^\sigma \leq \lambda'\}} (q^u - q^\sigma)^2 dx. \tag{5.79}$$

Now we observe that the definition of q^σ implies

$$B_r = \{q^\sigma < q^u\} \cup (B_r \cap \{q^\sigma = q^u\}), \text{ and } \{q^\sigma < q^u\} = \{q^h < q^u\}.$$

Therefore,

$$\int_{B_r} (W(\sigma) - W(u))\mathrm{d}x = \int_{\{q^h < q^u\}} (W(h) - W(u))\mathrm{d}x,$$

$$\int_{B_r} \Delta q^\sigma (q^u - q^\sigma)\mathrm{d}x = \int_{\{q^h < q^u\}} c_0 q^h (q^u - q^h)\mathrm{d}x, \tag{5.80}$$

$$\int_{B_r \cap \{q^u - q^\sigma \le \lambda'\}} (q^u - q^\sigma)^2 \mathrm{d}x = \int_{\{0 < q^u - q^h \le \lambda'\}} (q^u - q^h)^2 \mathrm{d}x$$

where we have used $\Delta q^h = c_0 q^h$ in B_r (cf. (5.58)). Note also that, under the standing assumption $\lambda \le q_0$, $\mathbf{H}$ implies that, for $x \in \{q^h < q^u\} \cap \{q^u \le \lambda\}$ we have

$$W(u) - W(h) = \int_{q^h}^{q^u} q \int_0^1 D_{qq} W(a + q\mathbf{n}^u)\mathrm{d}s\mathrm{d}q \ge \frac{c_0}{2}\left((q^u)^2 - (q^h)^2\right).$$

Therefore, if we take $A = \frac{1}{\sqrt{c_0}}$, on the subset $\{q^h < q^u\} \cap \{q^u \le \lambda\}$ we have

$$2A\left(W(h) - W(u) + c_0 q^h (q^u - q^h)\right) + \frac{1}{A}(q^u - q^h)^2$$

$$\le \left(\frac{2}{\sqrt{c_0}}\left(-\frac{c_0}{2}(q^u + q^h) + c_0 q^h\right) + \sqrt{c_0}(q^u - q^h)\right)(q^u - q^h) = 0.$$

From this (5.80) and (5.47) we obtain

$$\left(\int_{B_r} \beta^{\frac{2n}{n-1}} \mathrm{d}x\right)^{\frac{n-1}{n}} \le \frac{2C}{\sqrt{c_0}} \int_{B_r \cap \{q^h < q^u\} \cap \{q^u > \lambda\}} \left(W(h) - W(u) + c_0 q^h (q^u - q^h)\right)\mathrm{d}x$$

$$+ C\sqrt{c_0} \int_{B_r \cap \{0 < q^u - q^h \le \lambda'\} \cap \{q^u > \lambda\}} (q^u - q^h)^2 \mathrm{d}x. \tag{5.81}$$

Set $r = r_0 + pT$ and define $B_0 = B_{r_0}$, $B_j = B_{r_0 + jT}$, $j = 1, \ldots, p$ and

$$\omega_0 = \mathscr{L}^n (B_0 \cap \{q^u > \lambda\}),$$
$$\omega_j = \mathscr{L}^n ((B_j \setminus B_{j-1}) \cap \{q^u > \lambda\}). \tag{5.82}$$

Lemma 5.2 *There exists a constant $C^\circ > 0$, $C^\circ = C^\circ(W, M, \lambda, r_0)$, such that*

$$C^\circ\Big(\sum_{j=0}^{p-1} \omega_j\Big)^{\frac{n-1}{n}} \leq \sum_{j=1}^{p} \epsilon^j \omega_{p-j} + \omega_p \tag{5.83}$$

where $\epsilon := \mathrm{e}^{-c_1 T}$.

Proof From (5.82) we have $\sum_{j=0}^{p-1} \omega_j = \mathcal{L}^n(B_{r-T} \cap \{q^u > \lambda\})$, and therefore (5.78) implies

$$(\lambda')^2\Big(\sum_{j=0}^{p-1} \omega_j\Big)^{\frac{n-1}{n}} \leq \Big(\int_{B_r} \beta^{\frac{2n}{n-1}}\,\mathrm{d}x\Big)^{\frac{n-1}{n}}. \tag{5.84}$$

Set for the moment $\mathcal{A}^1 = B_r \cap \{q^h < q^u\} \cap \{q^u > \lambda\}$, $\mathcal{A}^2 = B_r \cap \{0 < q^u - q^h \leq \lambda'\} \cap \{q^u > \lambda\}$ and let I^1, I^2 stand for the integrals on the right-hand side of (5.81). Denote by I^i_{p-j}, $j = 0, \ldots, p-1$, $i = 1, 2$ the part of the integral I^i over $(B_{p-j} \setminus B_{p-j-1}) \cap \mathcal{A}^i$ and let also I^i_0 be the part of the integral I^i over $B_0 \cap \mathcal{A}^i$. Then we have

$$\frac{2}{\sqrt{c_0}} I^1_p + \sqrt{c_0} I^2_p \leq \Big(\frac{2}{\sqrt{c_0}}(W_M + c_0 M^2) + \sqrt{c_0}(\lambda')^2\Big)\omega_p. \tag{5.85}$$

This, using also **H**, follows from

$$x \in (B_p \setminus B_{p-1}) \cap \{q^h < q^u\} \cap \{q^u > \lambda\}$$
$$\implies W(h) - W(u) + c_0 q^h(q^u - q^h) \leq W(h) + c_0(q^u)^2 \leq W_M + c_0 M^2,$$

$$x \in (B_p \setminus B_{p-1}) \cap \{0 < q^u - q^h \leq \lambda'\} \cap \{q^u > \lambda\} \implies (q^u - q^h)^2 \leq (\lambda')^2.$$

Now observe that (5.77) and $B_{p-1} = B_{r-T}$ imply

$$I^2_{p-j} = 0, \quad j = 1, \ldots, p. \tag{5.86}$$

Moreover, recalling (5.76) and **H**, which imply

$$q^h \leq M\mathrm{e}^{-c_1 T} \leq q_0, \quad \text{and} \quad W(h) \leq \frac{c_0'}{2}(q^h)^2,$$

we obtain for $x \in (B_{p-j} \setminus B_{p-j-1}) \cap \{q^h < q^u\} \cap \{q^u > \lambda\}$, $j = 1, \ldots, p-1$:

$$W(h) - W(u) + c_0 q^h(q^u - q^h) \leq W(h) + c_0 q^h q^u \leq (c_0 + \frac{c_0'}{2})q^h M \leq (c_0 + \frac{c_0'}{2})M^2\epsilon^j,$$

and therefore

$$\frac{2}{\sqrt{c_0}} I^1_{p-j} \le \frac{2}{\sqrt{c_0}}(c_0 + \frac{c_0'}{2})M^2 \epsilon^j \omega_{p-j}, \quad j = 1, \ldots, p, \tag{5.87}$$

where we have observed that the estimate is also valid for $j = p$. Now (5.85) and (5.86) imply that the right-hand side of (5.81) is bounded by the constant $\tilde{C} = C\left(\frac{2}{\sqrt{c_0}}(W_0 + (c_0 + \frac{c_0'}{2})M^2) + \sqrt{c_0}(\lambda')^2\right)$ times the right hand side of (5.83). Therefore, recalling (5.84), (5.83) follows with $C^\circ = \frac{(\lambda')^2}{C}$. The constant C° depends on λ, r_0, M and on W through c_0, c_0' and W_M. The proof is complete. $\quad\square$

We are now in the position to conclude the proof of Theorem 5.2. We first show that if $\epsilon = e^{-c_1 T}$ is sufficiently small, then there is a constant $c^* > 0$, $c^* = c^*(W, M, \lambda, r_0)$ such that

$$\omega_{p-1} \ge c^* p^{n-1}, \quad \text{for } p = 1, \ldots \tag{5.88}$$

From the assumption in Theorem 5.2 this is true for $p = 1$ with any choice of $c^* \in (0, \mu_0]$. Therefore, we proceed by induction and show that, if (5.88) holds for $j = 1, \ldots, p$, then it also holds for $p + 1$. From the induction assumption we have

$$\frac{c^*}{n} p^n = c^* \int_0^p s^{n-1} ds \le c^* \sum_{j=1}^p j^{n-1} \le \sum_{j=0}^{p-1} \omega_j \Rightarrow \left(\frac{c^*}{n}\right)^{\frac{n-1}{n}} p^{n-1} \le \left(\sum_{j=0}^{p-1} \omega_j\right)^{\frac{n-1}{n}}. \tag{5.89}$$

On the other hand, assuming $T \ge r_0$, we have the obvious inequalities

$$\omega_j \le \gamma_n (r_0 + jT)^{n-1} T \le \gamma_n T^n p^{n-1}, \quad \text{for } j = 0, \ldots, p-1; \ p \ge 1,$$

where γ_n is the measure of the unit sphere in $\mathbb{R}^n$, and consequently:

$$\sum_{j=1}^p \epsilon^j \omega_{p-j} \le \gamma_n T^n p^{n-1} \sum_{j=1}^p \epsilon^j \le \gamma_n T^n \frac{e^{-c_1 T}}{1 - e^{-c_1 T}} p^{n-1} \le \frac{C^\circ}{2}\left(\frac{c^*}{n}\right)^{\frac{n-1}{n}} p^{n-1} \tag{5.90}$$

where for the last inequality, we have also assumed that T in $\epsilon = e^{-c_1 T}$ is sufficiently large. From this, (5.89) and Lemma 5.2 we obtain

$$\frac{C^\circ}{2^n}\left(\frac{c^*}{n}\right)^{\frac{n-1}{n}}(p+1)^{n-1} \le \frac{C^\circ}{2}\left(\frac{c^*}{n}\right)^{\frac{n-1}{n}}\left(\frac{p}{p+1}\right)^{n-1}(p+1)^{n-1}$$

$$\le \frac{C^\circ}{2}\left(\frac{c^*}{n}\right)^{\frac{n-1}{n}} p^{n-1} \le \omega_p \tag{5.91}$$

which, provided $c^* > 0$ is chosen so that $c^* \leq \frac{C^\circ}{2^n}\left(\frac{c^*}{n}\right)^{\frac{n-1}{n}}$, implies $\omega_p \geq c^*(p + 1)^{n-1}$ and establishes (5.88). Like C°, c^* depends on W, M, λ and r_0. Let $[\frac{r-r_0}{T}]$ the integer part of $\frac{r-r_0}{T}$. Then (5.88) implies

$$
\mathcal{L}^n(B_r \cap \{q^u > \lambda\}) \geq \mathcal{L}^n\left(B_{r_0+[\frac{r-r_0}{T}]T} \cap \{q^u > \lambda\}\right)
$$

$$
= \sum_{j=1}^{[\frac{r-r_0}{T}]+1} \omega_{j-1} \geq c^* \sum_{j=1}^{[\frac{r-r_0}{T}]+1} j^{n-1}
$$

$$
\geq \frac{c^*}{n}\left(\left[\frac{r-r_0}{T}\right]+1\right)^n = \frac{c^*}{nT^n}\left(\left[\frac{r-r_0}{T}\right]T + T\right)^n \geq \frac{c^*}{nT^n}r^n, \tag{5.92}
$$

which concludes the proof with $C = \min\{\frac{c^*}{nT^n}, \frac{\mu_0}{r_0^n}\}$. The proof of Theorem 5.2 is complete. $\qquad\square$

Remark 5.9 $(n = 1)$ We note that (5.78), (5.83) are replaced by

$$
|\lambda'|^2 \leq \int_{B_r} |\nabla(\beta^2)|\mathrm{d}x \leq \sum_{j=1}^{p} \epsilon^j \omega_{p-j} + \omega_p.
$$

Therefore, taking T sufficiently large we obtain (5.88) with $n = 1$. From this (5.92) for $n = 1$ follows and thus the theorem also holds for $n = 1$.

5.4   Pointwise Estimates via the Density Theorem

The lack of the classical maximum principle is a major obstruction to the derivation of pointwise bounds for solutions of systems. We show below that, in the case of minimizers, this obstruction can be handled to a certain extent by energy considerations. Indeed, a direct corollary of the Density Theorem 5.2 is the following pointwise estimate

Theorem 5.3 *Assume W and $u \in C^\beta(\mathcal{O}; \mathbb{R}^m)$ as in Theorem 5.2. Assume moreover that either*

1. $\{W = 0\} = \{a\}$, *or*
2. $d(u(\mathcal{O}), \mathcal{Z}_a) \geq \delta$, *for some $\delta > 0$, where $\mathcal{Z}_a := \{W = 0\} \setminus \{a\}$, and $d(\cdot, \cdot)$ stands for the Euclidean distance.*

Then, given $q \in (0, M)$, there is $r_q > 0$ such that

$$
B_{r_q}(x_0) \subset \mathcal{O} \quad implies \quad |u(x_0) - a| < q.
$$

The radius r_q depends on W and M in case 1. *and also on δ in case* 2.

Remark 5.10 The major hypothesis of the theorem is that on a certain subset $\mathcal{O}$ in its domain, the solution u is bounded away from all but one minimum of W. A typical situation where such an assumption can be verified is when symmetry conditions are satisfied. We refer the reader to Chap. 6 and to the concept of *positive map* for this purpose.

Proof (Theorem 5.3) Given $q \in (0, M)$, assume that

$$|u(x_0) - a| \geq q. \tag{5.93}$$

Then the continuity of u implies that the condition (5.33) in Theorem 5.2 is satisfied for $\lambda = \frac{q}{2}$ with $r_0 = \frac{q}{2M}$ and $\mu_0 = \mathcal{L}^n(B_{r_0})$. Therefore the Density Theorem implies

$$\mathcal{L}^n(B_r(x_0) \cap \{|u - a| > q/2\}) \geq Cr^n, \quad \text{for} \quad r \geq r_0. \tag{5.94}$$

Observe now that, given $p \in (0, M)$, assumptions 1. and 2. imply respectively

$$0 < w_p := \begin{cases} \min\{W(z) : |z - a| \in [p, M]\}, \\ \min\{W(z) : |z - a| \in [p, M], \ d(z, \mathcal{Z}_a) \geq \delta\}. \end{cases} \tag{5.95}$$

From this, (5.94), and the basic estimate in Lemma 5.1 it follows that

$$w_{\frac{q}{2}} Cr^n \leq J_{B_r}(u) \leq C_0 r^{n-1}$$

which is impossible for $r > \frac{C_0}{Cw_{\frac{q}{2}}}$. Therefore, if we set $r_q = \frac{2C_0}{Cw_{\frac{q}{2}}}$, then $B_{r_q}(x_0) \subset \mathcal{O}$ is incompatible with (5.93). The proof is concluded. $\square$

Theorem 5.3 implies a *Liouville* theorem for minimizers

Corollary 5.2 *Assume W and u are as in Theorem 5.3 and assume that $\mathcal{O} = \mathbb{R}^n$. Then*

$$u \equiv a.$$

Proof We have $B_{r_q}(x) \subset \mathbb{R}^n$ for all $x \in \mathbb{R}^n$ and all $q \in (0, M)$. Therefore, Theorem 5.3 implies

$$|u(x) - a| < q, \quad \text{for all} \quad x \in \mathbb{R}^n, \ q \in (0, M).$$

$\square$

Under the assumptions of Corollary 5.2, a minimizer reduces to a constant if $\mathcal{O} = \mathbb{R}^n$. For a general open set $\mathcal{O}$, a good example of which is the half-space, we have

Proposition 5.2 *Assume W and u are as in Theorem 5.3, and $\alpha = 2$. Then there exist constants $k_0, K_0 > 0$ depending on W and M (and on δ if 2. holds in Theorem 5.3) such that*

$$|u(x) - a| \leq K_0 e^{-k_0 d(x, \partial \mathcal{O})}, \quad \text{for } x \in \mathcal{O}. \tag{5.96}$$

Proof With q_0 the constant in **H** and r_{q_0} the corresponding radius of the ball in Theorem 5.3, set $k_0 = c_1(c_0, r_{q_0})$, $K_0 = Me^{2k_0 r_{q_0}}$, where c_1 is as in (5.59). It suffices to prove the inequality (5.96) for $x \in \mathcal{O}$ with $d(x, \partial \mathcal{O}) > 2r_{q_0}$, since

$$|u(x) - a| \leq M \leq K_0 e^{-k_0 d(x, \partial \mathcal{O})}, \quad \text{for } d(x, \partial \mathcal{O}) \leq 2r_{q_0}.$$

If $d(x, \partial \mathcal{O}) > 2r_{q_0}$ we have $r(x) := d(x, \partial \mathcal{O}) - r_{q_0} > r_{q_0}$, and Theorem 5.3 implies

$$|u - a| \leq q_0, \quad \text{on } \partial B_{r(x)}(x).$$

Therefore, Lemma 4.4 for $A = B_{r_{q_0}}(x)$ and (5.59) yield

$$|u(x) - a| \leq \varphi(x, r(x)) q_0 = \Phi(0; r(x)) q_0$$
$$\leq q_0 e^{-k_0 r(x)} = q_0 e^{k_0 r_{q_0}} e^{-k_0 d(x, \partial \mathcal{O})} \leq K_0 e^{-k_0 d(x, \partial \mathcal{O})}.$$

The proof is complete. $\square$

5.5   The Proof of Theorem 5.3 Without the Density Estimate

In this section we present the original proof of Theorem 5.3 for $\alpha = 2$. This proof does not rely on the density estimates in Theorem 5.2 and can be generalised to cover the case of degenerate potentials with a C^∞ contact with 0 at a [15]. The regularity hypothesis on W in Theorem 5.3 implies that, by reducing the value of the constant q_0 if necessary, we can assume that

$$\begin{cases} W(a + q\xi) < W(a + q'\xi), \text{ for } q \in [0, q_0], \ q < q', \ |\xi| = 1, \\ \text{provided } a + q'\xi \in \{|z - a| \leq M\} \cap \{d(z, \mathscr{Z}_a) \geq \delta\}. \end{cases} \tag{5.97}$$

Lemma 5.3 *Assume W as in Theorem 5.3, and let $\tilde{u} : \mathcal{O} \to \mathbb{R}^m$ a $C^{0,1}$ map (not necessarily a solution) that satisfies the bounds $|\tilde{u} - a| < M$, $|\nabla \tilde{u}| < M$ on $\mathcal{O}$. Given $r_0 > 0$, $\eta > 0$ and $\bar{q} \in (0, q_0]$, assume that, for some x_0 and $r \geq r_0 + \eta$, $B_r(x_0) \subset \mathcal{O}$ and*

$$q^{\tilde{u}} = |\tilde{u} - a| \leq \bar{q}, \quad \text{on } B_r(x_0). \tag{5.98}$$

Then there exists a $C^{0,1}$ map $v : \mathcal{O} \to \mathbb{R}^m$ that coincides with $\tilde{u}$ on $\mathcal{O} \setminus B_r(x_0)$ and satisfies

$$J_{B_r}(\tilde{u}) - J_{B_r}(v) \geq k\mathcal{L}^n(B_{r-\eta} \cap \{\bar{q} = q^{\tilde{u}}\}), \tag{5.99}$$

where $k > 0$, $k = k(W, r_0, \eta, \bar{q})$ is a constant that does not depend on $r \geq r_0 + \eta$.

Proof Let v be the solution to

$$\begin{cases} \Delta v = W_u(v), & \text{in } B_r(x_0), \\ v = \tilde{u}, & \text{on } \partial B_r(x_0), \end{cases} \tag{5.100}$$

extended by $v = \tilde{u}$ on $\mathcal{O} \setminus B_r(x_0)$. That such a solution exists follows by minimization of $\int_{B_r(x_0)} \left(\frac{1}{2}|\nabla u|^2 + W(u)\right)dx$ with $v = \tilde{u}$ on $\partial B_r(x_0)$. Moreover by Theorem 4.3, it is unique and satisfies the estimate (5.98): $|v - a| \leq \bar{q}$ on $B_r(x_0)$. Proceeding exactly as in the proof of Theorem 4.3 (with $B_r = A$ and $\tilde{u}, v$ replacing v, u respectively), we arrive at

$$J_{B_r(x_0)}(\tilde{u}) - J_{B_r(x_0)}(v) \geq \frac{c_0}{2} \int_{B_r(x_0)} |\tilde{u} - v|^2 dx \geq \frac{c_0}{2} \int_{B_{r-\eta}(x_0)} |\tilde{u} - v|^2 dx. \tag{5.101}$$

On the other hand, assuming $A = B_r$ and arguing as in the proof of Lemma 4.4, we obtain

$$|v(x) - a| \leq \sqrt{\varphi(x)}\bar{q}, \quad x \in B_r(x_0)$$

and therefore in $B_{r-\eta}(x_0) \cap \{|\tilde{u} - a| = \bar{q}\}$ one has

$$|\tilde{u} - v| \geq ||\tilde{u} - a| - |v - a|| \geq (1 - \sqrt{\varphi(x)})\bar{q}. \tag{5.102}$$

To estimate $1 - \sqrt{\varphi(x)}$, we recall that $\varphi(x) = \varphi(x; r) = \Phi(|x - x_0|; r)$ with $\Phi(s; r)$ strictly increasing in $(0, r)$. Moreover, by Lemma A.1,

$$r_1 < r_2, \quad t \in (0, r_1) \implies \Phi(r_1 - t; r_1) > \Phi(r_2 - t; r_2).$$

therefore, for $x \in B_{r-\eta}(x_0)$, we have

$$\varphi(x, r) = \Phi(|x - x_0|; r) \leq \Phi(r - \eta; r) \leq \Phi(r_0; r_0 + \eta).$$

It follows that

$$1 - \sqrt{\varphi(x)} \geq 1 - \sqrt{\Phi(r_0; r_0 + \eta)}, \quad \text{for } x \in B_{r-\eta}, \quad r \geq r_0 + \eta.$$

From this, (5.102) and (5.101) we obtain

$$J_{B_r}(\tilde{u}) - J_{B_r}(v) \geq \frac{c_0}{2}\bar{q}^2(1 - \sqrt{\Phi(r_0; r_0 + \eta)})^2 \mathcal{L}^n(B_{r-\eta} \cap \{q^{\tilde{u}} = \bar{q}\})$$

that concludes the proof with $k = \frac{c_0}{2}\bar{q}^2(1 - \sqrt{\Phi(r_0; r_0 + \eta)})^2$. $\square$

Lemma 5.3 suggests a strategy for the proof of Theorem 5.3. From Theorem 4.1 and Lemma 4.4 we could conclude that

$$q^u \leq \sqrt{\varphi}\bar{q} < \bar{q}, \quad \text{for} \ \ x \in B_r$$

provided we knew that the minimizer u satisfies the boundary condition

$$q^u \leq \bar{q}, \quad \text{on} \ \ \partial B_r.$$

In general, u will not satisfy this condition and so we cannot conclude that $q^u < \bar{q}$ in B_r. However, with the quantitative estimate (5.99) at hand, we are allowed to *invest* a certain amount of energy to *deform* u into a map $\tilde{u}$ that satisfies (5.98) and, if the energy spent in the deformation is less than the difference $J_{B_r}(\tilde{u}) - J_{B_r}(v)$ estimated by (5.99) we reach a contradiction with the minimality of u and we can conclude the strict inequality $q^u < \bar{q}$ on B_r. We show that this is indeed the case, provided r is sufficiently large. We need the estimate (5.103) below

Lemma 5.4 *Assume W and $u : \mathcal{O} \to \mathbb{R}^m$ are as in Theorem 5.3. Let r_0, η and $\bar{q}$ as in Lemma 5.3 and assume that $B_{r+\eta} = B_{r+\eta}(x_0) \subset \mathcal{O}$ for some $r \geq r_0 + \eta$. Then there exist $K > 0$, $K = K(W, M, \eta, \bar{q})$ independent of $r \geq r_0 + \eta$, and a $C^{0,1}$ map $\tilde{u} : \mathcal{O} \to \mathbb{R}^m$ which coincides with u in $(\mathcal{O} \setminus B_{r+\eta}) \cup \{q^u \leq \bar{q}\}$, satisfies (5.98), and*

$$J_{B_{r+\eta}}(\tilde{u}) - J_{B_{r+\eta}}(u) \leq K\mathcal{L}^n\Big((B_{r+\eta} \setminus B_{r-\eta}) \cap \{q^u > \bar{q}\}\Big). \tag{5.103}$$

Before giving the proof of the lemma we complete the proof of Theorem 5.3. Set $r_h = r_0 + 2h\eta$ for $h = 0, 1, \dots$ and let $\tilde{u}_h$ the map $\tilde{u}$ given by Lemma 5.4 for $r = r_h + \eta$, $h = 0, 1, \dots$. Let v_h the map v given by Lemma 5.3 with $\tilde{u} = \tilde{u}_h$ and $r = r_h + \eta$. Then, the minimality of u implies

$$0 \geq J_{B_{r_h+2\eta}}(u) - J_{B_{r_h+2\eta}}(v_h)$$

$$= J_{B_{r_h+2\eta}}(u) - J_{B_{r_h+2\eta}}(\tilde{u}_h) + J_{B_{r_h+2\eta}}(\tilde{u}_h) - J_{B_{r_h+2\eta}}(v_h)$$

$$= J_{B_{r_h+2\eta}}(u) - J_{B_{r_h+2\eta}}(\tilde{u}_h) + J_{B_{r_h+\eta}}(\tilde{u}_h) - J_{B_{r_h+\eta}}(v_h).$$

This together with (5.103) and (5.99) yield,

$$k\mathcal{L}^n(B_{r_h} \cap \{q^u > \bar{q}\}) \leq K\mathcal{L}^n\Big(B_{r_h+2\eta} \setminus B_{r_h}) \cap \{q^u > \bar{q}\}\Big), \quad \text{for} \ \ h = 0, \dots \tag{5.104}$$

If we set $\omega_h := \mathscr{L}^n(B_{r_h} \cap \{q^u > \bar{q}\})$, for $h = 0, 1, \ldots$ we can recast (5.104) as

$$\frac{k}{K}\omega_h \leq \omega_{h+1} - \omega_h, \qquad \text{for} \quad h = 0, 1, \ldots \tag{5.105}$$

We now show that, if $d(x_0, \partial\mathcal{O})$ is sufficiently large, this inequality leads to a contradiction with the assumption

$$q^u(x_0) = |u(x_0) - a| \geq q > 0,$$

for some $x_0 \in \mathcal{O}$. Set

$$\bar{q} = \min\left\{q_0, \frac{q}{2}\right\},$$

then the gradient bound $|\nabla u| \leq M$ implies

$$q^u > \bar{q}, \qquad \text{for} \quad x \in B_{r_0}(x_0), \quad r_0 = \frac{q - \bar{q}}{M},$$

and we have $\omega_0 = \mathscr{L}^n(B_{r_0}) > 0$. This and (5.105) implies $\omega_h \geq (1 + \frac{k}{K})^h \omega_0$, and therefore

$$\omega_0 \frac{k}{K}\left(1 + \frac{k}{K}\right)^{h+1} \leq \omega_{h+1} - \omega_h \leq \mathscr{L}^n(B_{r_h+2\eta} \setminus \overline{B_{r_h}}), \qquad \text{for} \quad h = 0, 1, \ldots.$$

This inequality cannot hold for large h. Indeed, the left-hand side grows exponentially in h, while the right-hand side only algebraically. Let $\bar{h} \geq 0$ the smallest value of h that violated this inequality and set $r_q := r_0 + 2\bar{h}\eta$. Then we conclude that

$$B_{r_q}(x_0) \subset \mathcal{O} \qquad \text{implies} \qquad |u(x_0) - a| < q.$$

For fixed W, M and η, the constants k and K depend only on r_0 and $\bar{q}$. Since $\bar{q} = \min\{q_0, \frac{q}{2}\}$ and $r_0 = \frac{q-\bar{q}}{M}$ depend on q_0, q and M, we have that for fixed W, M and η, r_q depends only on q and eventually on δ if q_0 has been reduced in order for (5.97) to hold. The proof of Theorem 5.3 is complete.

Proof (Lemma 5.4) Set $p^u(x) := q^u(x) - (q^u(x) - \bar{q})^+$ and define $\tilde{u}$ by

$$\tilde{u} = a + q^{\tilde{u}}\mathbf{n}^u,$$

with

$$q^{\tilde{u}}(x) := \begin{cases} q^u(x), & \text{for } x \in \mathcal{O} \setminus B_{r+\eta}, \\ p^u(x) + g(x)(q^u(x) - \bar{q})^+, & \text{for } x \in B_{r+\eta}, \end{cases}$$

and

$$g(x) := \begin{cases} 0, & \text{for } x \in B_r, \\ \frac{|x-x_0|-r}{\eta}, & \text{for } x \in \overline{B_{r+\eta}} \setminus B_r. \end{cases}$$

Inspecting (5.106) we see that $\tilde{u}$ is continuous, that $q^{\tilde{u}} \leq q^u \leq M$ and that $\nabla \tilde{u}$ is defined a.e. and bounded. To prove (5.103), we begin by estimating the difference of energy between $\tilde{u}$ and u in B_r. The definition of $q^{\tilde{u}}$ implies

$$J_{B_r}(\tilde{u}) - J_{B_r}(u) = J_{B_r \cap \{q^u > \bar{q}\}}(\tilde{u}) - J_{B_r \cap \{q^u > \bar{q}\}}(u)$$

and

$$q^{\tilde{u}} = \bar{q}, \quad \nabla q^{\tilde{u}} = 0, \quad \text{on} \quad B_r \cap \{q^u > \bar{q}\}. \tag{5.106}$$

From (5.35), (5.106) and (5.97) it follows

$$J_{B_r \cap \{q^u > \bar{q}\}}(\tilde{u}) - J_{B_r \cap \{q^u > \bar{q}\}}(u)$$

$$= \int_{B_r \cap \{q^u > \bar{q}\}} \left(\frac{1}{2}(-|\nabla q^u|^2 + (\bar{q}^2 - (q^u)^2)|\nabla \mathbf{n}^u|^2) + W(a + \bar{q}\mathbf{n}^u) - W(a + q^u \mathbf{n}^u) \right) dx$$

$$\leq 0$$

and therefore

$$J_{B_r}(\tilde{u}) \leq J_{B_r}(u), \quad \text{on} \quad B_r. \tag{5.107}$$

It remains to evaluate the difference of energy in $B_{r+\eta} \setminus B_r$. The definition of $q^{\tilde{u}}$ in (5.106) implies that

$$|\nabla q^{\tilde{u}}| \leq |\nabla g|(q^u - \bar{q}) + |g||\nabla q^u| \leq \left(\frac{1}{\eta} + 1 \right) M, \quad \text{a.e. on} \quad (B_{r+\eta} \setminus B_r) \cap \{q^u > \bar{q}\}.$$

From this and $q^{\tilde{u}} \leq q^u$ it follows that

$$|\nabla \tilde{u}|^2 - |\nabla u|^2 \leq \left(\frac{1}{\eta} + 1 \right)^2 M^2, \quad \text{a.e. on} \quad (B_{r+\eta} \setminus B_r) \cap \{q^u > \bar{q}\}. \tag{5.108}$$

On the other hand, (5.3) yields

$$W(a + q^{\tilde{u}}\mathbf{n}^u) - W(a + q^u \mathbf{n}^u) \leq W(a + q^{\tilde{u}}\mathbf{n}^u) \leq W_M, \quad \text{on} \quad (B_{r+\eta} \setminus B_r) \cap \{q^u > \bar{q}\}.$$

From this, (5.108) and (5.35), recalling also that $q^{\tilde{u}} \leq q^u$, we obtain

$$J_{(B_{r+\eta} \setminus B_r) \cap \{q^u > \bar{q}\}}(\tilde{u}) - J_{(B_{r+\eta} \setminus B_r) \cap \{q^u > \bar{q}\}}(u) \leq K \mathcal{L}^n((B_{r+\eta} \setminus B_r) \cap \{q^u > \bar{q}\})$$

$$(5.109)$$

where we have set $K = \frac{1}{2}M^2(1 + \frac{1}{\eta})^2 + W_M$. From this estimate and (5.107) we deduce that

$$\begin{aligned}
J_{B_{r+\eta}}(\tilde{u}) - J_{B_{r+\eta}}(u) &\leq J_{(B_{r+\eta} \setminus B_r) \cap \{q^u > \bar{q}\}}(\tilde{u}) - J_{(B_{r+\eta} \setminus B_r) \cap \{q^u > \bar{q}\}}(u) \\
&\leq K \mathcal{L}^n((B_{r+\eta} \setminus B_r) \cap \{q^u > \bar{q}\}) \\
&\leq K \mathcal{L}^n((B_{r+\eta} \setminus B_{r-\eta}) \cap \{q^u > \bar{q}\}).
\end{aligned}$$

The proof is complete. $\qquad\qquad\qquad\qquad\qquad\qquad\qquad\qquad\qquad\qquad\qquad\square$

In certain cases, for instance in the context of equivariant maps that we consider in Chap. 6, the following slight generalization of Theorem 5.3 is useful.

Corollary 5.3 *Assume $W : \mathbb{R}^m \to \mathbb{R}$ and $u : \mathcal{O} \subset \mathbb{R}^n \to \mathbb{R}^m$ are as in Theorem 5.3. Assume moreover that either*

1. $\{W = 0\} = \{a\}$, or
2. $d(u(\mathcal{O}), \mathcal{L}_a) \geq \delta$, for some $\delta > 0$.

Then, given $q \in (0, M)$, and $x_0, x_1 \in \mathcal{O}$, there is $r_{q,\rho} > 0$, $\rho \equiv |x_0 - x_1|$, such that

$$B_{r_{q,\rho}}(x_0) \subset \mathcal{O} \quad \text{implies} \quad |u(x_1) - a| < q.$$

The radius $r_{q,\rho}$ depends on W and M in case 1. and also on δ in case 2.

Proof In both proofs of Theorem 5.3 given above we have related the value of r_0 to the value of q by setting $r_0 := \frac{q-\bar{q}}{M}$. But, all the arguments developed in the proofs work exactly well when we define $r_0 := \frac{q-\bar{q}}{M} + |x_0 - x_1|$. $\qquad\qquad\square$

Remark 5.11 A situation where Corollary 5.3 finds a natural application is when the minimizer u is equivariant with respect to a reflection group $\mathscr{G}$ generated by reflections with respect to hyperplanes through x_0. In this case, for the validity of Corollary 5.3, it is necessary to check that, like u, also all the maps σ, h or $\tilde{u}, v$ considered in the proof of Theorem 5.3 are $\mathscr{G}$-equivariant. To see that this is indeed the case, we recall that $u = a + q^u \mathbf{n}^u$ is $\mathscr{G}$-equivariant if and only if $\mathbf{n}^u$ is equivariant and q^u is $\mathscr{G}$-invariant. Therefore, since all the functions $f = \sigma, h, \ldots$ we have introduced in the proofs satisfy the condition $\mathbf{n}^f = \mathbf{n}^u$, it suffices to verify that $q^f = q^\sigma, q^h, \ldots$ are $\mathscr{G}$-invariant. This follows from the fact that q^f is either radially symmetric like $q^f = q^\sigma, \varphi$, or is defined via operations like $q^f = \min\{q^\sigma, q^u\}$ that range inside the class of maps invariant under $\mathscr{G}$. In certain cases $\mathscr{G}$ is a subgroup of a reflection group $\widetilde{\mathscr{G}}$ and u is a minimizer in a class of $\widetilde{\mathscr{G}}$-equivariant maps. In this situation, Corollary 5.3 can be applied after verifying that the constructions developed in $B_{r_{q,\rho}}(x_0)$ in the proofs of Theorem 5.3 extend by equivariance to $g B_{r_{q,\rho}}(x_0)$ for $g \in \widetilde{\mathscr{G}}$. A necessary and sufficient condition for this is that, for each $g \in \widetilde{\mathscr{G}}$ one has that either $g B_{r_{q,\rho}}(x_0) = B_{r_{q,\rho}}(x_0)$, or $g B_{r_{q,\rho}}(x_0) \cap B_{r_{q,\rho}}(x_0) = \emptyset$. An equivalent condition is that

$$B_{r_{q,\rho}}(x_0) \cap \pi_\gamma = \emptyset,$$

for all projections $\gamma \in \widetilde{\mathscr{G}} \setminus \mathscr{G}$, where π_γ is the plane associated to γ.

5.6 Linking

In this subsection we assume hypothesis **H** (cf. (5.32)) and moreover we take the zero set of W to be a finite set of points, the *phases*

$$\{W = 0\} = \{a_1, \ldots, a_N\} \subset \mathbb{R}^m, \tag{5.110}$$

$\Omega \subset \mathbb{R}^n$, open, connected, with possibly nonempty C^1 boundary. We assume that $v_\epsilon : \Omega \to \mathbb{R}^m$, $\epsilon \in (0, \epsilon_0)$, $\epsilon_0 > 0$ is a minimizer of

$$J_\epsilon(v) = \int_\Omega \left(\frac{\epsilon}{2} |\nabla_y v|^2 + \frac{1}{\epsilon} W(v) \right) dy, \tag{5.111}$$

and that, as $\epsilon \to 0$, v_ϵ converges in $L^1(\Omega; \mathbb{R}^m)$ to a step map

$$v_\epsilon \xrightarrow{L^1} v_0 = \sum_{i=1}^{N} a_i \mathbb{1}_{\Omega_i}, \tag{5.112}$$

that corresponds to a minimizing partition $\mathscr{P} = \{\Omega_j\}_{j=1}^{N}$ of Ω by disjoint sets of finite perimeter, $\mathscr{H}^n(\Omega \setminus \bigcup_{i=1}^{N} \Omega_i) = 0$. $\mathscr{P}$ is associated to an appropriate perimeter functional

$$E(\mathscr{P}) = \sum_{0 < i < j \leq N} \sigma_{ij} \mathscr{H}^{n-1}(\partial \Omega_i \cap \partial \Omega_j) \tag{5.113}$$

that represents the *energy* of the partition, with $\partial \mathscr{P} = \bigcup_{0 < i < j \leq N}(\partial \Omega_i \cap \partial \Omega_j)$, and σ_{ij} the surface tension coefficients. We can and in fact will assume that $\partial \Omega_i \subset \partial \mathscr{P} \cup \partial \Omega$. By classical results of Almgren's improved and simplified in White [24] for minimizing partitions with surface tension coefficients $\sigma_{i,j}$ satisfying the strict triangle inequality (cf. Remark 5.13), Ω_j can be taken open with $\partial \Omega_j$ real analytic except possibly for a singular part with Hausdorff dimension at most $n - 2$. We assume a uniform bound for v_ϵ

$$\|v_\epsilon\|_{L^\infty(\Omega;\mathbb{R}^m)} < M', \tag{5.114}$$

where $M' > 0$ is independent of $\epsilon \in (0, \epsilon_0)$. For

$$\gamma \in (0, \min_{\substack{i \neq j \\ i \leq i,j \leq N}} |a_i - a_j|) \tag{5.115}$$

we define the *diffuse interface* as the set

$$I_{\gamma,\epsilon} := \{y \in \Omega : \min_{j=1,\dots,N} |v_\epsilon(y) - a_j| \geq \gamma\}. \tag{5.116}$$

We now show that the density estimates in Theorem 5.2 refine significantly the L^1 convergence of v_ϵ to v_0. In the following $d(x, y)$ stands for the Euclidean distance, and if $V \subset \mathbb{R}^n$ is an open set and $\delta > 0$ a small number we denote $V^\delta := \{x \in V : d(x, \partial V) > \delta\}$.

Proposition 5.3 *Assume W, v_ϵ and v_0 as before. Given V, open, bounded, with $\overline{V} \subset \Omega$, we have*

$$\max \left\{ d(y, \partial \mathscr{P}) : y \in I_{\gamma,\epsilon} \cap V^{C_\gamma(g(\epsilon))^{1/n}} \right\} \leq C_\gamma(g(\epsilon))^{1/n}, \tag{5.117}$$

where $g(\epsilon) = \|v_\epsilon - v_0\|_{L^1(V;\mathbb{R}^m)}$ and C_γ depends on γ. Moreover, if Ω_j, for some j, is open and with piecewise smooth boundary, and if $2\gamma < \min_{i \neq j} |a_i - a_j|$, then we have for $\epsilon > 0$ small

$$|v_\epsilon(y) - a_j| < \gamma \text{ on } \Omega_{j,\epsilon} = \left\{ y \in \Omega_j \cap V : d(y, \partial\Omega_j \cup \partial V) > C_\gamma(g(\epsilon))^{1/n} \right\}. \tag{5.118}$$

Remark 5.12 We do not assume that $\mathscr{P}$ is a minimizing partition.

Proof (Blow-Up, cf. [5, Theorem 2]) We proceed by contradiction. Thus, suppose there exist sequences $\{\epsilon_k\}$, $\{C_k\}$ and $\{y_k\} \subset V$ such that $\epsilon_k \to 0$, $C_k \to +\infty$ and

$$\min_j |v_{\epsilon_k}(y_k) - a_j| \geq \gamma, \quad d(y_k, \partial\mathscr{P} \cup \partial V) > C_k g(\epsilon_k)^{\frac{1}{n}}, \quad k = 1, 2, \dots \tag{5.119}$$

By passing to a subsequence we may assume that $\{y_k\} \subset \Omega_h \cap V$ for some $h \in \{1, \dots, N\}$. The minimality of $v_{\epsilon_k}(y)$ implies the minimality of $u_k(x) := v_{\epsilon_k}(\epsilon_k x)$ for the rescaled functional

$$J_{\Omega^k}(u) = \int_{\Omega^k} \left(\frac{1}{2}|\nabla_x u|^2 + W(u)\right) dx,$$

where $\Omega^k := \{x \in \mathbb{R}^n : x = y/\epsilon_k, \ y \in \Omega\}$. Since the bound (5.114) is equivalent to $\|u_k\|_{L^\infty(\Omega^k;\mathbb{R}^m)} \leq M'$, if W is sufficiently smooth, elliptic theory yields a gradient bound

$$|\nabla u_k| \leq M,$$

for some $M > 0$ independent of k. If W satisfies the assumptions $\mathbf{H}$ (cf. (5.32)) corresponding to $0 < \alpha \leq 1$, we only have that u_k is bounded in C^β (cf. Remark 5.2) for some $\beta \in (0, 1)$ with a bound independent of k. In any case, (5.119) implies the existence of $r_0 > 0$ such that

$$|u_k(x) - a_h| \geq \frac{\gamma}{2}, \quad x \in B_{r_0}(y_k/\epsilon_k), \ k = 1, 2, \ldots$$

This and the density estimate (5.34) imply that there is $c^* > 0$ independent of k such that

$$\mathscr{L}^n\left(\{|u_k - a_h| \geq \frac{\gamma}{2}\} \cap B_r(y_k/\epsilon_k)\right) \geq c^* r^n, \ r \geq r_0, \ B_r(y_k/\epsilon_k) \subset \Omega^k, \ k = 1, 2, \ldots$$

or equivalently

$$\mathscr{L}^n\left(\{|v_{\epsilon_k} - a_h| \geq \frac{\gamma}{2}\} \cap B_{\epsilon_k r}(y_k)\right) \geq c^*(\epsilon_k r)^n, \ r \geq r_0, \ B_{\epsilon_k r}(y_k) \subset \Omega, \ k = 1, 2, \ldots$$

Now choose $r = r_k$ by setting

$$\epsilon_k r_k = \left(\frac{4g(\epsilon_k)}{c^*\gamma}\right)^{\frac{1}{n}}, \ k = 1, 2, \ldots$$

Since $C_k \to +\infty$, for k sufficiently large $\epsilon_k r_k < C_k g(\epsilon_k)^{\frac{1}{n}}$ and therefore (5.119) implies $B_{\epsilon_k r_k}(y_k) \subset V \cap \Omega_h$. It follows that

$$g(\epsilon_k) = \|v_{\epsilon_k} - v_0\|_{L^1(V;\mathbb{R}^m)} \geq \int_{V \cap \Omega_h} |v_{\epsilon_k}(y) - a_h|dy \geq \int_{B_{\epsilon_k r_k}(y_k)} |v_{\epsilon_k}(y) - a_h|dy$$

$$\geq \frac{\gamma}{2}c^*(\epsilon_k r_k)^n = 2g(\epsilon_k).$$

This contradiction proves (5.117). Next we establish (5.118). Let

$$\Omega_\epsilon = \{y \in V \subset\subset \Omega : d(y, \partial\mathscr{P} \cup \partial V) > C_\gamma (g(\epsilon))^{1/n}\}$$

From (5.117) it follows that, for each $y \in \Omega_\epsilon$, there exists $a(y) \in \{a_1, \ldots, a_N\}$ such that

$$|v_\epsilon(y) - a(y)| < \gamma, \quad y \in \Omega_\epsilon.$$

From the hypothesis on γ it follows that the mapping $\Omega_\epsilon \ni y \mapsto a(y)$ is continuous. Indeed, if $\{y_k\} \subset \Omega_\epsilon$ is a sequence that converges to some $y \in \Omega_\epsilon$, the continuity of v_ϵ, provided k is sufficiently large, implies

$$|a(y_k) - a(y)| \leq |v_\epsilon(y_k) - a(y_k)| + |v_\epsilon(y_k) - v_\epsilon(y)| + |v_\epsilon(y) - a(y)|$$

$$\leq 2\gamma + |v_\epsilon(y_k) - v_\epsilon(y)| \leq \min_{i \neq j} |a_i - a_j|.$$

This implies $a(y_k) = a(y)$ for k large and so $a(y)$ is constant on each connected component of Ω_ϵ. On the other hand, by (5.112), along a subsequence we have $\lim_{\epsilon \to 0} v_\epsilon(y) = a_j$ a.e. for $y \in \Omega_{j,\epsilon}$ and therefore $a(y) = a_j$ on $\Omega_{j,\epsilon}$ for $\epsilon > 0$ small. The proof of Proposition 5.3 is complete. $\qquad\square$

Remark 5.13 For global minimizers, if Ω is bounded, we have the well-known and easy to establish estimate $J_\epsilon(v_\epsilon) < C$ from which, via

$$\frac{w_\gamma}{\epsilon} \mathscr{L}^n(I_{\gamma,\epsilon}) \leq \frac{1}{\epsilon} \int_\Omega W(v_\epsilon) dy \leq C,$$

with $w_\gamma := \min\{w(z) : \min_V |z - a_j| \geq \gamma\}$, we obtain $\mathscr{L}^n(I_{\gamma,\epsilon}) < C\epsilon$. This in turn suggests that

$$\max\{d(y, \partial P) : y \in I_{\gamma,\epsilon} \cap V\} \leq C_\gamma \epsilon. \tag{5.120}$$

The estimate (5.120) was established by Cecon et al. [8] for a special scalar problem, and should follow for the general scalar bistable nonlinearity from work of Caffarelli and Córdoba [6, Proposition 1]. In the equivariant class that we are considering in Chap. 6, this estimate is valid and actually holds all the way up to the boundary $\partial \Omega$. We refer the reader to (5.132) and to (1.32), and also to (6.29). Baldo [4] (see also Alberti [1]) studied the problem with mass constraints, for Ω bounded. He established that J_ϵ Γ-converges to E, that the sequence of minimizers $\{v_\epsilon\}$ under (5.114) is relatively compact in $L^1(\Omega; \mathbb{R}^m)$, and that the partition $\mathscr{P}$ in (5.112) minimizes the perimeter functional (5.113) with

$$\sigma_{ij} = d_W(a_i, a_j), \tag{5.121}$$

where $d_W(\cdot, \cdot)$ stands for the geodesic distance defined by

$$d_W(z_i, z_j) =$$

$$\inf \left\{ \int_0^1 \sqrt{2W(\zeta(s))} |\zeta'(s)| ds, \ \zeta \in C^1([0, 1]; \mathbb{R}^m), \ \zeta(0) = z_i, \ \zeta(1) = z_j \right\}.$$

$$\tag{5.122}$$

We recall from [24] (see Sect. 3.4) that to establish the existence of a minimizing partition for (5.113) the surface tension coefficients need to satisfy the triangle inequality $\sigma_{ij} \le \sigma_{ik} + \sigma_{kj}$, while to obtain the smoothness of the partition one needs the strict inequalities $\sigma_{ik} < \sigma_{ij} + \sigma_{jk}$. Alternatively, one can consider minimizers of (5.111) subject to Dirichlet conditions $v_\epsilon = \phi_\epsilon$ on $\partial\Omega$ with $\phi_\epsilon \to \phi : \partial\Omega \to \{a_1, \ldots, a_N\}$, where ϕ determines a partition $B = \{B_j\}_{j=1}^N$ of disjoint sets of $\partial\Omega$ analogously defined and require $\mathscr{P}$ to satisfy the Dirichlet conditions B.

5.7 A Lower Bound for the Potential Energy

We consider entire solutions of

$$\Delta u - W_u(u) = 0, \quad u : \mathbb{R}^n \to \mathbb{R}^m, \tag{5.123}$$

$u \in W_{\mathrm{loc}}^{1,2}(\mathbb{R}^n; \mathbb{R}^m) \cap L^\infty(\mathbb{R}^n; \mathbb{R}^m)$. We have seen in Sect. 3.2 that for arbitrary continuous W, $W \ge 0$, and any solution $u(x)$ of (5.123) we have the monotonicity formula

$$\frac{\mathrm{d}}{\mathrm{d}r}\left(r^{-(n-2)} J_{B_r}(u)\right) \ge 0, \quad \text{for } r > 0, \tag{5.124}$$

which is a consequence of the algebraic structure of (5.123). An immediate corollary is the lower bound

$$J_{B_r(x_0)}(u) \ge cr^{n-2}, \text{ for } r \ge 0, \tag{5.125}$$

with $c = J_{B_1(x_0)}(u)$, and $r \ge 1$ (cf. (3.17)). In this generality of potentials and solutions, (5.125) is sharp, as has been established by Farina [12] for the potential $W(u) = \frac{1}{4}(1 - |u|^2)^2$. On the other hand, for potentials with a finite set of global minima

$$\{W = 0\} = \{a_1, \ldots, a_N\} \subset \mathbb{R}^m, \quad N \ge 2, \tag{5.126}$$

and for minimal bounded u, we have the following improvement.

Proposition 5.4 *Let W satisfy H (cf. (5.32)) and (5.126) above, and let $u : \mathbb{R}^n \to \mathbb{R}^m$ be nonconstant and minimal, with*

$$\|u\|_{L^\infty(\mathbb{R}^n;\mathbb{R}^m)} < \infty, \quad \|\nabla u\|_{L^\infty(\mathbb{R}^n;\mathbb{R}^m)} < \infty. \tag{5.127}$$

Then

$$\int_{B_r(x_0)} W(u)\mathrm{d}x \ge C'r^{n-1}, \quad r \ge r(x_0), \tag{5.128}$$

with $C' > 0$ a constant independent of x_0.

Remark 5.14 In light of the basic estimate (5.4), (5.128) is sharp. Moreover, since the Modica estimate $|\nabla u|^2 \leq 2W(u(x))$ is generally false for solutions of (5.123), even if they are minimal (cf. Sect. 3.3), (5.128) is significantly stronger than the lower bound $J_{B_r(x_0)} \geq Cr^{n-1}$, derived in [2]. Sourdis [22] obtained independently (5.128) for $n = 2$.

Definition 5.1 Let $2\gamma_0 < \min_{i \neq j} |a_i - a_j|$, $\gamma_0 > 0$ fixed, and let $0 < \gamma < \gamma_0$. Assume that $2\gamma_0 \leq r_0$, where r_0 is as in the cut-off Lemma 4.1. Let $\delta(x) = d(u(x), \{W = 0\})$, where d is the Euclidean distance, and assume that $\gamma_0 < \sup_{\mathbb{R}^n} \delta$. We define the *diffuse interface* as

$$I_\gamma = \{x \in \mathbb{R}^n : \delta(x) \geq \gamma\}.$$

We note that $\overline{I}_{\gamma_1} \subset I_{\gamma_2}$ if $\gamma_1 > \gamma_2$.

Proposition 5.4 will be obtained as a corollary of the following lemma, which provides estimates on the size of the diffuse interface.

Lemma 5.5 *Under the hypotheses in Proposition 5.4, we have*

(i) *$I_\gamma \neq \emptyset$, I_γ unbounded, for all $\gamma \leq \gamma_0$.*
(ii) *Suppose $I_\gamma \neq \emptyset$. Then $c_1(\gamma)r^{n-1} \leq \mathscr{L}^n(I_\gamma \cap B_r(x_0)) \leq c_2(\gamma)r^{n-1}$, $r \geq r(x_0)$, $x_0 \in \mathbb{R}^n$ arbitrary, $c_i(\gamma) > 0$ independent of x_0, r.*

Remark 5.15 Uniform continuity for u is all that is required in the proof of the lower bound, and this is how the gradient bound in (5.127) is utilized. Thus the lower bound (5.128) holds also for the singular potentials $0 < \alpha < 2$ (see **H** and Remark 5.2). Assertion (ii) is independent of (i), and implies the unboundedness of I_γ if $I_\gamma \neq \emptyset$ (which holds for γ small because $u \not\equiv \text{Const}$). Assertion (i) establishes $I_{\gamma_0} \neq \emptyset$. However it requires possibly more regularity on u (cf. Theorem 4.5).

Proof Assume $I_\gamma \subset B_r$ for some $r > 0$. Then $\gamma \in (0, \gamma_0)$ and the continuity of u implies

$$|u(x) - a| < \gamma \quad \text{on } \mathbb{R}^n \setminus B_r,$$

for some $a \in \{W = 0\}$. This and Theorem 5.3 implies

$$\lim_{|x| \to +\infty} |u(x) - a| = 0,$$

and Theorem 4.1 (Maximum Principle) yields $u \equiv a$, contradicting the assumption that u is not a constant. This proves (i) (see also Theorem 4.5). Next we prove (ii). From (5.127), $x_0 \in I_\gamma \neq \emptyset$ and $a \in \{W = 0\}$ imply

$$|u(x) - a| \geq \frac{\gamma}{2} \quad \text{on } B_{r_\gamma}(x_0), \qquad r_\gamma = \frac{\gamma}{2\|\nabla u\|_{L^\infty}}.$$

Then the density estimate (5.34) yields

$$\mathscr{L}^n(B_r(x_0) \setminus A_{\gamma/2}(a)) \geq c(\gamma)r^n, \quad r \geq r_\gamma, \tag{5.129}$$

where

$$A_\gamma(a) = \{x \in \mathbb{R}^n : |u(x) - a| < \gamma\}.$$

Let $w_\gamma = \min\{W(z) : \min_j |z - a_j| \geq \gamma\} > 0$. Since the basic estimate (5.4) implies

$$w_{\gamma/2}\mathscr{L}^n(B_r(x_0) \setminus \cup_j A_{\gamma/2}(a_j)) \leq Cr^{n-1},$$

relation (5.129) yields

$$\mathscr{L}^n(B_r(x_0) \cap \cup_{a_j \neq a} A_{\gamma/2}(a_j)) \geq \frac{c(\gamma)}{2}r^n, \quad \text{for } r \geq r'_\gamma = \frac{2C}{c(\gamma)w_{\gamma/2}}.$$

It follows that

$$\mathscr{L}^n(B_r(x_0) \cap A_{\gamma/2}(a^+)) \geq \frac{c(\gamma)}{2(N-1)}r^n, \quad r \geq r'_\gamma, \tag{5.130}$$

for some $a^+ \in \{W = 0\}$, $a^+ \neq a$, and by repeating the argument with $a = a^+$ we obtain $a^- \in \{W = 0\}$, $a^- \neq a^+$, that satisfies the estimate corresponding to (5.130). Now we observe that the definition of γ_0 implies that

$$A_{\gamma/2}(a^+) \subset A_t(a^+), \qquad A_{\gamma/2}(a^-) \subset \mathbb{R}^n \setminus A_t(a^+), \quad t \in (\gamma, \gamma_0).$$

From these inclusions, (5.130), and the relative isoperimetric inequality (see [11, p. 190]) we obtain

$$\mathscr{H}^{n-1}(B_r(x_0) \cap \{|u(x) - a^+| = t\}) \geq c'(\gamma)r^{n-1}, \quad t \in (\gamma, \gamma_0),$$

with $c'(\gamma) = \frac{1}{2C_2}\left(\frac{c(\gamma)}{2(N-1)}\right)^{\frac{n-1}{n}}$. Therefore the co-area formula yields

$$\mathscr{L}^n(B_r(x_0) \cap I_\gamma) \geq \int_{B_r(x_0) \cap A_{\gamma,\gamma_0}(a^+)} dx \geq \frac{1}{\|\nabla u\|_{L^\infty}} \int_{B_r(x_0) \cap A_{\gamma,\gamma_0}(a^+)} |D|u - a^+||dx$$

$$\geq \int_\gamma^{\gamma_0} \mathscr{H}^{n-1}(B_r(x_0) \cap \{|u(x) - a^+| = t\})dt \geq (\gamma_0 - \gamma)c'(\gamma)r^{n-1},$$

where $A_{\gamma,\gamma_0}(a^+) = A_{\gamma_0}(a^+) \setminus A_\gamma(a^+)$. This proves the lower bound. The upper bound follows from (5.4). The proof is complete. $\qquad\square$

5.8   Comments

5.8.1   *First Comment*

In Chap. 6, under symmetry hypotheses on W, we construct entire solutions to $\Delta u - W_u(u) = 0$, $u : \mathbb{R}^n \to \mathbb{R}^m$, connecting as $|x| \to \infty$ the minima $\{W = 0\} = \{a_1, \dots, a_N\}$, along rays contained in certain sectors partitioning $\mathbb{R}^n$. In this final section we comment on the possibility of utilizing Density Estimates for constructing such entire solutions without any symmetry requirements on W.

Let $u_R : B_R(0) \to \mathbb{R}^m$ be a minimizer of $J_{B_R(0)}(u)$ with Dirichlet conditions $u = \phi_R$ on ∂B_R, $\phi_R \to \phi_\infty$ as $R \to \infty$. A priori the only thing we know on the family $\{u_R\}$ is that the set where $W(u_R)$ is small has full measure. This does not give any information on the structure of the regions where u_R is close to a_i. The plan is to construct u by taking the limit $\lim_{R \to \infty} u_R$ along a sequence $R_k \to \infty$. The two issues involved are

  (i) establishing the nontriviality of u;
 (ii) establishing the desired asymptotic behavior.

The limit above exists on compacts, C^2_{loc}, which by itself is too weak to be of any value. We set $\epsilon = R^{-1}$, $v_\epsilon(y) = u_{\epsilon^{-1}}(\epsilon^{-1} y)$, $h_\epsilon(y) = \phi_{\epsilon^{-1}}(\epsilon^{-1} y)$, and assume $h_\epsilon \to h : \partial B_1 \to \{a_1, \dots, a_N\}$, with h inducing a partition $\{B_j\}$ of ∂B_1 via $B_j = h^{-1}(\{a_j\})$. Along a sequence $v_\epsilon \to v_0 = \sum_{j=1}^N a_j \mathbb{1}_{D_j}$, where $\mathscr{P} = \{D_1, \dots, D_N\}$ is a partition of B_1 with Dirichlet values B_j, $\partial \mathscr{P} = \bigcup_{i \neq j}(\partial D_i \cap \partial D_j)$. A prominent example is the triple junction solution $u : \mathbb{R}^2 \to \mathbb{R}^2$, with $\{W = 0\} = \{a_1, a_2, a_3\}$, and h corresponding to the singular cone on the plane. Proposition 5.3 can be restated in the following form

Proposition 5.5 *There exists a constant $C_\gamma > 0$ depending on γ such that*

$$\max \left\{ d(y, \partial \mathscr{P}) : \ y \in I_{\gamma,\epsilon} \cap B_{1 - C_\gamma(g(\epsilon))^{1/n}} \right\} < C_\gamma (g(\epsilon))^{1/n}. \tag{5.131}$$

Thus if we set $d(R) = C_\gamma R(g(R^{-1}))^{1/n}$, we have

$$x \in D_{j,R} := \{x \in D_j : \ d(x, \partial D_j \cup \partial B_R) \geq d(R)\} \implies |u_R(x) - a_j| < \gamma \tag{5.132}$$

(for $2\gamma < \min_{i \neq j} |a_i - a_j|$, cf. Proposition 5.3)

To obtain pointwise estimates we would need to show that at least for a subsequence $\{R_k\} \to \infty$, $d(R_k) \leq d_0 < \infty$. This is accomplished in the symmetric case, in Chap. 6.

5.8.2 Second Comment

Let $u \in W_{loc}^{1,2}(\mathbb{R}^n; \mathbb{R}^m) \cap L^\infty(\mathbb{R}^n; \mathbb{R}^m)$ be a minimal solution to $\Delta u - W_u(u) = 0$, $\{W = 0\} = \{a_1, \ldots, a_N\}$. Set $u_\epsilon(y) = u(y/\epsilon)$, $y \in \mathbb{R}^n$, $\epsilon \in (0, 1)$. Then there is a sequence $\{\epsilon_i\}$, $\epsilon_i \to 0$ as $i \to +\infty$, such that

$$u_{\epsilon_i} \xrightarrow{L_{loc}^1} u_0 = \sum_{j=1}^{\bar{N}} \bar{a}_j \mathbb{1}_{D_j},$$

where $1 \leq \bar{N} \leq N$, $\bar{a}_1, \ldots, \bar{a}_{\bar{N}} \in \{a_1, \ldots, a_N\}$, $\bar{a}_j \neq \bar{a}_k$ for $j \neq k$, with $\mathscr{P} = \{D_1, \ldots, D_{\bar{N}}\}$ a minimal partition of $\mathbb{R}^n$ such that $\partial \mathscr{P} = \bigcup_{0 < j < k \leq \bar{N}} (\partial D_j \cap \partial D_k)$.

For $m = 1$ (scalar case), $N = 2$, Modica has established that $\partial \mathscr{P}$ is a minimal cone. This conclusion is derived from the strong monotonicity formula

$$\frac{d}{dr} \left(r^{-(n-1)} J_{B_r}(u) \right) \geq 0, \text{ for } r > 0,$$

which holds for all solutions of the scalar equation; this is simply not true for the vector case (cf. Sect. 3.3). For uniformly bounded nontrivial minimal solutions we have, by the basic estimate (5.4) and the lower bound (5.128),

$$0 < C_1 < r^{-(n-1)} J_{B_r}(u) < C_2 < \infty. \tag{5.133}$$

If moreover we assume that the limit

$$\lim_{r \to \infty} r^{-(n-1)} J_{B_r}(u) = \Theta \tag{5.134}$$

exists, then

$$\mathscr{H}^{n-1}(\partial \mathscr{P} \cap B_r) = \Theta r^{n-1}, \ \forall r > 0, \tag{5.135}$$

from which it follows that $\partial \mathscr{P}$ is a cone, hence a minimal cone, since $\mathscr{P}$ is a minimal partition. For $m = 1$, (5.134) holds by Modica's monotonicity formula (cf. Corollary 3.2 and [16]). This suggests the study of the following

Conjecture 5.1 For each $j = 1, \ldots, \bar{N}$ and for each unit vector $\xi \in D_j$,

$$\lim_{s \to +\infty} u(s\xi) = \bar{a}_j,$$

with the convergence being uniform for ξ in compact sets of $\mathbb{S}^{n-1} \setminus \partial \mathscr{P}$.

Assuming that $\partial \mathscr{P}$ is a cone, we establish next a result which supports the correctness of this conjecture.

Proposition 5.6 *Let W satisfy $\mathbf{H}$ for $\alpha = 2$, $\{W = 0\} = \{a_1, \ldots, a_N\}$, and let $u : \mathbb{R}^n \to \mathbb{R}^m$ be nonconstant and minimal with $\|u\|_{L^\infty(\mathbb{R}^n;\mathbb{R}^m)} < \infty$. Then there exists a sequence $\{r_k\}_{k=1}^\infty \to +\infty$, such that for each $j = 1, \ldots, \bar{N}$, and for each vector $\xi \in D_j$*

$$\lim_{k \to +\infty} \frac{1}{r_k} \int_0^{r_k} u(s\xi)\mathrm{d}s = \bar{a}_j, \tag{5.136}$$

and the convergence is uniform for ξ in compact sets of $\mathbb{S}^{n-1} \setminus \partial \mathscr{P}$.

Proof For each $\epsilon \in (0, 1)$ and each $\gamma \in (0, \min_{i \neq j} |\bar{a}_i - \bar{a}_j|)$, introduce the *diffuse interface* as the set

$$I_{\gamma,\epsilon} = \{y \in \mathbb{R}^n : \min_{j=1,\ldots,\bar{N}} |u_\epsilon(y) - \bar{a}_j| \geq \gamma\},$$

Then we have

Lemma 5.6 *For any $R > 0$,*

$$\lim_{i \to +\infty} \max\{d(y, \partial\mathscr{P}) : y \in I_{\gamma,\epsilon_i} \cap B_R(0)\} = 0.$$

This lemma follows from (5.118). Set $\mathscr{K} := \partial\mathscr{P}$, and let $\mathscr{K}_\delta = \bigcup_{y \in \mathscr{K}} B_\delta(y)$ denote a δ-neighborhood. By Lemma 5.6, for each $k = 1, 2, \ldots$ there exists i_k such that

$$I_{\gamma,\epsilon_i} \cap B_1 \subset \mathscr{K}_{1/k}, \quad \text{for } i \geq i_k, \ k = 1, 2, \ldots$$

From this, provided γ is sufficiently small ($2\gamma < \min_{i \neq j} |a_i - a_j|$), it follows that

$$|u_{\epsilon_i}(y) - \bar{a}_j| < \gamma, \quad \text{for } y \in (D_j \setminus \mathscr{K}_{1/k}) \cap B_1, \ i \geq i_k. \tag{5.137}$$

In particular, if we set $i = i_k$ and $r_k = \epsilon_{i_k}^{-1}$ and recall the definition of u_ϵ, then (5.137) yields

$$|u(x) - \bar{a}_j| < \gamma, \quad \text{for } x \in (D_j \setminus \mathscr{K}_{r_k/k}) \cap B_{r_k}, \ j = 1, \ldots \bar{N}, \ k = 1, 2, \ldots. \tag{5.138}$$

Given a unit vector $\xi \in D_j$, set $d_\xi = d(\xi, \mathscr{K})$. Then we have

$$d(s\xi, \mathscr{K}) > \frac{r_k}{k}, \quad \text{for } s > \frac{r_k}{k d_\xi}$$

$$\frac{r_k}{k d_\xi} < r_k, \quad \text{for } k > \frac{1}{d_\xi}.$$

Hence for $k > \frac{1}{d_\xi}$ we have

$$s\xi \in (D_j \setminus \mathscr{K}_{r_k/k}) \cap B_{r_k} \quad \text{for} \ \ s \in \Big(\frac{r_k}{kd_\xi}, r_k\Big).$$

This, (5.138) and linear theory (cf. Proposition 6.4), provided $k > 1/d_\xi$ is sufficiently large, imply

$$|u(s\xi) - \bar{a}_j| \le C_0 e^{-c_0 d(s\xi, \partial D_j^k)} = C_0 e^{-c_0 \min\{sd_\xi - \frac{r_k}{k}, r_k - s\}}, \quad \text{for} \ \ s \in \Big(\frac{r_k}{kd_\xi}, r_k\Big).$$

$$(5.139)$$

It follows that

$$\frac{1}{r_k} \int_{\frac{r_k}{kd_\xi}}^{r_k} |u(s\xi) - \bar{a}_j| \mathrm{d}s \le C_0 \int_{\frac{1}{kd_\xi}}^{1} e^{-c_0 r_k \min\{t d_\xi - \frac{i}{k}, 1 - t\}} \mathrm{d}t$$

and therefore $\frac{1}{r_k} \int_{\frac{r_k}{kd_\xi}}^{r_k} |u(s\xi) - \bar{a}_j| \mathrm{d}s$ converges to zero as $k \to +\infty$. On the other hand, by the uniform bound $|\nabla u(x)| < M$,

$$\frac{1}{r_k} \int_0^{\frac{r_k}{kd_\xi}} |u(s\xi) - \bar{a}_j| \mathrm{d}s \le \frac{M}{kd_\xi}.$$

The proof of Proposition 5.6 is complete. $\qquad\square$

5.9 Scholia on Chap. 5

The Caffarelli-Córdoba density estimates [5] (1995) played a major role in the resolution of the De Giorgi conjecture [17]. Other extensions of the density estimates in different contexts have been provided by Farina and Valdinoci [13], Savin and Valdinoci [19], [20], Sire and Valdinoci [21] and very recently by Cesaroni et al. [9]. We benefited from Valdinoci's improved presentation in [23]. The method of proof of Theorem 5.3 in Sect. 5.5 was developed in [15] (see also [14]). It is an alternative to the Caffarelli-Córdoba method for establishing a pointwise estimate in situations where the restriction of the minimizer u to a certain set A is bounded away from all but a single minimum a of W.

References

1. Alberti, G.: Variational models for phase transitions, an approach via Gamma convergence. In: Ambrosio, L., Dancer, N. (eds.) Calculus of Variations and Partial Differential Equations, pp. 95–114. Springer, Berlin (2000)

2. Alikakos, N.D., Fusco, G.: Density estimates for vector minimizers and application. Discrete Contin. Dynam. Syst, **35**(12), 5631–5663 (2015)
3. Alikakos, N.D., Zarnescu, A.: in preparation
4. Baldo, S.: Minimal interface criterion for phase transitions in mixtures of Cahn-Hilliard fluids. Ann. Inst. Henri Poincaré **7**(2), 67–90 (1990)
5. Caffarelli, L., Córdoba, A.: Uniform convergence of a singular perturbation problem. Commun. Pure Appl. Math. **48**, 1–12 (1995)
6. Caffarelli, L., Córdoba, A.: Phase transitions: uniform regularity of the intermediate layers. J. Reine Angew. Math. **593**, 209–235 (2006)
7. Caffarelli, L., Salsa, S.: A Geometric Approach to Free Boundary Problems. Graduate Studies in Mathematics, vol. 68. American Mathematical Society, Providence (2005)
8. Cecon, B., Paolini, M., Romeo, M.: Optimal interface error estimates for a discrete double obstacle approximation to the prescribed curvature problem. Math. Models Methods Appl. Sci. **9**, 799–823 (1999)
9. Cesaroni, A., Muratov, C.M., Novaga, M.: Front propagation and phase field models of stratified media. Arch. Ration. Mech. Anal. **216**(1), 153–191 (2015)
10. Evans, L.C.: Partial Differential Equations. Graduate Studies in Mathematics, vol. 19, 2nd edn. American Mathematical Society, Providence (2010)
11. Evans, L.C., Gariepy, R.F.: Measure Theory and Fine Properties of Functions. CRC Press, Boca Raton (1992)
12. Farina, A.: Two results on entire solutions of Ginzburg–Landau system in higher dimensions. J. Funct. Anal. **214**(2), 386–395 (2004)
13. Farina, A., Valdinoci, E.: Geometry of quasiminimal phase transitions. Calc. Var. Partial Differ. Equ. **33**(1), 1–35 (2008)
14. Fusco, G.: Equivariant entire solutions to the elliptic system $\Delta u - W_u(u) = 0$ for general G-invariant potentials. Calc. Var. Partial Differ. Equ. **49**(3), 963–985 (2014)
15. Fusco, G.: On some elementary properties of vector minimizers of the Allen-Cahn energy. Commun. Pure Appl. Anal. **13**(3), 1045–1060 (2014)
16. Modica, L.: Γ-convergence to minimal surfaces problem and global solutions of $\Delta u = 2(u^3 - u)$. In: Proceedings of the International Meeting on Recent Methods in Nonlinear Analysis (Rome, 1978), Pitagora, Bologna, pp. 223–244 (1979)
17. Savin, O.: Regularity of flat level sets in phase transitions. Ann. Math. **169**, 41–78 (2009)
18. Savin, O.: Minimal surfaces and minimizers of the Ginzburg Landau energy. Cont. Math. Mech. Anal. **526**, 43–58 (2010)
19. Savin, O., Valdinoci, E.: Density estimates for a variational model driven by the Gagliardo norm. J. Math. Pures Appl. **101**(1), 1–26 (2014)
20. Savin, O., Valdinoci, E.: Density estimates for a nonlocal variational model via the Sobolev inequality. SIAM J. Math. Anal. **43**(6), 2675–2687 (2011)
21. Sire, Y., Valdinoci, E.: Density estimates for phase transitions with a trace. Interfaces Free Bound. **14**, 153–165 (2012)
22. Sourdis, C.: Optimal energy growth lower bounds for a class of solutions to the vectorial Allen-Cahn Equation. arXiv:1402.3844
23. Valdinoci, E.: Plane-like minimizers in periodic media: jet flows and Ginzburg-Landau-type functionals. J. Reine Angew. Math. **574**, 147–185 (2004)
24. White, B.: Existence of least energy configurations of immiscible fluids. J. Geom. Anal. **6**, 151–161 (1996)

Chapter 6
Symmetry and the Vector Allen–Cahn Equation: The Point Group in $\mathbb{R}^n$

Abstract In this chapter we begin the study of entire solutions $u : \mathbb{R}^n \to \mathbb{R}^n$ of the vector Allen–Cahn equation (6.1) that describe the coexistence of different phases in a neighborhood of a point. We work in a symmetry context where a finite reflection group G is acting both on the domain space $\mathbb{R}^n_x$ and on the target space $\mathbb{R}^n_u$, which are assumed to be of the same dimension. The scope of this chapter is to introduce the main ideas involved in the proof of Theorem 1.2 which invokes estimate (1.34) or alternatively the density estimate (1.28), but otherwise is self-contained. In Chap. 7 we present a systematic study of all symmetric entire solutions that can be obtained by a variational approach.

6.1 Notation

We denote by B_R the ball of radius $R > 0$ centered at the origin, by $\cdot$ the Euclidean inner product, by $|\cdot|$ the Euclidean norm, and by $d(x, \partial D)$ the distance from x to ∂D. We also denote the functional associated to

$$\Delta u - W_u(u) = 0, \ \text{for } u : \mathbb{R}^n \to \mathbb{R}^n, \tag{6.1}$$

by

$$J_\Omega(u) = \int_\Omega \left\{ \frac{1}{2} |\nabla u|^2 + W(u) \right\} \mathrm{d}x.$$

A *Coxeter group*, or more simply a *reflection group G*, is a finite subgroup of the orthogonal group $O(\mathbb{R}^n)$, generated by a set of reflections. The notation $|G|$ stands for the order of G, that is, the number of elements of G. In this chapter we assume that the same reflection group G acts both on the domain space $\mathbb{R}^n$ or $B_R \subset \mathbb{R}^n$ and on the target space $\mathbb{R}^m$, and take $n = m$. A map $u : B_R \subset \mathbb{R}^n \to \mathbb{R}^n$ is said to be

© Springer Nature Switzerland AG 2018
N. D. Alikakos et al., *Elliptic Systems of Phase Transition Type*,
Progress in Nonlinear Differential Equations and Their Applications 91,
https://doi.org/10.1007/978-3-319-90572-3_6

equivariant with respect to the action of G, simply equivariant, if

$$u(gx) = gu(x), \ \forall g \in G, \ \forall x \in B_R. \tag{6.2}$$

We denote by $W_{\mathrm{E}}^{1,2}(B_R; \mathbb{R}^n) \subset W^{1,2}(B_R; \mathbb{R}^n)$ the subspace of equivariant maps. If G is a reflection group acting on $\mathbb{R}^n$, a *reflection* $\gamma \in G$ is a map $\gamma : \mathbb{R}^n \to \mathbb{R}^n$ of the form

$$\gamma x = x - 2(x \cdot \eta_\gamma)\eta_\gamma, \ \text{ for } x \in \mathbb{R}^n,$$

for some unit vector $\eta_\gamma \in \mathbb{S}^{n-1}$ which, aside from its orientation, is uniquely determined by γ. The hyperplane

$$\pi_\gamma = \{x \in \mathbb{R}^n : x \cdot \eta_\gamma = 0\},$$

is the set of the points that are fixed by γ. The open half space $S_\gamma^+ = \{x \in \mathbb{R}^n : x \cdot \eta_\gamma > 0\}$ depends on the orientation of η_γ. We let $\Gamma \subset G$ denote set of all reflections in G. Every finite subgroup of $O(\mathbb{R}^n)$ has a *fundamental region*, that is, a subset $F \subset \mathbb{R}^n$ with the following properties:

1. F is open and convex,
2. $F \cap gF = \varnothing$, for $I \neq g \in G$, where I is the identity,
3. $\mathbb{R}^n = \bigcup \{g\overline{F} : g \in G\}$.

The set $\bigcup_{\gamma \in \Gamma} \pi_\gamma$ divides $\mathbb{R}^n \setminus \bigcup_{\gamma \in \Gamma} \pi_\gamma$ in exactly $|G|$ congruent conical regions. Each one of these regions can be identified with the fundamental region F for the action of G on $\mathbb{R}^n$. We assume that the orientation of η_γ is such that $F \subset S_\gamma^+$ and we have

$$F = \bigcap_{\gamma \in \Gamma} S_\gamma^+.$$

Given $a \in \mathbb{R}^n$, the *stabilizer* of a, denoted by $G_a \subset G$, is the subgroup of the elements $g \in G$ that fix a:

$$G_a = \{g \in G : ga = a\}.$$

To give a simple example, consider the action on $\mathbb{R}^2$ of $G = D_3$, the group of symmetries of the equilateral triangle. D_3 has order 6 and, if we assume that the center of the triangle coincides with the origin and that one of the axes of the triangle is aligned with the x_1 axis, we can take as generators the reflections γ_1 and γ_2 defined by the vectors $\eta_{\gamma_1} = (0, 1)$ and $\eta_{\gamma_2} = (\sqrt{3}/2, -1/2)$. In this case $\mathbb{R}^2 \setminus \bigcup_{\gamma \in \Gamma} \pi_\gamma$ is the union of 6 congruent sectors of angle $\pi/3$ and F can be chosen to be the sector

$$F = \{x \in \mathbb{R}^2 : 0 < x_2 < \sqrt{3}x_1, \ x_1 > 0\}.$$

If $a \in F$, then $G_a = \{I\}$. If $a \in \pi_{\gamma_1} \setminus \{0\}$, we have $G_a = \{I, \gamma_1\}$. Finally, if $a = 0$, then $G_a = G$.

6.2 The Hypotheses of the Theorem

H_1 (N nondegenerate global minima) The potential W is of class C^2 and satisfies $W(a_i) = 0$, for $i = 1, \ldots, N$, and $W > 0$ on $\mathbb{R}^n \setminus \{a_1, \ldots a_N\}$. Furthermore, there holds $\xi^\top W_{uu}(u)\xi \geq 2c^2|\xi|^2$, for $\xi \in \mathbb{R}^n$ and $|u - a_i| \leq \bar{q}$, for some c, $\bar{q} > 0$, and for $i = 1, \ldots, N$.

H_2 (Symmetry) The potential W is invariant under a finite reflection group G acting on $\mathbb{R}^n$, that is,

$$W(gu) = W(u), \quad \text{for all } g \in G \text{ and } u \in \mathbb{R}^n.$$

Moreover, there exists $M > 0$ such that $W(su) \geq W(u)$, for $s \geq 1$ and $|u| = M$. We seek *equivariant* solutions of system (6.1), that is, solutions satisfying

$$u(gx) = gu(x), \quad \text{for all } g \in G \text{ and } x \in \mathbb{R}^n.$$

H_3 (Location and number of global minima) Let $F \subset \mathbb{R}^n$ be a fundamental region of G. We assume that $\overline{F}$ (the closure of F) contains a single global minimum of W, say a_1, and let G_{a_1} be he stabilizes of a_1. Setting $D := \mathrm{Int}\left(\bigcup_{g \in G_{a_1}} g\overline{F}\right)$, a_1 is also the unique global minimum of W in the region D.

Notice that, by the invariance of W, Hypothesis H_3 implies that the number of minima of W is

$$N = \frac{|G|}{|G_{a_1}|}.$$

We recall several examples of groups. For $G = D_3$, the group of symmetries of the equilateral triangle on the plane, we can take as F the $\frac{\pi}{3}$ sector. If $a_1 \in F$, then $N = 6$, while if a_1 is on the walls, then $N = 3$. In higher dimensions we have more options, since we can place a_1 in the interior of $\overline{F}$, in the interior of a face, on an edge, and so on. For cxample, if $G = \mathscr{K}$, the group of symmetries of the cube in three-dimensional space, then $|G| = 48$. If the cube is situated with its center at the origin and its vertices at the eight points $(\pm 1, \pm 1, \pm 1)$, then we can take as F the simplex generated by $s_1 = e_1 + e_2 + e_3$, $s_2 = e_2 + e_3$, and $s_3 = e_3$, where the e_i's are the standard basis vectors. We then have the following options:

1. On the edge s_3, $N = 6$.
2. On the edge s_1, $N = 8$.
3. On the edge s_2, $N = 12$.
4. In the interior of a face, $N = 24$.

5. In the interior of the fundamental region, $N = 48$.
6. At the origin, $N = 1$.

6.3   Examples of Potentials

Assume $n = 2$ and let $a_i \in \mathbb{R}^2$, $i = 1, 2, 3$ be the vertices of the equilateral triangle $\mathscr{T}$ with center at the origin and one of the vertices in $a_1 = (1, 0)$. Then $W(u) = \prod_{i=1}^3 |u - a_i|^2$ satisfies Hypotheses $\mathbf{H}_1$–$\mathbf{H}_3$, with G the symmetry group of $\mathscr{T}$. We can take $F = \{u \in \mathbb{R}^2 : 0 < u_2 < \sqrt{3}u_1\}$ and $D = \{u \in \mathbb{R}^2 : 0 < |u_2| < \sqrt{3}u_1\}$. In this case G_{a_1} is the subgroup of order 2 generated by the reflection in the u_1 axis.

As another example for the equilateral triangle symmetry group on the plane, consider again, F the $\frac{\pi}{3}$ sector, $a_1 = (1, 0)$, and the triple-well potential

$$W(u_1, u_2) = |u|^4 + 2u_1u_2^2 - \frac{2}{3}u_1^3 - |u|^2 + \frac{2}{3}.$$

This potential has the additional property that $Q_u(u) \cdot W_u(u) \geq 0$, in $D \setminus \{a_1\}$, where $Q(u) = |u - a_1|$, $u = (u_1, u_2)$, and $D = \{u \in \mathbb{R}^2 : 0 < |u_2| < \sqrt{3}u_1\}$.

As another example for $n = 3$, consider the group of symmetries of the tetrahedron, $G = \mathscr{T}^3$, with F the cone generated by $(\sqrt{2/3}, 0, 1/\sqrt{3})$, $(0, \sqrt{2/3}, 1/\sqrt{3})$, $(0, 0, 1/\sqrt{3})$, and $a_1 = (\sqrt{2/3}, 0, 1/\sqrt{3})$. We can take the quadruple-well potential

$$W(u_1, u_2, u_3) = |u|^4 - \frac{4}{\sqrt{3}}(u_1^2 - u_2^2)u_3 - \frac{2}{3}|u|^2 + \frac{5}{9},$$

where $u = (u_1, u_2, u_3)$, and D is the cone generated by $(0, \sqrt{2/3}, 1/\sqrt{3})$, $(0, -\sqrt{2/3}, 1/\sqrt{3})$, $(\sqrt{2/3}, 0, -1/\sqrt{3})$. It can be checked that W satisfies Hypotheses $\mathbf{H}_1$–$\mathbf{H}_3$, and that it has the additional property that $Q_u(u) \cdot W_u(u) \geq 0$, in $D \setminus \{a_1\}$, where $Q(u) = |u - a_1|$.

More generally, for each choice of a reflection group G and of a vector $a_1 \in \overline{F}$ the polynomial $W : \mathbb{R}^n \to \mathbb{R}$

$$W(u) = \frac{1}{2} \prod_{g \in G} |u - ga_1|^2, \quad u \in \mathbb{R}^n, \tag{6.3}$$

satisfies Hypotheses $\mathbf{H}_1$–$\mathbf{H}_3$. Indeed a_1 is obviously a nondegenerate point of minimum for W. Moreover, we have

$$W(\tilde{g}u) = \frac{1}{2} \prod_{g \in G} |\tilde{g}u - ga_1|^2 = \frac{1}{2} \prod_{g \in G} |u - \tilde{g}^{-1}ga_1|^2 = W(u), \quad u \in \mathbb{R}^n, \tilde{g} \in G,$$

and for $|u| \geq |a_1|$

$$
W_u(u) \cdot \frac{u}{|u|} = \left(\sum_{\tilde{g} \in G} (u - \tilde{g}a_1) \cdot \frac{u}{|u|} \right) \prod_{g \in G \setminus \{\tilde{g}\}} |u - ga_1|^2
$$

$$
\geq (|u| - |a_1|) \sum_{\tilde{g} \in G} \prod_{g \in G \setminus \{\tilde{g}\}} |u - ga_1|^2 \geq 0.
$$

A potential of physical interest arising in the study of magnetism (cf. [5] and [6]) is

$$
W^\mu(u) = \frac{1}{2} \sum_{i \neq j} u_i u_j + \sum_i u_i (\ln u_i - \mu), \quad u_i > 0. \tag{6.4}
$$

Here $u = (u_1, \ldots, u_n)$ is an order parameter that describes the magnetization of the material and $\mu \in \mathbb{R}$ is a rescaled parameter corresponding to the chemical potential. Note that W^μ is only defined in the positive cone, therefore we need to extend W^μ to $\mathbb{R}^n$. From (6.4) it follows that W^μ is invariant under the symmetry group G of order $n!$ of the $n - 1$ dimensional hyper-tetrahedron $\mathscr{T}^n$ (of side $\sqrt{2}$), which can be identified with the convex hull of the n standard unit vectors e_i, $i = 1, \ldots, n$, in $\mathbb{R}^n$. The group G is generated by the reflections $\gamma_{ij} e_h = e_h$, $h \notin \{i, j\}$; $\gamma_{ij} e_i = e_j$. The critical points of W^μ are the solutions of the system

$$
\sum_{j \neq i} u_j + \ln u_i + 1 - \mu = 0, \quad i = 1, \ldots, n. \tag{6.5}
$$

These equations can be rewritten as $u_i - \ln u_i = f(u)$ with the obvious definition for f. Since f does not depend on i all solutions of (6.5) must be of the form $u_i \in \{p, q\}$ for some $0 < p \leq 1 \leq q$. By analyzing the Hessian matrixs of W^μ one sees that the local minimizers of W^μ are necessarily of one of two types: $u = pe_0$, or $u = pe_0 + (q - p)e_i$, $i = 1, \ldots, n$, where $e_0 = (1, \ldots, 1)$. In [5, Theorem 2.3] it is shown that there exists a critical value μ_c of μ such that $u = pe$ is the unique global minimizer of W^μ for $\mu < \mu_c$, while for $\mu > \mu_c$, W^μ has n global nondegenerate minimizers of the form $u = pe_0 + (q - p)e_i$, $i = 1, \ldots, n$. Set $u = \lambda v$ with $\sum_i v_i = 1$, $v_i > 0$, $i = 1, \ldots, n$. Then we have

$$
W_u^\mu(u) \cdot v = \sum_i v_i (\lambda(1 - v_i) + \ln v_i + \ln \lambda + 1 - \mu)
$$

$$
\geq \sum_i v_i \ln v_i + \ln \lambda + 1 - \mu \geq -\frac{n}{e} + \ln \lambda + 1 - \mu.
$$

It follows $W_u(u) \cdot v \geq 0$ for all v provided $\lambda > 0$ is sufficiently large. Therefore, W^μ satisfies Hypothesis $\mathbf{H_3}$ for some $M > 0$ and u in the positive cone. Now observe that, if $|u| \leq M$, we have

$$-W_u^\mu(u) \cdot e_i = -\sum_{j \neq i} u_j - \ln u_i - 1 + \mu \geq -\sqrt{n}\, M - \ln u_i - 1 + \mu.$$

Therefore $|u| \leq M$ implies the existence of $\delta \in (0, 1)$ such that

$$- W_u^\mu(u) \cdot e_i > 0, \quad \text{for } u_i \in (0, \delta), \ i = 1, \ldots, n. \tag{6.6}$$

These observations show that we can restrict to maps with range in the compact subset C of the positive cone defined by

$$C := \{u : |u| \leq M, \ \delta \leq u_i, \ i = 1, \ldots, n\}.$$

It follows that we can extend the definition of W^μ from C to $\mathbb{R}^n$ in such a way that the extended potential satisfies Hypothesis $\mathbf{H}_1$, for fixed $\mu > \mu_{\mathrm{c}}$, with $a_1 = pe_0 + (q - p)e_1$ and $D := \{u \in \mathbb{R}^n : u_i < u_1, \ i = 2, \ldots, n\}$.

6.4   Statement of the Theorem

Theorem 6.1 *Under Hypotheses* $\mathbf{H}_1$–$\mathbf{H}_3$*, there exists an equivariant classical solution to system* (6.1) *such that*

1. $|u(x) - a_1| \leq K e^{-kd(x, \partial D)}$, *for* $x \in D := \mathrm{Int}\left(\bigcup_{g \in G_{a_1}} g\overline{F}\right)$, *and for positive constants* k, K,
2. $u(\overline{F}) \subset \overline{F}$ *and* $u(D) \subset D$.

In particular, u connects the $N = |G|/|G_{a_1}|$ *global minima of W in the sense that*

$$\lim_{\lambda \to +\infty} u(\lambda g\eta) = ga_1, \quad \text{for all } g \in G,$$

uniformly for η *in compact subsets of* $D \cap \mathbb{S}^{n-1}$.

6.5   Outline of the Proof

The proof is based on minimizing

$$J_{B_R}(u) = \int_{B_R} \left(\tfrac{1}{2}|\nabla u|^2 + W(u)\right) \mathrm{d}x,$$

over balls B_R centered at the origin, and then taking the limit

$$u(x) = \lim_{R \to \infty} u_R(x),$$

along subsequences of minimizers u_R. Minimizing over compact sets is forced by the fact that the action evaluated over $\mathbb{R}^n$ is finite only for trivial, constant maps (cf. Theorem 3.5). Minimizing in the equivariant class does not affect the Euler–Langrange equation (by classical facts) and relatively easily renders the estimate $J_{B_r}(u_R) \leq Cr^{n-1}$, $0 < r < R - 1$. This estimate implies the existence of a nontrivial solution $u(x)$ in the equivariant class under only Hypotheses $\mathbf{H}_1$ and $\mathbf{H}_2$, and very mild regularity assumptions on W, and also very mild nondegeneracy hypotheses on a_i. To obtain information on the asymptotic behavior of the solution, we introduce the notion of *positivity*, $u(\overline{F}) \subset \overline{F}$, as a constraint in the minimization process. This, in principle, could affect the Euler–Langrange equation. We show that the associated gradient flow with Neumann condition on B_R preserves positivity, and since it reduces J_{B_R}, we conclude that positivity is a removable constraint.

By Hypothesis $\mathbf{H}_3$, there is a unique minimum a_1 of W in $\overline{F}$. Thus, the aforementioned estimate $J_{B_r}(u_R) \leq Cr^{n-1}$, with $r \in (0, R - 1)$ (which also holds under the positivity constraint), implies easily that $\mathcal{L}^n(A_{\bar{q}} \cap B_r) \leq Kr^{n-1}$, where $A_{\bar{q}} = \{x \in F : |u(x) - a_1| \geq \bar{q}\}$, $\bar{q} > 0$, and arbitrary otherwise. This estimate says that the solution in most of D is close to a_1. Obtaining however the pointwise statement in Theorem 6.1 is considerably more involved, and requires the pointwise estimates developed in Chap. 5. Minimality of u in the equivariant class is sufficient for this purpose.

6.6   Proof of an Easy Fact: The Existence of a Nontrivial Equivariant Solution

Before presenting the proof of Theorem 6.1, we would like to show that establishing existence of a nontrivial solution to (6.1) in the equivariant class is not hard. Consider the minimization problem

$$\min_{W_E^{1,2}(B_R;\mathbb{R}^n)} J_{B_R}, \quad \text{where } J_{B_R}(u) = \int_{B_R} \left\{ \frac{1}{2}|\nabla u|^2 + W(u) \right\} dx.$$

We will argue first that the minimizer exists. We redefine $W(u)$ for $|u| \geq M + 1$, so that the modified W is C^2, satisfies $W(u) \geq c^2|u|^2$ for $|u| \geq M+1$ and a constant c, and also $W(gu) = W(u)$, for all $g \in G$. We still denote the modified potential by W and the modified functional by J_{B_R}. The modified functional J_{B_R} satisfies all the properties required by the direct method and, as a result, a minimizer v_R exists.

Next we will show that as a consequence of Hypothesis $\mathbf{H}_2$ we can produce a minimizer u_R, which in addition satisfies the estimate

$$|u_R(x)| \leq M. \tag{6.7}$$

Due to this estimate, the values of W outside $\{|u| \leq M\}$ will not matter in the considerations that follow and, therefore, the equation that will be solved is (6.1) with the original unmodified potential W. Set

$$u_R(x) = Pv_R(x),$$

where Pv equals the projection on the sphere $\{v \in \mathbb{R}^n : |v| = M\}$, for points outside the sphere ($Pv = Mv/|v|$), and equals the identity inside the sphere. Since P is a contraction with respect to the Euclidean norm in $\mathbb{R}^n$, it follows that $u_R \in W^{1,2}(B_R; \mathbb{R}^n)$, with $|\nabla u_R(x)| \leq |\nabla v_R(x)|$. Furthermore,

$$u_R(gx) = Pv_R(gx) = Pgv_R(x) = gPv_R(x) = gu_R(x),$$

hence $u_R \in W_{\mathrm{E}}^{1,2}(B_R; \mathbb{R}^n)$. Clearly $|u_R(x)| \leq M$, for $x \in B_R$.

The fact that u_R is also a minimizer follows from $|\nabla u_R(x)| \leq |\nabla v_R(x)|$ and from Hypothesis $\mathbf{H}_2$, which implies

$$W(u_R) = W(v_R), \quad \text{if } |v_R| \leq M,$$

$$W(u_R) = W\left(\frac{M}{|v_R|}v_R\right) \leq W(v_R), \quad \text{if } |v_R| > M.$$

We will construct the solution by taking the limit, possibly along a subsequence,

$$u(x) = \lim_{R \to \infty} u_R(x).$$

From the considerations above we have that

$$\Delta u_R - W_u(u_R) = 0 \quad \text{in } W_{\mathrm{loc}}^{1,2}(B_R; \mathbb{R}^n). \tag{6.8}$$

Indeed, the equivariance constraint can be removed (cf. [11]), and thus it does not affect the Euler–Lagrange equation (6.8).

By elliptic theory, u_R satisfies the equation classicaly in $\overline{B}_{R-1}$, and also

$$|\nabla u_R(x)| \leq M', \text{ in } \overline{B}_{R-1}, \text{ for some constant } M' \text{ depending only on } M. \tag{6.9}$$

Proposition 6.1

$$J_{B_r}(u_R) \leq Cr^{n-1}, \ \forall r \in (0, R-1), \ \textit{with a constant C depending only on M.} \tag{6.10}$$

Proof By (6.7) and (6.9), it is clear that $J_{B_r}(u_R) \le Cr^n$, $\forall r \in (0, R-1) \cap (0, 1]$, and with a constant C depending only on M. Thus, (6.10) holds when $r \in (0, R-1) \cap (0, 1]$. Next, we examine the case where $1 < r < R-1$, and define

$$u_{\mathrm{aff}}(x) = \begin{cases} d(x, \partial D)a_1 & \text{for } x \in D_R, \text{ and } d(x, \partial D) \le 1, \\ a_1 & \text{for } x \in D_R, \text{ and } d(x, \partial D) \ge 1, \end{cases} \tag{6.11}$$

where $D_R = D \cap B_R$, and extend equivariantly in B_R. Since u_{aff} vanishes on ∂D, the extended map is also continuous. As it is well known, the distance function is 1-Lipschitz, and therefore in $W^{1,\infty}(B_R)$ (cf. [7]). Fix now a number $h \in (0, 1)$, and for $r \in (1, R-1)$ define (Fig. 6.1)

$$\hat{u}_R(x) = \chi\left(1 - \frac{|x| - (r-h)}{h}\right)u_{\mathrm{aff}}(x) + \chi\left(\frac{|x| - (r-h)}{h}\right)u_R(x), \tag{6.12}$$

where $\chi : \mathbb{R} \to [0, 1]$ is a fixed C^1 function such that $\chi(s) = 0$, for $s \le 0$, and $\chi(s) = 1$, for $s \ge 1$. Note that $\hat{u}_R \in W_E^{1,2}(B_R; \mathbb{R}^n)$, and most importantly $\hat{u}_R = u_R$ on ∂B_r. Moreover, $\hat{u}_R = u_{\mathrm{aff}}$ in B_{r-h} and $\hat{u}_R = u_R$ on $B_R \setminus B_r$, and $u_{\mathrm{aff}}(x) = a_1$ if $d(x, \partial D) \ge 1$. By the minimality of u_R, we have

$$J_{B_r}(u_R) \le J_{B_r}(\hat{u}_R)$$

$$= \int_{B_{r-h} \cap \{d(x, \partial D) \le 1\}} \left\{ \frac{|\nabla \hat{u}_R|^2}{2} + W(\hat{u}_R) \right\} + \int_{B_r \setminus B_{r-h}} \left\{ \frac{|\nabla \hat{u}_R|^2}{2} + W(\hat{u}_R) \right\}$$

$$\le C_1(r-h)^{n-1} + C_2 r^{n-1}, \text{ with constants } C_i \text{ depending only on } M,$$

$$\le Cr^{n-1}, \text{ for } 1 < r < R-1 \text{ and a constant } C \text{ depending only on } M. \tag{6.13}$$

This completes the proof of (6.10). $\square$

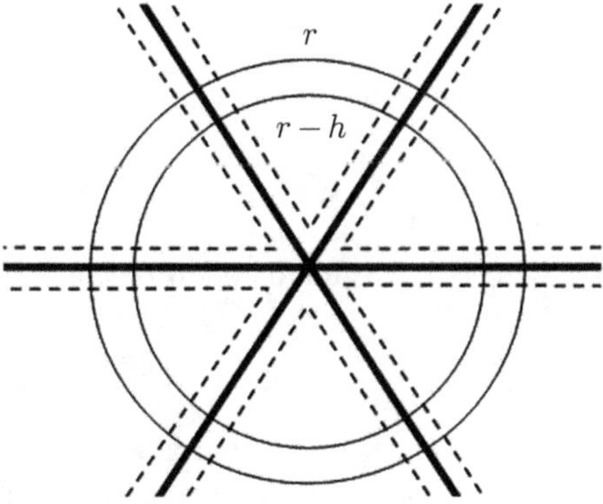

Fig. 6.1 The comparison map $\hat{u}_R$

Corollary 6.1 *There is a classical equivariant solution $u : \mathbb{R}^n \to \mathbb{R}^n$ to (6.1), satisfying the estimate*

$$J_{B_r}(u) \leq Cr^{n-1}, \quad \forall r > 0. \tag{6.14}$$

Proof From (6.8) and elliptic estimates we obtain by a diagonal argument a subsequence $\{u_{R_k}\}$ such that

$$u(x) = \lim_{R_k \to \infty} u_{R_k}(x), \text{ for the } C^1_{\mathrm{loc}}(\mathbb{R}^n; \mathbb{R}^n) \text{ convergence.}$$

Estimate (6.14) follows by taking the limit in (6.10). $\qquad\qquad\square$

Remark 6.1 Note that if $W(0) > 0$, then estimate (6.14) guarantees that u is not identically 0. Hence, $u : \mathbb{R}^n \to \mathbb{R}^n$ is a nontrivial equivariant solution to (6.1).

6.7   Proof of Theorem 6.1

6.7.1   *The Gradient Flow and Positivity*

We define the set of *positive maps* (in the class of equivariant Sobolev maps)

$$\mathscr{A}^R := \left\{u \in W_{\mathrm{E}}^{1,2}(B_R; \mathbb{R}^n) : u(\overline{F_R}) \subset \overline{F}\right\}, \tag{6.15}$$

where $F_R = F \cap B_R$. Here $R > 0$ and clearly the sets $\mathscr{A}^R$ depend on R.
  We will utilize the L^2-gradient flow of the functional J_{B_R}:

$$\begin{cases} \dfrac{\partial u}{\partial t} = \Delta u - W_u(u), & \text{in } B_R \times (0, \infty), \\[2mm] \dfrac{\partial u}{\partial n} = 0, & \text{on } \partial B_R \times (0, \infty), , \\[2mm] u(x, 0) = u_0(x), & \text{in } B_R, \end{cases} \tag{6.16}$$

where $\partial/\partial n$ is the normal derivative. We note that, by Hypothesis $\mathbf{H}_2$,

$$- W_u(u) \cdot u \leq 0, \text{ for } |u| = M. \tag{6.17}$$

We will consider initial conditions in (6.16) satisfying in addition

$$\|u_0\|_{L^\infty(B_R; \mathbb{R}^n)} \leq M. \tag{6.18}$$

Since W is C^2 (cf. Hypothesis $\mathbf{H}_1$), the results in [10, Ch. 3, §3.3, §3.5] apply and provide a unique solution to (6.16) in $C([0, \infty); W_{\mathrm{E}}^{1,2}(B_R; \mathbb{R}^n))$, which for $t > 0$, as a function of x, is in $C^{2+\alpha}(\overline{B_R}; \mathbb{R}^n))$, for some $0 < \alpha < 1$. Moreover, the solution satisfies the estimate

$$\|u(\cdot, t)\|_{L^\infty(B_R;\mathbb{R}^n)} \leq M, \text{ for } t \geq 0.$$

This follows from (6.17), (6.18), and by well-known invariance results [12, Ch. 14, §B], and therefore the solution exists globally by well known facts for semilinear equations (cf. [10]).

Proposition 6.2 *Let W be a C^2 potential satisfying Hypothesis $\mathbf{H}_2$. If $u_0 \in \mathscr{A}^R$ and $\|u_0\|_{L^\infty(B_R;\mathbb{R}^n)} \leq M$, then*

$$u(\cdot, t; u_0) \in \mathscr{A}^R, \quad \text{for } t \geq 0.$$

Proof Let $u : B_R \to \mathbb{R}^n$ be an equivariant map. We prove that u is a positive map if and only if

$$u(\overline{(S_\gamma^+)_R}) \subset \overline{S_\gamma^+}, \text{ for all } \gamma \in \Gamma, \tag{6.19}$$

where $(S_\gamma^+)_R = S_\gamma^+ \cap B_R$.

Suppose first that (6.19) holds. Then

$$u(\overline{F_R}) = u(\bigcap_{\gamma \in \Gamma} \overline{(S_\gamma^+)_R}) \subset \bigcap_{\gamma \in \Gamma} u(\overline{(S_\gamma^+)_R}) \subset \bigcap_{\gamma \in \Gamma} \overline{S_\gamma^+} = \overline{F}.$$

Hence, u is positive. Conversely, suppose that u is a positive equivariant map on B_R. Then, equivalently, u_e defined by

$$u_e(x) := \begin{cases} u(x), & \text{for } x \in B_R \\ 0, & \text{for } x \in \mathbb{R}^n \setminus B_R \end{cases}$$

is a positive equivariant map on $\mathbb{R}^n$. For any $g \in G$, equivariance and positivity ensure that

$$u_e(g(\overline{F})) = g(u_e(\overline{F})) \subset g(\overline{F}). \tag{6.20}$$

Now pick a $\gamma \in \Gamma$ and take an $x \in S_\gamma^+$ and fix it. There is a $g \in G$, denoted by g_x, such that $x \in g_x(\overline{F})$ and $g_x(F)$ is also a fundamental region. Since for each fundamenal region F' and for each reflection γ we have either $F' \subset S_\gamma^+$ or $F' \subset -S_\gamma^+$, we conclude that

$$g_x(\overline{F}) \subset \overline{S_\gamma^+}.$$

This, (6.20), and the fact that $x \in g_x(\overline{F})$ imply

$$u_e(x) \in u_e(g_x(\overline{F})) \subset \overline{S_\gamma^+}.$$

Since this is true for every $x \in S_\gamma^+$, (6.19) follows.

Now consider (6.16) with $u_0 \in \mathscr{A}^R$. First we note that the solution is global in time and regular. Indeed by the regularizing property of the equation, the solution is classical for $t > 0$ and by (6.17) and the uniform L^∞ bound in t, it exists globally in time as it was noted above and belongs to $C([0, +\infty); W_{\mathrm{E}}^{1,2}(B_R; \mathbb{R}^n)) \cap C^1((0, +\infty); C^{2+\alpha}(\overline{B_R}; \mathbb{R}^n))$, for some $0 < \alpha < 1$ (see [10]).

Consider a reflection $\gamma \in \Gamma$ and set

$$\zeta(x, t) = u(x, t, u_0) \cdot \eta_\gamma, \qquad \text{on } B_R \times (0, \infty),$$

$$\zeta_0(x) = u_0(x) \cdot \eta_\gamma, \qquad \text{on } B_R.$$

By taking the inner product of Eq. (6.16) with η_γ, we obtain

$$\begin{cases} \dfrac{\partial \zeta}{\partial t} = \Delta\zeta + c\zeta, & \text{in } B_R \times (0, \infty), \\[2mm] \dfrac{\partial \zeta}{\partial n} = 0, & \text{on } \partial B_R \times (0, \infty), \\[2mm] \zeta(\cdot, 0) = \zeta_0, & \end{cases} \tag{6.21}$$

where we have set

$$c(x, t) = \frac{W_u(u(x, t, u_0)) \cdot \eta_\gamma}{\zeta(x, t)}.$$

From the equivariance of $u(\cdot, t, u_0)$ and $W_u(\gamma u) = \gamma W_u(u)$ it follows that

$$\zeta(x, t) = -\zeta(\gamma x, t), \qquad \text{in } B_R \times (0, \infty), \tag{6.22}$$

$$c(x, t) = c(\gamma x, t), \qquad \text{in } B_R \times (0, \infty). \tag{6.23}$$

By the symmetry of W, we also have that $u \in \pi_\gamma$ implies $W_u(u) \in \pi_\gamma$. From this we deduce

$$W_u(u) \cdot \eta_\gamma = (u \cdot \eta_\gamma) \left(\int_0^1 W_{uu}\big(u + (s-1)(u \cdot \eta_\gamma)\eta_\gamma\big)\eta_\gamma \, ds \right) \cdot \eta_\gamma.$$

Thus, the coefficient $c(x, t)$ of ζ in (6.21) is bounded (actually continuous) on $B_R \times (0, \infty)$. Since u_0 is a positive map, we have $\zeta_0 \geq 0$ for $x \cdot \eta_\gamma \geq 0$. Therefore, to establish positivity it is sufficient to show that $\zeta(x, t) \geq 0$, for $x \in B_R^+ = \{x \in B_R : x \cdot \eta_\gamma > 0\}$ and $t \geq 0$. We note that by (6.22) there holds $\zeta(x, t) = 0$ for $x \in \pi_\gamma \times [0, \infty)$, hence if ζ is a classical solution of (6.21), then $\zeta(x,t)$ is nonnegative on $B_R^+ \times [0,\infty)$ by the maximum principle. For general $u_0 \in \mathscr{A}_R$ we approximate via mollification as in [7, §4.2, Thm. 2], and note that positivity and symmetry are preserved by the approximation process, rendering $u_0^\varepsilon \in C^\infty(\overline{B_R}; \mathbb{R}^n) \cap L^\infty(B_R; \mathbb{R}^n)$, with $u_0^\varepsilon \to u_0$ in $W^{1,2}(B_R; \mathbb{R}^n)$. For the convenience of the reader we detail below the construction and the properties of the sequence u_0^ε. We first consider an extension of u_0 to the whole space, still called u_0, such that:

- $u_0 \in W_{\mathrm{E}}^{1,2}(\mathbb{R}^n; \mathbb{R}^n) \cap L^\infty(\mathbb{R}^n; \mathbb{R}^n)$ (equivariance),
- $u_0(\overline{F}) \subset \overline{F}$ (positivity).

By the definition of the standard mollifier $\theta_\epsilon(x) := \epsilon^{-n}(\int_{\mathbb{R}^n} \theta)^{-1}\theta(x/\epsilon)$, $\epsilon > 0$, where

$$\theta(x) = \begin{cases} e^{(|x|^2-1)^{-1}} & \text{for } x \in \mathbb{R}^n, \ |x| < 1, \\ 0 & \text{for } x \in \mathbb{R}^n, \ |x| \geq 1, \end{cases}$$

one can check that $u_0^\varepsilon := u_0 * \theta_\epsilon$ is positive and equivariant. Indeed, for every $x \in \mathbb{R}^n$, and $g \in G$,

$$\begin{aligned}
u_0^\varepsilon(gx) &= \int_{B_\epsilon(gx)} \theta_\epsilon(gx - y)u_0(y)\mathrm{d}y \\
&= \int_{B_\epsilon(x)} \theta_\epsilon(gx - gz)u_0(gz)\mathrm{d}z \\
&= g \int_{B_\epsilon(x)} \theta_\epsilon(x - z)u_0(z)\mathrm{d}z = gu_0^\varepsilon(x),
\end{aligned}$$

which proves the equivariance. On the other hand, let $x \in \mathbb{R}^n$ be such that $x \cdot \eta_\gamma \geq 0$, and define $B_\epsilon^1(x) := B_\epsilon(x) \cap (-S_\gamma^+)$, $B_\epsilon^2(x) := \gamma B_\epsilon^1(x)$, and $B_\epsilon^3(x) := B_\epsilon(x) \setminus (B_\epsilon^1(x) \cup B_\epsilon^2(x))$. For $y \in B_\epsilon^1(x)$, we have $|x - y| \geq |x - \gamma y|$ and $0 \leq \theta_\epsilon(x - y) \leq \theta_\epsilon(x - \gamma y)$, while $u_0(y) \cdot \eta_\gamma \leq 0$. As a consequence,

$$\begin{aligned}
\int_{B_\epsilon^1(x)} \theta_\epsilon(x - y)u_0(y) \cdot \eta_\gamma \, \mathrm{d}y &\geq \int_{B_\epsilon^1(x)} \theta_\epsilon(x - \gamma y)u_0(y) \cdot \eta_\gamma \, \mathrm{d}y \\
&= -\int_{B_\epsilon^1(x)} \theta_\epsilon(x - \gamma y)u_0(\gamma y) \cdot \eta_\gamma \, \mathrm{d}y \\
&= -\int_{B_\epsilon^1(x)} \theta_\epsilon(x - z)u_0(z) \cdot \eta_\gamma \, \mathrm{d}z.
\end{aligned}$$

Finally, since $\int_{B_\epsilon^3(x)} \theta_\epsilon(x - y)u_0(y) \cdot \eta_\gamma \, dy \geq 0$, we deduce that

$$u_0^\epsilon(x) \cdot \eta_\gamma = \int_{B_\epsilon^1(x) \cup B_\epsilon^2(x) \cup B_\epsilon^3(x)} \theta_\epsilon(x - y)u_0(y) \cdot \eta_\gamma \, dy \geq 0,$$

from which the positivity of u_0^ϵ follows.

By construction, $u_0^\epsilon \in C^\infty(\overline{B_R}; \mathbb{R}^n) \cap L^\infty(B_R; \mathbb{R}^n)$, with $u_0^\epsilon \to u_0$ in $W^{1,2}(B_R; \mathbb{R}^n)$. Applying the classical maximum principle with initial condition u_0^ϵ, there holds that $\zeta^\epsilon(x, t) := u^\epsilon(x, t) \cdot \eta_\gamma \geq 0$ on $B_R^+ \times [0, \infty)$, and by continuous dependence for (6.21) in $W^{1,2}(B_R)$ [10, Thm. 3.4.1], we have that $\zeta^\epsilon(\cdot, t) \to \zeta(\cdot, t)$ a.e. in B_R along subsequences $\varepsilon_n \to 0$, hence $\zeta(x, t) \geq 0$ a.e.. $\qquad\square$

6.7.2   The Minimization

We consider the minimization problem

$$\min_{\mathscr{A}^R} J_{B_R}, \quad \text{where } J_{B_R}(u) = \int_{B_R} \left\{ \frac{1}{2}|\nabla u|^2 + W(u) \right\} dx.$$

Proceeding exactly as in Sect. 6.6, and observing that the convexity of $\overline{F}$ implies that $\mathscr{A}^R$ is convex and closed in $W_{\mathrm{E}}^{1,2}(B_R; \mathbb{R}^n)$, we deduce the existence of a minimizer u_R satisfying the estimate (6.7). As before, we will construct the solution by taking the limit, possibly along a subsequence,

$$u(x) = \lim_{R \to \infty} u_R(x).$$

For this purpose, we will need to show that the positivity constraint built in $\mathscr{A}^R$ does not affect the Euler–Lagrange equation.

Lemma 6.1 *Let u_R be as above. Then, for every $R > 0$,*

$$\Delta u_R - W_u(u_R) = 0 \quad \text{in } W_{\mathrm{loc}}^{1,2}(B_R; \mathbb{R}^n). \tag{6.24}$$

Proof By Proposition 6.2, we have $u(\cdot, t; u_R) \in \mathscr{A}^R$, for $t \geq 0$. Since u_R is a global minimizer of J_{B_R} in $\mathscr{A}^R$, and since $u(\cdot, t; u_R) \in C^1(0, \infty); C^{2+\alpha}(\overline{B_R}))$, a classical solution to (6.16) for $t > 0$, we conclude from

$$\frac{\mathrm{d}}{\mathrm{d}t} J_{B_R}(u(\cdot, t)) = -\int_{B_R} |u_t|^2 \, dx \tag{6.25}$$

that $|u_t(x, t)| = 0$, for all $x \in B_R$ and $t > 0$. Hence, for $t > 0$, $u(\cdot, t)$ satisfies

$$\Delta u(x, t) - W_u(u(x, t)) = 0. \tag{6.26}$$

Taking $t \to 0+$ and using the continuity of the flow in $W^{1,2}(B_R; \mathbb{R}^n)$ at $t = 0$, $u(\cdot, \cdot; u_R) \in C([0, \infty); W^{1,2}(B_R; \mathbb{R}^n))$, we obtain the lemma. $\qquad\qquad\square$

Remark 6.2 In some situations[1] one can show that

$$\min_{\mathscr{A}^R} J_{B_R} = \min_{W_{\mathrm{E}}^{1,2}(B_R; \mathbb{R}^n)} J_{B_R},$$

from which (6.24) follows immediately without resorting to the gradient flow. Let us explain in the case where $n = 2$, and G is the group generated by the reflections with respect to the coordinate axes, how we can produce a minimizer in $W_{\mathrm{E}}^{1,2}(B_R; \mathbb{R}^n)$ which is positive. Let $v_R(x_1, x_2) = (f(x_1, x_2), g(x_1, x_2))$ be the minimizer of J_{B_R} in $W_{\mathrm{E}}^{1,2}(B_R; \mathbb{R}^n)$. We consider the restriction of v_R to F_R, with $F = \{(x_1, x_2) \in \mathbb{R}^2 : x_1 > 0, \ x_2 > 0\}$, and define the map

$$F_R \ni (x_1, x_2) \longmapsto u_R(x_1, x_2) = (|f(x_1, x_2)|, |g(x_1, x_2)|).$$

Clearly, $u_R(\overline{F}_R) \subset \overline{F}$. Since v_R is equivariant, we can check that the image under u_R of a point belonging to a coordinate axis, remains in the same coordinate axis. This implies that u_R can be extended equivariantly to a $W_{\mathrm{E}}^{1,2}(B_R; \mathbb{R}^n)$ map, still called u_R. In addition, by symmetry we have for every $x \in F_R$:

$$W(u_R(x)) = W(v_R(x)), \text{ and } |\nabla u_R(x)| \le |\nabla v_R(x)| \implies J_{B_R}(u_R) = J_{B_R}(v_R).$$

To describe intuitively the whole construction, we can say that we have 'folded' the image of F_R under v_R, once with respect to each coordinate axis. For general reflection groups this technique does not apply, since foldings do not always preserve the boundary conditions: a point of $\overline{F}_R$ belonging to a reflection plane may not remain after a folding in the same reflection plane.

Remark 6.3 We note that a slight modification of the argument in Proposition 6.1 produces a nontrivial entire positive solution to (6.1). We explain below.

Proposition 6.3 *Let u_R be the minimizer in $\mathscr{A}^R$. Then,*

(i) *$J_{B_r}(u_R) \le C r^{n-1}$, for $r \in (0, R - 1)$, and for a constant C depending only on M.*

(ii) *Let $A_{\bar{q}}^R = \{x \in F_R : |u_R(x) - a_1| \ge \bar{q}\}$, where $\bar{q} > 0$, and arbitrary otherwise. Then, $\mathscr{L}^n(A_{\bar{q}}^R \cap B_r) \le K r^{n-1}$, for $r \in (0, R - 1)$, $R \ge 1$, and with a constant K independent of R.*

---

[1]This is true in particular for all the reflection groups G acting on $\mathbb{R}^2$ and containing the antipodal map $\sigma : u \mapsto -u$, but the proof is somewhat more involved.

Proof

(i) The proof is identical to that of Proposition 6.1. By Lemma 6.1,

$$\Delta u_R - W_u(u_R) = 0 \quad \text{in } W^{1,2}_{\text{loc}}(B_R; \mathbb{R}^n),$$

and also $|u_R(x)| \leq M$ in B_R. By elliptic theory, u_R satisfies the equation classicaly in $\overline{B}_{R-1}$, and also we have the estimate $|\nabla u_R(x)| \leq M'$, in $\overline{B}_{R-1}$, for some constant M' depending only on M. The competitor $\hat{u}_R$ in (6.11) is positive, by the convexity of $\overline{F}$, and hence in $\mathscr{A}^R$. As before, $\hat{u}_R = u_R$ on ∂B_r. Hence (i) follows.

(ii) The positivity of u_R implies that $u_R : \overline{D} \to \overline{D}$, and since a_1 is the unique zero of W in $\overline{D}$, we conclude that

$$W(u_R(x)) \geq \bar{w} > 0 \text{ in } A^R_{\bar{q}}, \quad \bar{w} = \bar{w}(\bar{q}) > 0, \text{ independent of } R. \tag{6.27}$$

This immediately implies the lower bound

$$J_{B_r}(u_R) \geq N\bar{w}\mathscr{L}^n(A^R_{\bar{q}} \cap B_r), \tag{6.28}$$

hence (ii) follows from (i).   □

Corollary 6.2 *Let $u(x) = \lim_{R_k \to \infty} u_{R_k}(x)$, $x \in \mathbb{R}^n$, where the convergence is in C^1_{loc}. Note that u is equivariant, positive, and in $W^{1,2}_{\text{loc}}(\mathbb{R}^n; \mathbb{R}^n)$. Set*

$$A_{\bar{q}} = \{x \in F : |u(x) - a_1| \geq \bar{q}\},$$

with $\bar{q} > 0$, arbitrary otherwise. Then $\mathscr{L}^n(A_{\bar{q}} \cap B_r) \leq K r^{n-1}, \forall r > 0$.

Proof Follows from Proposition 6.3 (ii) above.   □

6.7.3 *Minimality*

Lemma 6.2 *Let $u : \mathbb{R}^n \to \mathbb{R}^n$ be an equivariant solution to (6.1), as established in Corollary 6.2. Then*

$$J(u; \Omega) \leq J(v; \Omega),$$

for every $\Omega \subset \mathbb{R}^n$, open, bounded, with Lipschitz boundary, and for every $v \in C^1(\overline{\Omega}; \mathbb{R}^n)$ such that $v = u$ on $\partial\Omega$, and v is the restriction on $\overline{\Omega}$ of a positive and equivariant map.

Proof The idea is to intersect Ω with every fundamental domain, and use the fact that the energy of an equivariant map is $|G|$ times its energy restricted to a fundamental domain. Let $\mu := \max_{x \in \Omega} |x|$. For any fundamental domain F, we define the map

$$\psi(x) := \begin{cases} (v - u)(x) & \text{for } x \in \overline{F} \cap \Omega, \\ 0 & \text{for } x \in \overline{F} \setminus \Omega. \end{cases}$$

Clearly, $\psi \in C(\overline{F}; \mathbb{R}^m) \cap W^{1,2}(F; \mathbb{R}^m)$. We also notice that if π_γ is a hyperplane bounding F, then $x \in \pi_\gamma \cap \overline{F} \Rightarrow \psi(x) \in \pi_\gamma$. As a consequence, ψ can be extended to an equivariant, continuous and Sobolev map defined in $\mathbb{R}^n$, which we still call ψ. If ψ is positive, the proof is staightforward, since we have successively

$$J_{B_R}(u_R + \psi) \geq J_{B_R}(u_R), \forall R,$$

$$J_{B_\mu}(u_R + \psi) = J_{B_R}(u_R + \psi) - J_{B_R \setminus B_\mu}(u_R)$$

$$\geq J_{B_R}(u_R) - J_{B_R \setminus B_\mu}(u_R) = J_{B_\mu}(u_R), \ \forall R > \mu,$$

and by the C^1 convergence on compacts,

$$J_{B_\mu}(u + \psi) \geq J_{B_\mu}(u).$$

Thus, by equivariance,

$$J_{B_\mu \cap F}(u + \psi) \geq J_{B_\mu \cap F}(u) \Rightarrow J_{\Omega \cap F}(v) \geq J_{\Omega \cap F}(u),$$

and since we have similar inequalities for the other fundamental domains, we obtain the desired result.

In the general case, we utilize a correcting term $\lambda \phi$, where $\lambda > 0$ is fixed, and $\phi(x) = \rho(|x|)x$, with $\rho : [0, \infty) \to [0, 1]$ a smooth function such that

$$\rho(\alpha) := \begin{cases} 1 & \text{for } 0 \leq \alpha \leq \mu, \\ 0 & \text{for } \alpha \geq \mu + 1. \end{cases}$$

We claim that for R large enough $u_R + \psi + \lambda \phi \in \mathscr{A}^R$. Indeed, if π_γ are the hyperplancs bounding F with normal vectors η_γ, then by the C^1 convergence on compacts, there exists $R_0 > \mu + 1$, such that

$$\frac{\partial(u_R - u + \lambda \phi)}{\partial \eta_\gamma}(x) \cdot \eta_\gamma \geq 0, \quad \forall \gamma, \ \forall x \in \overline{B_\mu}, \ \forall R > R_0.$$

Consequently, $u_R - u + \lambda \phi \in \mathscr{A}^\mu$, and since $u + \psi \in \mathscr{A}^\mu$ we obtain that $u_R + \psi + \lambda \phi = (u_R - u + \lambda \phi) + (u + \psi) \in \mathscr{A}^\mu$. Then, we easily see that for $R > R_0$, $u_R + \psi + \lambda \phi \in \mathscr{A}^R$ since $(u_R + \psi + \lambda \phi) = u_R + \lambda \phi$ on $\overline{B_R} \setminus \overline{B_\mu}$).

To conclude, we proceed as before. We have successively

$$J_{B_R}(u_R + \psi + \lambda\phi) \geq J_{B_R}(u_R), \quad \forall R > R_0,$$

$$J_{B_{\mu+1}}(u_R + \psi + \lambda\phi) = J_{B_R}(u_R + \psi + \lambda\phi) - J_{B_R \setminus B_{\mu+1}}(u_R)$$

$$\geq J_{B_R}(u_R) - J_{B_R \setminus B_{\mu+1}}(u_R) = J_{B_{\mu+1}}(u_R), \quad \forall R > R_0,$$

and by the C^1 convergence on compacts,

$$J_{B_{\mu+1}}(u + \psi + \lambda\phi) \geq J_{B_{\mu+1}}(u).$$

Letting $\lambda \to 0$, we obtain

$$J_{B_{\mu+1}}(u + \psi) \geq J_{B_{\mu+1}}(u),$$

and by equivariance,

$$J_{B_{\mu+1} \cap F}(u + \psi) \geq J_{B_{\mu+1} \cap F}(u) \implies J_{\Omega \cap F}(v) \geq J_{\Omega \cap F}(u).$$

Adding the corresponding inequalities for the other fundamental domains, we obtain the desired result. $\qquad\square$

6.7.4 Exponential Decay

Proposition 6.4 *Assume* $\mathbf{H}_1$–$\mathbf{H}_3$ *and let* u *as in Corollary 6.2. Then we have the estimate*

$$|u(x) - a_1| \leq K\mathrm{e}^{-kd(x,\partial D)},$$

where $K = K(M)$ *and* $k = k(c)$ *are positive constants.*

Proof In view of Lemma 4.5, it suffices to establish that given $q \in (0, r_0)$, there is $d_0 > 0$ such that

$$|u(x) - a_1| \leq q, \quad \forall x \in D, \ d(x, \partial D) \geq d_0. \tag{6.29}$$

In proving this we need to consider that u is a minimizer in the space of positive equivariant maps. Since all the comparison maps $\sigma, \ldots$ considered in the proof of Theorem 5.3 satisfy $q^\sigma \leq q^u \ldots$, the positivity of u and the convexity of F imply that $\sigma, \ldots$ are positive maps (Fig. 6.2). Therefore for the proof of (6.29) we can utilize Corollary 5.3 under the conditions specified in Remark 5.11 for equivariant minimizers. We proceed by contradiction. So assume that there is $\{x_k\} \subset D$ such that

$$|u(x_k) - a_1| > q, \quad d(x_k, \partial D) \to \infty \text{ as } k \to \infty. \tag{6.30}$$

Fig. 6.2 The positivity of σ

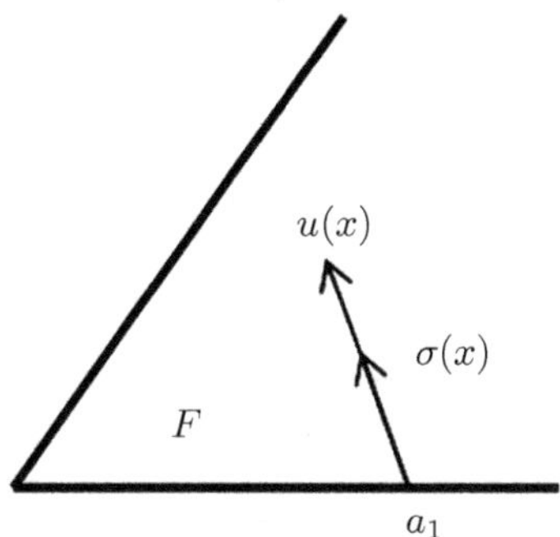

By passing to a subsequence and by a an appropriate choice of a fundamental domain $F \subset D$ we can assume that $\{x_k\} \subset \bar{F}$. Let $\pi_1, \ldots, \pi_l$ be the hyperplanes that correspond to the reflections in the stabilizer G_{a_1} of a_1. Consider first the case

$$\lim_{k \to +\infty} d(x_k, \pi_i) = +\infty, \quad i = 1, \ldots, l.$$

This and $\lim_{k \to +\infty} d(x_k, \partial D) = +\infty$ imply $\lim_{k \to +\infty} d(x_k, \partial F) = +\infty$. This case fits with the situation in Remark 5.11 for $\mathscr{G} = I$ and $\widetilde{\mathscr{G}} = G$. Therefore, we can apply Corollary 5.3 with $\rho = 0$ and, in contradiction with (6.30), we get $|u(x_k) - a_1| \leq q$ as soon as k is so large that $B_{r_q}(x_k) \subset F$. Next consider the general case where there exists some $1 \leq i \leq l$ such that

$$\lim_{k \to +\infty} d(x_k, \pi_i) \leq \text{Const.} \tag{6.31}$$

After a renumbering of the π_i we can assume that the ones that satisfy (6.31) are the first $\bar{l}$ for $1 \leq \bar{l} \leq l$. Let $x_{0,k}$ the orthogonal projection of x_k on $\bigcap_{i=1}^{\bar{l}} \pi_i$. From (6.31) it follows that there is a constant $\rho > 0$ such that $|x_k - x_{0,k}| \leq \rho$. Let $r_{q,\rho}$ the radius in Corollary 5.3 and observe that $\lim_{k \to +\infty} d(x_k, \pi_i) = +\infty$ for $\bar{l} < i \leq l$ together with $\lim_{k \to +\infty} d(x_k, \partial D) = +\infty$ imply that, for k sufficiently large, $B_{r_{q,\rho}}(x_{0,k})$ has empty intersection with all hyperplanes corresponding to reflections in G which are not associated to $\pi_1, \ldots, \pi_{\bar{l}}$. Therefore on the basis of Remark 5.11 we can apply Corollary 5.3 to conclude again $|u(x_k) - a_1| \leq q$ for k large in contradiction with (6.30). This establish (6.29) and concludes the proof of Proposition 6.4 and with it the proof of the theorem.

Next we give a second proof that employs Theorem 5.2, but otherwise is self-contained, at the expense of being repetitions.

Part A   We will first establish that given $c_0 > 0$, there is $d_0 > 0$, depending on c_0, such that

$$|u(x) - a_1| \leq c_0, \quad \forall x \in D, \ d(x, \partial D) \geq d_0. \tag{6.32}$$

We proceed by contradiction. So assume that there is $\{x_k\} \subset D$ such that

$$|u(x_k) - a_1| > c_0, \, d(x_k, \partial D) \to \infty \text{ as } k \to \infty. \tag{6.33}$$

Let Π_{a_1} be the union of all the reflection planes that contain a_1, that is, all the reflections in the stabilizer G_{a_1}. We distinguish two cases.

Case 1 $d(x_k, \Pi_{a_1}) \to \infty$ **as** $k \to \infty$.

It follows that $d(x_k, \partial F) \to \infty$ as $k \to \infty$, and by passing, of necessary, to a subsequence of $\{x_k\}$, we may assume that $\{x_k\} \subset F$. From the bound $|u(x)| \leq M$, via elliptic theory, we obtain as above that $|\nabla u(x)| \leq M'$, some constant. Thus $u(\cdot)$ is uniformly continuous. Therefore, by (6.33), we conclude the existence of a $\mu_0 > 0$ independent of k, such that

$$\mathscr{L}^n(B_1(x_k) \cap \{x : |u(x) - a_1| \geq c_0/2\}) \geq \mu_0. \tag{6.34}$$

At this point we want to invoke the density estimate in Theorem 5.2 and conclude that

$$\mathscr{L}^n(B_R(x_k) \cap \{x : |u(x) - a_1| \geq c_0/2\}) \geq CR^n, \, \forall R \geq 1, \quad B_R(x_k) \subset F. \tag{6.35}$$

To justify this step, we need a couple of observations. Firstly, by Lemma 6.2, u is minimal in the class of equivariant positive maps. For utilizing this, we note that in the proof of the density estimate (cf. (5.36)), the energy comparison maps have the required regularity and are obtained by reducing the modulus of the map, and leaving the angular part unchanged, $u(x) = a_1 + q^u(x)\eta^u(x)$, $\sigma(x) = a_1 + q^\sigma(x)v^u(x)$, $0 \leq q^\sigma(x) \leq q^u(x)$. Therefore, by the convexity of F, the comparison map $\sigma(x)$ is also positive, that is $\sigma(\overline{F}) \subset \overline{F}$. Now, clearly the comparison map can be extended equivariantly from F to $\mathbb{R}^n$, since $B_R(x_k) \subset \overline{F}$. Hence, the proof of the density estimate works without modification and yield the estimate (6.35). Utilizing the positivity of u, and that a_1, by Hypothesis $\mathbf{H}_3$, is the unique zero of W in $\overline{F}$, we obtain from (6.35):

$$W(u(x)) \geq \text{Const.} > 0, \quad \text{on } B_R(x_k) \cap \{x : |u(x) - a_1| \geq c_0/2\} =: E_k. \tag{6.36}$$

This in turn implies the lower bound

$$\int_{B_R(x_k)} W(u(x))dx \geq (\text{Const.})\mathscr{L}^n(E_k) \geq (\text{Const.})R^n, \text{ as long as } B_R(x_k) \subset F. \tag{6.37}$$

We now proceed to derive an easy upper bound on the energy that will contradict (6.37). Let $\chi \in C^1((-\infty, 1], [0, 1])$ be a map such that $\chi(s) = 0$, for $s \leq 0$, and $\chi(1) = 1$, and consider the map

$$\hat{u}(x) = a_1 + \chi\left(|x - x_k| - (R - 1)\right)q^u(x)\eta^u(x), \qquad \text{on } B_R(x_k). \qquad (6.38)$$

This is in $W^{1,2}$ and satisfies $J_{B_R(x_k)}(\hat{u}) \leq CR^{n-1}$, for $R \geq 2$, with C depending only on $\|u\|_{L^\infty}$. In addition, it is a positive map, by the discussion before, that coincides with u on $\partial B_R(x_k)$, and can clearly be extended equivariantly on $\mathbb{R}^n$. By the minimality of u in the positive equivariance class (cf. Lemma 6.2 above) we have

$$J_{B_R(x_k)}(u) \leq J_{B_R(x_k)}(\hat{u}) \leq CR^{n-1}, \qquad R \geq 2. \qquad (6.39)$$

We can now choose a sequence $R = R_k$, with $R_k \to \infty$, $B_{R_k}(x_k) \subset F$, and note that (6.39) and (6.37) are in contradiction for $R = R_k$ and k large. Thus (6.29) is established in Case 1.

Case 2 $d(x_k, \Pi_{a_1}) < \text{Const.}, \ \forall k$.

The obstruction in this case is that we cannot take arbitrarily large balls $B_R(x_k)$, inside F. However, as we will see, we can take large balls $B_R(x_k')$ in D, with x_k' an appropriate projection of x_k, lying at a uniformly bounded distance from x_k. The equivariant extension now is less trivial, but again doable since $B_R(x_k')$ will be an equivariant set (its center x_k' will be on a reflection plane in Π_{a_1}). In the rest, the argument proceeds as before. We now give the details. Consider the subset of the reflection planes that contain a_1 with the property that the distance of x_k from each such plane is uniformly bounded in k. By the hypothesis above, there is at least one such plane. If there is only one such plane, we will take x_k' to be the projection of x_k on this plane. If there are more such planes, we will take as x_k' the projection of x_k on their intersection. By the uniform continuity of u, and the uniform boundedness of $|x_k - x_k'|$ in k, we have that

$$\mathscr{L}^n\left(B(x_k', |x_k - x_k'| + 1) \cap \{x : |u(x) - a_1| \geq c_0/2\}\right) \geq \mu_0 > 0, \mu_0 \text{ as before.} \qquad (6.40)$$

We now proceed as before with $B_R(x_k')$ in the place of $B_R(x_k)$. Finally, we conclude by taking a sequence of balls $B_{R_k}(x_k) \subset D$, with $R_k \to \infty$, and using that a_1 is the unique zero of W in D. The proof of Case 2 is complete, and with it the proof of Part A.

Part B: Comparison Argument   To complete the proof, we need to use the fact that for c_0 small enough, W is strictly convex in the ball $B(a_1, c_0)$. For completeness we give the details of this rather well-known comparison argument.

We note that if we take $c_0 \leq \bar{q}$ in (6.32), then in $D_{d_0} = \{x \in D : d(x, \partial D) \geq d_0\}$, we have by Hypothesis $\mathbf{H}_1$,

$$(u - a_1) \cdot W_u(u) \geq c^2|u - a_1|^2 \Rightarrow \Delta|u - a_1|^2 \geq c^2|u - a_1|^2. \qquad (6.41)$$

To finish, we need an O.D.E. estimate for the radial solution to

$$\begin{cases} \Delta\varphi = c^2\varphi, & \text{in } B_r = B(0; 1), \\ \varphi = 1, & \text{on } \partial B_r, \end{cases} \tag{6.42}$$

(see Appendix A.1).

Conclusion

We will now use a comparison argument on (6.41), (6.42). For $x \in D_{d_0}$, consider the ball with center x and radius $r = d(x, \partial D_{d_0})$, and notice that by (A.3):

$$c_0^2\varphi(0) \leq c_0^2 e^{-kd(x,\partial D_{d_0})}, \quad \text{for } d(x, \partial D_{d_0}) \geq 1,$$

where $k = h(1)$. Therefore, by the maximum principle,

$$|u(x) - a_1|^2 \leq \varphi(0) \leq c_0^2 e^{-kd(x,\partial D_{d_0})}, \quad \text{for } d(x, \partial D_{d_0}) \geq 1,$$

and so the proof of Proposition 6.4 is complete, and with it the proof of the theorem.

$$\square$$

6.8 Heteroclinic Connections for Symmetric Potentials

Before closing this chapter, we shall examine the properties that symmetric potentials induce on heteroclinic connections.

Proposition 6.5 *Let W be a potential satisfying $\mathbf{H}_1$–$\mathbf{H}_3$, and let $a^- := a_1$ be a minimum of W. Then, there exist $a^+ \in \{a_2, \ldots, a_N\}$ and a heteroclinic connection $\bar{u} : \mathbb{R} \to \mathbb{R}^m$, $\bar{u}'' - W_u(\bar{u}) = 0$, $\lim_{t\to\pm\infty} \bar{u}(t) = a^\pm$ with the following properties:*

(i) *$\bar{u}$ is a minimizer of the action J in the class*

$$\mathscr{A} = \{v \in W_{\text{loc}}^{1,2}((l_-^v, l_+^v); \mathbb{R}^m) : -\infty \leq l_-^v < l_+^v \leq +\infty,$$

$$\lim_{t\to l_-^u} v(t) = a_1, \ \lim_{t\to l_+^v} v(t) \in \{a_2, \ldots, a_N\}, \ u((l_-^u, l_+^u)) \subset \mathbb{R}^m \setminus \{a_1, \ldots, a_N\}\}. \tag{6.43}$$

(ii) *$a^+ = \gamma a^-$, where $\gamma \in G$ is a reflection with respect to a hyperplane π_γ bounding the domain D containing a_1 (cf. Hypothesis $\mathbf{H}_3$).*

(iii) *For every $t < 0$, $\bar{u}(t) \in D$, and $\bar{u}(-t) = \gamma\bar{u}(t) \in \gamma D$. On the other hand, π_γ is the unique reflection plane of the group G containing $\bar{u}(0)$.*

Proof By Theorem 2.1, there exists a minimizer $\bar{u} : \mathbb{R} \to \mathbb{R}^m$ of J in the class $\mathscr{A}$, connecting a_1 to a minimum $a^+ \in \{a_2, \dots, a_N\}$. To prove that $\bar{u}$ satisfies (ii) and (iii), let $t_0 = \min\{t \in \mathbb{R} : \bar{u}(t) \in \partial D\}$, and let π_γ be a reflection plane bounding D and containing $\bar{u}(t_0)$. By the translation invariance of J, we may assume without loss of generality that $t_0 = 0$. We claim that

$$J_{(-\infty,0]}(\bar{u}) = \frac{1}{2} J_{\mathbb{R}}(\bar{u}). \tag{6.44}$$

Indeed, if $J_{(-\infty,0]}(\bar{u}) < \frac{1}{2} J_{\mathbb{R}}(\bar{u})$, the map

$$\bar{v}(s) = \begin{cases} \bar{u}(s) & \text{for } s \leq 0, \\ \gamma \bar{u}(-s) & \text{for } s \geq 0, \end{cases} \tag{6.45}$$

belongs to $\mathscr{A}$, and satisfies $J_{\mathbb{R}}(\bar{v}) < \frac{1}{2} J_{\mathbb{R}}(\bar{u})$, which is impossible. Similarly, if $J_{(-\infty,0]}(\bar{u}) > \frac{1}{2} J_{\mathbb{R}}(\bar{u})$, we can construct another competitor in $\mathscr{A}$ with smaller action than $J_{\mathbb{R}}(\bar{u})$. Now that (6.44) is established, it is clear that the map $\bar{v}$ defined in (6.45) is a minimizer of J in $\mathscr{A}$, and also a heteroclinic connection. Since $\bar{u}$ and $\bar{v}$ coincide on the interval $(-\infty, 0]$, it follows by the uniqueness result for O.D.E. that $\bar{u} \equiv \bar{v}$. To complete the proof, it remains to show that π_γ is the unique reflection plane of G containing $\bar{u}(0)$. But if $\bar{u}(0) \in \pi_{\gamma'}$ for another reflection $\gamma' \in G$, then we could construct as in (6.45), a connection between a_1 and $\gamma' a_1$ coinciding with $\bar{u}$ on $(-\infty, 0]$. Again by the uniqueness result for O.D.E., this is a contradiction. $\qquad\square$

Remark 6.4 For every $g \in G_{a_1}$ we also obtain a minimizer in the class $\mathscr{A}$ connecting a_1 and ga_1. More precisely, if G_{a_1} contains k distinct reflections, then a_1 is connected to k distinct minima of W. In the particular cases of the triple and the quadruple junction, we have $n = k = N - 1$, thus any pair of minima (a_i, a_j), $a_i \neq a_j$ is connected by a minimal orbit satisfying the property (iii) of Proposition 6.5 (symmetry and positivity).

6.9   Scholia on Chap. 6

Theorem 6.1 first appeared in [2] in 2011, and was established under the extra hypothesis

$\mathbf{H_4}$ (Q-monotonicity): the potential W is such that there is a continuous function $Q : \mathbb{R}^n \to \mathbb{R}$ that satisfies

$$Q(u + a_1) = |u| + H(u), \tag{6.46a}$$

where $H : \mathbb{R}^n \to \mathbb{R}$ is a C^2 function such that $H(0) = 0$, $H_u(0) = 0$, and

$$Q \text{ is convex,} \tag{6.46b}$$

$$Q(u) > 0 \quad \text{on } \mathbb{R}^n \setminus \{a_1\}, \tag{6.46c}$$

$$Q_u(u) \cdot W_u(u) \geq 0 \quad \text{in } D \setminus \{a_1\}. \tag{6.46d}$$

First observe that (6.46d) holds in D, not in $\mathbb{R}^n$. Very roughly Q-monotonicity implies a certain monotonicity for each well. It allows for nontrivial W's as is explained in Sect. 6.3. The proof in [2] employs the minimization of

$$J_{B_R}(u) = \int_{B_R} \left(\frac{1}{2} |\nabla u|^2 + W(u) \right) dx,$$

under two constraints,

$$u_R(\overline{F}_R) \subset \overline{F} \text{ (positivity)},$$

$$|u_R - a_1| \leq q_0 \text{ for } x \in \omega^R, \text{ a certain subset of } D \cap B_R.$$

Both of these are potentially dangerous for the Euler–Lagrange equation. The positivity constraint is removed by using the gradient as in Sect. 6.7.1 above. The pointwise constraint that was introduced along the lines of the method in Sect. 2.4 is removed by a comparison argument based on

$$|u - a_1| \leq M, \quad x \in \mathbb{R}^n, \tag{6.47a}$$

$$\Delta Q(u) \geq 0, \quad x \in D \text{ (by the convexity of } Q). \tag{6.47b}$$

$$\Delta Q(u) \geq c^2 Q(u), \quad |u(x) - a_1| \leq q_0, \tag{6.47c}$$

which also, after some work, yields the exponential estimate. The proof in [2] is also complicated partly because of the use of a polar representation of the energy based on $Q(u)$. This requires involved arguments near $Q(u) = 0$. Subsequently in [1], in 2012, a different simpler proof was given which imposed only the positivity constraint in the minimization of $J_{B_R}(u)$, and which also localized the arguments in D, thus avoiding equivariant extensions. Also it avoided the Q-polar form. Since the gradient flow $u_t = \Delta u - W_u(u)$, $\frac{\partial u}{\partial n} = 0$ on ∂B_R, preserves positivity and reduces the energy J_{B_R}, it follows that the minimizer u_R is an equilibrium, i.e., $\Delta u_R - W_u(u_R) = 0$, $\frac{\partial u_R}{\partial n} = 0$. On the other hand, the easy estimate

$$\int_{B_R} W(u_R) dx \leq J_{B_R}(u_R) \leq C R^{n-1} \tag{6.48}$$

implies that u is close to a_1 on a set of large measure in B_R, as $R \to \infty$. Using (6.47b) and applying (iteratively) the De Giorgi oscillation lemma (cf. [4, p. 195]), we can convert the integral estimate into pointwise information and deduce that in a smaller ball $B_{R_*}(x_R)$, $|u-a| < q_0$, hence recovering the pointwise estimate previously imposed as a constraint. The exponential estimate is then obtained as in [2] by combining (6.47a), (6.47b), (6.47c), and showing that $|u - a| < q_0$, except possibly on a strip of width l_0 around ∂D. In [8] in 2014 (and subsequently in [9]), the Q-monotonicity hypothesis was eliminated. The proof of the theorem proceeds in two steps. Firstly, it is observed that (6.48) can be upgraded to $J_{B_r}(u_R) \leq Cr^{n-1}$, for all $r \in (0, R]$, with C independent of R. From this, by fixing r and letting $R \to \infty$, we obtain existence of an equivariant nontrivial, minimal, positive u. In the second step the exponential estimate was obtained by a very close variant of the second method presented in Sect. 5.5. Finally in [3] in 2015, the density estimates of Caffarelli and Córdoba were extended to the vector case and as a by product we obtained one more proof of the theorem under optimal hypotheses.

References

1. Alikakos, N.D.: A new proof for the existence of an equivariant entire solution connecting the minima of the potential for the system $\Delta u - W_u(u) = 0$. Commun. Partial Diff. Equ. **37**(12), 2093–2115 (2012)
2. Alikakos, N.D., Fusco, G.: Entire solutions to equivariant elliptic systems with variational structure. Arch. Rat. Mech. Anal. **202**(2), 567–597 (2011)
3. Alikakos, N.D., Fusco, G.: Density estimates for vector minimizers and applications. Discrete Cont. Dyn. Syst. **35**(12), 5631–5663 (2015)
4. Caffarelli, L., Salsa, S.: A Geometric Approach to Free Boundary Problems. Graduate Studies in Mathematics, vol. 68. American Mathematical Society, Providence (2005)
5. De Masi, A., Merola, I., Presutti, E., Vignaud, Y.: Potts models in the continuum. Uniqueness and exponential decay in the restricted ensembles. J. Stat. Phys. **133**, 281–345 (2008)
6. De Masi, A., Merola, I., Presutti, E., Vignaud, Y.: Coexistence of ordered and disordered phases in Potts models in the continuum. J. Stat. Phys. **134**, 243–345 (2009)
7. Evans, L.C., Gariepy, R.F.: Measure Theory and Fine Properties of Functions. CRC Press, Boca Raton (1992)
8. Fusco, G.: Equivariant entire solutions to the elliptic system $\Delta u - W_u(u) = 0$ for general G-invariant potentials. Calc. Var. Part Diff. Equ. **49**(3), 963–985 (2014)
9. Fusco, G.: On some elementary properties of vector minimizers of the Allen-Cahn energy. Commun. Pure Appl. Anal. **13**(3), 1045–1060 (2014)
10. Henry, D.: Geometric Theory of Semilinear Parabolic Equations. Lecture Notes in Mathematics, vol. 840. Springer, Berlin (1981)
11. Palais, R.S.: The principle of symmetric criticality. Commun. Math. Phys. **69**(1), 19–30 (1979)
12. Smoller, J.: Shock Waves and Reaction-Diffusion Equations. Grundlehren der Mathematischen Wissenschaften, vol. 258, 2nd edn. Springer, Berlin (1994)

Chapter 7
Symmetry and the Vector Allen–Cahn Equation: Crystalline and Other Complex Structures

Abstract We present a systematic study of entire symmetric solutions $u : \mathbb{R}^n \to \mathbb{R}^m$ of the vector Allen–Cahn equation $\Delta u - W_u(u) = 0$, $x \in \mathbb{R}^n$, where $W : \mathbb{R}^m \to \mathbb{R}$ is smooth, symmetric, nonnegative with a finite number of zeros, and $W_u := (\partial W/\partial u_1, \ldots, \partial W/\partial u_m)^\top$. We assume that W is invariant under a finite reflection group Γ acting on target space $\mathbb{R}^m$ and that there is a finite or discrete reflection group G acting on the domain space $\mathbb{R}^n$. G and Γ are related by a homomorphism $f : G \to \Gamma$ and a map u is said to be equivariant with respect to f if

$$u(gx) = f(g)u(x), \quad \text{for } g \in G, \ x \in \mathbb{R}^n.$$

We prove two abstract theorems, concerning the cases of G finite and G discrete, on the existence of equivariant solutions. Our approach is variational and based on a mapping property of the parabolic vector Allen–Cahn equation and on a pointwise estimate for vector minimizers. The abstract results are then applied for particular choices of G, Γ and $f : G \to \Gamma$, and solutions with complex symmetric structure are described.

7.1 Introduction

A symmetric nonnegative function $W : \mathbb{R}^m \to \mathbb{R}$, $m > 1$, with a finite number of zeros can model the bulk free energy density of an alloy that can exist in several equally preferred crystalline phases corresponding to the zeros $a_1, \ldots, a_N \in \mathbb{R}^m$ of W. The symmetry of W reflects the symmetry of the underlining microscopic crystal lattice of the alloy [6].

   W depends smoothly on a vector parameter $u \in \mathbb{R}^m$ which describes the fraction of the components of the alloy in each of m sublattices of the microscopic crystal lattice and $u = a_j$, $j = 1, \ldots, N$ corresponds to a pure phase.

   Under the simplifying assumption that the interfacial energy density can be modeled by the isotropic quantity $\frac{1}{2}|\nabla u|^2$, the free energy of the alloy in a bounded

© Springer Nature Switzerland AG 2018
N. D. Alikakos et al., *Elliptic Systems of Phase Transition Type*,
Progress in Nonlinear Differential Equations and Their Applications 91,
https://doi.org/10.1007/978-3-319-90572-3_7

region $\Omega \subset \mathbb{R}^n$, $n \geq 1$ has the form

$$J_\Omega(u) = \int_\Omega \left(\frac{1}{2}|\nabla u|^2 + W(u)\right)dx, \tag{7.1}$$

where $u : \Omega \to \mathbb{R}^m$ describes the distribution of the order parameter in Ω. This assumption is probably adequate for computing surface tension for interfaces parallel to symmetry planes of the microscopic lattice. For general oriented interfaces non-isotropic surface energy densities are necessary [6]. We expect that, in spite of the above simplifying assumption, the analysis that we develop here can be extended to more general energy functionals.

Owing to the symmetry of W, a natural mathematical question is the classification of symmetric bounded entire solution of the Euler–Lagrange equation associated to the free energy (7.1), the vector Allen–Cahn equation

$$\Delta u - W_u(u) = 0, \quad x \in \mathbb{R}^n. \tag{7.2}$$

This question is also relevant from the physical point of view, given that there are situations where different phases of the same alloy organize in space in regular patterns, in particular, lamellar and similar structures of high interfacial energy, which, in spite of their instability with respect to nonsymmetric perturbations, are observed in nature as metastable states.

While the set of general bounded entire solutions of (7.2) is largely unknown and, in particular, there is no established method for describing the geometry of the sets where a minimal solutions of (7.2) is near to one or another of the zeros of W, we show that the pointwise estimates derived in Chap. 5 in combination with the symmetry allow for a fairly complete and systematic study of bounded symmetric entire solutions of (7.2). Indeed, by exploiting the symmetry we prove the existence of minimizers $u : \mathbb{R}^n \to \mathbb{R}^m$ that map a *fundamental domain* in the domain x-space into a *fundamental domain* in the target u-space. A basic consequence of this is that, provided W has a unique zero in the closure of each fundamental domain, a minimizer $u : \mathbb{R}^n \to \mathbb{R}^m$, when restricted to a fundamental domain, remains at a distance from all the zeros of W but one. This allows for the use of Theorem 5.3 and its consequences and a precise understanding of the structure of u becomes possible.

We assume that W is invariant under a finite reflection group Γ acting on $\mathbb{R}^m$ and that there is a reflection group G acting on the domain space $\mathbb{R}^n$. Since we intend to include also periodic patterns, we consider both the cases where G is a finite or an infinite (discrete) reflection group.

We assume that G and Γ are related by a homomorphism $f : G \to \Gamma$ and define a map $u : \mathbb{R}^n \to \mathbb{R}^m$ to be f-equivariant if

$$u(gx) = f(g)u(x), \quad \text{for } g \in G, \ x \in \mathbb{R}^n. \tag{7.3}$$

We characterize the homomorphisms which allow for the existence of f-equivariant maps that send a fundamental domain F for the action of G on $\mathbb{R}^n$ into a fundamental domain Φ for the action of Γ on $\mathbb{R}^m$:

$$u(\overline{F}) \subset \overline{\Phi}. \tag{7.4}$$

We refer to such homomorphisms and to the maps that satisfy (7.4) as *positive*. Positive homomorphisms (see Definition 7.1 below) have certain mapping properties that relate the reflections associated to the walls of a fundamental domain F to the reflections associated to the walls of a corresponding region Φ. These properties are instrumental for showing that minimizing in the class of f-equivariant maps that satisfy (7.4) does not affect the Euler–Lagrange equation and yields a smooth solution of (7.2). The proof of this fact is based on a quite sophisticated use of the maximum principle for parabolic equations that was first introduced in [13] and [3]. We prove (see Lemma 7.2) that, provided f is a positive homomorphism, the L^2 gradient flow associated to the functional (7.1) preserves the positivity condition (7.4). By a careful choice of certain scalar projections of the vector parabolic equation that describes the above mentioned gradient flow, we show that this fact is indeed a consequence of the maximum principle.

Based on this and on a pointwise estimate from Chap. 5 we prove two abstract existence results: Theorem 7.1, which concerns the case where G is a finite reflection group and Theorem 7.2, which treats the case of a discrete (infinite) group G.

From (7.4) and the f-equivariance of u it follows that

$$u(g\overline{F}) \subset f(g)\overline{\Phi}, \quad \text{for } g \in G. \tag{7.5}$$

Therefore, besides its importance for the proofs of Theorems 7.1 and 7.2, the mapping property (7.4) is a source of information on the geometric structure of the vector valued map u. The fact that (7.5) holds true in general in the abstract setting that we consider can perhaps be regarded as one of the significant results of the analysis that we present in this chapter. Indeed, due to the variety of choices for n and m, the dimensions of the domain and target spaces, of the possible choices of the reflection groups G and Γ, and of the homomorphism $f : G \to \Gamma$, we will deduce from Theorem 7.1 and Theorem 7.2 the existence of various complex multi-phase solutions of (7.2), including several types of lattice solutions. A characterization of all homomorphisms between reflection groups in general dimensions is not available. For the special case $n = m = 2$, in Sect. 7.7, we determine all positive homomorphisms between finite reflection groups and the corresponding solutions of (7.2) system.

7.2 Equivariance with Respect to a Group Homomorphism

We begin with some examples of f-equivariant maps. Let I_k be the identity map of $\mathbb{R}^k$, $k \geq 1$. As a first example we observe that, in the particular case where $G = \Gamma$, $n = m$, and the homomorphism f is the identity, f-equivariance reduces to the notion considered in Chap. 6:

$$u(gx) = gu(x), \quad \text{for } g \in G, \ x \in \mathbb{R}^n.$$

The next example is a genuine f-equivariant map. In [4], under the assumption that W is invariant under the group Γ of the equilateral triangle, a solution $u : \mathbb{R}^2 \to \mathbb{R}^2$ to system (7.2) was constructed such that

(i) $u(\gamma x) = \gamma u(x)$ for all $\gamma \in \Gamma$ (which is the dihedral group D_3);
(ii) $u(-x) = u(x)$ for all $x \in \mathbb{R}^2$.

If we incorporate the additional symmetry (ii) in a group structure, this solution can be seen as an f-equivariant map. Indeed, the regular hexagon reflection group $G = D_6$ contains $\Gamma = D_3$, and the antipodal map $\sigma : \mathbb{R}^2 \to \mathbb{R}^2$ given by $\sigma(x) = -x$. Since σ commutes with the elements of D_3, G is isomorphic to the group product $\{I_2, \sigma\} \times D_3$. Furthemore, we can define a homomorphism $f : D_6 = \{I_2, \sigma\} \times D_3 \to D_3$, by setting $f(\gamma) = \gamma$ and $f(\sigma\gamma) = \gamma$, for every $\gamma \in D_3$. Then, the above conditions (i) and (ii) express the f-equivariance of the solution u in [4].

Similarly, we can consider the action on $\mathbb{R}^2$ of the discrete reflection group G' generated by the reflections s_1, s_2 and s_3 with respect to the correponding lines $P_1 := \{x_2 = 0\}$, $P_2 := \{x_2 = x_1/\sqrt{3}\}$, and $P_3 := \{x_2 = -\sqrt{3}(x_1 - 1)\}$ (the dashed lines in Fig. 7.1). These three lines bound a triangle F' with angles 30°, 60° and 90°, which is a fundamental domain of G'. The discrete group G' contains also all the reflections with respect to the lines drawn in Fig. 7.1, which partition the plane into triangles congruent to F'.

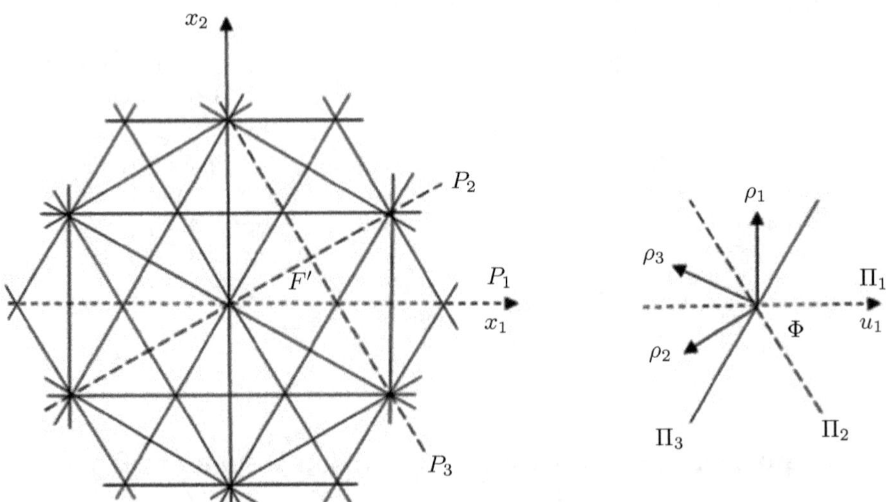

Fig. 7.1 The discrete reflection group G' on the left and the finite reflection group $\Gamma = D_3$ on the right

The point group of G', which is the stabilizer of the origin: $\{g \in G' : g(0) = 0\}$, is the group $G = D_6$, and we have $G' = TG$, where T is the translation group of G'; T is generated by the translations $t^{\pm}$ by the vectors $(\frac{3}{2}, \pm\frac{\sqrt{3}}{2})$. Now, if we compose the canonical homomomorphism $p : G' \to G$ such that $p(tg) = g$ for every $t \in T$ and $g \in G$, with the homomorphism $f : D_6 \to D_3$ defined in the previous paragraph, we obtain a homomorphism $f' : G' \to D_3$. We have, in particular,

$$f'(s_1) = f(p(s_1)) = s_1,$$
$$f'(s_3) = f(p(s_3)) = p(s_3), \tag{7.6}$$
$$f'(s_2) = f'(\sigma p(s_3)) = f(\sigma p(s_3)) = p(s_3),$$

where $p(s_3)$ is the reflection in the line $\Pi_2 = \{u_2 = -\sqrt{3}u_1\}$. We note that the image of the line $P_1 = \{x_2 = 0\}$ under an f'-equivariant map $u : \mathbb{R}^2 \to \mathbb{R}^2$ is contained in the line $\Pi_1 := \{u_2 = 0\}$, while the images of the lines $P_2 = \{x_2 = x_1/\sqrt{3}\}$ and $P_3 = \{x_2 = -\sqrt{3}(x_1 - 1)\}$ are contained in the line $\Pi_2 := \{u_2 = -\sqrt{3}u_1\}$. Indeed

$$x = s_1 x \quad \Longrightarrow \quad u(x) = u(s_1 x) = f'(s_1)u(x) = s_1 u(x),$$
$$x = s_2 x \quad \Longrightarrow \quad u(x) = u(s_2 x) = f'(s_2)u(x) = p(s_3)u(x), \tag{7.7}$$
$$x = s_3 x \quad \Longrightarrow \quad u(x) = u(s_3 x) = f'(s_3)u(x) = p(s_3)u(x).$$

The lines Π_1 and Π_2 define a $60°$ sector Φ which is a fundamental domain of the finite reflection group D_3. At a later stage, we will prove the existence of a solution to (7.2) that maps the triangle F' in this sector.

7.3   The Notion of Positive Homomorphism

Now we return to the general setting and discuss the notion of positive homomorphism $f : G \to \Gamma$ between reflection groups G and Γ. As in Chap. 6, we are interested in constructing f-equivariant solutions $u : \mathbb{R}^n \to \mathbb{R}^m$ to (7.2) that satisfy (7.4), that is, map a fundamental domain F for the action of G on $\mathbb{R}^n$ into a fundamental domain Φ for the action of Γ on $\mathbb{R}^m$. We still call (cf. Sect. 6.5) property (7.4) *positivity* (in analogy with the case where $m = 1$, $\Gamma = \{I_1, \sigma_1\}$, and $\Phi = (0, \infty)$), and characterize the homomorphisms which allow for the existence of such maps. For coherence of terminology, we refer to these homomorphism as *positive*. Positive homomorphisms (see Definition 7.1 below) have certain mapping properties that relate the reflections associated to the walls of a fundamental domain F to the reflections associated to the walls of a correponding domain Φ. These properties are instrumental for showing that minimizing in the class of f-equivariant maps that satisfy (7.4) does not affect the Euler–Lagrange equation and renders a smooth solution of (7.2). Before giving the definition, we observe that if $s \in G$ is a reflection we have $I_m = f(I_n) = f(s)f(s)$. This and the fact that $f(s)$ is an orthogonal transformation imply that $f(s)$ is symmetric. Thus $f(s)$ has m orthonormal eigenvectors $v_1, \ldots, v_m$, and $v_j = f(s)f(s)v_j = \mu_j^2 v_j$ implies that $|\mu_j| = 1$ for the corresponding eigenvalues μ_j, $j = 1, \ldots, m$. Therefore, if we let $E \subset \mathbb{R}^m$ be the span of the eigenvectors corresponding to the eigenvalue $\mu = 1$, that is, $E = \ker(f(s) - I_m)$, the set of the points fixed by $f(s)$, we have $\mathbb{R}^m = E \oplus E^\perp$ and

$$f(s)u = f(s)(u_E + (u - u_E)) = u_E - (u - u_E) = -u + 2u_E, \qquad (7.8)$$

where we have used the decomposition $u = u_E + (u - u_E)$ with $u_E \in E$ and $u - u_E \in E^\perp$. We can interpret (7.8) by saying that $f(s)$ is a symmetry with respect to the subspace E, or that $f(s)$ coincides with the identity on E and with the antipodal map on $E^\perp$.

Definition 7.1 Let F be a fundamental domain of G, bounded by the hyperplanes $P_1, \ldots, P_l$, correponding to the reflections $s_1, \ldots, s_l$. We say that a homomorphism $f : G \to \Gamma$ is *positive* if there exists a fundamental domain Φ of Γ, bounded by the hyperplanes $\Pi_1, \ldots, \Pi_k$, such that for every $i = 1, \ldots, l$, there is an index $1 \le k_i \le k$ and $\tilde{\Pi}_1, \ldots, \tilde{\Pi}_{k_i} \in \{\Pi_1, \ldots, \Pi_k\}$ such that

$$\ker(f(s_i) - I_m) = \bigcap_{j=1}^{k_i} \tilde{\Pi}_j. \qquad (7.9)$$

That is, the set of points fixed by the orthogonal map $f(s_i)$ is one of the hyperplanes Π_j, or the intersection of several of them.

The property of being positive for a homomorphism f is independent of the choice of F. Indeed, if we take $\widehat{F} = gF$, with $g \in G$, then $\widehat{F}$ is bounded by the hyperplanes $gP_1, \ldots, gP_l$, correponding to the reflections $gs_1g^{-1}, \ldots, gs_lg^{-1}$. In addition, $\ker(f(gs_ig^{-1}) - I_m) = f(g)\ker(f(s_i) - I_m)$, thus, the fundamental domain $\widehat{\Phi} = f(g)\Phi$ can be associated with $\widehat{F}$, in accordance with the definition.

Note that the choice of Φ is not unique, since the homomorphism f can associate F to Φ, or to $-\Phi$, with no preference.

The homomorphism $f' : G' \to D_3$ defined in Sect. 7.2 is an example of a positive homomorphism. Indeed, if we identify F with the triangle F' and Φ with the $60°$ sector $\{-\sqrt{3}u_1 < u_2 < 0, \ u_1 > 0\}$ bounded by the lines Π_1 and Π_2, then (7.6) expresses the positivity of f'.

It is not true in general that a homomorphism $f : G \to \Gamma$ between reflection groups G and Γ must be positive. For example, the canonical projection p of a discrete reflection group G' onto its point group G does not, in general, fulfill this requirement. To see this, let us revisit the discrete reflection group G' depicted in

Fig. 7.1. We have

$$p(s_1) = \text{reflection in the line } \Pi_1,$$

$$p(s_2) = \text{reflection in the line } \{u_2 = u_1/\sqrt{3}\},$$

$$p(s_3) = f(s_3) = \text{reflection in the line } \{u_2 = -\sqrt{3}u_1\};$$

then $p(s_i), i = 1, 2, 3$ are reflections with respect to three distinct lines intersecting at the origin. Thus, the canonical projection $p : G' \to G = D_6$ cannot associate F' to any fundamental domain of D_6 (a 30° sector).

A characterization of all homomorphisms between reflection groups in general dimensions is not known. For the special case $n = m = 2$, in Sect. 7.7, we determine all positive homomorphisms between finite reflection groups and the corresponding solutions of (7.2).

7.4 The Theorems

We assume:

H$_1$ (Homomorphism) There exist: a finite (or discrete) reflection group G acting on $\mathbb{R}^n$, a finite reflection group Γ acting on $\mathbb{R}^m$, and a homomorphism $f : G \to \Gamma$ which is positive in the sense of Definition 7.1. We denote by Φ the fundamental domain of Γ that f associates with the fundamental domain F of G.

H$_2$ The potential $W : \mathbb{R}^m \to [0, \infty)$, of class C^3, is invariant under the finite reflection group Γ, that is,

$$W(\gamma u) = W(u), \text{ for all } \gamma \in \Gamma \text{ and } u \in \mathbb{R}^m. \tag{7.10}$$

Moreover, we assume that there exists $M > 0$ such that $W(su) \geq W(u)$, for $s \geq 1$ and $|u| = M$.

H$_3$ There exists $a \in \overline{\Phi}$, the closure of Φ, such that:

(i) $0 = W(a) < W(u)$, for $u \in \overline{\Phi} \setminus \{a\}$, and
(ii) The Hessian matrix $W_{uu}(a)$ is positive definite.

Hypotheses **H$_2$** and **H$_3$** determine the number N of minima of W. From Hypothesis **H$_2$** we have $W(\gamma a) = 0, \forall \gamma \in \Gamma$. Therefore, if $a \in \Phi$, that is, a is in the interior of Φ, then from the fact that $\gamma \Phi \neq \Phi$ for $\gamma \in \Gamma \setminus \{I_m\}$ it follows that W has exactly $N = |\Gamma|$ distinct minima, where $|\mathscr{G}|$ denotes the order of a group $\mathscr{G}$. If $a \in \partial\Phi$, then the stabilizer $\Gamma_a = \{\gamma \in \Gamma : \gamma a = a\}$ of a is nontrivial, and we have $N = |\Gamma|/|\Gamma_a| < |\Gamma|$. In addition, a is the unique minimum of W in the cone $\mathscr{D} \subset \mathbb{R}^m$ defined by $\mathscr{D} = \text{Int}\left(\bigcup_{\gamma \in \Gamma_a} \gamma\overline{\Phi}\right)$. The set $\mathscr{D}$ satisfies

$$\text{for } \gamma \in \Gamma : \text{ either } \gamma\mathscr{D} \cap \mathscr{D} = \varnothing, \text{ or } \gamma\mathscr{D} = \mathscr{D}. \tag{7.11}$$

It follows that $\mathbb{R}^m = \bigcup_{\gamma \in \Gamma} \left(\gamma \overline{\mathscr{D}} \right)$, that is, $\mathbb{R}^m$ is partitioned into $N = |\Gamma|/|\Gamma_a|$ cones congruent to $\mathscr{D}$. The cone $\mathscr{D} \subset \mathbb{R}^m$ has its counterpart in the set $D \subset \mathbb{R}^n$ given by

$$D = \mathrm{Int} \left(\bigcup_{g \in f^{-1}(\Gamma_a)} g \overline{F} \right), \qquad (7.12)$$

which is mapped into $\overline{\mathscr{D}}$ by any positive f-equivariant map $u : \mathbb{R}^n \to \mathbb{R}^m$. Indeed, for such a map (7.5) implies that $u(g\overline{F}) \subset \overline{\mathscr{D}}$ if and only if $f(g) \in \Gamma_a$, or equivalently $g \in f^{-1}(\Gamma_a)$. In the case where $G = \Gamma$ and f is the identity, the expression for D reduces to the set $D = \mathrm{Int}\left(\bigcup_{g \in \Gamma_a} g\overline{F} \right)$ defined in Chap. 6.

The set D satisfies the analog of (7.11). Therefore, also the domain space $\mathbb{R}^n$ is partitioned into $N = |\Gamma|/|\Gamma_a|$ sets congruent to D.

Let us consider a few examples:

(i) if $G = \Gamma$, f is the identity and a is in the interior of Φ, then $\mathscr{D} = \Phi$ and $D = F$.

(ii) if $m = n = 2$, $G = \Gamma = D_3$, $\Phi = \{u : 0 < u_2 < \sqrt{3}u_1, \ u_1 > 0\}$, f is the identity and $a = (1, 0)$, then $\Gamma_a = \{I_2, g_1\}$, where g_1 is the reflection in the line $\{u_2 = 0\}$. Therefore, (7.12) yields $D = \mathrm{Int}\left(\overline{F} \cup g_1 \overline{F} \right) = \{x : |x_2| < \sqrt{3}x_1, \ x_1 > 0\}$.

(iii) if $n = m = 2$, $G = D_6 = \{I_2, \sigma\} \times D_3$, $\Gamma = D_3$, $f(\gamma) = \gamma$ and $f(\sigma\gamma) = \gamma$ for every $\gamma \in D_3$, and $a \in \Phi = \{u : 0 < u_2 < \sqrt{3}u_1, \ u_1 > 0\}$, then $\Gamma_a = \{I_2\}$ and $f^{-1}(\Gamma_a) = \{I_2, \sigma\}$. Therefore, $\mathscr{D} = \Phi$ and $D = \mathrm{Int}\left(\overline{F} \cup \sigma \overline{F} \right) = \{x : 0 < -\frac{x_1 x_2}{|x_1|} < \frac{|x_1|}{\sqrt{3}}, \ x_1 \neq 0\}$. $(F = \{x : 0 < -x_2 < \frac{x_1}{\sqrt{3}}, \ x_1 > 0\}.)$

(iv) If in the previous example we take $a = (1, 0) \in \overline{\Phi}$, we have $\Gamma_a = \{I_2, g_1\}$ and $f^{-1}(\Gamma_a) = \{I_2, \sigma, g_1, \sigma g_1\}$. It follows that $\mathscr{D} = \mathrm{Int}\left(\overline{\Phi} \cup g_1 \overline{\Phi} \right) = \{0 < |u_2| < \sqrt{3}u_1, \ u_1 > 0\}$, and $D = \mathrm{Int}\left(\overline{F} \cup \sigma \overline{F} \cup g_1 \overline{F} \cup \sigma g_1 \overline{F} \right) = \{x : 0 < |x_2| < \frac{|x_1|}{\sqrt{3}}\}$.

If G is a discrete (infinite) group, then D has infinitely many connected components. As examples (iii) and (iv) above show, even when G is a finite group, D does not need to be connected. To characterize one of the connected components of D, let $G^a \subset f^{-1}(\Gamma_a)$ be the subgroup generated by $f^{-1}(\Gamma_a) \cap \{s_1, \ldots, s_l\}$, and define

$$D_0 := \mathrm{Int} \left(\bigcup_{g \in G^a} g \overline{F} \right). \qquad (7.13)$$

Since G^a is a reflection group generated by a subset of $\{s_1, \ldots, s_l\}$, the set of reflections in the planes that bound F, the set D_0 is connected. To show that D_0 is one of the connected components of D, we show that D_0 and $D \setminus D_0$ are

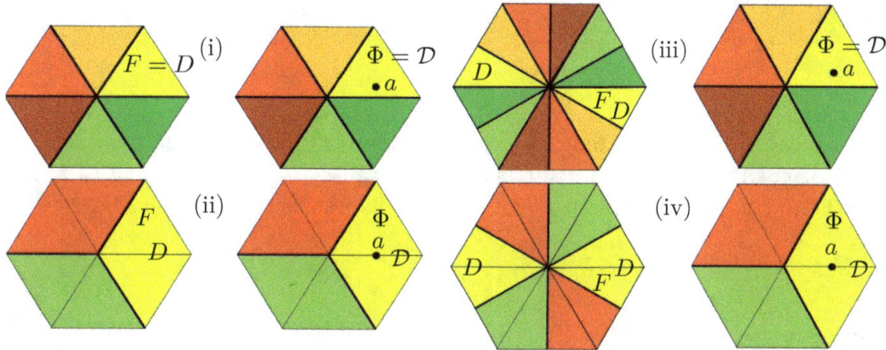

Fig. 7.2 The sets F, Φ, D and $\mathscr{D}$ and their correspondence under an f-equivariant map in the examples (i)–(iv)

disconnected. This is equivalent to proving that if s is the reflection in one of the planes that bound D_0, then $s \notin f^{-1}(\Gamma_a)$. The definition of D_0 implies that $s = s_i s^0 s_i$, for some $s_i \in G^a$ and some reflection $s^0 \in \{s_1, \ldots, s_l\} \setminus f^{-1}(\Gamma_a)$. Therefore, $s^0 = s_i s s_i \;\Rightarrow\; s^0 \in G^a$ if $s \in f^{-1}(\Gamma_a)$.

For examples (i)–(iv), we have: $F = D_0 = D$ in (i); $F \subsetneqq D_0 = D$ in (ii); $F = D_0 \subsetneqq D = D_0 \cup \sigma D_0$ in (iii); $F \subsetneqq D_0 = \mathrm{Int}\left(\overline{F} \cup g_1 \overline{F}\right) \subsetneqq D = D_0 \cup \sigma D_0$ in (iv). Figure 7.2 illustrates these properties.

We are now in a position to state the main results.

Theorem 7.1 *Under Hypotheses* $\mathbf{H}_1$–$\mathbf{H}_3$, *and assuming that G is a finite reflection group, there exist an f-equivariant classical solution u to system (7.2), and positive constants k, K such that*

(i) $u(\overline{F}) \subset \overline{\Phi}$ *and* $u(\overline{D}) \subset \overline{\mathscr{D}}$,
(ii) $|u(x) - a| \le K e^{-kd(x,\partial D)}$, *for* $x \in D$.

Theorem 7.2 *Under Hypotheses* $\mathbf{H}_1$–$\mathbf{H}_3$, *and assuming that G is a discrete reflection group, there exists for every $R > R_0$ (a positive constant), a nontrivial f-equivariant classical solution u_R to the system*

$$\Delta u_R - R^2 W_u(u_R) = 0, \quad \text{for } x \in \mathbb{R}^n, \tag{7.14}$$

such that

(i) $u_R(\overline{F}) \subset \overline{\Phi}$ *and* $u_R(\overline{D}) \subset \overline{\mathscr{D}}$,
(ii) $|u_R(x) - a| \le K e^{-kR d(x,\partial D)}$, *for* $x \in D$, *for positive constants k, K.*

The solution u_R of (7.14) given by Theorem 7.2 is periodic. We describe this periodic structure of u_R under the assumption that the positive homomorphism f

satisfies[1]

$$f(t) = I_m, \quad \text{for } t \in T. \tag{7.15}$$

This covers the examples that we present below. On the other hand, we are not aware of positive homomorphisms that do not satisfy (7.15). Assuming (7.15), if $G = TG_0$ with G_0 the point group of G and T its translation group, we have

$$u_R(tx) = u_R(x), \quad \text{for } t \in T, \; x \in C$$

where

$$C = \text{Int} \bigcup_{g \in G_0} g\overline{F}$$

is the elementary *cell*. C is a convex polytope that satisfies

$$tC \cap C = \varnothing, \quad \text{for } t \in T \setminus \{I_n\}$$

and defines a tessellation of $\mathbb{R}^n$ as the union of the translates tC, $t \in T$ of C: $\mathbb{R}^n = \overline{\bigcup_{t \in T} tC}$. In this sense we can say that u_R has a *crystalline* structure and that C is the elementary *crystal*.

Let us illustrate Theorem 7.2 with the help of the example described in Sect. 7.2, where the discrete reflection group G' acts on the domain x-plane while the finite reflection group $\Gamma = D_3$ acts on the target u-plane. We have already verified that the homomorphism $f' : G' \to \Gamma$ is positive and therefore Hypothesis $\mathbf{H}_1$ is satisfied. When Hypotheses $\mathbf{H}_2$–$\mathbf{H}_3$ also hold, Theorem 7.2 ensures the existence, for every R sufficiently large, of a nontrivial f'-equivariant solution u_R to (7.14), such that $u_R(\overline{F'}) \subset \overline{\Phi}$. By f'-equivariance, the other fundamental domains of G' are mapped into fundamental domains of Γ as in Fig. 7.3. Properties (ii) and (iii) state that for every $x \in D$, $u_R(x)$ approaches, as R grows, the unique minimum a of W in $\overline{\Phi}$, with a speed that depends on $d(x, \partial D)$. If for instance the potential W has six minima (one in the interior of each fundamental domain of D_3), then the set D is the union of the fundamental domains of G' with the same colour (cf. Fig. 7.3 left) and $\mathscr{D}$ is the sector with the same color of D (cf. Fig. 7.3 right). If a lies on the boundary of

---

[1] For an example of a homomorphism that does not satisfy (7.15), take $\Gamma = \{I_m, \gamma\}$ with γ the reflection in the plane $\{u_1 = 0\}$, and $G = \langle s_j \rangle_{j \in \mathbb{Z}} = TG_0$, where s_j is the reflection in the plane $\{x_1 = j\}$, T is the translation group generated by the translation t_0 by the vector $(2, 0, \dots, 0)$, and $G_0 = \{I_n, s_0\}$. Define $f : G \to \Gamma$ by

$$f(t_0) = f(s_{2j}) = \gamma,$$

$$f(s_{2j+1}) = I_m.$$

Fig. 7.3 Fundamental domains for the actions on $\mathbb{R}^2$ of G' (left) and D_3 (right). The f'-equivariant solution u_R of (7.2) given by Theorem 7.2 maps fundamental domains into fundamental domains with the same color

the fundamental domain of D_3, for instance $a = (1, 0)$, then $\mathscr{D}$ is the $120°$ sector that contains a and D is the union of all fundamental domains (triangles) with the same colors of the two sectors that compose $\mathscr{D}$. For this example condition (7.15) is satisfied and the elementary crystal C is the hexagon determined by the fundamental domains (triangles) whose closure contains $0 \in \mathbb{R}^2$.

To give a first application of Theorem 7.1, consider the particular case where $G = D_6 = \{I_2, \sigma\} \times D_3$, $\Gamma = D_3$, and f is the positive homomorphism defined by $f(\gamma) = \gamma$, $f(\sigma\gamma) = \gamma$, for every $\gamma \in D_3$. Figure 7.2 (iii) and (iv) shows the correspondence of the fundamental domains by f- equivariant solutions u of (7.2) when the potential W has respectively six and three minima. The case of W with three minima when u has a sixfold structure, see Fig. 7.2 (iv) was first considered in [4]. Other examples of application of Theorem 7.1 and Theorem 7.2 are given in Sects. 7.6 and 7.7.

7.5   Proofs of Theorems 7.1 and 7.2

The proofs of Theorems 7.1 and 7.2 proceed along lines similar to the proof of Theorem 6.1. We shall recall the differents steps, insisting mainly on the new elements.

7.5.1   Minimization

In the case where G is a finite reflection group, we first construct the solution in a ball $B_R \subset \mathbb{R}^n$ of radius $R > 0$ centered at the origin. We set $F_R = F \cap B_R$ and define the class

$$\mathscr{A}^R := \left\{ u \in W^{1,2}(B_R, \mathbb{R}^m) : u \text{ is } f\text{-equivariant and } u(\overline{F}_R) \subset \overline{\Phi} \right\}$$

in which we have imposed the positivity constraint $u(\overline{F}_R) \subset \overline{\Phi}$. Then, we consider the minimization problem

$$\min_{\mathscr{A}^R} J_{B_R}, \quad \text{where } J_{B_R}(u) = \int_{B_R} \left(\frac{1}{2} |\nabla u|^2 + W(u) \right) dx.$$

Since $\mathscr{A}^R$ is convex and hence weakly closed in $W^{1,2}(B_R, \mathbb{R}^m)$, a minimizer u_R exists, and because of Hypothesis $\mathbf{H}_2$ we can assume that

$$|u_R(x)| \leq M. \tag{7.16}$$

In the case where G is a discrete reflection group, we can work directly in the fundamental domain F. Suppose first that F is bounded. Then, we consider the class

$$\mathscr{A} := \{ u \in W^{1,2}_{\text{loc}}(\mathbb{R}^n, \mathbb{R}^m) : u \text{ is } f\text{-equivariant and such that } u(\overline{F}) \subset \overline{\Phi} \},$$

and choose a minimizer u_R of the problem

$$\min_{\mathscr{A}} J_F^R, \quad \text{where } J_F^R(u) = \int_F \left(\frac{1}{2} |\nabla u|^2 + R^2 W(u) \right) dx,$$

satisfying the estimate (7.16).

Now, suppose that F is not bounded. This implies that all the reflection hyperplanes of G are parallel to a subspace $\{0\}^\nu \times \mathbb{R}^d \subset \mathbb{R}^n$ (with $\nu + d = n$, $d \geq 1$), and that G also acts in $\mathbb{R}^\nu$. Since $F = F_\nu \times \mathbb{R}^d$, with $F_\nu \subset \mathbb{R}^\nu$ bounded, we have, according to the preceding argument, a minimizer $v_R : \mathbb{R}^\nu \to \mathbb{R}^m$ of

$$J_{F_\nu}^R(v) = \int_{F_\nu} \left(\frac{1}{2} |\nabla_{x_\nu} v|^2 + R^2 W(v) \right) dx_\nu,$$

in the analog of $\mathscr{A}$ with n replaced by ν, that is, the class of $W^{1,2}_{\text{loc}}(\mathbb{R}^\nu, \mathbb{R}^m)$ maps v that are f-equivariant and satisfy $v(\overline{F}_\nu) \subset \overline{\Phi}$. Then, we set $u_R(x) := v_R(x_\nu)$, where $x = (x_\nu, x_d) \in \mathbb{R}^n$.

7.5.2 Removing the Positivity Constraint with the Gradient Flow

To show that the positivity constraint built in $\mathscr{A}^R$ (or $\mathscr{A}$) does not affect the Euler–Lagrange equation we will utilize the gradient flow associated to the elliptic system. In the case where G is a finite reflection group we consider

$$
\begin{cases}
\dfrac{\partial u}{\partial t} = \Delta u - W_u(u), & \text{in } B_R \times (0, \infty), \\[2mm]
\dfrac{\partial u}{\partial n} = 0, & \text{on } \partial B_R \times (0, \infty), \\[2mm]
u(x, 0) = u_0(x), & \text{in } B_R,
\end{cases}
\tag{7.17}
$$

where $\partial/\partial n$ is the normal derivative.

In the case where G is a discrete reflection group, we consider

$$
\begin{cases}
\dfrac{\partial u}{\partial t} = \Delta u - R^2 W_u(u), & \text{in } \mathbb{R}^n \times (0, \infty), \\[2mm]
u(x, 0) = u_0(x), & \text{in } \mathbb{R}^n.
\end{cases}
\tag{7.18}
$$

Since W is C^3, the results in [11] apply and provide a unique solution to (7.17) (or (7.18)) which is smooth if we assume that u_0 is globally Lipschitz. In the next two lemmas we will establish that the gradient flow preserves the f-equivariance and the positivity of a smooth initial condition. The properties of the positive homomorphism are instrumental for proving the latter result. In the point reflection group case treated in Chap. 6, one could reduce the problem of positivity to a half-space determined by a reflection plane. Now, one has to deal with the fundamental domain Φ, and chose carefully certain scalar projections of the vector equation (7.17) or (7.18), to which the maximum principle can be applied. We mention that these ideas were first introduced in [3].

Lemma 7.1 *Under Hypothesis* $\mathbf{H}_2$, *if the initial condition* u_0 *is a smooth, f-equivariant map, then for every* $t > 0$, *the solution* $u(\cdot, t)$ *of problem (7.17) (or (7.18)) is also f-equivariant.*

Proof We only present the proof for (7.17), since it is identical for (7.18). Let $g \in G$ and $\gamma := f(g) \in \Gamma < O(\mathbb{R}^m)$. We are going to check that for every $x \in B_R$ and every $t > 0$, we have $u(gx, t) = \gamma u(x, t)$, or equivalently $u(x, t) = \gamma^\top u(gx, t)$. Let us set $v(x, t) := \gamma^\top u(gx, t)$. Since g is an isometry, $\Delta v(x, t) = \gamma^\top (\Delta u)(gx, t)$. On the other hand, we have $\frac{\partial v}{\partial t}(x, t) = \gamma^\top \frac{\partial u}{\partial t}(gx, t)$, and by the symmetry of the potential, $W_u(v(x, t)) = \gamma^\top W_u(u(gx, t))$. Finally, we see that for $x \in \partial B_R$ and $t > 0$, $\dfrac{\partial v}{\partial n}(x, t) = \gamma^\top \dfrac{\partial u}{\partial gn}(gx, t) = 0$. Thus, v is also a smooth solution of (7.17) with initial condition $v_0(x) = \gamma^\top u_0(gx) = u_0(x)$, and by uniqueness $u \equiv v$. $\qquad\square$

Lemma 7.2 *Under Hypotheses* $\mathbf{H}_1$–$\mathbf{H}_2$, *and assuming that the initial condition* u_0 *is a smooth map, we have:*

1. $u_0 \in \mathscr{A}^R \Rightarrow u(\cdot, t) \in \mathscr{A}^R$, $\forall t > 0$, *when* G *is a finite reflection group.*
2. $u_0 \in \mathscr{A} \Rightarrow u(\cdot, t) \in \mathscr{A}$, $\forall t > 0$, *when* G *is a discrete reflection group.*

Proof The idea of the proof is to consider the restriction of the gradient flow to a fundamental domain and assume by contradiction that the above statement 1. or 2. does not hold. According to the value of the gradient flow which maximizes the distance to $\overline{\Phi}$, we project the flow in an appropriate direction ρ. The equivariance and the fact that the angles of Φ are acute, are both essential to locate the point $\tilde{x}$ where this maximum is attained. Next, by considering if necessary the extension of the gradient flow to the union of several fundamental domains, one shows that $\tilde{x}$ is an interior point, and then the maximum principle is applied to reach a contradiction. We first present the proof in a specific case where the argument can be described with simpler notation and then consider the abstract situation and give the proof for the general case.

The case we discuss first is the example of Sect. 7.2, where we have the discrete reflection group G' acting on the domain, $\Gamma = D_3$ acting on the target, and the homomorphism $f' : G' \to \Gamma$. We preserve the notation of Sect. 7.2 and refer to Fig. 7.1 and to the comments following Theorem 7.2. In particular, we still denote by F' the fundamental domain of G' and by f' the homomorphism $G' \to \Gamma$ (which are denoted by F and f in the statement of Theorem 7.2 and in Sect. 7.5.1). We also denote by ρ_1 and ρ_2 the outward unit normals to the lines Π_1 and Π_2 that bound the fundamental domain Φ of Γ. Let $\Pi_3 = \{u_2 = \sqrt{3}u_1\}$ be the third reflection line of Γ and let $\rho_3 := (-\sqrt{3}/2, 1/2) \perp \Pi_3$. From (7.18) and the symmetry of W given by (7.10), it follows that for every $j = 1, 2, 3$, the projection $h_j(x, t) := u(x, t) \cdot \rho_j$ satisfies the linear scalar equation:

$$\Delta h_j + c_j^* h_j - \frac{\partial h_j}{\partial t} = 0, \quad \text{in } \mathbb{R}^2 \times (0, +\infty),$$

with $c_j^* = R^2 c_j$ and c_j (cf. (7.22) below) continuous and bounded on $\mathbb{R}^2 \times [0, T]$, for every $T > 0$.

Now, suppose that for some $t_0 > 0$, $u(\cdot, t_0)$ does not belong to the class $\mathscr{A}$.

In order to have equations with nonpositive coefficients, we perform the standard transformation and set $\tilde{h}_j(x, t) := e^{-\lambda t} h_j(x, t)$, where the constant λ is chosen larger than $\sup\{c_j^*(x, t) : x \in \mathbb{R}^2, \ t \in [0, t_0], \ j = 1, 2, 3\}$. Then, we have

$$\Delta \tilde{h}_j + \tilde{c}_j^* \tilde{h}_j - \frac{\partial \tilde{h}_j}{\partial t} = 0, \quad \text{in } \mathbb{R}^2 \times (0, t_0], \text{ with } \tilde{c}_j^* = c_j^* - \lambda \leq 0. \tag{7.19}$$

Let $\mu := \max\{d(e^{-\lambda t}u(x,t),\overline{\Phi}) : x \in \overline{F'},\ t \in [0,t_0]\} > 0$, and suppose that this is achieved at $\tilde{x} \in \overline{F'}$ at time $\tilde{t} \in (0,t_0]$ (since $u_0 \in \mathscr{A}$). Define

$$\tilde{u} := e^{-\lambda\tilde{t}}u(\tilde{x},\tilde{t}),\quad \rho := \frac{\tilde{u} - \tilde{v}}{|\tilde{u} - \tilde{v}|},$$

where $\tilde{v}$ is the unique point of $\partial\Phi$ (since Φ is convex) such that $d(\tilde{u},\tilde{v}) = \mu$. According to the direction of ρ, we distinguish the following cases:

(i) If $\rho = \rho_1$, then $\tilde{v} \in \Pi_1 \cap \overline{\Phi}$ and we define

$$\omega := \{(x,t) \in \mathbb{R}^2 \times (0,t_0] : e^{-\lambda t}u(x,t) \cdot \rho_1 > 0\}.$$

Clearly, $(\tilde{x},\tilde{t}) \in \omega$, which is relatively open in $\mathbb{R}^2 \times (0,t_0]$.

(ii) If $\rho = \rho_2$, then $\tilde{v} \in \Pi_2 \cap \overline{\Phi}$. Similarly, define

$$\omega := \{(x,t) \in \mathbb{R}^2 \times (0,t_0] : e^{-\lambda t}u(x,t) \cdot \rho_2 > 0\},$$

and we have $(\tilde{x},\tilde{t}) \in \omega$, which is relatively open in $\mathbb{R}^2 \times (0,t_0]$.

(iii) If $\rho = \rho_3$, then $\tilde{v} = 0$. We check that

$$(\tilde{x},\tilde{t}) \in \omega := \{(x,t) \in \mathbb{R}^2 \times (0,t_0] : e^{-\lambda t}u(x,t) \cdot \rho_3 > 0\},$$

which is relatively open in $\mathbb{R}^2 \times (0,t_0]$.

(iv) If $\rho = \alpha_1\rho_1 + \alpha_3\rho_3$ with $\alpha_1,\alpha_3 > 0$, then $\tilde{v} = 0$ and we define

$$\omega := \{(x,t) \in \mathbb{R}^2 \times (0,t_0] : e^{-\lambda t}u(x,t) \cdot \rho_j > 0 \text{ for } j = 1 \text{ and } j = 3\}.$$

Thanks to the fact that $\rho_1 \cdot \rho_3 \geq 0$, we have again $(\tilde{x},\tilde{t}) \in \omega$ which is relatively open in $\mathbb{R}^2 \times (0,t_0]$.

(v) If $\rho = \alpha_2\rho_2 + \alpha_3\rho_3$ with $\alpha_2,\alpha_3 > 0$, then $\tilde{v} = 0$ and we define ω in a similar way.

We want to apply the maximum principle to $\tilde{h}(x,t) := e^{-\lambda t}u(x,t) \cdot \rho$ in a neighborhood of $(\tilde{x},\tilde{t})$. In the cases (i), (ii) and (iii) above, the Eq. (7.19) trivially holds in ω. In the cases (iv) and (v), we have the inequality

$$\Delta\tilde{h} + \tilde{c}^*\tilde{h} - \frac{\partial\tilde{h}}{\partial t} \geq 0 \text{ in } \omega,\quad \text{with } \tilde{c}^* = \max\{\tilde{c}_j^* : j = 1,2,3\} \leq 0.$$

Indeed, we can check that for instance in the case (iv):

$$\Delta\tilde{h} + \tilde{c}^*\tilde{h} - \frac{\partial\tilde{h}}{\partial t} = \alpha_1(\tilde{c}^* - \tilde{c}_1^*)\tilde{h}_1 + \alpha_3(\tilde{c}^* - \tilde{c}_3^*)\tilde{h}_3 \geq 0,\quad \forall(x,t) \in \omega.$$

At this stage, the fact that Φ is an acute angle sector (i.e., $\rho_1 \cdot \rho_2 \le 0$) is essential to conclude the proof. This property implies that

$$\tilde{u} \in \Pi_j \ (\text{with } j = 1, 2) \Longrightarrow \tilde{v}, \ \rho \in \Pi_j, \tag{7.20}$$

and as a consequence

$$f'(s_i) \notin \Gamma_\rho \Longrightarrow \tilde{x} \notin P_i, \tag{7.21}$$

where P_i $(i = 1, 2, 3)$ is a line bounding F', corresponding to the reflection $s_i \in G'$. To show (7.21), suppose that $\tilde{x} \in P_i$. Then, $s_i(\tilde{x}) = \tilde{x}$, and by f'-equivariance

$$u(\tilde{x}, \tilde{t}) = u(s_i(\tilde{x}), \tilde{t}) = f'(s_i)u(\tilde{x}, \tilde{t}) \Longleftrightarrow u(\tilde{x}, \tilde{t}) \in \ker(f'(s_i) - I_2).$$

Since $\ker(f'(s_i) - I_2)$ is either the line Π_1 or the line Π_2 (cf. Hypothesis $\mathbf{H}_1$), we deduce thanks to (7.20) that

$$\rho \in \ker(f'(s_i) - I_2) \Longleftrightarrow f'(s_i) \in \Gamma_\rho.$$

Property (7.21) will enable us to locate $\tilde{x}$ in $\overline{F'}$. Indeed, if $\Gamma_\rho = \{I_2\}$, we immediately see that $\tilde{x}$ is an interior point of F'. Otherwise we extend by reflection $\tilde{h}$ to the union of several fundamental domains. Note that the acute angle property of Φ allows both for proving (7.21) and reflecting. Let $\widetilde{G}'$ be the subgroup of G' generated by the reflections s_i $(i = 1, 2, 3)$ such that $f'(s_i) \in \Gamma_\rho$. By f'-equivariance,

$$\mu = \tilde{h}(\tilde{x}, \tilde{t}) = \max\{\tilde{h}(x, t) : x \in \bigcup_{g \in \widetilde{G}'} g\overline{F'}, \ t \in [0, t_0]\},$$

and now $\tilde{x} \in \mathrm{Int}\left(\bigcup_{g \in \widetilde{G}'} g\overline{F'}\right)$. Thus, thanks to the maximum principle for parabolic equations applied in ω, we see that $\tilde{h}(x, \tilde{t}) \equiv \mu$, for $x \in B_\delta(\tilde{x}) \cap \overline{F'}$ (where $\delta > 0$). To finish the proof, we show that the set $S := \{y \in \overline{F'} : \tilde{h}(y, \tilde{t}) = \mu\}$ is relatively open. Indeed, let $y \in S$ and let w be the projection of $e^{-\lambda \tilde{t}} u(y, \tilde{t})$ on $\overline{\Phi}$. We have $e^{-\lambda \tilde{t}} u(y, \tilde{t}) - w = \mu\rho$, and repeating the above argument we find $\tilde{h}(x, \tilde{t}) \equiv \mu$, for $x \in B_{\delta'}(y) \cap \overline{F'}$ (where $\delta' > 0$). Thus, by connectedness, $\tilde{h}(\cdot, \tilde{t}) \equiv \mu > 0$ on $\overline{F'}$, which is a contradiction since $\tilde{h}(0, \tilde{t}) = 0$.

Let us now give the proof of the lemma for arbitrary groups. We just present it when G is a finite reflection group, since it is similar for discrete reflection groups. We will need to apply the maximum principle to some projections of the solution u. We denote by $\rho_1, \ldots, \rho_k$ the outward unit normals to the hyperplanes $\Pi_1, \ldots, \Pi_k$ that bound the fundamental domain Φ (see Definition 7.1). We also consider the collection $\Pi_1, \ldots, \Pi_q$ $(k \le q)$ of all the reflection hyperplanes of Γ, and choose for $j > k$, a unit normal ρ_j to Π_j. Since the potential W is symmetric, we know that for every $j = 1, \ldots, q$, the projection $h_j(x, t) := u(x, t) \cdot \rho_j$ satisfies the linear scalar equation

$$\Delta h_j + c_j h_j - \frac{\partial h_j}{\partial t} = 0, \quad \text{in } B_R \times (0, +\infty),$$

with

$$c_j = -\left(\int_0^1 W_{uu}\big(u + (s-1)(u \cdot \rho_j)\rho_j\big)\rho_j \, ds\right) \cdot \rho_j. \tag{7.22}$$

This formula, shows that c_j is continuous and bounded on $B_R \times [0, T]$, for every $T > 0$.

Now, suppose that for some $t_0 > 0$, $u(\cdot, t_0)$ does not belong to the class $\mathscr{A}^R$.

In order to have an equation with a nonpositive coefficient, we again perform the standard transformation and set $\tilde{h}_j(x,t) := \mathrm{e}^{-\lambda t} h_j(x,t)$, where the constant λ is chosen bigger than $\sup\{c_j(x,t) : x \in B_R,\ t \in [0, t_0],\ j = 1, \ldots, q\}$. Then, we have

$$\Delta \tilde{h}_j + \tilde{c}_j \tilde{h}_j - \frac{\partial \tilde{h}_j}{\partial t} = 0, \quad \text{in } B_R \times (0, t_0], \text{ with } \tilde{c}_j = c_j - \lambda \le 0.$$

Let $\mu := \max\{d(\mathrm{e}^{-\lambda t} u(x,t), \overline{\Phi}) : x \in \overline{F_R},\ t \in [0, t_0]\} > 0$, and suppose that this is achieved at $\tilde{x} \in \overline{F_R}$ at time $\tilde{t} \in (0, t_0]$ (since $u_0 \in \mathscr{A}_R$). Define

$$\tilde{u} := \mathrm{e}^{-\lambda \tilde{t}} u(\tilde{x}, \tilde{t}), \quad \rho := \frac{\tilde{u} - \tilde{v}}{|\tilde{u} - \tilde{v}|},$$

where $\tilde{v}$ is the unique point of $\partial \Phi$ (since Φ is convex) such that $d(\tilde{u}, \tilde{v}) = \mu$. We will apply the maximum principle to $\tilde{h}(x,t) := \mathrm{e}^{-\lambda t} u(x,t) \cdot \rho$ in a neighborhood of $(\tilde{x}, \tilde{t})$ in $\overline{B_R} \times (0, t_0]$. To do this, in analogy to what was done in the special case considered above we need to consider various possibilities for the unit vector ρ. If $\tilde{v}$ belongs to the interior of an $m - p$ dimensional face $\Pi_1 \cap \ldots \cap \Pi_p \cap \overline{\Phi}$ ($1 \le p \le k$) of Φ, then, using also that $\rho_1, \ldots, \rho_k$ are linearly independent, we have

$$\rho \perp E, \quad E := \Pi_1 \cap \ldots \cap \Pi_p, \text{ that is, } \rho \in E^\perp = \mathbb{R}\rho_1 \oplus \cdots \oplus \mathbb{R}\rho_p$$

where $\mathbb{R}\rho_j = \{x : x = t\rho_j, t \in \mathbb{R}\}$ and $E^\perp$ is the orthogonal complement of E. Let $\widetilde{\Gamma}$ be the subgroup of Γ generated by the reflections with respect to the hyperplanes $\Pi_1, \ldots, \Pi_p$. The elements of $\widetilde{\Gamma}$ leave invariant the subspace E, and actually $\widetilde{\Gamma}$ acts in $E^\perp$. Let $N \supset \{\pm\rho_1, \ldots, \pm\rho_p\}$ be the set of all the unit normals to the reflection hyperplanes of $\widetilde{\Gamma}$. We claim that

$$\begin{cases} \rho = \alpha_1 \nu_1 + \ldots + \alpha_{\tilde{p}} \nu_{\tilde{p}}, \text{ with } 1 \le \tilde{p} \le p,\ \alpha_1, \ldots, \alpha_{\tilde{p}} > 0,\ \nu_1, \ldots, \nu_{\tilde{p}} \in N, \\ \text{and } \nu_i \cdot \nu_j \ge 0, \text{ for } 1 \le i, j \le \tilde{p}. \end{cases}$$

$$\tag{7.23}$$

Given p linearly independent vectors $v_1, \ldots, v_p \in N$ we denote by $C(v_1, \ldots, v_p)$ the cone

$$C(v_1, \ldots, v_p) := \{\alpha_1 v_1 + \cdots + \alpha_p v_p : \alpha_1, \ldots, \alpha_p \geq 0\}. \qquad (7.24)$$

To prove the claim we start by observing that (since $\rho_1, \ldots, \rho_p$ are linearly independent) we have $\rho \in \mathscr{C}$, where

$$\mathscr{C} = C(\rho_1', \ldots, \rho_p'), \quad \text{with } \rho_j' = \rho_j \text{ or } -\rho_j, \ j = 1, \ldots, p.$$

To conclude the proof we show that $\mathscr{C}$ can be partitioned into cones of the form (7.24) that satisfy the condition

$$C(v_1, \ldots, v_p) \cap N = \{v_1, \ldots, v_p\}. \qquad (7.25)$$

Note that (7.25) and the fact that $\widetilde{\Gamma}$ is a reflection group automatically imply

$$v_i \cdot v_j \geq 0, \quad \text{for } 1 \leq i, j \leq p.$$

Indeed, N is invariant under $\widetilde{\Gamma}$ (N is the root system of $\widetilde{\Gamma}$) and therefore $nu_i \cdot v_j < 0$ implies that the reflection $v \notin \{v_i, v_j\}$ of v_j in the hyperplane orthogonal to v_i belongs to both N and $C(v_1, \ldots, v_p)$, in contradiction with (7.25). If $\mathscr{C}$ does not satisfy (7.25), then there exists $v \in N$ that (possibly after a renumbering the vectors $\rho_1', \ldots, \rho_p'$) has the form $v = \alpha_1 \rho_1' + \cdots + \alpha_{\hat{p}} \rho_{\hat{p}}'$ with $2 \leq \hat{p} \leq p, \alpha_1, \ldots, \alpha_{\hat{p}} > 0$ and we can partition $\mathscr{C}$ into the $\hat{p}$ cones $\mathscr{C}_i = C(\rho_1', \ldots, \rho_{i-1}', v, \rho_{i+1}', \ldots, \rho_p'), i = 1, \ldots, \hat{p}$ defined by the linearly independent vectors $\rho_1', \ldots, \rho_{i-1}', v, \rho_{i+1}', \ldots, \rho_p'$, $i = 1, \ldots, \hat{p}$. If $\mathscr{C}_i$ does not satisfy (7.25), we partition $\mathscr{C}_i$ in the same fashion used for $\mathscr{C}$ and continue in this way. Note that at each step (if some of the cones of the partition do not satisfy (7.25)) the number of vectors in N used to generate the cones of the partition increases by one. Therefore, since N is a finite set, the process terminates after a finite number of steps exactly when all the cones of the partition satisfy (7.25). This concludes the proof of the claim.

Since $\tilde{v} \in E$, it follows that (with $v_1 \ldots, v_{\tilde{p}}$ the vectors in (7.23))

$$(\tilde{x}, \tilde{t}) \in \omega := \{(x, t) \in \overline{B_R} \times (0, t_0] : e^{-\lambda t} u(x, t) \cdot v_j > 0, \ \forall j = 1, \ldots, \tilde{p}\},$$

which is relatively open in $\overline{B_R} \times (0, t_0]$, and in addition we have

$$\Delta \tilde{h} + \tilde{c} \tilde{h} - \frac{\partial \tilde{h}}{\partial t} \geq 0 \text{ in } \omega, \text{ with } \tilde{c} = \max\{\tilde{c}_j : j = 1, \ldots, q\} \leq 0.$$

At this stage, the fact that Φ has acute angles (i.e., $\rho_i \cdot \rho_j \leq 0$ for $1 \leq i < j \leq k$) is essential to conclude the proof. This property implies that

$$\tilde{u} \in \Pi_j \text{ (with } 1 \leq j \leq k) \Longrightarrow \tilde{v}, \rho \in \Pi_j, \qquad (7.26)$$

and as a consequence

$$f(s_i) \notin \Gamma_\rho \implies \tilde{x} \notin P_i, \tag{7.27}$$

where P_i is, as in Definition 7.1, a hyperplane bounding F corresponding to the reflection $s_i \in G$. To show (7.27), suppose that $\tilde{x} \in P_i$. Then

$$u(\tilde{x}, \tilde{t}) \in \ker(f(s_i) - I_m) \quad \text{by } f \text{-equivariance,}$$

and thanks to Hypothesis $\mathbf{H}_1$ and (7.26),

$$\rho \in \ker(f(s_i) - I_m) \implies f(s_i) \in \Gamma_\rho.$$

Property (7.27) will enable us to locate $\tilde{x}$ in $\overline{F}_R$. Let $\widetilde{G}$ be the subgroup of G generated by the reflections s_i ($i = 1, \ldots, l$) such that $f(s_i) \in \Gamma_\rho$. By f-equivariance, we have

$$\mu = \tilde{h}(\tilde{x}, \tilde{t}) = \max\{\tilde{h}(x, t) : x \in \cup_{g \in \widetilde{G}} g\overline{F_R}, \ t \in [0, t_0]\}.$$

But now either $\tilde{x} \in \text{Int}\left(\bigcup_{g \in \widetilde{G}} g\overline{F_R}\right)$, or $\tilde{x} \in \text{Int}\left(\bigcup_{g \in \widetilde{G}} g\overline{F}\right) \cap \partial B_R$. In both cases, thanks to the maximum principle for parabolic equations applied in ω and thanks to Hopf's Lemma, $\tilde{h}(x, \tilde{t}) \equiv \mu$, for $x \in B_\delta(\tilde{x}) \cap \overline{F_R}$ (where $\delta > 0$). To finish the proof we show that the set $S := \{y \in \overline{F_R} : \tilde{h}(y, \tilde{t}) = \mu\}$ is relatively open. Indeed, let $y \in S$ and let w be the projection of $e^{-\lambda \tilde{t}} u(y, \tilde{t})$ on $\overline{\Phi}$. We have $e^{-\lambda \tilde{t}} u(y, \tilde{t}) - w = \mu\rho$ and according to the preceding argument $\tilde{h}(x, \tilde{t}) \equiv \mu$ for $x \in B_{\delta'}(y) \cap \overline{F_R}$ (where $\delta' > 0$). Thus, by connectedness, $\tilde{h}(\cdots, \tilde{t}) \equiv \mu > 0$ on $\overline{F}_R$, which is a contradiction since $\tilde{h}(0, \tilde{t}) = 0$. $\qquad\square$

If the initial condition in (7.17) (respectively (7.18)) is a $W^{1,2}(B_R, \mathbb{R}^m)$ (respectively, $W_{\text{loc}}^{1,2}(\mathbb{R}^n, \mathbb{R}^m)$) bounded map, then the solution to (7.17) (respectively, (7.18)) belongs to $C([0, \infty), W^{1,2}(B_R, \mathbb{R}^m))$ (respectively, $C([0, \infty), W_{\text{loc}}^{1,2}(\mathbb{R}^n, \mathbb{R}^m))$), and is smooth for $t > 0$. We are now going to take the minimizer u_R constructed in Step 1 as the initial condition in (7.17) (respectively, (7.18)). Thanks to Lemma 7.3 below (and to its analog for discrete reflection groups), we can construct a sequence of smooth, f-equivariant, and positive maps (u_k) which converges to u_R in the $W^{1,2}$ norm, as $k \to \infty$. Applying then Lemmas 7.1 and 7.2 to u_k, and using the continuous dependence of the flow on the initial condition, we obtain that the solution to (7.17) (respectively, (7.18)), with initial condition u_R, is f-equivariant and positive, that is, $u(\cdot, t; u_R) \in \mathscr{A}^R$, for $t \geq 0$ (respectively, $u(\cdot, t; u_R) \in \mathscr{A}$, for $t \geq 0$).

Lemma 7.3 *Let* $u \in W^{1,2}(B_R, \mathbb{R}^m) \cap L^\infty(B_R, \mathbb{R}^m)$ *be an* f-*equivariant map such that* $u(\overline{F_R}) \subset \overline{\Phi}$. *Then, there exists a sequence* $(u_k) \subset C(\overline{B_R}, \mathbb{R}^m)$ *of globally*

Lipschitz maps with the following properties:

(i) u_k is f-equivariant;
(ii) $\|u_k\|_{L^\infty(B_R,\mathbb{R}^m)} \le \|u\|_{L^\infty(B_R,\mathbb{R}^m)}$;
(iii) $u_k(\overline{F_R}) \subset \overline{\Phi}$ (positivity);
(iv) u_k converges to u in $W^{1,2}(B_R,\mathbb{R}^m)$, as $k \to \infty$.

Proof See the end of the proof of Proposition 6.2, and also [1] or [3, Proposition 5.2]. $\qquad\square$

Since u_R is a global minimizer of J_{B_R} (respectively, J_F^R) in $\mathscr{A}^R$ (respectively, $\mathscr{A}$), and since $u(\cdot, t; u_R)$ is a classical solution to (7.17) (respectively (7.18)) for $t > 0$, we conclude from the calculation

$$\frac{\mathrm{d}}{\mathrm{d}t} J_{B_R}(u(\cdot, t)) = -\int_{B_R} |u_t|^2 \, \mathrm{d}x \left(\text{respectively}, \; \frac{\mathrm{d}}{\mathrm{d}t} J_F^R(u(\cdot, t)) = -\int_F |u_t|^2 \, \mathrm{d}x\right)$$

$$(7.28)$$

that $|u_t(x, t)| = 0$, for all $x \in B_R$ (respectively, $x \in F$) and $t > 0$. Hence, for $t > 0$, $u(\cdot, t)$ satisfies

$$\Delta u(x, t) - W_u(u(x, t)) = 0 \;\; (\text{respectively } \Delta u(x, t) - R^2 W_u(u(x, t)) = 0).$$

$$(7.29)$$

Letting $t \to 0+$ and using the continuity of the flow in $W^{1,2}(B_R,\mathbb{R}^m)$ (respectively $W_{\mathrm{loc}}^{1,2}(\mathbb{R}^n,\mathbb{R}^m)$) at $t = 0$, we obtain that u_R is an f-equivariant, classical solution to system (7.2) (respectively (7.14)) satisfying also $u_R(\overline{F_R}) \subset \overline{\Phi}$ (respectively $u_R(\overline{F}) \subset \overline{\Phi}$).

If G is a finite reflection group, then the family of solutions $u_R \in C^3(B_R,\mathbb{R}^m)$, $R \ge 1$, yields an entire f-equivariant classical solution $u \in C^3(\mathbb{R}^n,\mathbb{R}^m)$ to system (7.2) defined by

$$u(x) = \lim_{j \to +\infty} u_{R_j}(x),$$

$$(7.30)$$

where $R_j \to +\infty$ is a suitable subsequence and the convergence is in the C^2 sense on compact subsets of $\mathbb{R}^n$. This follows from the fact that u_R satisfies the bound

$$\|u_R\|_{C^{2+\alpha}(\overline{B_R},\mathbb{R}^m)} \le M',$$

$$(7.31)$$

for some $\alpha \in (0, 1)$ and $M' > 0$ independent of $R \ge 1$. The estimate (7.31) follows by elliptic regularity from (7.16), from the assumed smoothness of W and from the fact that ∂B_R is uniformly smooth for $R \ge 1$. The solution u satisfies also: $u(\overline{F}) \subset \overline{\Phi}$.

7.5.3   Pointwise Estimates

Continuing Sect. 7.5.2, to complete the proof of Theorems 7.1 and 7.2 it remains to prove that the entire solution u to system (7.2) and the solution u_R to system (7.14) satisfy the pointwise estimates stated in Theorems 7.1 and 7.2, respectively. To establish these estimates we resort to Theorem 5.3, which holds under Hypothesis $\mathbf{H}_3$ (ii).

This theorem is proved for the case of a generic potential and covers all cases where $D_0 = F$. To treat the general case $F \subset D_0$ we need to show that Theorem 5.3 holds true in the case of symmetric potentials and f-equivariant local minimizers. In [9] the validity of Theorem 5.3 is established for symmetric potential and $f = I$, but the arguments therein extend naturally to cover the general case of f-equivariance.

If G is a finite reflection group we apply Theorem 5.3 to u_R with $\Omega = D_{0,R} := D_0 \cap B_R$ and $\mathscr{L}_a = \{\gamma a\}_{\gamma \in \Gamma} \setminus \{a\}$. From Sect. 7.5.2, we have $u_R(\overline{F}_R) \subset \overline{\Phi}$ and therefore, by f-equivariance,

$$u_R(\overline{D}_{0,R}) \subset u_R(\overline{D}_R) \subset \cup_{\gamma \in \Gamma_a} \gamma \overline{\Phi} = \overline{\mathscr{D}}. \tag{7.32}$$

Since by Hypothesis $\mathbf{H}_3$, as we have seen, a is the unique minimizer of W in $\mathscr{D}$, it follows that

$$u_R(D_{0,R}) \cap \Big(\bigcup_{z \in Z_a} B_{z,\delta} \Big) = \varnothing, \quad \text{for } \delta = d(a, \partial \mathscr{D}) > 0. \tag{7.33}$$

Thus, the bound (7.31) and Theorem 5.3 imply

$$|u_R(x) - a| < q_0, \ \ B_{r_{q_0}}(x) \subset D_{0,R}, \tag{7.34}$$

where q_0 is the constant in the assumption $\mathbf{H}$ of Theorem 5.3. From this and Proposition 5.2 it follows that there are positive constants k_0, K_0 independent of R such that

$$|u_R(x) - a| \leq K_0 e^{-k_0 d(x \partial D_{0,R})} q_0, \quad x \in D_{0,R}, \tag{7.35}$$

Therefore, in inequality (7.35) we can pass to the limit along the sequence u_{R_j} that defines the entire solution u. This and the f-equivariance of u establish (ii). The proof of Theorem 7.1 is complete.

Assume now that G is a discrete group. In this case the fundamental domain can be bounded, as for the group G' considered in Sect. 7.2, or unbounded, as for the group G generated by the reflections in the plane $\{x_1 = 0\}$ and $\{x_1 = 1\}$ of $\mathbb{R}^n$, $n > 1$ where $F = \{x \in \mathbb{R}^n : 0 < x_1 < 1\}$. To establish the estimate (ii) in

Theorem 7.2 it suffices to consider the case where F is bounded. For $R > 0$ define

$$v_R(y) = u_R(\tfrac{y}{R}), \quad y \in \mathbb{R}^n,$$

$$F^R = \{y \in \mathbb{R}^n : \tfrac{y}{R} \in F\},$$

(7.36)

and let G^R denote the discrete reflection group generated by the reflections in the planes $P_1^R, \ldots, P_l^R$ that bound F^R. There is an obvious group isomorphim $\eta^R : G^R \to G$ between G^R and G, and the minimality of u_R implies that $v_R \in W_{\mathrm{loc}}^{1,2}(\mathbb{R}^n, \mathbb{R}^m)$ is a local minimizers in the class of positive f^R-equivariant maps, where $f^R := f \circ \eta^R$. Therefore v_R is a solution of

$$\Delta v - W_u(v) = 0, \quad \text{in } \mathbb{R}^n.$$

(7.37)

This, (7.16) and elliptic regularity implies

$$|\nabla v_R| \le M', \quad \text{in } \mathbb{R}^n,$$

(7.38)

for some $M' > 0$ independent of R. As before, we have

$$u_R(\overline{D}_0) \subset \bigcup_{\gamma \in \Gamma_a} \gamma \overline{\Phi} = \overline{\mathscr{D}},$$

or equivalently

$$v_R(\overline{D}_0^R) \subset \overline{\mathscr{D}},$$

(7.39)

where $\overline{D}_0^R = \cup_{g \in G^a}(\eta^R)^{-1}(g)\overline{F}^R$. From (7.39) it follows

$$v_R(D_0^R) \cap \Big(\bigcup_{z \in Z_a} B_{z,\delta} \Big) = \varnothing, \quad \text{for } \delta = d(a, \partial \mathscr{D}) > 0.$$

(7.40)

Therefore, we can apply Theorem 5.3 to v_R with $\mathscr{O} = D_0^R$ and deduce, for $R \ge R_0 := \min\{R : B_{r_{q0}}(y) \subset D_0^R, \ \text{for some } y \in D_0^R\}$, that

$$|v_R(y) - a| \le q_0, \quad B_{r_{q0}}(y) \subset D_0^R,$$

(7.41)

and hence, by Proposition 5.2, that

$$|v_R(y) - a| \le K_0 e^{-k_0 d(y, \partial D_0^R)}; \ y \in D_0^R,$$

which is equivalent to

$$|u_R(x) - a| \le K_0 e^{-k_0 R d(x, \partial D_0)}, \quad x \in D_0. \tag{7.42}$$

The rest of the proof is as in the case of G finite discussed before. The proof of Theorem 7.2 is complete.

The approach in Sect. 6.7.4 applies also in the context of this chapter. Let us mention the adjustments needed, for instance, in the case where G is a discrete reflection group. We consider as above the solutions v_R of the original system (7.2) in the blown up lattice. The minimality of these solutions in the class of positive f^R-equivariant maps is evident by construction (cf. Sect. 7.5.1). Now let slow that given $c_0 > 0$, there is $d_0 > 0$, depending on c_0, such that

$$|v_R(x) - a| \le c_0, \forall x \in D_0^R, \quad d(x, \partial D_0^R) \ge d_0, \quad \text{and } \forall R > R_0. \tag{7.43}$$

Again, we proceed by contradiction, and assume that there are squences $\{x_k\} \subset D_0^{R_k}$ and R_k, such that

$$|v_{R_k}(x_k) - a| > c_0, \quad d(x_k, \partial D_0^R) \to \infty, \quad \text{and } R_k \to \infty, \text{ as } k \to \infty. \tag{7.44}$$

Next, we define Π_a to be the union of all the reflection planes of the domain space corresponding to reflections s such that $f^R(s) \in \Gamma_a$. In both cases considered in Sect. 6.7.4, we check the positivity and the f^R-equivariance of the comparison maps σ and $\hat{u}$ (cf. (6.38)), thanks to the choice of the center x_k (respectively, x_k') of the balls. Finally, the density estimates of Theorem 5.2 apply to v_R with constants independent of R, since these solutions are uniformly bounded.

7.6   Three Detailed Examples Involving the Reflection Group of the Tetrahedron

7.6.1   Preliminaries

In this subsection we recall some properties of the symmetry group of a regular tetrahedron and the symmetry group of a cube.

Let $\mathscr{T}$ be the symmetry group of a regular tetrahedron $A_1 A_2 A_3 A_4$ (with $A_1 = (1, 1, 1)$, $A_2 = (-1, -1, 1)$, $A_3 = (1, -1, -1)$, $A_4 = (-1, 1, -1)$), which can be inscribed in a cube centered at the origin O (see Fig. 7.4). The order of $\mathscr{T}$ is $|\mathscr{T}| = 24$ and $\mathscr{T}$ is isomorphic to the permutation group S_4. The 24 elements of $\mathscr{T}$ are associated to the following elements of S_4:

- the identity map I_3 of $\mathbb{R}^3$ corresponds to the unit of S_4 (I_3 fixes the 4 vertices),
- the 8 rotations of angle $\pm 2\pi/3$ with respect to the axes OA_i correspond to the 3-cycles $(j\ k\ l)$ (only the vertex A_i is fixed),

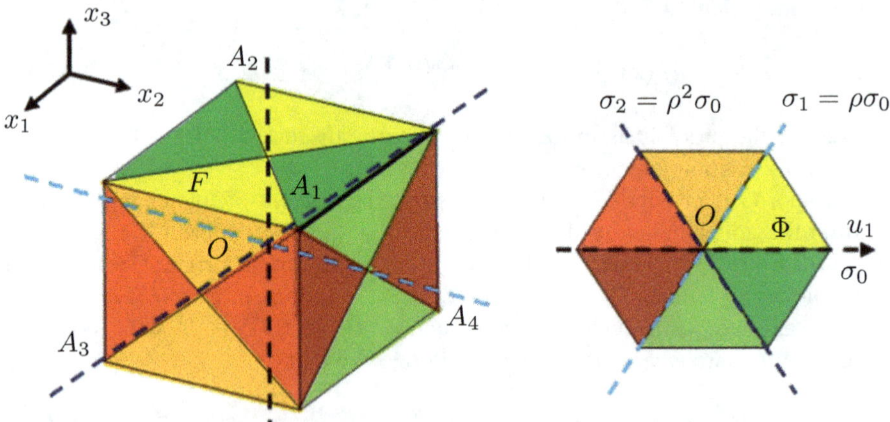

Fig. 7.4 Fundamental domains for the action of $\mathscr{T}$ on $\mathbb{R}^3$ (left) and for the action of D_3 on $\mathbb{R}^2$ (right). The ϕ-equivariant solution $u : \mathbb{R}^3 \to \mathbb{R}^2$ of (7.2) given by Theorem 7.1 maps fundamental domains into fundamental domains with the same color. In particular, u maps the infinite double cone (union of four fundamental domains) generated by O and by the two yellow triangles into the sector Φ

- the 6 reflections with respect to the planes OA_iA_j correspond to the transpositions $(k\,l)$ (the vertices A_i and A_j are fixed),
- the 3 symmetries with respect to the coordinate axes correspond to the permutations $(i\,j)(k\,l)$ (no vertex fixed),
- a reflection with respect to the plane OA_kA_l composed with a rotation around the axis OA_i corresponds to one of the six 4-cycles $(i\,j\,k\,l)$ (no vertex fixed).

Since there exists a homomorphism between the permutation groups S_4 and S_3, and since S_3 is isomorphic to the dihedral group D_3, which is the symmetry group of an equilateral triangle in the plane, we also have a homomorphism $\phi : \mathscr{T} \to D_3$. This homomorphism associates:

- I_3 and the symmetries with respect to the coordinate axes to I_2 (i.e., $\ker \phi = \{I_3$ and the three symmetries with respect to the coordinate axes$\}$),
- the rotation of angle $2\pi/3$ around the axis OA_1 to the rotation ρ of angle $2\pi/3$ in the plane,
- the reflections with respect to the planes OA_1A_2 and OA_3A_4 to the reflection $\sigma_0 : \mathbb{R}^2 \to \mathbb{R}^2$ in the u_1 axis (cf. Fig. 7.4),
- the reflections with respect to the planes OA_1A_4 and OA_2A_3 to the reflection $\sigma_1 = \rho\sigma_0$,
- the reflections with respect to the planes OA_1A_3 and OA_2A_4 to the reflection $\sigma_2 = \rho^2\sigma_0$.

Let $\mathscr{K}$ be the symmetry group of the cube centered at the origin O with vertices at the points $(\pm 1, \pm 1, \pm 1)$. The order of $\mathscr{K}$ is $|\mathscr{K}| = 48$, and $\mathscr{K}$ contains the symmetry group of the regular tetrahedron as a subgroup (i.e., $\mathscr{T} < \mathscr{K}$). Let σ :

$\mathbb{R}^3 \to \mathbb{R}^3$, $\sigma(x) = -x$, be the antipodal map. Clearly, σ is an element of $\mathscr{K}$ of order 2 which does not belong to $\mathscr{T}$, and $\mathscr{K} = \mathscr{T} \cup \sigma \mathscr{T}$. Furthermore, σ commutes with the reflections with respect to the planes OA_iA_j, thus σ commutes with every element of $\mathscr{K}$. As a consequence, the correspondence

$$\{I_3, \sigma\} \times \mathscr{T} \ni (\alpha, \beta) \longmapsto \alpha\beta \in \mathscr{K},$$

defines an isomorphism of the group product $\{I_3, \sigma\} \times \mathscr{T}$ onto $\mathscr{K}$, and we can define a homomorphism $\psi : \mathscr{K} \to \mathscr{T}$ by setting $\psi(\beta) = \beta$, and $\psi(\sigma\beta) = \beta$, for every $\beta \in \mathscr{T}$. By definition, ψ leaves invariant the elements of $\mathscr{T}$. We also mention that $\mathscr{K}$ contains the 3 reflections with respect to the coordinate planes $x_i = 0$ (which are the symmetries with respect to the coordinate axes Ox_i composed with σ).

7.6.2   A Solution $u : \mathbb{R}^3 \to \mathbb{R}^2$ to (7.2) with the Reflection Group of the Tetrahedron Acting on the Domain and the Reflection Group of the Equilateral Triangle Acting on the Target

In this example we consider the aforementioned homomorphism $\phi : \mathscr{T} \to D_3$. Let F be the fundamental domain of $\mathscr{T}$ bounded by the planes OA_1A_2, OA_3A_4 and OA_1A_4, and Φ be the fundamental domain of D_3 bounded by the lines $u_2 = 0$ and $u_2 = \sqrt{3}u_1$ corresponding to the reflections σ_0 and σ_1. By the foregoing discussion, the image under ϕ of the reflections with respect to the planes OA_1A_2 and OA_3A_4 is σ_0, while the image under ϕ of the reflection with respect to the plane OA_3A_4 is σ_1. Thus ϕ is a positive homomorphism that associates F to Φ, and Hypothesis $\mathbf{H}_1$ is satisfied. If Hypotheses $\mathbf{H}_2$–$\mathbf{H}_3$ also hold, Theorem 7.1 ensures the existence of a ϕ-equivariant solution u to (7.2). In particular (see Fig. 7.4), ϕ-equivariance implies that u maps

- the coordinate axes into the reflection lines (with the same colour),
- the planes OA_1A_2 and OA_3A_4 into the reflection line $u_2 = 0$,
- the planes OA_1A_4 and OA_2A_3 into the reflection line $u_2 = \sqrt{3}u_1$,
- the planes OA_1A_3 and OA_2A_4 into the reflection line $u_2 = -\sqrt{3}u_1$,
- the diagonals of the cube into the origin O.

In addition, Theorem 7.1 implies that the solution u is positive (i.e. $u(\overline{F}) \subset \overline{\Phi}$), and therefore u maps each fundamental domain of $\mathscr{T}$ into a fundamental domain of D_3, as in Fig. 7.4. If for instance the potential W has 6 minima (one in the interior of each fundamental domain of D_3), then the domain space $\mathbb{R}^3$ also splits into six regions as in Fig. 7.4. Properties (ii) and (iii) of Theorem 7.1 state that for every x in such a region D, $u(x)$ converges as $d(x, \partial D) \to \infty$ to the corresponding minimum a of W.

7.6.3   A Solution $u : \mathbb{R}^3 \to \mathbb{R}^3$ to (7.2) with the Reflection Group of the Cube Acting on the Domain and the Reflection Group of the Tetrahedron Acting on the Target

In this example we consider the homomorphism $\psi : \mathscr{K} \to \mathscr{T}$ defined previously. Now, we denote by F the fundamental domain of $\mathscr{K}$ bounded by the planes OA_1A_2, OA_1A_4, and $x_2 = 0$, and by Φ be the fundamental domain of $\mathscr{T}$ bounded by the planes OA_1A_2, OA_1A_4, and OA_2A_3 (cf. Fig. 7.5). By the foregoing discussion, ψ leaves invariant the reflections with respect to the planes OA_1A_2 and OA_1A_4, while the image under ψ of the reflection with respect to the plane $x_2 = 0$ is the coordinate axis x_2, that is, the intersection of the planes OA_1A_4 and OA_2A_3. Thus ψ is a positive homomorphism that associates F to Φ, and Hypothesis $\mathbf{H_1}$ is satisfied. If Hypotheses $\mathbf{H_2}$–$\mathbf{H_3}$ also hold, Theorem 7.1 ensures the existence of a ψ-equivariant solution u to (7.2). More precisely, ψ-equivariance implies that u maps

- every plane OA_iA_j into itself,
- every diagonal of the cube into itself,
- the coordinate planes into the perpendicular coordinate axes,
- the coordinate axes at the origin O.

In addition, the solution u is positive (i.e., $u(\overline{F}) \subset \overline{\Phi}$), and in fact u maps each fundamental domain of $\mathscr{K}$ into a fundamental domain of $\mathscr{T}$ as in Fig. 7.5.

If the potential W has for instance 4 minima (located at the vertices A_1, A_2, A_3 and A_4) of the tetrahedron, then the stabilizer Γ_a of $a = A_1$ in $\Gamma = \mathscr{T}$ has six

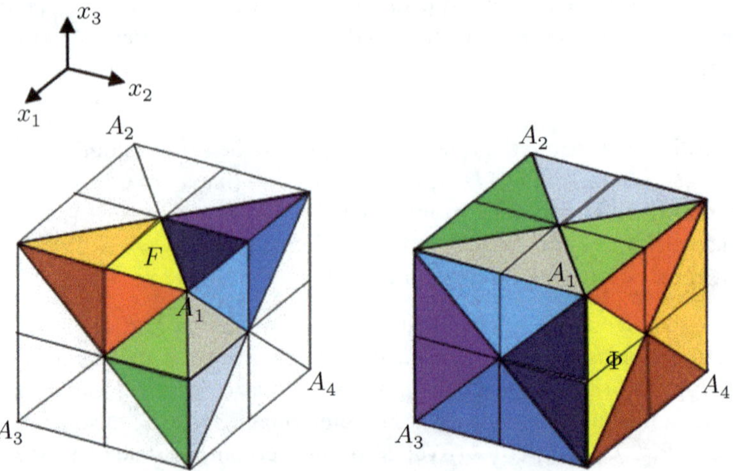

Fig. 7.5 Fundamental domains for the action on $\mathbb{R}^3$ of $\mathscr{K}$ (left) and $\mathscr{T}$ (right). The ψ-equivariant solution $u : \mathbb{R}^3 \to \mathbb{R}^3$ of (7.2) given by Theorem 7.1 maps fundamental domains into fundamental domains with the same color. Note, in particular, that u maps $\overline{F} \cup \sigma\overline{F}$ into $\overline{\Phi}$

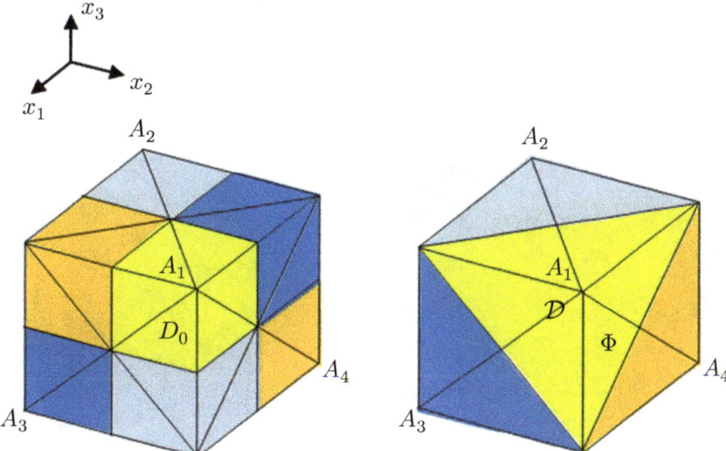

Fig. 7.6 The sets $\mathscr{D} = \{u = \alpha_1(1, 1, -1) + \alpha_2(-1, 1, 1) + \alpha_3(1, -1, 1), \alpha_i > 0, i = 1, 2, 3\}$, $D_0 = \{x_i > 0, \forall i = 1, 2, 3\}$ and $D = D_0 \cup \sigma D_0$ when W has four minima at the vertices of the tetrahedron. In this case the solution u of (7.2) given by Theorem 7.1 satisfies: $u(x) \to A_1$, for $\min_i |x_i| \to +\infty$, $x \in D$

elements: I_3, the reflections with respect to the planes OA_1A_2, OA_1A_3 and OA_1A_4, and the rotations of angle $\pm 2\pi/3$ around the axis OA_1. Thus $\mathscr{D}$ is the (interior of the closure) of the union of the six fundamental domains that have A_1 on their boundary; the group $\psi^{-1}(\Gamma_a) = \Gamma_a \cup \sigma \Gamma_a$ has 12 elements, and the set D has two connected components: the solid right angle $D_0 = \{x_i > 0, \ \forall i = 1, 2, 3\}$ and $\sigma D_0 = -D_0$ (cf. Fig. 7.6, and also note that the group G^a is in this particular case the group Γ_a). According to Theorem 7.1, if $x \in D_0$ and $d(x, \partial D_0) \to \infty$ (that is, if $x_i \to +\infty$ for every $i = 1, 2, 3$) then $u(x) \to a$. Of course, the same result is true when $x \in -D_0$ and $d(x, \partial(-D_0)) \to \infty$, and the solution also converges in the other solid right angle cones to the corresponding minima of W, as in Fig. 7.6.

7.6.4   A Crystalline Structure in $\mathbb{R}^3$

Now, let us consider the discrete reflection group $\mathscr{K}'$ acting in $\mathbb{R}^3$ which is generated by the reflections s_1, s_2, s_3 and s_4 with respect to the corresponding planes $P_1 := OA_1A_2$, $P_2 := OA_1A_4$, $P_3 := \{x_2 = 0\}$ and $P_4 := \{x_1 + x_3 = 2\}$. These planes bound the fundamental domain F' of $\mathscr{K}'$ with vertices at the points O, A_1, $I := (1, 0, 1)$ and $B := (0, 0, 2)$ (cf. Fig. 7.7). The point group of $\mathscr{K}'$, that is the stabilizer of the origin, is the group $\mathscr{K}$, and we have $\mathscr{K}' = T\mathscr{K}$, where T is the translation group of $\mathscr{K}'$. T is generated by the translations given by the vectors $t_1 := (2, 0, 2)$, $t_2 := (0, 2, 2)$ and $t_3 := (0, -2, 2)$. By composing the canonical homomorphism $p : \mathscr{K}' \to \mathscr{K}$ such that $p(tg) = g$ for every $t \in T$ and $g \in \mathscr{K}$, with the homomorphism $\psi : \mathscr{K} \to \mathscr{T}$ defined previously, we

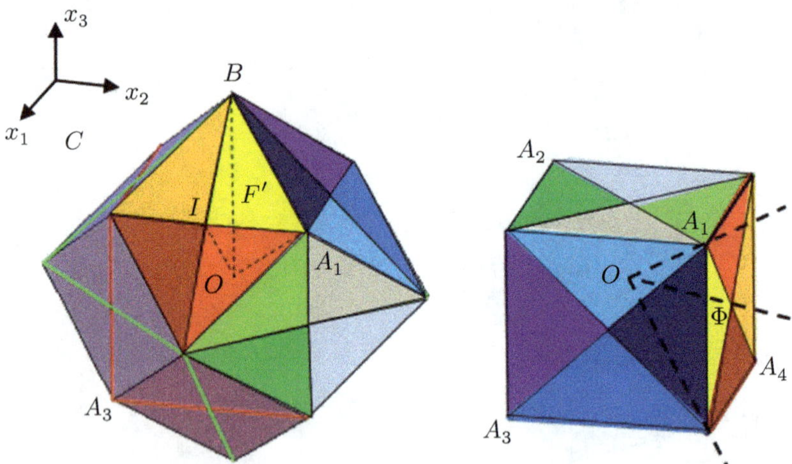

Fig. 7.7 Fundamental domains for the action on $\mathbb{R}^3$ of $\mathcal{K}'$ (left) and $\mathcal{T}$ (right). The fundamental domain F' of $\mathcal{K}'$ is a pyramid with base the triangle $A_1 B I$ and vertex in O. Under the action of the point group $\mathcal{K}$, F' generates the rhombic dodecahedron C (left) which tiles the domain space $\mathbb{R}^3$ when translated by the elements of T. The ψ'-equivariant solution $u : \mathbb{R}^3 \to \mathbb{R}^3$ of (7.2) given by Theorem 7.2 maps fundamental domains into fundamental domains with the same color. Note, in particular, that u maps $\bigcup_{t \in T}(\overline{F'} \cup \sigma\overline{F'})$ into $\overline{\Phi}$

obtain a homomorphism $\psi' : \mathcal{K}' \to \mathcal{T}$. We note that $\psi'(s_4)$ is the image under ψ of the reflection with respect to the plane $O A_2 A_3$, which is the reflection with respect to $O A_2 A_3$. Thus, ψ' is a positive homomorphism which associates F' to the fundamental domain Φ of $\mathcal{T}$ bounded by the planes $O A_1 A_2$, $O A_1 A_4$ and $O A_2 A_3$. In this case the elementary crystal $C = \bigcup_{g \in \mathcal{K}} g F'$ is a *rhombic dodecahedron* (cf. Fig. 7.7) that tiles the three-dimensional space when translated by the elements of T.[2] Several structures are possible for the solution u_R given by Theorem 7.2, depending on the position of $a \in \overline{\Phi}$.

For instance, if $a \in \Phi$ we have $\mathcal{D} = \Phi$ and $D_0 = F'$. If $a = (0, 1, 0)$ (cf. Fig. 7.8) then $\mathcal{D} = \{u : \max\{|u_1|, |u_3|\} < u_2\}$ and D_0 is the pyramid with base the rhombus defined by the points A_1, B, $(1, -1, 1)$, $(2, 0, 0)$, and vertex in O. Finally, if $a = A_1$, $\mathcal{D}$ is the cone which has vertex in O and is generated by the triangle with vertices at the points $(1, 1, -1)$, $(1, -1, 1)$, $(-1, 1, 1)$, while D_0 is the polyhedron (union of two pyramids) with vertices at the points O, $(2, 0, 0)$, $(0, 2, 2)$, B, A_1 and $D = \bigcup_{t \in T}(D_0 \cup \sigma D_0)$.

---

[2]Space filling tessellation with rhombic dodecahedra is the crystal structure often found in garnets and other minerals, such as pyrite and magnetite.

Fig. 7.8 The sets D_0 and $\mathscr{D}$ when W has six minima (one in the middle of each side of the tetrahedron). In this case the ψ'-equivariant solution $u_R : \mathbb{R}^3 \to \mathbb{R}^3$ given by Theorem 7.2 satisfies $\lim_{R \to +\infty} u_R(x+t) = (0, 1, 0)$ for $x \in D_0 \cup \sigma D_0$, $t \in T$

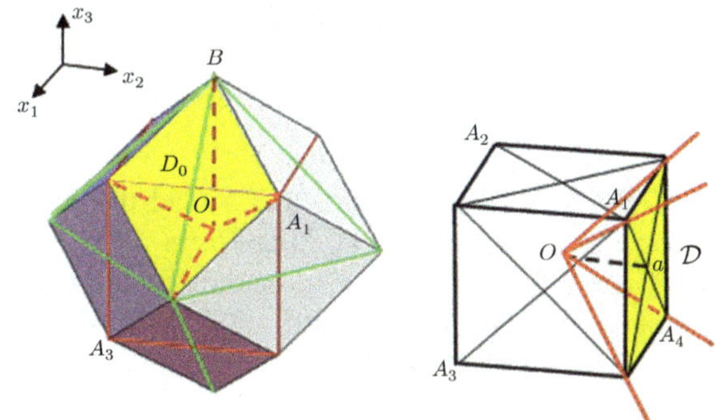

7.7   Other Examples in Lower Dimension

7.7.1   Positive Homomorphisms Between Finite Reflection Groups of the Plane

The finite reflection groups of the plane are the dihedral groups D_n with $n \geq 1$. The group D_n contains $2n$ elements: the rotations $r_n^0 = I_2, r_n^1, \ldots, r_n^{n-1}$ (where I_2 is the identity map of the plane, and r_n is the rotation of angle $2\pi/n$), and the reflections $r_n^0 s = s, r_n^1 s, \ldots, r_n^{n-1} s$ (where s is the reflection with respect to the x_1 coordinate axis). Similarly, the elements of D_{nk} (with $k \geq 1$) are the rotations $r_{nk}^0 = I_2, r_{nk}^1, \ldots, r_{nk}^{nk-1}$ (where r_{nk} is the rotation of angle $2\pi/nk$) and the reflections $r_{nk}^0 s = s, r_{nk}^1 s, \ldots, r_{nk}^{nk-1} s$. In the two propositions below we have determined all the positive homomorphisms between finite reflection groups of the plane (up to an isomorphism). From the list of homomorphisms between dihedral groups established in [12], we have extracted the positive ones.

For $m = \pm 1$ we define the homomorphism $f_m : D_{nk} \to D_n$ by setting $f_m(r_{nk}^p) = r_n^{mp}$ and $f_m(r_{nk}^p s) = r_n^{mp} s$, for every integer p. Thanks to the property $sr = r^{-1}s$ (which holds for every reflection s and every rotation r), it is easy to check that f_m is a homomorphism from D_{nk} onto D_n. We can also define the homomorphism $g : D_{2k} \to D_2$ by setting $g(r_{2k}^p) = s^p$ and $g(r_{2k}^p s) = s^p \sigma$ for every integer p, where σ denotes the the antipodal map $\sigma u = -u$.

Proposition 7.1 *If $n \geq 2$, $k \geq 1$, $G = D_{nk}$ acts on the domain plane $\mathbb{R}^2$, and $\Gamma = D_n$ on the target plane $\mathbb{R}^2$, then for every $m = \pm 1$, f_m is a positive homomorphism*

Fig. 7.9 The correspondence of the fundamental domains for a solution to (7.2) equivariant with respect to the homomorphism $f_1 : D_4 \to D_2$

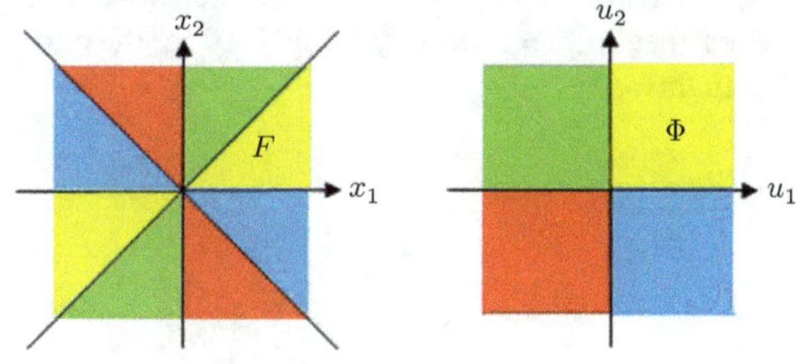

which associates the fundamental domain $F := \{re^{it} : 0 < r, \ 0 < t < \pi/nk\}$ of G to the fundamental domain $\Phi := \{re^{it} : 0 < r, \ 0 < mt < \pi/n\}$ of Γ. In addition, the homomorphism f_m leaves invariant the elements of $D_n < D_{nk}$ if and only if $mk = 1 \mod n$.

Proof By definition, the lines that bound the fundamental domain F correspond to the reflections s and $r_{nk}s$. Since for $m = \pm 1$ the fixed points of the reflections $f_m(s) = s$ and $f_m(r_{nk}s) = r_n^m s$ are the lines that bound the fundamental domain Φ, the homomorphism f_m can associate F to Φ. Also, f_m leaves invariant the elements of $D_n < D_{nk}$ if and only if $f_m(r_n) = r_n$, that is, if and only if

$$f_m(r_n) = f_m(r_{nk}^k) = r_n^{mk} = r_n \iff r_n^{mk-1} = I_2 \iff mk = 1 \mod n.$$

$\square$

Proposition 7.2 *If $k \geq 1$, $G = D_{2k}$ acts on the domain plane $\mathbb{R}^2$, and $\Gamma = D_2$ on the target plane $\mathbb{R}^2$, then g is a positive homomorphism which associates the fundamental domain $F = \{re^{it} : 0 < r, \ 0 < t < \pi/2k\}$ of G to the fundamental domain $\Phi = \{re^{it} : 0 < r, \ 0 < t < \pi/2\}$ of Γ.*

Proof As before, we see that $g(s) = \sigma$ fixes only the origin, while $g(r_{2k}s) = s\sigma$ fixes the u_2 coordinate axis. Thus, the homomorphism g can associate F to Φ (and in fact it can associate F to any of the four fundamental domains of D_2). $\square$

To illustrate the propositions above, let us give some examples.

- The homomorphism $f : D_6 \to D_3$ that was mentioned at the beginning of Sect. 7.2 (cf. also [4]) coincides with the homomorphism f_{-1} of Proposition 7.1 with $n = 3$, $k = 2$ and $m = -1$. Since $mk = -2 = 1 \mod 3$, we see again that it leaves invariant the elements of D_3.
- Taking $n = 3$, $k = 5$ and $m = -1$, we check that $mk = 1 \mod n$, and we obtain a new homomorphism $f_{-1} : D_{15} \to D_3$ that leaves invariant the elements of D_3. The kernel of this homomorphism is the cyclic group generated by the rotation r_5.
- Taking $n = 2$, $k = 2$ and $m = 1$, we obtain the positive homomorphism $f_1 : D_4 \to D_2$. When Hypotheses $\mathbf{H_2}$–$\mathbf{H_3}$ also hold, Theorem 7.1 ensure the existence of a f_1-equivariant solution to (7.2) which maps $\overline{F}$ into $\overline{\Phi}$, and the other fundamental domains of $G = D_4$ as in Fig. 7.9.
- Considering the homomorphism $g : D_4 \to D_2$ of Proposition 7.2 (with $k = 2$) we can also construct a g-equivariant solution u to (7.2). This solution has the particularity that the coordinate axes are mapped at the origin. Indeed, by g-equivariance, if $x, y \in \mathbb{R}^2$ are symmetric with respect to one of the coordinate axes, then $u(x) = -u(y)$ (cf. Fig. 7.10 for the correspondence of the fundamental domains).

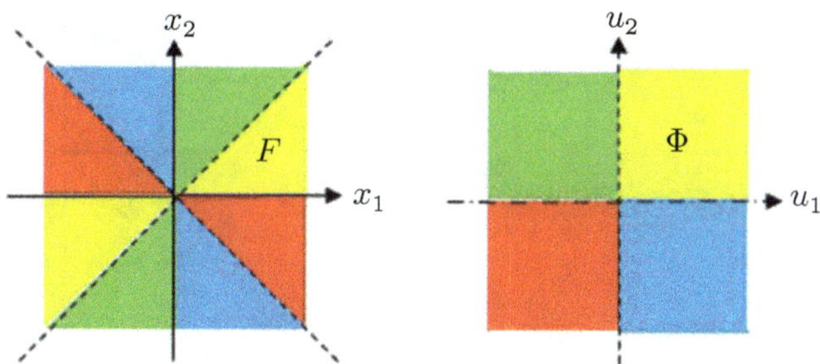

Fig. 7.10 The correspondence of the fundamental domains for a solution to (7.2) equivariant with respect to the homomorphism $g : D_4 \to D_2$

7.7.2 Saddle Solutions

In this subsection we are going to construct, as an application of Theorems 7.1 and 7.2, scalar solutions to (7.2) and (7.14) having particular symmetries. The only finite reflection group that acts on the target space $\mathbb{R}$ is the dihedral group $\Gamma = D_1$ with two elements: the identity I_1 and the antipodal map $\sigma u = -u$. We assume that a finite or a discrete reflection group G acts on the domain space $\mathbb{R}^n$, and that $W : \mathbb{R} \to \mathbb{R}$ satisfies Hypotheses $\mathbf{H}_2$–$\mathbf{H}_3$, that is,

- W is a nonnegative and even function,
- there exists $M > 0$ such that $W(u) \geq W(M)$, for $u \geq M$,
- $W(u) = 0 \Leftrightarrow u = \pm a$, with in addition $a > 0$ and $W''(a) > 0$.

Clearly, the map ϵ which sends the orientation-preserving motions to I_1, and the orientation-reversing motions to σ, is a positive homomorphism from G onto Γ. Thus, Theorems 7.1 and 7.2 ensure the existence of classical solutions $u : \mathbb{R}^n \to \mathbb{R}$ to (7.2) and (7.14) with the following properties:

(i) ϵ-equivariance implies that if $x, y \in \mathbb{R}^n$ are symmetric with respect to a reflection plane of G, then $u(x) = -u(y)$. In particular, u vanishes on the reflection planes of G. If G is a discrete reflection group, then u is periodic in the sense that $u(x + t) = u(x)$, for every $x \in \mathbb{R}^n$, and every translation t in the translation group $T < G$.

(ii) Positivity means that either $u \geq 0$, or $u \leq 0$ in each fundamental domain F of G.

(iii) In each fundamental domain F, $u(x)$ approaches either a or $-a$, as $x \in F$ and $d(x, \partial F)$ increases.

We also give another example, more elaborated, when $G = D_{2k}$ (with $k \geq 1$) acts on $\mathbb{R}^2$. Let us consider the homomorphism $h : D_{2k} \to D_1$ such that $h(r_{2k}^p) = \sigma^p$ and $h(r_{2k}^p s) = \sigma^p$, for every integer p (see the previous subsection for the notation). In this particular set-up, we can again construct a h-equivariant solution $u : \mathbb{R}^2 \to \mathbb{R}$

Fig. 7.11 The symmetries of
a solution $u : \mathbb{R}^2 \to \mathbb{R}$
to (7.2) equivariant with
respect to the homomorphism
$h : D_4 \to D_1$

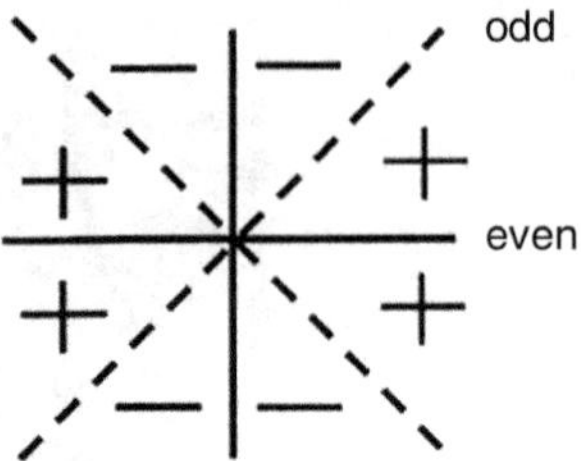

to (7.2) which has in each fundamental domain of D_{2k} alternatively even and odd symmetries. Figure 7.11 represents the symmetries of such a solution for $k = 2$.

7.7.3   Other Examples Involving Discrete Reflection Groups

To finish, we give some more examples illustrating Theorem 7.2. Let us assume again that the discrete reflection group G' acts on the domain x-plane as in Sect. 7.2 (cf. also the end of Sect. 7.5), but let us consider now a new reflection group acting on the target u-plane: the group $\Gamma = D_2$. We construct a homomorphism $f'' : G' \to D_2$ by composing the canonical projection $p : G' \to D_6$ with the homomorphism $g : D_6 \to D_2$ defined in Sect. 7.7.1 (that is, $f'' = g \circ p$). As we did before for the homomorphism f', we can check that f'' is a positive homomorphism. Thus, once again, Theorem 7.2 allows us to construct f''-equivariant solutions u_R to (7.14). Figure 7.12 represents the correspondence of the fundamental domains of G' with the fundamental domains of D_2 for such solutions (compare with Fig. 7.3). The f''-equivariant solutions have the special property that some reflection lines of the group G' are mapped at the origin.

Let us also mention a last example involving the discrete reflection group of the plane H, generated by the reflections with respect to the lines $x_2 = 0$, $x_2 = x_1$ and $x_1 = 1$. The point group associated to H is the group D_4, and we can compose the canonical projection $H \to D_4$ either with the homomorphism $f_1 : D_4 \to D_2$, or with the homomorphism $g : D_4 \to D_2$ defined in Sect. 7.7.1, to construct positive homomorphisms from H onto D_2.

7.8   Scholia on Chap. 7

The study of bounded symmetric entire solution of the vector Allen–Cahn equation initiated with the work of Bronsard and Reitich [7] and Bronsard, Gui and Schatzman [8] for $n = m = 2$ and $G = \Gamma = D_3$, D_3 the reflection group of the equilateral triangle. Later Gui and Schatzman [10] considered the case $n = M = 3$ and $G = \Gamma$, with Γ the symmetry group of the regular tetrahedron. The case of

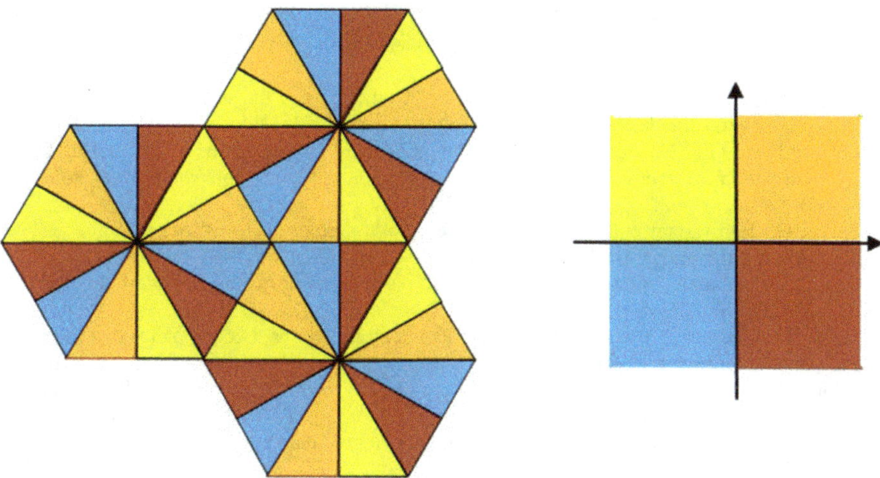

Fig. 7.12 The correspondence of the fundamental domains by the f''-equivariant solution u_R: on the right the fundamental domains of the discrete reflection group G' and on the left the fundamental domains of the finite reflection group D_2

general $n = m$ and general $G = \Gamma$ was studied in [2] under a monotonicity condition of W along rays emanating from the unique zero of W in $\overline{F}$. This condition was removed in [9]. Discrete groups where first considered in Smyrnelis' thesis [13] where, in a different vein from what we have discussed before, it was assumed that also $\Gamma = G$ is a discrete infinite group. The idea for the proof of the basic Lemma 7.2 is also from Smyrnelis' thesis, see also [3]. The first paper that considered different groups $G \neq \Gamma$ was [4] where, assuming $n = m = 2$, $G = D_6$ the dihedral group of symmetries of the hexagon and $\gamma = D_3$, the existence of a *sixfold* solution was established. Our presentation follows [5].

References

1. Alikakos, N.D.: A new proof for the existence of an equivariant entire solution connecting the minima of the potential for the system $\Delta u - W_u(u) = 0$. Commun. Partial Diff. Equ. **37**(12), 2093–2115 (2012)
2. Alikakos, N.D., Fusco, G.: Entire solutions to equivariant elliptic systems with variational structure. Arch. Ration. Mech. Anal. **202**(2), 567–597 (2011)
3. Alikakos, N.D., Smyrnelis, P.: Existence of lattice solutions to semilinear elliptic systems with periodic potential. Electron. J. Diff. Equ. **15**, 1–15 (2012)
4. Bates, P.W., Fusco, G., Smyrnelis, P.: Entire solutions with six-fold junctions to elliptic gradient systems with triangle symmetry. Adv. Nonlinear Stud. **13**(1), 1–12 (2013)
5. Bates, P.W., Fusco, G., Smyrnelis, P.: Multiphase solutions to the vector Allen-Cahn equation: crystalline and other complex symmetric structures. Arch. Ration. Mech. Anal. **225**(2), 685–715 (2017)

6. Braun, R.J., Cahn, J.W., McFadden, G.B., Wheeler, A.A.: Anisotropy of interfaces in an ordered alloy: a multiple-order-parameter model. Philos. Trans. R. Soc. Lond. A **355**, 1787–1833 (1997)
7. Bronsard, L., Reitich, F.: On three-phase boundary motion and the singular limit of a vector-valued Ginzburg- Landau equation. Arch. Ration. Mech. Anal. **124**(4), 355–379 (1993)
8. Bronsard, L., Gui, C., Schatzman, M.: A three-layered minimizer in $\mathbb{R}^2$ for a variational problem with a symmetric three-well potential. Commun. Pure. Appl. Math. **49**(7), 677–715 (1996)
9. Fusco, G.: Equivariant entire solutions to the elliptic system $\Delta u - W_u(u) = 0$ for general G-invariant potentials. Calc. Var. Part. Diff. Equ. **49**(3), 963–985 (2014)
10. Gui, C., Schatzman, M.: Symmetric quadruple phase transitions. Ind. Univ. Math. J. **57**(2), 781–836 (2008)
11. Henry, D.: Geometric Theory of Semilinear Parabolic Equations. Lecture Notes in Mathematics, vol. 840. Springer, Berlin (1981)
12. Johnson, J.W.: The number of group homomorphisms from D_m into D_n. Collage Math. J. **44**(3), 191–192 (2013)
13. Smyrnelis, P.: Solutions to elliptic systems with mixed boundary conditions. Phd Thesis (2012)

Chapter 8
Hierarchical Structure—Stratification

Abstract In this chapter we extend the density estimate in Theorem 5.2 by replacing the constant solution a with a symmetric, minimal, hyperbolic connection e, (and more generally with any equivariant minimal hyperbolic solution), and then derive Liouville theorems and asymptotic information for minimal solutions under symmetry hypotheses. Utilizing the extended density estimate we give a proof of a result of Alama et al. (Calc Var 5:359–390, 1997) on the existence of stationary layered solutions in $\mathbb{R}^2$. The Alama, Bronsard and Gui example is revisited in Chap. 9 under no symmetry hypotheses. Our results were originally obtained by a different method in Alikakos and Fusco (Annali della Scuola Normale Superiore di Pisa XV:809–836, 2016).

8.1 Introduction

Consider a triple-well potential with the symmetries of the equilateral triangle, and satisfying the hypotheses in Theorem 6.1 (Fig. 8.1). Then by that theorem there exists a solution $u_{\mathrm{tr}} : \mathbb{R}^2_x \to \mathbb{R}^2_u$ of

$$\Delta u - W_u(u) = 0, \tag{8.1}$$

equivariant with respect to the symmetry group of the equilateral triangle, connecting the minima of W at infinity, that is, along rays l contained in the interior of each component D_i of the partition, the solution approaches exponentially the corresponding minimum a_i:

$$\lim_{x \in l, |x| \to \infty} u_{\mathrm{tr}}(x) = a_1 \quad \text{(cf. Fig. 8.2).} \tag{8.2}$$

Theorem 6.1 gives no information on the asymptotic behavior of the solution as $|x| \to \infty$ along lines parallel to the 'walls' of the partition. On the other hand, Bronsard et al. [7] established the existence of such an equivariant solution

N. D. Alikakos et al., *Elliptic Systems of Phase Transition Type*,
Progress in Nonlinear Differential Equations and Their Applications 91,
https://doi.org/10.1007/978-3-319-90572-3_8

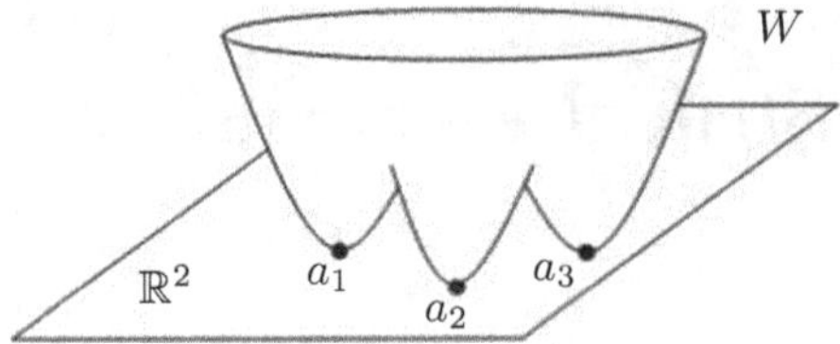

Fig. 8.1 The potential W

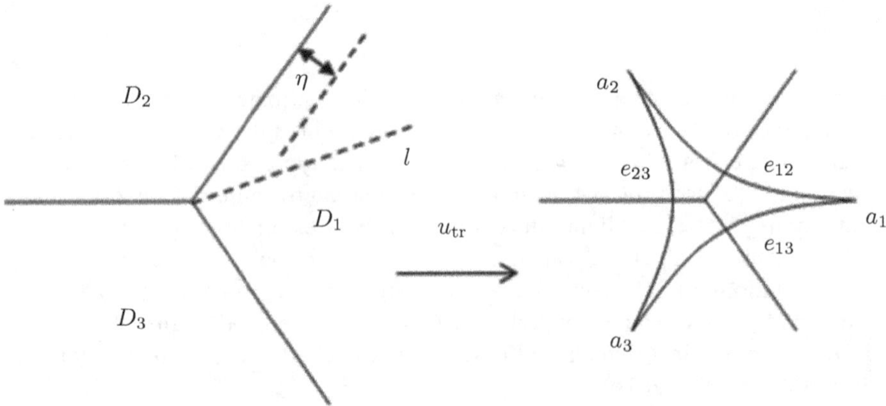

Fig. 8.2 $u_{\mathrm{tr}} : \mathbb{R}^2_x \to \mathbb{R}^2_u$, the ray l, $d(x, \partial D_1) = \eta$

satisfying (8.2) above, with the additional property that

$$\lim_{d(x,\partial D_i)=\eta, |x|\to\infty} u_{\mathrm{tr}}(x) = e_{ij}(\eta) \quad \text{(cf. Fig. 8.2)}, \tag{8.3}$$

where $e_{ij} : \mathbb{R} \to \mathbb{R}^2$ is a solution of $e'' - W_u(e) = 0$ connecting the minima a_i, a_j, and $d(x, \partial D_i)$ is the signed distance. By invoking Theorem 2.1 it can be shown that all three connections exist and are minimal for general perturbations.

Thus, we see that in this example the solution has stratified structure consisting of lower-dimensional minimal solutions of (8.1), $\{a_1, a_2, a_3\}$ and $\{e_{12}, e_{13}, e_{23}\}$ (zero- and one-dimensional respectively). To establish (8.2) in Chap. 6, we had to study the measure of the sets $B_r(x_0) \cap \{|u - a_i| > \lambda\}$ for large r. An appropriate tool for that purpose is the density estimate in Theorem 5.2. We recall the analogy behind it with minimal surface theory, introduced by Caffarelli and Córdoba (cf. Sect. 5.2.1), that is, the identification of the area and volume with

$$A(r) = \int_{B^n_r(x_0)\cap\{x:|u-a|\leq\lambda\}} W(u)\mathrm{d}x, \tag{8.4a}$$

$$V(r) = \mathscr{L}^n(B^n_r(x_0) \cap \{x : |u - a| > \lambda\}), \tag{8.4b}$$

where B_r^n is an open ball in $\mathbb{R}^n$. In relation to (8.3) above it is useful to introduce a further such identification of area and volume. By introducing suitable coordinates we can restate (8.3) as

$$\lim_{y \to +\infty} u_{\mathrm{tr}}(s, y) = e_{ij}(s), \quad x = (s, y). \tag{8.5}$$

We can adopt the point of view of u as a map

$$\mathbb{R} \ni y \longmapsto u(\cdot, y) \in L^2_\#(\mathbb{R}; \mathbb{R}^2),$$

where $L^2_\#$ is a function space made up of curves in $\mathbb{R}^2$, parametrized by s, $u(s, y) =: \tilde{u}(y)(s)$. We define the *effective potential* on $L^2_\#(\mathbb{R}; \mathbb{R}^2)$ for $e = e_{12}$:

$$\mathscr{W}(v(\cdot)) = J_\mathbb{R}(v) - J_\mathbb{R}(e). \tag{8.6}$$

We note that $\mathscr{W} \geq 0$ (cf. (8.18)). Note also that $J_\mathbb{R}(e)$ is finite, but for a higher-dimensional solution this term will be infinite. Nevertheless, in the appropriate set-up the difference defining the effective potential will be finite. The expression $\mathscr{W}(\xi(\cdot)) = J_\mathbb{R}(e + \xi) - J_\mathbb{R}(e)$ is an alternative and suggestive way of writing (8.6). Then by analogy to (8.4), we introduce the area and volume by

$$\mathscr{A}(r) = \int_{B_r^{n-1}(y^0) \cap \{y : \|u(\cdot, y) - e(\cdot)\| \leq \lambda\}} \mathscr{W}(u(\cdot, y)) \mathrm{d}y, \tag{8.7a}$$

$$\mathscr{V}(r) = \mathscr{L}^{n-1}(B_r^{n-1}(y^0) \cap \{y : \|u(\cdot, y) - e(\cdot)\| > \lambda\}), \tag{8.7b}$$

where $\| \cdot \|$ is the $L^2(\mathbb{R})$ norm. Clearly, we can continue in this manner and replace $e(\cdot)$ with a more general lower-dimensional solution and establish appropriate analogs of the density Theorem 5.2. These will be useful for studying the asymptotic behavior of the solution as $|x| \to \infty$. More precisely, we are interested in the limits

$$\lim_{\lambda \to \infty} u(x' + \lambda v) = \tilde{u}(x'), \quad \mathbb{R}^n \ni x = x' + \lambda v, \ |v| = 1, \ x' \perp v,$$

where v is in the interior of intersections of reflection planes $\Pi_1, \ldots, \Pi_k$ making up pieces of the walls of D. Take for example the tetrahedral cone in Fig. 1.2. We have

$$\lim_{\lambda \to \infty} u(x' + \lambda v) = a, \ x' \in \mathbb{R}^0, \ \forall v \in \mathrm{Int}\, D, \tag{8.8a}$$

$$\lim_{\lambda \to \infty} u(x' + \lambda v) = e(x'), \ x' \in \mathbb{R}^1, \ \forall v \in \mathrm{Int}\, \Pi_i, \tag{8.8b}$$

$$\lim_{\lambda \to \infty} u(x' + \lambda v) = u_{\mathrm{tr}}(x'), \ x' \in \mathbb{R}^2, \ \forall v \in \mathrm{Int}(\Pi_i \cap \Pi_j). \tag{8.8c}$$

8.2   The Density Estimate for a Connection

We begin with the simplest possible result of this kind, in the simplest possible setting, namely, the whole space.

The Polar Form (cf. Sect. 4.1, Sect. 5.3.1)

We introduce the *polar form* of a vector map $u \in W^{1,2}_{\mathrm{loc}}(\mathbb{R}^n; \mathbb{R}^m) \cap L^\infty(\mathbb{R}^n; \mathbb{R}^m)$ with respect to the connection e. We write

$$u(s, y) = e(s) + q^u(y)\boldsymbol{n}^u(s, y), \quad x = (s, y) \in \mathbb{R}^n, \ y = (y_1, \ldots, y_{n-1}), \qquad (8.9)$$

with

$$q^u(y) := \|u(\cdot, y) - e(\cdot)\|, \qquad (8.10)$$

where we have dropped the subscript $\| \cdot \|_{L^2(\mathbb{R})}$, and

$$\boldsymbol{n}^u(\cdot, y) = \begin{cases} \dfrac{u(\cdot, y) - e(\cdot)}{\|u(\cdot, y) - e(\cdot)\|}, & \text{if } q^u(y) \neq 0, \\ 0, & \text{otherwise} . \end{cases} \qquad (8.11)$$

We also introduce the horizontal cylinders $C_r(y^0) = \mathbb{R} \times B^{n-1}_r(y^0)$, with cross-section $B^{n-1}_r(y^0)$, the open ball in $\mathbb{R}^{n-1}$. From (8.9) we have

$$u_{y_i} = q^u_{y_i} \boldsymbol{n}^u + q^u \boldsymbol{n}^u_{y_i}.$$

and therefore observing that

$$\|\boldsymbol{n}^u(\cdot, y)\| = 1, \ \left\langle \boldsymbol{n}^u(\cdot, y), \boldsymbol{n}^u_{y_i}(\cdot, y) \right\rangle = 0, \ i = 1, \ldots, n - 1,$$

we obtain the following *polar representation* of the energy:

$$J_{C_r(y^0)}(u) = \int_{C_r(y^0)} \left(\frac{1}{2}|\nabla u|^2 + W(u) \right) dx$$

$$= \int_{B^{n-1}_r(y^0)} \left(\frac{1}{2}\left(|\nabla_y q^u|^2 + (q^u)^2 \sum_{i=1}^{n-1} \|\boldsymbol{n}^u_{y_i}\|^2\right) + \mathscr{W}(u) + J_{\mathbb{R}}(e) \right) dy.$$

$$(8.12)$$

Since only differences of the energy J are involved, we can disregard the term $J_{\mathbb{R}}(e)$ in (8.12) and replace $J_{C_r(y^0)}$ with

$$
\begin{aligned}
\widehat{J}_{C_r(y^0)}(u) &= \int_{C_r(y^0)} \left(\left[\frac{1}{2}|\nabla u|^2 + W(u) \right] - J_{\mathbb{R}}(e) \right) dx \\
&= \int_{B_r^{n-1}(y^0)} \left(\frac{1}{2} \|\nabla_y u(\cdot, y)\|^2 + \mathscr{W}(u) \right) dy \\
&= \int_{B_r^{n-1}(y^0)} \left(\frac{1}{2} \left(|\nabla_y q^u|^2 + (q^u)^2 \sum_{i=1}^{n-1} \|\mathbf{n}_{y_i}^u\|^2 \right) + \mathscr{W}(u) \right) dy. \qquad (8.13)
\end{aligned}
$$

Note that $\widehat{J}_{C_r(y^0)}(e) = 0$.

The Set-Up

A. For $z \in \mathbb{R}^d$, $d \geq 1$, we denote by $\hat{z}$ the reflection of z in the plane $\{z_1 = 0\}$, $\hat{z} = (-z_1, z_2, \ldots, z_d)$. We consider $W : \mathbb{R}^m \to \mathbb{R}$, a C^3 double-well potential, symmetric with respect to the reflection above, $W(u) = W(\hat{u})$, with nondegenerate[1] global minima at $a^+ \neq a^-$, $0 = W(a^+) = W(a^-)$, $W(u) > 0$ for $u \in \mathbb{R}^m \setminus \{a^+, a^-\}$. We consider symmetric, minimal (with respect to equivariant perturbations) solutions to

$$
\Delta u - W_u(u) = 0, \ u : \mathbb{R}^n \to \mathbb{R}^m, \qquad (8.14)
$$

which are also *positive*, $u(\mathbb{R}_+^n) \subset \mathbb{R}_+^m$, $\mathbb{R}_+^d := \{z \in \mathbb{R}^d : z_1 \geq 0\}$. Theorem 6.1 (or Theorem 7.1, which covers also the case $m \neq n$), $D = F = \mathbb{R}_+^m$, ensure the existence of such a solution satisfying the estimate

$$
|u(x) - a^+| + |\nabla u(x)| \leq K e^{-kx_1}, \quad x \in \mathbb{R}_+^n. \qquad (8.15)
$$

B. Our hypotheses on W imply the existence of a connection $e : \mathbb{R} \to \mathbb{R}^m$, that is a global minimizer of the *action* in the general class of v's satisfying $\lim_{s \to \pm\infty} v(s) = a^\pm$. We can assume that e is symmetric, *positive*, and satisfies the estimate

$$
|e(s) - a^+| + |e_s(s)| \leq K e^{-ks}, \quad s \geq 0. \qquad (8.16)
$$

All these properties can be deduced from Theorem 2.1 and Proposition 2.4. We require in addition that e is *hyperbolic*:

$$
\langle Tv, v \rangle \geq \eta \|v\|^2, \quad v \in W_s^{1,2}(\mathbb{R}; \mathbb{R}^m), \text{ for some } \eta > 0, \qquad (8.17)
$$

---

[1] i.e., with positive definite Hessian at $a^\pm$.

where $Tv = -v_{ss} + W_{uu}(e)v$, $W_S^{1,2}(\mathbb{R}; \mathbb{R}^m) \subset W_{\text{loc}}^{1,2}(\mathbb{R}; \mathbb{R}^m)$ is the subspace of the symmetric maps, $\langle \cdot, \cdot \rangle$ the inner product in $L^2(\mathbb{R}; \mathbb{R}^m)$, and $\|\cdot\|$ the associated norm. Note that (8.17) excludes zero from being an eigenvalue of T. Here symmetry is crucial because it excludes translations. Nevertheless, in the vector case simplicity of eigenvalues is not generally true, and so (8.17) is an extra requirement. We denote by $E^{xp} \subset C^1(\mathbb{R}; \mathbb{R}^m)$ the set of symmetric maps $v : \mathbb{R} \to \mathbb{R}^m$ that satisfy the estimate $|v(s) - a^+| + |v_s(s)| \leq K e^{-ks}$, $\forall s \geq 0$, with k, K as in (8.15), (8.16) above. Note that (8.15) and the minimizing property of e imply that

$$\mathscr{W}(u) \geq 0, \quad u \in E^{xp}. \tag{8.18}$$

The Analog of the Basic Estimate (cf. Sect. 5.1)

Lemma 8.1 *Under the hypotheses in* A. *above (symmetry, minimality), we have the estimate*

$$0 \leq J_{C_r(y^0)}(u) = \int_{C_r(y^0)} \left(\left[\frac{1}{2} |\nabla u|^2 + W(u) \right] - J_{\mathbb{R}}(e) \right) dx$$

$$= \int_{B_r^{n-1}(y^0)} \left(\frac{1}{2} \|\nabla_y u(\cdot, y)\|^2 + \mathscr{W}(u(\cdot, y)) \right) dy \leq C_1 r^{n-2}, \tag{8.19}$$

with $C_1 = C_1(k, K)$, $\forall y^0 \in \mathbb{R}^{n-1}$.

Proof The inequality on the left follows from (8.18). The estimate will follow from minimality via an appropriate test function argument. Minimality states that u minimizes the energy subject to its Dirichlet values on the boundary of compact sets. Note that this class of sets can be extended to include infinite horizontal cylinders for $u \in E^{xp}$.

Step 1 (Minimality Over Infinite Cylinders)
Let $O \subset \mathbb{R}^{n-1}$, open, bounded. Then

$$\widehat{J}_{\mathbb{R} \times O}(u) = \min_{v \in u + W_{0S}^{1,2}(\mathbb{R} \times O; \mathbb{R}^m)} \widehat{J}_{\mathbb{R} \times O}(v), \tag{8.20}$$

where $W_{0S}^{1,2}(\mathbb{R} \times O; \mathbb{R}^m)$ is the closure in $W_S^{1,2}(\mathbb{R} \times O; \mathbb{R}^m)$ of the smooth maps v that satisfy $v = 0$ on $\mathbb{R} \times \partial O$.

We proceed by contradiction. Thus assume there are $\delta > 0$ and $v \in u + W_{0S}^{1,2}(\mathbb{R} \times O; \mathbb{R}^m)$ such that

$$\widehat{J}_{\mathbb{R} \times O}(u) - \widehat{J}_{\mathbb{R} \times O}(v) \geq \delta. \tag{8.21}$$

For each $l > 0$ define $\tilde{v} \in u + W_{0S}^{1,2}(\mathbb{R} \times O; \mathbb{R}^m)$ by

$$
\tilde{v} = \begin{cases}
v, & \text{for } s \in [0, l], \ y \in O, \\
(1 + l - s)v + (s - l)u, & \text{for } s \in [l, l+1], \ y \in O, \\
u, & \text{for } s \in [l+1, +\infty), \ y \in O.
\end{cases}
$$

The minimality of u implies

$$
0 \geq \widehat{J}_{[-l-1,l+1]\times O}(u) - \widehat{J}_{[-l-1,l+1]\times O}(\tilde{v}) = \widehat{J}_{[-l-1,l+1]\times O}(u) - \widehat{J}_{[-l,l]\times O}(v) + O(e^{-kl}),
$$

where we have taken into account that both u and v belong to E^{xp}. Taking the limit as $l \to +\infty$ yields

$$
0 \geq \widehat{J}_{\mathbb{R}\times O}(u) - \widehat{J}_{\mathbb{R}\times O}(v),
$$

in contradiction to (8.21).

Step 2
Let

$$
v(\cdot, y) = \begin{cases}
e(\cdot), & \text{for } y \in B_{r-1}^{n-1}(y^0), \\
e(\cdot) + (|y - y^0| - r + 1)q^u(y)n^u(\cdot, y), & \text{for } y \in B_r^{n-1}(y^0) \setminus B_{r-1}^{n-1}(y^0).
\end{cases}
$$

From (8.20) we have

$$
\begin{aligned}
\widehat{J}_{C_r(y^0)}(u) \leq \widehat{J}_{C_r(y^0)}(v) &= \widehat{J}_{\mathbb{R}\times(B_r^{n-1}(y^0)\setminus B_{r-1}^{n-1}(y^0))}(v) \\
&\leq C\mathscr{L}^{n-1}(B_r^{n-1}(y^0) \setminus B_{r-1}^{n-1}(y^0)) \\
&\leq Cr^{n-2}.
\end{aligned}
$$

$\square$

Consequences of Hyperbolicity

Lemma 8.2 *Under the hypothesis of hyperbolicity in B. above, there is $\bar{q} > 0$ such that*

$$
D_{qq}\mathscr{W}(e + qn) \geq \frac{\eta}{2}, \ \text{for } q \in [0, \bar{q}], \ n \in \mathbb{S}, \tag{8.22}
$$

where $\mathbb{S} = W^{1,2}(\mathbb{R}; \mathbb{R}^m) \cap \{\|n\| = 1\}$, for every $v(s) = e(s) + qn(s) \in \mathrm{E}^{\mathrm{xp}}$.

Proof First note that $\|v - e\|_{W^{1,2}(\mathbb{R};\mathbb{R}^m)} \leq M_1$, for $v \in E^{\mathrm{xp}}$, for some $M_1 > 0$. We begin by differentiating twice $\mathscr{W}(e + qn)$ with respect to q. We obtain

$$D_{qq}\mathscr{W}(e + qn) = \|n_s\|^2 + \int_{\mathbb{R}} W_{uu}(e + qn)n \cdot n \, ds$$

$$= D_{qq}\mathscr{W}(e + qn)|_{q=0} + \int_{\mathbb{R}} (W_{uu}(e + qn) - W_{uu}(e))n \cdot n \, ds.$$

From the interpolation inequality

$$\|f\|_{L^\infty(\mathbb{R};\mathbb{R}^m)} \leq \sqrt{2}\|f\|^{\frac{1}{2}}\|f_s\|^{\frac{1}{2}} \ (f \in W^{1,2}(\mathbb{R}; \mathbb{R}^m))$$

$$\leq \sqrt{2}\|f\|_{W^{1,2}(\mathbb{R};\mathbb{R}^m)},$$

applied to qn we obtain via the second inequality

$$\|qn\|_{L^\infty(\mathbb{R};\mathbb{R}^m)} \leq \sqrt{2}M_1,$$

and via the first

$$\|n\|_{L^\infty(\mathbb{R};\mathbb{R}^m)} \leq \sqrt{2}M_1^{\frac{1}{2}}q^{-\frac{1}{2}},$$

since $\|qn\| = q$ and $\|qn_s\| \leq M_1$. Therefore we have

$$|W_{u_iu_j}(e(s) + qn(s)) - W_{u_iu_j}(e(s))| \leq \sqrt{2}M_1^{\frac{1}{2}}\overline{W}'''q^{\frac{1}{2}},$$

where $\overline{W}'''$ is defined by

$$\overline{W}''' := \max_{\substack{1 \leq i, j, k \leq m \\ s \in \mathbb{R}, |\tau| \leq 1}} |W_{u_iu_ju_k}(e(s) + \tau\sqrt{2}M_1)|.$$

Thus we obtain

$$\left| \int_{\mathbb{R}} (W_{uu}(e + qn) - W_{uu}(e))n \cdot n \, ds \right| \leq C_1 q^{\frac{1}{2}}\langle n, n \rangle = C_1 q^{\frac{1}{2}},$$

where $C_1 > 0$ is a constant that depends on M_1. We now observe that

$$D_{qq}\mathscr{W}(e + qn)|_{q=0} = \langle Tn, n \rangle \geq \eta\|n\|^2 = \eta,$$

where we have also used (8.17) above. Consequently,

$$D_{qq}\mathscr{W}(e + q\boldsymbol{n}) \geq c_0 := \frac{\eta}{2}, \quad \text{for } q \in [0, \bar{q}],$$

where $\bar{q} = \frac{1}{4}\frac{\eta^2}{C_1^2}$. This concludes the proof of the lemma. $\square$

Inequality (8.22) implies that e is isolated in $\{\mathscr{W} = 0\} \cap \mathrm{E^{xp}}$. Hence,

$$\hat{d}_0 = \inf_v \{\|e - v\| : v \neq e, \{\mathscr{W}(v) = 0\} \cap \mathrm{E^{xp}}\} > 0.$$

Theorem 8.1 ([4])   *Assume the hypotheses* A. *and* B. *(nondegeneracy of W, symmetry, minimality of u, hyperbolicity of e),* $u : \mathbb{R}^n \to \mathbb{R}^m$, $n \geq 2$. *Then for any* $\mu_0 > 0$ *and any* $\lambda \in (0, \hat{d}_0)$, *the condition*

$$\mathscr{L}^{n-1}(B_{r_0}^{n-1}(y^0) \cap \{y : \|u(\cdot, y) - e(\cdot)\| \geq \lambda\}) \geq \mu_0 > 0$$

implies the estimate

$$\mathscr{L}^{n-1}(B_r^{n-1}(y^0) \cap \{y : \|u(\cdot, y) - e(\cdot)\| \geq \lambda\}) \geq Cr^{n-1}, \quad \text{for } r \geq r_0,$$

where $C = C(W, \mu_0, \lambda, r_0, K, k)$ *is independent of* y^0 *and* u.

Remark 8.1   It is a simple consequence of (8.19) that the validity of the theorem for any value of $\lambda \in (0, \hat{d}_0)$ implies its validity for all $\lambda' \in (0, \hat{d}_0)$. The argument is analogous to that in Remark 5.4, relying on the estimate

$$\min_{\lambda \leq q \leq \lambda'} \mathscr{W}(e + q\boldsymbol{n}) \geq w_\lambda^{\lambda'}, \quad \|\boldsymbol{n}\| = 1,$$

which follows in $\mathrm{E^{xp}}$ by strong L^2 compactness, and the lower semicontinuity of $\mathscr{W}$ (cf. argument in the conclusion of the proof of Lemma 8.4 below).

Before giving the proof of Theorem 8.1, we present as a corollary a Liouville type result.

Corollary 8.1   *Assume the hypotheses in Theorem* 8.1 *above, and moreover assume that the connection e is unique. Then*

$$u(x) = e(x_1), \quad x = (x_1, \ldots, x_n).$$

Proof   Assume, by contradiction, that for some y

$$\|u(\cdot, y) - e(\cdot)\| \geq 2\lambda > 0, \quad \text{with some } \lambda > 0. \tag{8.23}$$

By the membership of $u(\cdot, y)$ in E^{xp}, and by the continuity of u, we obtain that

$$\mathcal{L}^{n-1}(B_1^{n-1}(y) \cap \{\|u(\cdot, y) - e(\cdot)\| \geq \lambda\}) \geq \mu_0 > 0,$$

and so, via Theorem 8.1 above,

$$\mathcal{L}^{n-1}(B_r^{n-1}(y) \cap \{\|u(\cdot, y) - e(\cdot)\| \geq\}) \geq Cr^{n-1}, \quad \text{for } r \geq 1. \tag{8.24}$$

From the assumed uniqueness and hyperbolicity of e, by (8.42) with $l = \infty$, we have

$$\|u(\cdot, y) - e(\cdot)\| \geq \lambda \quad \Rightarrow \quad \mathcal{W}(u(\cdot, y)) \geq \bar{w}(\lambda) > 0.$$

This fact together with (8.24) imply the lower bound

$$\bar{w}(\lambda)Cr^{n-1} \leq J_{C_r(y)}(u), \quad \text{for } r \geq 1,$$

which contradicts Lemma 8.1 for large r. This concludes the proof of Corollary 8.1.
$$\square$$

Proof (Theorem 8.1) (cf. proof of Theorem 5.2)

1. In the present framework the comparison maps $h = e + q^h \boldsymbol{n}^u$ and $\sigma = e + q^\sigma \boldsymbol{n}^u$, $q^\sigma = \min\{q^h, q^u\}$, $\sigma = u$ on ∂B_r^{n-1}, should be regarded as maps from B_r^{n-1} into the function space E^{xp}. The minimality of u and the polar form (8.13) of the energy imply the inequality

$$\frac{1}{2}\int_{B_r^{n-1}}(|\nabla q^u|^2 - |\nabla q^\sigma|^2)dy = \widehat{J}_{C_r}(u) - \widehat{J}_{C_r}(\sigma) + \int_{B_r^{n-1}}(\mathcal{W}(\sigma) - \mathcal{W}(u))dy$$

$$+ \frac{1}{2}\int_{B_r^{n-1}}\left((q^\sigma)^2 - (q^u)^2\right)\sum_{i=1}^{n-1}\left\|\frac{\partial \boldsymbol{n}^u}{\partial y_i}\right\|^2 dy$$

$$\leq \int_{B_r^{n-1}}(\mathcal{W}(\sigma) - \mathcal{W}(u))dy. \tag{8.25}$$

Indeed, minimality and (8.20) imply $\widehat{J}_{C_r}(u) - \widehat{J}_{C_r}(\sigma) \leq 0$, and the third term is also nonpositive because $0 \leq q^\sigma \leq q^u$.

2. As in the proof of the case $\alpha = 2$ in Theorem 5.2, we let $\varphi : B_r^{n-1} \subset \mathbb{R}^{n-1} \to \mathbb{R}$ be the solution of the problem

$$\begin{cases} \Delta\varphi = c_1^2\varphi, & \text{on } B_r^{n-1}, \\ \varphi = 1, & \text{on } \partial B_r^{n-1}, \end{cases} \tag{8.26}$$

where $c_1 > 0$ will be chosen. We set

$$q_M = \sup_{y \in \mathbb{R}^{n-1}} \|u(\cdot, y) - e(\cdot)\|$$

and define

$$h = e + q^h \boldsymbol{n}^u, \quad q^h = \varphi q_M, \quad \text{and as before}$$
$$\sigma = e + q^\sigma \boldsymbol{n}^u, \quad q^\sigma = \min\{q^u, q^h\}, \tag{8.27}$$
$$\beta = \min\{q^u - q^\sigma, \lambda\},$$

where $\lambda \in (0, \bar{q})$ with $\bar{q}$ as in Lemma 8.2. We also recall from (5.59) that $\varphi(y) = \Phi(|y|, r)$ and the exponential estimate

$$\Phi(s, r) \le e^{-c_2(r-s)}, \quad \text{for } s \in [0, r], \ r \ge 1, \tag{8.28}$$

holds for some $c_2 > 0$.

Note that the definition of σ in (8.27) implies

$$q^\sigma = q^u, \quad \text{on } \partial B_r^{n-1}.$$

Proceeding as in the proof of Theorem 5.2 by applying the inequality (5.42) on $B_r^{n-1} \subset \mathbb{R}^{n-1}$ to β^2, we obtain

$$\left(\int_{B_r^{n-1}} \beta^{\frac{2(n-1)}{n-2}} dy \right)^{\frac{n-2}{n-1}} = \left(\int_{B_r^{n-1}} (\beta^2)^{\frac{n-1}{n-2}} dy \right)^{\frac{n-2}{n-1}}$$

$$\le C \int_{B_r^{n-1}} |\nabla(\beta^2)| dy \qquad (\beta = 0, \ \text{on } \partial B_r^{n-1})$$

$$\le 2C \int_{B_r^{n-1} \cap \{q^u - q^\sigma \le \lambda\}} |\nabla \beta| |\beta| dy$$

$$\le CA \int_{B_r^{n-1}} |\nabla(q^u - q^\sigma)|^2 dy + \frac{C}{A} \int_{B_r^{n-1} \cap \{q^u - q^\sigma \le \lambda\}} (q^u - q^\sigma)^2 dy$$

$$= CA \left(\int_{B_r^{n-1}} (|\nabla q^u|^2 - |\nabla q^\sigma|^2) dy - 2 \int_{B_r^{n-1}} \nabla q^\sigma \cdot \nabla(q^u - q^\sigma) dy \right)$$

$$+ \frac{C}{A} \int_{B_r^{n-1} \cap \{q^u - q^\sigma \le \lambda\}} (q^u - q^\sigma)^2 dy$$

where we have used $\nabla\beta = 0$ a.e. on $q^u - q^\sigma > \lambda$ and Young's inequality. Thus by (8.25) we derive that

$$\left(\int_{B_r^{n-1}} \beta^{\frac{2(n-1)}{n-2}}\,\mathrm{d}y\right)^{\frac{n-2}{n-1}} \leq \frac{C}{A}\int_{B_r^{n-1}\cap\{q^u-q^\sigma\leq\lambda\}}(q^u - q^\sigma)^2\mathrm{d}y$$

$$+ 2CA\left(\int_{B_r^{n-1}}(\mathscr{W}(\sigma) - \mathscr{W}(u))\mathrm{d}y - \int_{B_r^{n-1}}\nabla q^\sigma\cdot\nabla(q^u - q^\sigma)\mathrm{d}y\right). \qquad (8.29)$$

3. Conclusion

The inequality (8.29), aside from the fact that n is replaced by $n - 1$, B_r^{n-1} is the ball of radius r in $\mathbb{R}^{n-1}$, and W is replaced by $\mathscr{W}$, coincides with (5.79). Moreover, by Lemma 8.2, $\mathscr{W}$ has the properties required for W in **H**, for $\alpha = 2$, in (5.32). The inequality

$$W(h) - W(u) \leq W(h)$$

is now replaced by

$$\mathscr{W}(h) - \mathscr{W}(u) \leq \mathscr{W}(h).$$

Thus the arguments developed in the proof of Theorem 5.2 for the case $\alpha = 2$ can be repeated essentially verbatim to complete the proof. $\qquad\square$

Remark 8.2 Theorem 8.1 is a special case of Theorem 8.2 below, corresponding to the case $l = \infty$. In the proof of that theorem we give a very detailed account of the modifications of the proof of Theorem 5.2 that are required for proving Theorem 8.2.

8.3 Localization of the Density Estimate

Our purpose in this section is to obtain a localized version of Theorem 8.1 with $\mathbb{R}^n$ replaced by an open, symmetric-convex $\Omega \neq \mathbb{R}^n$, playing the role of $\mathscr{O}$ in Theorem 5.2. As an application we establish Corollary 8.2 below which originally was proved via the analog in the connection setting of the method in Sect. 5.5 [5, Theorem 1.4]. Such a localization is needed for establishing the hierarchical structure of solutions. In that set-up Ω is the intersection of half-spaces defined by reflection planes. The infinite cylinders are replaced by finite cylinders of increasing length, and roughly speaking the proofs are modified by an exponentially small term that can be absorbed.

To see the necessity for such a localization, let Ω be the upper half-plane, $\mathbb{R}^2 \cap \{x_2 > 0\}$, and take W and the symmetry group as in Sect. 8.2 above. Theorem 6.1 applies and produces an equivariant solution $u : \Omega \to \mathbb{R}^2$, $u(\hat{x}) = \hat{u}(x)$, with the estimate $|u(x) - a_1| \le Ke^{-kd(x,\partial\Omega^+)}$, $x \in \Omega^+ = \mathbb{R}^2_+ \cap \{x_2 > 0\}$. Notice that this estimate implies exponential decay along rays in the first quadrant emanating from the origin, but on the other hand gives no information along horizontal lines as $x_1 \to \infty$. Thus, infinite cylinders parallel to the x_1-axis are not appropriate test sets, since minimality over bounded sets does not imply minimality over the unbounded cylinder.

The Set-Up

A. $\Omega \subset \mathbb{R}^n$ is *symmetric-convex* if $x \in \Omega \Rightarrow (tx_1, x_2, \ldots, x_n) \in \Omega$ for $|t| \le 1$. We define $\Omega^+ := \Omega \cap \{x_1 > 0\}$. For $z \in \mathbb{R}^d$ we denote by $\hat{z}$ the reflection of z in the plane $\{z_1 = 0\}$, $\hat{z} = (-z_1, z_2, \ldots, z_d)$. For simplicity we will restrict ourselves to the case where $\Omega = \mathbb{R}^n \cap \{x_n > c|x_1|\}$, for some $c > 0$. We consider $W : \mathbb{R}^m \to \mathbb{R}_+$, a C^3 double-well potential, symmetric with respect to the reflection above, $W(u) = W(\hat{u})$, with nondegenerate minima at a^+, a^-. We consider symmetric minimal solutions to

$$\Delta u - W_u(u) = 0, \quad \mathbb{R}^n \supset \Omega : x \mapsto u(x) \in \mathbb{R}^m, \tag{8.30}$$

which are also *positive*, $u(\Omega^+) \subset \mathbb{R}^m_+$. By Theorem 6.1 (or Theorem 7.1, which also covers $m \ne n$), $D = F = \Omega^+$, and we have the estimate

$$|u(x) - a^+| + |\nabla u(x)| \le Ke^{-kd(x,\partial\Omega^+)}, \quad x \in \Omega^+. \tag{8.31}$$

We bisect Ω^+ with the $x_1 = \Lambda x_n$ plane $\Lambda^{-1} = c + \sqrt{1 + c^2}$ into Ω_I^+ and Ω_{II}^+, and define $\Omega_I = \Omega_I^+ \cup \widehat{\Omega_I^+}$ as in Fig. 8.3.

It follows from (8.31), since $d(x, \partial\Omega^+) = x_1$ for $x \in \Omega_I^+$, that

$$|u(x) - a^+| + |\nabla u(x)| \le Ke^{-kx_1}, \quad x \in \Omega_I^+. \tag{8.32}$$

B. Our hypotheses on W imply the existence of a connection $e : \mathbb{R} \to \mathbb{R}^m$, that is a global minimizer of the *action* in the class of v's satisfying $\lim_{s \to \pm\infty} v(s) = a^\pm$.

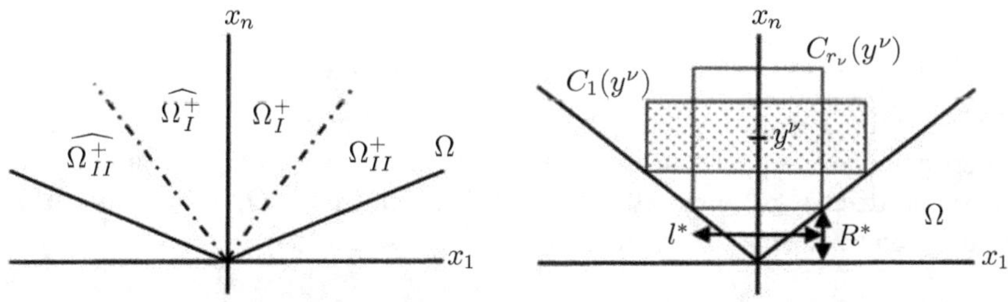

Fig. 8.3 Examples of Ω, Ω_I^+, $\Omega_I = \Omega_I^+ \cup \widehat{\Omega_I^+}$. In the second figure Ω is the upper half-plane. We exhibit a few horizontal cylinders $C_{r_\nu}(y^\nu)$, $r_\nu > 1$

We can assume that e is *positive*, symmetric, and satisfies the estimate

$$|e(s) - a^+| + |e_s(s)| \leq Ke^{-ks}, \quad s \geq 0. \tag{8.33}$$

All these can be deduced from Theorem 2.2 and Proposition 2.4. We require in addition that e is *hyperbolic* in the class of symmetric positive perturbations:

$$\langle Tv, v \rangle \geq \eta \|v\|^2, \quad v \in W_s^{1,2}(\mathbb{R}, \mathbb{R}^m), \text{ for some } \eta > 0,$$

where $Tv = -v_{ss} + W_{uu}(e)v$, $W_s^{1,2}(\mathbb{R}, \mathbb{R}^m) \subset W_{\text{loc}}^{1,2}(\mathbb{R}, \mathbb{R}^m)$ is the subspace of the symmetric maps, $\langle \cdot, \cdot \rangle$ the inner product in $L^2(\mathbb{R}, \mathbb{R}^m)$, and $\|\cdot\|$ the associated norm. We denote by $E_l^{\text{xp}} \subset C^1([-l, l]; \mathbb{R}^m)$, the symmetric maps $v : [-l, l] \to \mathbb{R}^m$ that satisfy the estimate $|v(s) - a^+| + |v_s(s)| \leq Ke^{-ks}$, $\forall s \in [0, l]$, with k, K as in (8.33) above. For $l \in (0, \infty]$, we let $\langle v, w \rangle_l = \int_{-l}^{l} v(s)w(s)ds$, $\|v\|_l = (\langle v, v \rangle_l)^{1/2}$, $\|v\| := \|v\|_\infty$, $E^{\text{xp}} := E_\infty^{\text{xp}}$.

C. We write $x = (s, y)$, $y = (y_1, \ldots, y_{n-1})$, and introduce the *cylinders* $C_r^l(y^0) = (-l, l) \times B_r^{n-1}(y^0)$ with cross-section $B_r^{n-1}(y^0)$, and height $2l$, and in particular we consider those cylinders in Ω_I which are 'sitting' on the bisector $x_1 = \Lambda x_n$ (cf. Fig. 8.3 above). Their height $2l$ is determined linearly by y^0 and r, and is denoted by $2l_r^{y^0}$. For example if $\Omega = \{x_n > 0\}$, then $l_r^{y^0} = y_{n-1}^0 - r$, $y = (y_1, \ldots, y_{n-1})$. We denote these cylinders by $C_r(y^0)$.

Theorem 8.2 *Under the Hypotheses in* A. *and* B. *(nondegeneracy of* W, *symmetry, minimality of* u, *hyperbolicity of* e), $u : \Omega_I \to \mathbb{R}^m$, $n \geq 2$, *there exists* $\lambda^* > 0$ *depending only on* K, k *in (8.31), such that for any* $\mu_0 > 0$ *and* $\lambda \in (0, \lambda^*)$, *there exists* $l^* = l^*(\lambda)$ *so that the condition* $l_1^{y^0} > l^*$ *and*

$$\mathscr{L}^{n-1}(B_1^{n-1}(y^0) \cap \{y : \|u(\cdot, y) - e(\cdot)\|_{l_1^{y^0}} \geq \lambda\}) \geq \mu_0 > 0,$$

imply

$$\mathscr{L}^{n-1}(B_r^{n-1}(y^0) \cap \{y : \|u(\cdot, y) - e(\cdot)\|_{l_r^{y^0}} \geq \lambda\}) \geq Cr^{n-1}, \quad 1 \leq r \leq y_{n-1}^0 - R^*,$$

$\frac{l^*}{\Lambda} = R^*$, *where* $C = C(\mu_0, \lambda, k, K)$ *is independent of* y^0, *and of* u *otherwise.*

What makes Theorem 8.2 possible is that the horizontal, but bounded now, cylinders provide an upper bound (Lemma 8.3) that differs from the optimal one by an error that decays exponentially with the distance from $\partial \Omega_I$. This is absorbable in the difference scheme that extracts the density estimates.

The proof of Theorem 8.2 is a modification of the proof of Theorem 8.1. We postpone it for a while and present instead

Corollary 8.2 *Let $\Omega = \mathbb{R}^n \cap \{x_n > c|x_1|\}$, for some $c \geq 0$, and assume the hypotheses in Theorem 8.2 above, and moreover assume that the connection e is unique. Then, there exist $\bar{k}$, $\bar{K}$, positive constants, such that*

$$|u(x) - e(x_1)| \leq \bar{K}e^{-\bar{k}d(x,\partial\Omega)}, \quad x = (x_1, \ldots, x_n) \in \Omega. \tag{8.34}$$

Proof For $x \in \Omega_{II}^+$, we have that $d(x, \partial\Omega) = d(x, \partial\Omega^+)$, and so the corollary follows easily for such x's by (8.31):

$$|u(s, y) - e(s)| \leq |u(s, y) - a^+| + |a^+ - e(s)|$$

$$\leq Ke^{-kd(x,\partial\Omega^+)} + Ke^{-ks}$$

$$\leq Ke^{-kd(x,\partial\Omega)} + Ke^{-kd(x,\partial\Omega)}.$$

Thus (8.34) has been established for $x \in \Omega_{II}^+$. Next, we consider $x \in \Omega_I^+$, where (8.32) holds. Using this, we can estimate u on the cups of the cylinder and obtain a modification of the bound (8.19):

$$\int_{B_r^{n-1}(y^0)} \left[\frac{1}{2}\|\nabla_y u(\cdot, y)\|_{l_r^{y^0}}^2 + \mathcal{W}_{l_r^{y^0}}(u(\cdot, y))\right] dy \leq C_1 r^{n-2} + C_2 e^{-kl_r^{y^0}} r^{n-1}, \tag{8.35}$$

for all $l_r^{y^0} \geq l^*$, with $C_i = C_i(l^*, k, K)$, $i = 1, 2$ (by Lemma 8.3), where $\mathcal{W}_{l_r^{y^0}}$ is the modified *effective potential*

$$\mathcal{W}_l(v) = J_{(-l,l)}(v) - J_{(-l,l)}(e) = \int_{-l}^{l} \left(\left[\frac{1}{2}|v_s|^2 + W(v)\right] - \left[\frac{1}{2}|e_s|^2 + W(e)\right]\right) ds. \tag{8.36}$$

This upper bound is complemented by the lower bound

$$\mathcal{W}_{l_r^{y^0}}(u(\cdot, y)) \geq C_3 \|u(\cdot, y) - e(\cdot)\|_{l_r^{y^0}}^2 - O(e^{-kl_r^{y^0}}), \quad \forall l_r^{y^0} \geq l^{**}, \tag{8.37}$$

with $C_3 = C_3(l^{**}, k, K) > 0$, by Lemma 8.4, which follows by the hyperbolicity and uniqueness of e.

To finish the proof it suffices to establish the following:

$$\forall \bar{q} \ll 1, \exists d_0 > 0 : d(x, \partial\Omega_I) \geq d_0 \Rightarrow |u(s, y) - e(s)| \leq \bar{q}, \ x = (s, y) \in \Omega_I. \tag{8.38}$$

To prove (8.38), we proceed by contradiction. So assume that there is $\bar{q}_0 > 0$ such that

$$|u(s_\nu, y_\nu) - e(s_\nu)| > \bar{q}_0, \quad d(x_\nu, \partial\Omega_I) \to \infty \quad \text{as } \nu \to \infty, \ x_\nu = (s_\nu, y_\nu) \in \Omega_I.$$

By (8.32) and (8.33), $|s_\nu| \le C$ for a constant depending only on k, K. Hence $|y_\nu| \to \infty$. By uniform continuity there is $0 < \delta_0 < 1$, independent of ν, such that $|u(s, y) - e(s)| \ge \frac{1}{2}\bar{q}_0$, for $|y - y_\nu| < \delta_0$, $|s - s_\nu| < \delta_0$. Since $l_1^{y_\nu} \to \infty$, we have

$$\int_{-l_1^{y_\nu}}^{l_1^{y_\nu}} |u(s, y) - e(s)|^2 \mathrm{d}s \ge \left(\frac{\bar{q}_0}{2}\right)^2 (2\delta_0),$$

for $|y - y_\nu| < \delta_0$, and ν sufficiently large. Hence,

$$\mathscr{L}^{n-1}\left(B_1^{n-1}(y_\nu) \cap \{y : \|u(\cdot, y) - e(\cdot)\|_{l_1^{y_\nu}} \ge \lambda := (q_0/2)(2\delta_0)^{1/2}\}\right) \ge \mu_0,$$

with $\mu_0 := (\delta_0)^{n-1}|\mathbb{S}^{n-1}|$, and so by Theorem 8.2

$$\mathscr{L}^{n-1}(B_{r_\nu}^{n-1}(y_\nu) \cap \{y : \|u(\cdot, y) - e(\cdot)\|_{l_{r_\nu}^{y_\nu}} \ge \lambda\}) \ge Cr_\nu^{n-1}, \tag{8.39}$$

provided $|y_\nu| \ge r_\nu + R^*$. Clearly, since $|y_\nu| \to \infty$, we can choose $r_\nu \to \infty$ so that $e^{-kl_{r_\nu}^{y_\nu}} r_\nu^{n-1} \to 0$. By (8.37), (8.39), we obtain

$$\int_{B_{r_\nu}^{n-1}(y_\nu)} \mathscr{W}_{l_{r_\nu}^{y_\nu}}(u(\cdot, y))\mathrm{d}y \ge [C_3\lambda^2 - O(e^{-kl_{r_\nu}^{y_\nu}})]Cr_\nu^{n-1}. \tag{8.40}$$

However this clashes with (8.35), and thus (8.38) is established.

Conclusion By linear theory, (8.38) implies

$$|u(s, y) - e(s)| \le \bar{K}e^{-\bar{k}d(x, \partial\Omega_I)}, \quad x \in \Omega_I.$$

The detailed argument is as follows: we have

$$u_{ss} + \Delta_y u - W_u(u) = 0,$$

$$e_{ss} - W_u(e) = 0.$$

Subtracting the second equation from the first, we obtain

$$(u - e)_{ss} + \Delta_y u - (W_u(u) - W_u(e)) = 0.$$

Multiplying by $(u - e)$ and integrating we get

$$\int_{-l}^{l} (u - e)_{ss} \cdot (u - e)\mathrm{d}s + \int_{-l}^{l} \Delta_y u \cdot (u - e)\mathrm{d}s - \int_{-l}^{l} (W_u(u) - W_u(e))(u - e)\mathrm{d}s = 0.$$

Integrating by parts the first term and using (8.32), (8.33) yields

$$\frac{1}{2} \int_{-l}^{l} \Delta_y(|u - e|^2)\mathrm{d}s - \int_{-l}^{l} |\nabla_y u|^2 \mathrm{d}s - \int_{-l}^{l} |u_s - e_s|^2 \mathrm{d}s$$

$$- \int_{-l}^{l} (W_u(u) - W_u(e))(u - e)\mathrm{d}s = O(\mathrm{e}^{-kl}).$$

Utilizing that $l = l(y)$ is a linear function and appealing once more to (8.32), (8.33) gives

$$\int_{-l}^{l} \Delta_y(|u - e|^2)\mathrm{d}s = \Delta_y \int_{-l}^{l} |u - e|^2 \mathrm{d}s + O(\mathrm{e}^{-kl}).$$

Taking $\bar{q}$ in (8.38) small, and using the mean value theorem on $(W_u(u) - W_u(e))$ $(u - e) = W_{uu}(\cdot)(u - e)^2$, we obtain

$$\int_{-l}^{l} [|u_s - e_s|^2 + W_{uu}(\cdot)(u - e)^2]\mathrm{d}s \geq \frac{\eta}{4}\|u - e\|_l^2 + O(\mathrm{e}^{-kl}).$$

This is a consequence of the fact that for $l^* > 0$ large, and $v \in \mathrm{E}_l^{\mathrm{xp}}$, $l \geq l^*$, we have

$$\int_{-l}^{l} [|v_s - e_s|^2 + W_{uu}(e)(v - e)^2]\mathrm{d}s \geq \frac{\eta}{2}\|v - e\|_l^2 + O(\mathrm{e}^{-kl}).$$

This in turn follows from the hyperbolicity of e, the exponential convergence of $W(e(s))$ to $W(a^{\pm})$, as $s \to \pm\infty$, and the uniform estimate defining $\mathrm{E}_l^{\mathrm{xp}}$. We refer to [1] and [11] for the relevant functional analysis. Finally, setting $\varphi(y) = \|u - e\|_l^2$ we conclude from above that

$$\Delta_y \varphi \geq c^2 \varphi - C\mathrm{e}^{-kl} \quad \text{on } B_r^{n-1}(y^0),$$

where $c^2 = \frac{\eta}{2}$, and C is a constant independent of l. Denote now by φ_h the solution of (A.1) in the Appendix, on $B_r^{n-1}(y^0)$, $\varphi_h = \Phi_h(|y|; r)$, and set $\Psi = \Phi_h + \frac{C}{c^2}\mathrm{e}^{-kl}$. Observe that

$$\Delta_y \Psi - c^2 \Psi = -C\mathrm{e}^{-kl} \quad \text{in } B_r^{n-1}(y^0),$$

$$\Psi = 1 + \frac{C}{c^2}\mathrm{e}^{-kl} \quad \text{on } \partial B_r^{n-1}(y^0),$$

$$\Psi'(0) = 1.$$

Notice that since φ is bounded, we can assume that $(\varphi - \Psi) \le 0$ on $\partial B_r^{n-1}(y^0)$. Thus by comparison we obtain the estimate

$$\|u(\cdot, y^0) - e\|_l^2 \le \Phi_h(0) + \frac{C}{c^2}e^{-kl} \le e^{-h(r)r} + \frac{C}{c^2}e^{-kl} \quad \text{(by (A.3), (A.4))}.$$

Therefore by taking $r = \frac{|y^0|}{2}$, we obtain an L^2-version of the desired estimate. To upgrade this estimate we can write

$$\Delta(u - e) = W_u(u) - W_u(e) = W_{uu}(\cdot)(u - e)$$

and employ L^p-local linear elliptic estimates [9, Theorem 9.11] and the Sobolev embedding. Notice that from the bound on $\|u - e\|_{L^\infty}$ it follows that

$$\|u(\cdot, y^0) - e\|_{L^p(-l,l)} = O(e^{-kl}), \quad \forall p < \infty.$$

We now continue the concluding argument. On the other hand by (8.32)

$$|u(s, y) - e(s)| \le K e^{-ks}, \quad x \in \Omega_I^+.$$

Therefore

$$|u(s, y) - e(s)|^2 \le K \bar{K} e^{-ks} e^{-\bar{k}d(x, \partial \Omega_I)}$$

$$\le K \bar{K} e^{-k_{\min}(s + d(x, \partial \Omega_I))} \quad (k_{\min} = \min(k, \bar{k}))$$

$$= K \bar{K} e^{-k_{\min}(s + \overline{xP}\cos\theta)}$$

$$\le K \bar{K} e^{-k_{\min}\cos\theta \, d(x, \partial \Omega)}.$$

The proof of the corollary is complete. $\qquad\qquad\qquad\qquad\qquad\square$

We now establish two lemmas.

Lemma 8.3 (*Upper bound*) *For cylinders $C_r(y^0)$ in Ω_I extending all the way to $\partial \Omega_I$, under the hypotheses in A above (symmetry, minimality), we have the estimate*

$$\int_{B_r^{n-1}(y^0)} \left[\frac{1}{2} \|\nabla_y u(\cdot, y)\|_{l_r^{y^0}}^2 + \mathscr{W}_{l_r^{y^0}}(u(\cdot, y)) \right] dy \le C_1 r^{n-2} + C_2 e^{-k(l_r^{y^0} - 1)} r^{n-1},$$

$$(8.41)$$

for all $l_r^{y^0} \in [l^*, \infty]$, with $C_i = C_i(l^*, k, K)$, where $\mathscr{W}_{l_r^{y^0}}$ is defined in (8.36) above and l^* is the fixed number given in Theorem 8.2.

Proof The argument is based on a modification of the test function in the proof of
[4, Lemma 6.3]. The top and bottom of the cylinders cannot be ignored anymore,
and their contribution is estimated via (8.32), (8.33). We use the modified polar form

$$u(\cdot, y) = e(\cdot) + q_r^u(y)\boldsymbol{n}_r^u(\cdot, y),$$

$$q_r^u(y) = \|u(\cdot, y) - e(\cdot)\|_{l_r^{y^0}},$$

$$\boldsymbol{n}_r^u(\cdot, y) = \frac{u(\cdot, y) - e(\cdot)}{\|u(\cdot, y) - e(\cdot)\|_{l_r^{y^0}}}.$$

Let $v(\cdot, y) =$

$$\begin{cases} e(\cdot), & \text{on } (-l_r^{y^0}, l_r^{y^0}) \times B_{r-1}^{n-1}(y^0), \\ e(\cdot) + (|y - y^0| - (r - 1)))q_r^u(y)\boldsymbol{n}_r^u, & \text{on } (-l_r^{y^0}, l_r^{y^0}) \times \left(B_r^{n-1}(y^0) \setminus B_{r-1}^{n-1}(y^0)\right). \end{cases}$$

$$z(\cdot, y) = \chi(s)(v(s, y) - u(s, y)) + u(s, y) \text{ on } C_r(y^0) = (-l_r^{y^0}, l_r^{y^0}) \times B_r^{n-1}(y^0),$$

where χ is a smooth cut-off function defined for $|s| \leq l_r^{y^0}$:

$$\chi(s) = \begin{cases} 0 & \text{for } |s| = l_r^{y^0}, \\ 1 & \text{for } |s| \leq l_r^{y^0} - 1, \end{cases}$$

and satisfying also $0 \leq \chi(s) \leq 1$, $|\chi'(s)| \leq 2$. We note that $z = u$ on $\partial C_r(y^0)$, and
thus by minimality it is sufficient to estimate

$$\int_{B_r^{n-1}(y^0)} \left[\frac{1}{2}\|\nabla_y z(\cdot, y)\|_{l_r^{y^0}}^2 + \mathscr{W}_{l_r^{y^0}}(z(\cdot, y))\right]dy.$$

The Gradient Term The integration is split into three regions as in Fig. 8.4. In the
interior $\nabla_y z = 0$. On the lateral part, by (8.32) and, (8.33):

$$\int_{B_r^{n-1} \setminus B_{r-1}^{n-1}} \|\nabla_y z(\cdot, y)\|_{l_r^{y^0}}^2 \, dy \leq C_1 r^{n-2}, \ C_1 = C_1(k, K).$$

Finally, the part near the cups, left and right, can be estimated via (8.32), (8.33) by
the second term on the right in (8.41).

The Effective Potential On the lateral part we have, as above,

$$\int_{B_r^{n-1} \setminus B_{r-1}^{n-1}} \mathscr{W}_{l_r^{y^0}}(z(\cdot, y))dy \leq C_1 r^{n-2}.$$

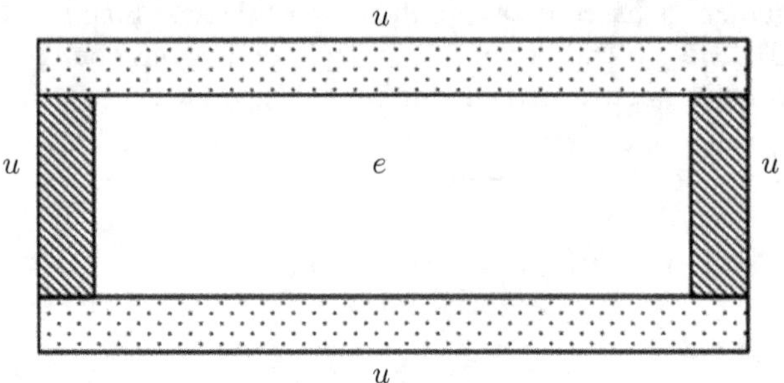

Fig. 8.4 The test function $z(\cdot, y)$

In the interior region $\mathcal{W}_{l_r^{y^0}}(z(\cdot, y)) = 0$, while the part near the cups can be estimated via (8.32), (8.33), by the second term on the right in (8.41). The proof of the lemma is complete. $\qquad\qquad\square$

Lemma 8.4 (*Lower bound*) *Under the hypotheses* B. *above (nondegeneracy, symmetry, hyperbolicity), and also under the assumption of uniqueness of the connection* e, *there is a large fixed number* $l^{**} > 0$, *and* $C_3 := C_3(l^{**}, k, K) > 0$, *such that*

$$\mathcal{W}_l(v) \geq C_3 \, \|v - e\|_l^2 + O(e^{-kl}), \quad \forall l \in [l^{**}, \infty], \ \forall v \in E_l^{\mathrm{xp}}. \tag{8.42}$$

Proof The plan for the proof is for v close to e to use hyperbolicity, while for v far, to invoke uniqueness. We first consider the case $l = +\infty$ and show that, given $p > 0$, there is $\varepsilon_p > 0$ such that

$$\|v - e\| \geq p \quad \text{implies } \mathcal{W}(v) \geq \varepsilon_p > 0, \quad v \in E^{\mathrm{xp}}. \tag{8.43}$$

Suppose, by contradiction, that there is a sequence $\{v_j\} \subset E^{\mathrm{xp}}$ that satisfies

$$\lim_{j \to +\infty} \mathcal{W}(v_j) = 0,$$
$$\|v_j - e\| \geq p. \tag{8.44}$$

Since E^{xp} is an equibounded and equicontinuous family, we can assume that

$$\lim_{j \to +\infty} v_j(s) = \bar{e}(s),$$

uniformly on compacts for some $\bar{e} \in E^{\mathrm{xp}}$. Since $v \in E^{\mathrm{xp}}$ implies $\int_{\mathbb{R}} |v'|^2 ds \leq \frac{K^2}{k}$, by passing to a subsequence (still labeled $\{v_j\}$) we have that v_j converges to $\bar{e}$

weakly in $W^{1,2}_{\text{loc}}(\mathbb{R}; \mathbb{R}^m)$. It follows that

$$\liminf_{j \to +\infty} \int_\mathbb{R} |v_j'|^2 ds \geq \int_\mathbb{R} |\bar{e}'|^2 ds. \tag{8.45}$$

Moreover, by Fatou's lemma,

$$\liminf_{j \to +\infty} \int_\mathbb{R} W(v_j) ds \geq \int_\mathbb{R} W(\bar{e}) ds.$$

This and (8.45) imply $\mathscr{W}(\bar{e}) = 0$ and therefore $\bar{e} = e$ by the uniqueness of e. Since both u_j and $\bar{e}$ belong to $\in \mathrm{E}^{\mathrm{xp}}$, there exists $\lambda_p > 0$ such that $\int_{\mathbb{R} \setminus (-\lambda_p, \lambda_p)} |v_j - e|^2 ds \leq \frac{p^2}{2}$. Then (8.44) yields

$$p^2 \leq \int_{-\lambda_p}^{\lambda_p} |v_j - e|^2 ds + \frac{p^2}{2},$$

which, since v_j converges uniformly to e in $(-\lambda_p, \lambda_p)$, is impossible for large j. This establishes (8.43). To complete the proof for the case $l = +\infty$ observe that $v \in \mathrm{E}^{\mathrm{xp}}$ implies

$$\|v - e\|^2 \leq \hat{q}^2 := 4\frac{K^2}{k},$$

and let $f : [0, \hat{q}] \to \mathbb{R}$ be defined by

$$f(t) = \begin{cases} \frac{\eta}{4} t^2, & \text{if } t \in [0, \bar{q}), \\ \varepsilon_{\bar{q}}, & \text{if } t \in [\bar{q}, \hat{q}], \end{cases}$$

where η and $\bar{q}$ are as in (8.22). Set $C_3 = \min\{\frac{\eta}{4}, \frac{\varepsilon_{\bar{q}}}{\hat{q}^2}\}$ and note that $f(t) \geq C_3 t^2$ for $t \in [0, \hat{q}]$. From (8.22) and (8.43) it follows

$$\mathscr{W}(v) \geq f(\|v - e\|) \geq C_3 \|v - e\|^2, \quad v \in \mathrm{E}^{\mathrm{xp}}, \tag{8.46}$$

which concludes the proof for the case $l = +\infty$. Assume now $l \in (0, +\infty)$ and observe that, possibly by reducing the value of $C_3 > 0$, (8.46) holds true for $v \in \widetilde{\mathrm{E}}^{\mathrm{xp}}$, where $\widetilde{\mathrm{E}}^{\mathrm{xp}}$ is defined as E^{xp} with $2K$ instead of K. Note also that each $v \in \mathrm{E}_l^{\mathrm{xp}}$ can be extended to a $\tilde{v} \in \widetilde{\mathrm{E}}^{\mathrm{xp}}$. Consider the estimate

$$|\mathscr{W}(\tilde{v}) - \mathscr{W}_l(v)| \leq \int_{\mathbb{R} \setminus (-l,l)} \left(\frac{1}{2} |\tilde{v}'|^2 + W(\tilde{v}) \right) ds$$

$$\leq (1 + 2C_a) \frac{4K^2}{k} e^{-2kl}, \quad l \geq l^{**} := \frac{1}{k} \ln \frac{2K}{\bar{r}}, \quad v \in \mathrm{E}^{\mathrm{xp}}, \tag{8.47}$$

where we have also used that $a^\pm$ nondegenerate implies $W(z) \le C_a|z - a^\pm|^2$, for $|z - a^\pm| \le \bar{r}$ for some constants $C_a > 0$ and $\bar{r} > 0$. From (8.46) and (8.47) it follows, with $C = (1 + 2C_a)\frac{4K^2}{k}$ that

$$\mathscr{W}_l(v) \ge \mathscr{W}(\tilde{v}) - Ce^{-2kl} \ge C^*\|\tilde{v} - e\|^2 - Ce^{-2kl}$$

$$\ge C^*\|v - e\|_l^2 - Ce^{-2kl}, \quad l \ge l^{**}, \ v \in \mathrm{E}^{\mathrm{xp}}.$$

The proof is complete. $\square$

We now present the proof of Theorem 8.2.

Proof (Theorem 8.2) We provide the necessary modifications in the proof of Theorem 8.1.

1. The polar form should be replaced by its relative version

$$u(s, y) = e(s) + q_r^u(y)\boldsymbol{n}_r^u(s, y),$$

$$q_r^u(y) = \|u(\cdot, y) - e(\cdot)\|_{l_r^{y^0}},$$

$$\boldsymbol{n}_r^u(\cdot, y) = \begin{cases} \dfrac{u(\cdot,y)-e(\cdot)}{\|u(\cdot,y)-e(\cdot)\|_{l_r^{y^0}}}, & \text{if } q_r^u(y) \ne 0, \\[2ex] 0, & \text{otherwise.} \end{cases}$$

The energy should be replaced by

$$\widehat{J}_{C_r(y^0)}(u) = \int_{B_r^{n-1}(y^0)} \left[\frac{1}{2}\left\|\nabla_y u(\cdot, y)\right\|_{l_r^{y^0}}^2 + \mathscr{W}_{l_r^{y^0}}(u(\cdot, y))\right] dy.$$

2. In (8.25), the term $\widehat{J}_{C_r}(u) - \widehat{J}_{C_r}(\sigma)$ is no longer nonpositive. Instead, we have

$$\frac{1}{2}\int_{B_r^{n-1}(y^0)} \left[|\nabla q_r^u|^2 - |\nabla q_r^\sigma|^2\right] dy \le \tilde{C}e^{-kl_r^{y^0}} + \int_{B_r^{n-1}(y^0)} (\mathscr{W}_{l_r^{y^0}}(\sigma) - \mathscr{W}_{l_r^{y^0}}(u)) dy,$$

where $q_r^\sigma = \min_{C_r(y^0)}\{q_r^h, q_r^u\}$, $q_r^\sigma = q_r^u$ on $\partial B_r^{n-1}(y^0)$.
3. Inequality (8.29) should be replaced by

$$\left(\int_{B_r^{n-1}} \beta^{\frac{2(n-1)}{n-2}} dy\right)^{\frac{n-2}{n-1}} \le 2\tilde{C}CAe^{-kl_r^{y^0}} + \frac{C}{A}\int_{B_r^{n-1}\cap\{q_r^u - q_r^\sigma \le \lambda'\}} (q_r^u - q_r^\sigma)^2 dy$$

$$+ 2CA\left(\int_{B_r^{n-1}} (\mathscr{W}_{l_r^{y^0}}(\sigma) - \mathscr{W}_{l_r^{y^0}}(u)) dy + \int_{B_r^{n-1}} \Delta q_r^\sigma (q_r^u - q_r^\sigma) dy\right).$$

4. Equation (8.18) and Lemma 8.2 should be replaced by

   (i) $\mathscr{W}_l(u) \geq -\frac{C}{k}e^{-kl}$,

   (ii) $D_{qq}\mathscr{W}_l(e + qv) \geq c_0, \ \forall l \geq l^*, \ q \in [0, \bar{q}], \ v \in \mathbb{S} = W^{1,2}((-l, l); \mathbb{R}^m) \cap$
   $\{\|v\|_l = 1\}$.

5. The analog of the inequality after (5.80) is here

$$\mathscr{W}_{l_r}(u) - \mathscr{W}_{l_r}(h) \geq \frac{c_0}{2}((q_r^u)^2 - (q_r^h)^2) + O(e^{-2kl_r}), \ 0 < q_r^h \leq q_r^u \leq \bar{q}.$$

6. The test function is defined as before in (8.26)

$$\begin{cases} \Delta\varphi = c_1^2\varphi, & \text{in } B_r^{n-1}, \\ \varphi = 1, & \text{on } \partial B_r^{n-1}, \end{cases} \tag{8.48}$$

with

$$q_M = \sup_{y,l} \|u(\cdot, y) - e(\cdot)\|_l, \ l \geq l^*,$$

$$q_r^h = \begin{cases} \varphi q_M & \text{on } B_r^{n-1} \\ q_M & \text{on } \mathscr{O} \setminus B_r^{n-1}, \end{cases}$$

$q_r^\sigma = \min\{q_r^u, q_r^h\}, h = e + q_r^h v, u = e + q_r^u v, q^\sigma \leq q^h \leq q_M e^{-c_1\tau}$.

7. The analog of (5.81) is

$$\left(\int_{B_r^{n-1}} \beta^{\frac{2(n-1)}{n-2}} dy\right)^{\frac{n-2}{n-1}} \leq \frac{\tilde{C}}{\sqrt{c_0}}e^{-kl_r^{y^0}} + C\sqrt{c_0}\int_{(\{0<q_r^u-q_r^h\leq\lambda'\}\cap\{q_r^u>\lambda\})\cap B_r^{n-1}} (q_r^u - q_r^h)^2 dy$$

$$+ 2\frac{C}{\sqrt{c_0}}\int_{(\{q_r^h<q_r^u\}\cap\{q_r^u>\lambda\})\cap B_r^{n-1}} (\mathscr{W}_{l_r^{y^0}}(h) - \mathscr{W}_{l_r^{y^0}}(u) + c_0 q_r^h(q_r^u - q_r^h)) dy.$$

8. *The difference scheme—the analog of Lemma 5.2*
   We set

$$\omega_0 = \mathscr{L}^{n-1}(B_0 \cap \{q^u > \lambda\}),$$

$$\omega_j = \mathscr{L}^{n-1}((B_j \setminus B_{j-1}) \cap \{q^u > \lambda\}), \tag{8.49}$$

with $B_0 := B_{r_0}^{n-1}$, $B_j := B_{r_0+j\tau}^{n-1}$, $r = r_0 + p\tau$, $j = 1, \ldots, p$, where for
notational simplicity we have dropped the subscripts in $q_r^u, q_r^h, l_r^{y^0} \geq l_\tau^*$, with
τ to be chosen large enough. The main difference with the whole space case is
that the cylinders now are not nested, the 'fatter' the cylinder, the 'shorter' it is,

i.e. $r \to l_r^{y^0}$ decreases (cf. Fig. 8.3), $y_{n-1}^0 \geq r_0 + p\tau + r^*$. Now there are two exponentials involved, one from φ and one from the E_l^{xp} class. In addition to the estimates in 4. above, we need

$$\mathscr{W}_{l_r}(h) \leq \frac{C(\lambda)}{2}(q_r^h)^2 + O(e^{-2kl_r}), \qquad \text{on } (B_{p-j} \setminus B_{p-j-1}) \cap \{q^h < q^u\} \cap \{q^u > \lambda\},$$

for $j = 1, \ldots, p$, which implies

$$\mathscr{W}_{l_r}(h) - \mathscr{W}_{l_r}(u) + c_0 q_r^h (q_r^u - q_r^h) \leq 2e^{-kl_r} + \left(c_0 + \frac{c_0'}{2}\right) q_r^h M. \tag{8.50}$$

Setting $l_\nu = \frac{1}{\Lambda}(y_{n-1}^0 - r_\nu)$, see Theorem 8.2, $r_\nu = r_0 + \nu\tau$, $\nu = p - j$, and noting that

$$e^{-\frac{k}{\Lambda}(y_{n-1}^0 - r_{p-j})} = e^{-\frac{k}{\Lambda}(y_{n-1}^0 - [r_0 + (p-j)\tau])} \leq e^{-\frac{k}{\Lambda}(r^* + j\tau)}, \qquad j = 1, \ldots, p,$$

we obtain that the right-hand side of (8.50) above can be estimated by

$$e^{-\frac{k}{\Lambda}(r^* + j\tau)} + \left(c_0 + \frac{c_0'}{2}\right) M^2 e^{-c_1 j\tau},$$

leading to the following modification of (5.87):

$$\frac{2}{\sqrt{c_0}} I_{p-j}^1 \leq \left[\frac{2}{\sqrt{c_0}} e^{-\frac{k}{\Lambda}(r^* + j\tau)} + \frac{2}{\sqrt{c_0}}\left(c_0 + \frac{c_0'}{2}\right) M^2 e^{-c_1 j\tau}\right]\omega_{p-j}, \qquad j = 1, \ldots, p.$$

Taking $\bar{c}_1 = \min\{k/\Lambda, c_1\}$, $\epsilon = e^{-\bar{c}_1\tau}$, we obtain (5.83). The proof of (5.88), from which the theorem follows, is based entirely on Lemma 5.2. In particular, the basic estimate (5.4) is not used for $\lambda \in (0, q_0)$, and so (8.41), its analog in the present set up is not needed for $\lambda \in (0, \bar{q})$. Thus Theorem 8.2 is established. $\square$

8.4   Application to the Singular Cone Solutions of $\Delta u - W_u(u) = 0$ in $\mathbb{R}^3$

We will consider equivariant solutions $u : \mathbb{R}^3 \to \mathbb{R}^3$ corresponding to the tetrahedral cone and the $\mathbb{R}^3$-triod in Fig. 1.2. For the tetrahedral solution we will examine how Theorem 8.2 and Corollary 8.2 and their higher-dimensional extensions complement the information provided by Theorem 6.1. Next, for the $\mathbb{R}^3$-triod we will conclude via an analog of Corollary 8.1 that it has cylindrical structure:

$$u(x_1, x_2, x_3) = u_{\text{tr}}(x_1, x_2), \tag{8.51}$$

where $u_{\mathrm{tr}} : \mathbb{R}^2 \to \mathbb{R}^3$ is a solution provided by Theorem 7.1 corresponding to the planar triod.

We begin with the tetrahedral solution. Recall that this goes with a four-well potential $W : \mathbb{R}^3 \to \mathbb{R}$ with nondegenerate minima at a_1, a_2, a_3, a_4 (Fig. 1.4). We assume that the hypotheses of Theorem 6.1 are in force. In addition, we assume existence of all six connections e_{ij} connecting a_i to a_j ($1 \leq i < j \leq 4$), and possessing the following properties:

  (i) minimality,
 (ii) positivity,
(iii) hyperbolicity,
(iv) and uniqueness,

(all these in the symmetry class). Existence of positive minimal connections in the symmetry class follows from Remark 6.4. Hyperbolicity and uniqueness are generic properties in the sense of C^2 symmetric perturbations for W (cf. [10, Appendix], and [7, Proposition 2.10]). Theorem 6.1 provides an equivariant (with respect to the tetrahedral group) positive solution u satisfying

$$\lim_{\lambda \to \infty} u(\lambda v) = a_i, \quad \forall v \in \mathrm{Int}\, D_i, \ |v| = 1, \tag{8.52}$$

where $\{D_1, D_2, D_3, D_4\}$ is the partition of $\mathbb{R}^3$, and with the convergence in (8.52) being exponential. None of the properties (i)–(iv) above is required for this. To establish (8.8b) and (8.8c), we first observe that the tetrahedral cone is made up of four ends, each of which is a 'half' $\mathbb{R}^3$-triod. Thus in order to study the asymptotic behavior, we can restrict the solution u to one such end. We begin with (8.8b), by considering the limit as we are approaching infinity moving parallel to a reflection plane, but not parallel to the spine of the $\mathbb{R}^3$-triod. We focus on Fig. 8.5 and proceed as follows.

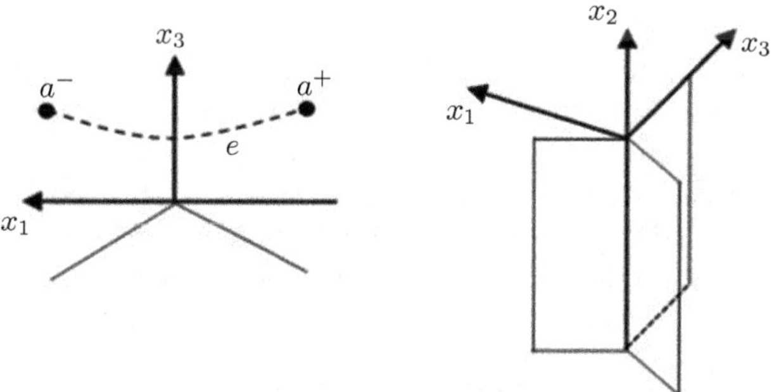

Fig. 8.5 Viewing the $\mathbb{R}^3$-triod from above (left). The spine is identified with the x_2-axis, $x = (x_1, x_2, x_3) = (s, y_1, y_2)$

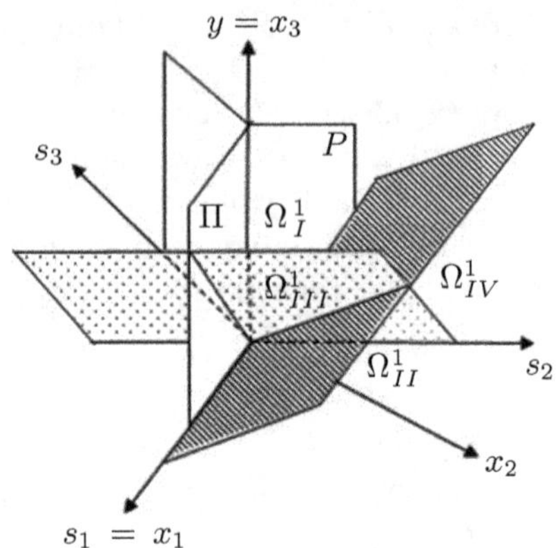

Fig. 8.6 Viewing the $\mathbb{R}^3$-triod from the side. The sets $\Omega_I^1, \Omega_{II}^1, \Omega_{III}^1, \Omega_{IV}^1$, making up Ω^1. $\Omega = \{y > 0\} = \bigcup_{i=1}^3 \Omega^i$, $\Omega^i = \Omega \cap D_i$, $s_1 = x_1$, $y = x_3$

We identify the reflection plane with $\{x_1 = 0\}$, and take $\Omega = \mathbb{R}^3 \cap \{x_3 > 0\}$, and thus deduce (8.8b) from Corollary 8.2. For this step the hyperbolicity and the uniqueness of the connections are needed. For (8.8c) we will have to extend Theorem 8.2 and its corollary by replacing $e(s)$ with $u_{\mathrm{tr}}(s_1, s_2)$, whose existence is provided by Theorem 7.1, with point group $\mathscr{T}$, corresponding to the symmetries of the equilateral triangle. We now focus on Fig. 8.6.

We begin by defining the counterparts of the sets Ω_I^+, Ω_{II}^+, which now are four sets. We restrict ourselves to the 'one third' of the $\mathbb{R}^3$-triod, $\Omega^1 = \Omega \cap D_1 = \{y > 0\} \cap D_1$ ($\{D_1, D_2, D_3\}$ is the partition corresponding to the $\mathbb{R}^3$-triod, $a_i \in D_i$), and coordinetize the plane $\{y = 0\}$ with the (nonorthogonal) (s_1, s_2) system $\{y = 0\} \cap \partial D_1$. We consider the two planes that are bisecting respectively the angle between the planes $(s_1 - s_2)$ and $(s_1 - y)$, and the angle between $(s_1 - s_2)$ and $(s_2 - y)$. These two planes partition Ω^1 into four parts: Ω_I^1 and Ω_{II}^1 above and below, and Ω_{III}^1 and Ω_{IV}^1 left and right. Our strategy for establishing Corollary 8.3 below, which measures the difference between $u(s_1, s_2, y)$ and $u_{\mathrm{tr}}(s_1, s_2)$ in terms of the distance from the $(s_1 - s_2)$ plane, is completely analogous to that of Corollary 8.2 and goes as follows. In Ω_{II}^1 this difference is estimated solely on the basis that both solutions converge to a_1. In Ω_{III}^1 the estimate follows from the fact that both solutions converge to the connection e_Π corresponding to the plane Π, and analogously in Ω_{IV}^1. Finally, in Ω_I^1 we employ the tool provided by Theorem 8.3. In Ω_{II}^1 we use (8.8a); in Ω_{III}^1 and Ω_{IV}^1 we use (8.8b), established above. Thus the proof is inductive and requires hyperbolicity and uniqueness for the connections, and also for u_{tr}. On the other hand, Theorem 8.3 does not require uniqueness for u_{tr}.

We begin with Ω_{II}^1. Note that

$$y = d(x, \partial\Omega) = d(x, \partial\Omega^1), \quad x \in \Omega_{II}^1. \tag{8.53}$$

Also note that $d(x, \partial D_1) \geq d(x, \partial\Omega)$ for $x \in \Omega_{II}^1$, and thus, by Theorem 6.1,

$$|u(x) - a_1| + |\nabla u(x)| \leq K e^{-kd(x, \partial D_1)} \leq K e^{-kd(x, \partial\Omega)}, \quad x \in \Omega_{II}^1. \tag{8.54}$$

Similarly, by Theorem 7.1,

$$|u_{\mathrm{tr}}(s) - a_1| + |\nabla u_{\mathrm{tr}}(s)| \le K e^{-kd(s, \partial D_1 \cap \{y=0\})}, \quad s = (s_1, s_2) \in D_1 \cap \{y = 0\}.$$

$$(8.55)$$

Therefore for $x = (s, y) \in \Omega_{II}^1$ we have the estimates

$$|u(s, y) - u_{\mathrm{tr}}(s)| \le |u(s, y) - a_1| + |u_{\mathrm{tr}}(s) - a_1|$$

$$\le K e^{-kd(x, \partial D_1)} + K e^{-kd(s, \partial D_1 \cap \{y=0\})} = 2K e^{-ky}. \quad (8.56)$$

Next, we consider Ω_{III}^1. We note $d(x, \Pi) \le d(x, \partial\Omega) = y \le d(x, P)$, where Π is the $(s_1 - y)$ plane, and P the $(s_2 - y)$ plane. We have

$$|u(x_1, x_2, y) - u_{\mathrm{tr}}(x_1, x_2)| \le |u(x_1, x_2, y) - e_\Pi(x_2)| + |u_{\mathrm{tr}}(x_1, x_2) - e_\Pi(x_2)|$$

$$\le 2K e^{-kx_1} \le 2K e^{-k\lambda y}, \quad (8.57)$$

where we utilized (8.8b) proved above, and the inequality $x_1 \ge \lambda d(x, P) \ge \lambda y$, where $\lambda > 0$ is a constant that can be determined by trigonometry and e_Π is the connection parametrized by x_2. In Ω_{IV}^1 the argument is completely analogous and yields (8.57).

On the other hand, note that $d(x, \partial\Omega^1) = d(x, \partial D_1)$, $x \in \Omega_I^1$, and thus as in (8.54) we have the estimate

$$|u(s, y) - a_1| + |\nabla u(s, y)| \le K e^{-kd(x, \partial D_1)}, \quad x \in \Omega^1. \quad (8.58)$$

and similarly we have (8.55). Therefore, for $x = (s, y) \in \Omega_I^1$, both $u(s, y)$ and $u_{\mathrm{tr}}(s)$ satisfy the same exponential estimate. We set $\Omega_I = \bigcup_{g \in \mathscr{G}} g(\Omega_I^1)$ and similarly for $\Omega_{II}, \Omega_{III}, \Omega_{IV}$.

We require in addition that u_{tr} is *hyperbolic* in its equivariance class:

$$\langle Tv, v \rangle \ge \eta \|v\|^2, \; v \in W_{Tr}^{1,2}(\mathbb{R}^2; \mathbb{R}^3), \text{ for some } \eta > 0,$$

where $Tv = -\left(\dfrac{\partial^2 v}{\partial x_1^2} + \dfrac{\partial^2 v}{\partial x_2^2}\right) + W_{uu}(u_{\mathrm{tr}})v$, $W_{\mathscr{G}}^{1,2}(\mathbb{R}^2; \mathbb{R}^3) \subset W_{\mathrm{loc}}^{1,2}(\mathbb{R}^2; \mathbb{R}^3)$ is the subspace of equivariant maps, $\langle \cdot, \cdot \rangle$ the inner product in $L^2(\mathbb{R}^2; \mathbb{R}^3)$, and $\| \cdot \|$ the associated norm. We consider the equilateral triangle T_l of side l, in the (s_1, s_2)-plane, with vertices on $s_1 = s_2$, $s_1 = s_3$, $s_2 = s_3$, and we denote by $E_l^{\mathrm{xp}} \subset C^1(T_l; \mathbb{R}^3)$, the $\mathscr{G}$-equivariant maps on T_l satisfying the estimate (8.55) with $v(s)$ replacing $u_{\mathrm{tr}}(s)$, $s \in T_l$, $s_1 \ge 0$, $s_2 \ge 0$, and $l \in [0, \infty]$. We set $\langle v, w \rangle_l = \int_{T_l} v(s) \cdot w(s) \mathrm{d}s$, $\|v\|_l = (\langle v, v \rangle_l)^{1/2}$, $\|v\| := \|v\|_\infty$, $\mathrm{E}^{\mathrm{xp}} := \mathrm{E}_\infty^{\mathrm{xp}}$.

We write $x = (s, y)$, $s = (s_1, s_2)$, $y \in \mathbb{R}$, and introduce the *cylinders* $C_r^l(y) = T_l \times B_r^{3-2}(y) = T_l \times (y - r, y + r)$, and in particular we consider those contained in

Fig. 8.7 The cylinder $C_r^{y^0}$ in Ω_I. Its cross-section $B_r^{3-2}(y^0)$ is the intersection with the y-axis

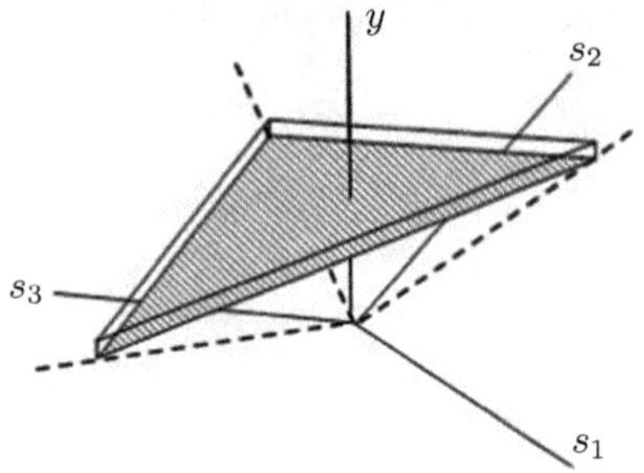

Ω_I which are 'sitting' on the three lines $s_1 = s_2 = y$, $s_2 = s_3 = y$, and $s_1 = s_3 = y$. The side l is determined linearly by y and r, and is denoted by l_r^y. We denote these cylinders by $C_r(y)$. Notice that the cross-section is $B_r^{3-2}(y) = (y - r, y + r)$; $n = 3$ is the dimension of the space, 2 the dimension of $u_{\mathrm{tr}}(x_1, x_2)$, hence $3 - 2$ the dimension of the cross-section of the cylinder (Fig. 8.7).

Theorem 8.3 *Under the hypotheses: nondegeneracy for the minima of W, equilateral triangle symmetry, minimality of $u : \Omega_I \to \mathbb{R}^3$, and hyperbolicity of u_{tr}, there exists $\lambda^* > 0$ depending only on K, k in (8.54), such that for any $\mu_0 > 0$ and $\lambda \in (0, \lambda^*)$, there exists $l^* = l^*(\lambda)$ so that the conditions $l_1^{y^0} > l^*$ and*

$$\mathscr{L}^{3-2}(B_1^{3-2}(y^0) \cap \{y : \|u(\cdot, y) - u_{\mathrm{tr}}(\cdot)\|_{l_1^{y^0}} \geq \lambda\}) \geq \mu_0 > 0,$$

imply

$$\mathscr{L}^{3-2}(B_r^{3-2}(y^0) \cap \{y : \|u(\cdot, y) - u_{\mathrm{tr}}(\cdot)\|_{l_r^{y^0}} \geq \lambda\}) \geq Cr^{3-2}, \quad 1 \leq r \leq y^0 - \frac{l^*}{\sqrt{3}}.$$

Corollary 8.3 *Let $\Omega = \{(s_1, s_2, y) \in \mathbb{R}^3 : y \geq 0\}$, and assume the hypotheses of Theorem 8.3, and moreover assume that u_{tr} is unique (in its equivariance class) and that all connections are hyperbolic and unique in their symmetry class. Then, there exist positive constants $\bar{k}$, $\bar{K}$, such that*

$$|u(s, y) - u_{\mathrm{tr}}(s)| \leq \bar{K} e^{-\bar{k}y}, \quad x = (s_1, s_2, y) \in \Omega,$$

where $u(x_1, x_2, x_3)$ is the tetrahedral solution provided by Theorem 6.1. s_1, s_2 are the coordinates in the oblique s_1-s_2 system, and $y = x_3$.

The proofs uses the following ingredients:

(a) The polar form:

$$u(s, y) = u_{\mathrm{tr}}(s) + q_r^u(y)\boldsymbol{n}_r^u(s, y), \quad (s, y) \in \mathbb{R}^n, \ (s, y) \in C_r(y^0),$$

with

$$q_r^u(y) := \|u(\cdot, y) - u_{\text{tr}}(\cdot)\|_{l_r^{y^0}},$$

and

$$\boldsymbol{n}_r^u(\cdot, y) = \begin{cases} \dfrac{u(\cdot,y)-u_{\text{tr}}(\cdot)}{\|u(\cdot,y)-u_{\text{tr}}(\cdot)\|_{l_r^{y^0}}}, & \text{if } q_r^u(y) \neq 0, \\ 0, & \text{otherwise.} \end{cases}$$

(b) The effective potential:

$$\mathscr{W}_l(v) = J_{T_l}(u_{\text{tr}} + v) - J_{T_l}(v).$$

(c) The upper bound:

$$\int_{B_r^{3-2}(y^0)} \left[\frac{1}{2} \left\| \nabla_y u(\cdot, y) \right\|_{l_r^{y^0}}^2 + \mathscr{W}_{l_r^{y^0}}(u(\cdot, y)) \right] dy \leq C_1 r^{3-2} + C_2 e^{-kl_r^{y^0}} r^2, \quad \forall l_r^{y^0} \geq l^*.$$

(d) The lower bound:

$$\mathscr{W}_{l_r^{y^0}}(v) \geq C_3 \|v\|_{l_r^{y^0}}^2 + O(e^{-kl_r^{y^0}}), \quad \forall l_r^{y^0} \geq l^{**}, \forall v \in E_l^{\text{xp}}.$$

The proofs of Theorem 8.3 and Corollary 8.3 are analogous to the proofs of Theorem 8.2 and Corollary 8.2, with appropriate modifications for the present setting.

We conclude this section with a theorem establishing the reduction of variables result in (8.51) above. We consider a symmetric three-well potential $W : \mathbb{R}^3 \to \mathbb{R}$, and assume the existence of an equivariant, minimal solution $u : \mathbb{R}^3 \to \mathbb{R}^3$, with $\|u\|_{L^\infty} < \infty$, corresponding to the triod.

Theorem 8.4 *Let $u(x_1, x_2, x_3)$ be a solution as above. Assume that the triod solution $u_{\text{tr}} : \mathbb{R}^2 \to \mathbb{R}^3$ provided by Theorem 7.1 is unique and hyperbolic (in its equivariance class). Then $u(x_1, x_2, x_3) = u_{\text{tr}}(x_1, x_2)$.*

The proof of this theorem is completely analogous to the proof of Corollary 8.1 and based on the density estimate provided in Theorem 8.3.

8.5   The Alama, Bronsard and Gui Example

In this section we present one of the first results for systems obtained in Alama, Bronsard and Gui [2], see also F. Alessio [3] which brings out some of the differences with the scalar case. First, the possibility of nonuniqueness of connections between the minima of the potential [6] shows that at the interface the transitions need not be one-dimensional. Higher dimensionality is a general phenomenon for systems that can be seen at the junctions. The paper [2] also was the first to indicate that for systems, entire minimal solutions even in low dimensions, need not be one-dimensional connections. The drawback here is that one symmetry is imposed, so it may be argued that the constructed solution is not minimal with respect to general perturbations. Schatzman [13] managed to remove the symmetry and establish the same theorem. Chapter 9 is dedicated to this issue. Our proof of [2] uses the density theorems 5.2 and 8.1, together with some ideas from the original derivation.

We consider

$$\Delta u - W_u(u) = 0, \quad u : \mathbb{R}^2 \to \mathbb{R}^m, \quad m \geq 2. \tag{8.59}$$

We assume that W is a double-well potential with exactly two minimizing connections (Fig. 8.8).

Finally, we assume only one reflection symmetry in the $\{z_1 = 0\}$ plane. More precisely our hypotheses are as follows:

H$_1$ (Two nondegenerate global minima of W) The potential $W : \mathbb{R}^m \to [0, \infty)$ is of class C^3, with $W(a^-) = W(a^+) = 0$, and $W > 0$ on $\mathbb{R}^m \setminus \{a^+, a^-\}$. Furthermore the Hessian is positive definite: $W_{uu}(a^\pm) \geq c^2 I$.

H$_2$ (Symmetry) For $z \in \mathbb{R}^d, d \geq 1$, denote by $\hat{z}$ the reflection of z in the $\{z_1 = 0\}$ plane, $\hat{z} = (-z_1, z_2, \ldots, z_d)$. We assume that

$$W(\hat{u}) = W(u).$$

Moreover, we assume that there exists $M > 0$ such that $W(su) \geq W(u)$, for $s \geq 1, |u| = M$.

We seek *equivariant* solutions to (1.1),

$$u(\hat{x}) = \hat{u}(x),$$

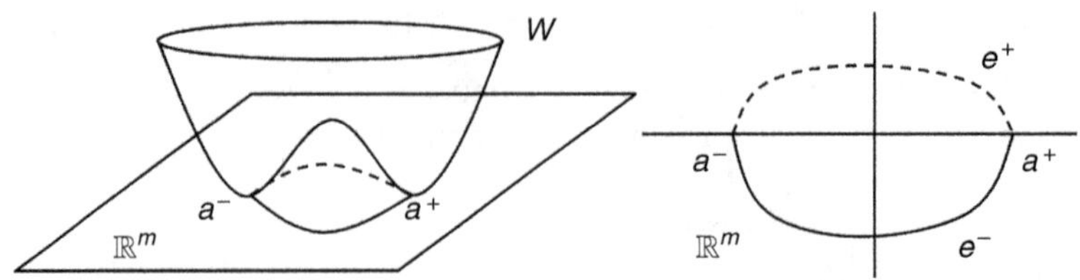

Fig. 8.8 The double-well potential W with the two minimizing connections

which are also *positive*:

$$u(\overline{\mathbb{R}}_+^2) \subset \overline{\mathbb{R}}_+^m, \quad \mathbb{R}_+^d = \{z \in \mathbb{R}_+^d : z_1 > 0\}.$$

H₃ (Two nondegenerate minimizing connections) We assume the existence of exactly two global (positive) minimizers $e^\pm : \mathbb{R} \to \mathbb{R}^m$ of the *action* functional

$$J_{\mathbb{R}}(v) = \int_{\mathbb{R}} \left\{ \frac{1}{2}|v_s|^2 + W(v) \right\} ds,$$

symmetric (i.e., $e(-s) = \hat{e}(s)$) and satisfying $\lim_{s \to \pm\infty} v(s) = a^\pm$. Thus, $e(\overline{\mathbb{R}}_+) \subset \overline{\mathbb{R}}_+^m$, and by the nondegeneracy in Hypothesis **H₁**, also the following estimate holds:

$$|e(s) - a^+| + |e_s(s)| \leq Ke^{-ks}, \quad s \geq 0, \, e = e^\pm. \tag{8.60}$$

Furthermore we assume that $e^\pm$ are *hyperbolic*,

$$\langle Tv, v \rangle \geq \eta \|v\|^2, \; v \in W_S^{1,2}(\mathbb{R}, \mathbb{R}^m), \text{ some } \eta > 0,$$

where $W_S^{1,2}(\mathbb{R}, \mathbb{R}^m) \subset W_{loc}^{1,2}(\mathbb{R}, \mathbb{R}^m)$ is the subspace of the symmetric maps, $\langle \cdot, \cdot \rangle$ the $L^2(\mathbb{R}, \mathbb{R}^m)$ inner product, and $\|\cdot\|$ the associated norm, and

$$Tv = -v_{ss} + W_{uu}(e)v, \quad e = e^\pm.$$

Theorem 8.5 ([2]) *Under Hypotheses* **H₁**-**H₃**, *there exists an equivariant classical solution to* (8.59), $u : \mathbb{R}^2 \to \mathbb{R}^m$, $m \geq 2$, *positive, and satisfying the following*

1. $|u(x) - a^+| \leq Ke^{-\bar{k}x_1}$, $x_1 \geq 0$, *with positive constants* k, K.

2. $|u(x) - e^+(x_1)| \leq \bar{K}e^{-\bar{k}x_2}$, $\forall x_2 \geq 0$, *and* $|u(x) - e^-(x_1)| \leq \bar{K}e^{\bar{k}x_2}$, $\forall x_2 \leq 0$, *with positive constants* $\bar{k}$, $\bar{K}$.

3. $u(\overline{\mathbb{R}}_+^2) \subset \overline{\mathbb{R}}_+^m$ *(positivity)*,

4. $\mathscr{W}(u(\cdot, x_2)) = \frac{1}{2}\int_{\mathbb{R}} |u_{x_2}(x_1, x_2)|^2 dx_1$, $\forall x_2 \in \mathbb{R}$ *(equipartition), where* $\mathscr{W}$ *is the effective potential:* $\mathscr{W}(v) := J_{\mathbb{R}}(v) - J_{\mathbb{R}}(e^+)$.

5. $\int_{\mathbb{R}^2} \frac{1}{2}|u_{x_2}(x_1, x_2)|^2 dx + \int_{\mathbb{R}} \mathscr{W}(u(\cdot, x_2)) dx_2 = \int_{\mathbb{R}} \left(\frac{1}{2}\|u_{x_2}(\cdot, x_2)\|^2 + \mathscr{W}(u(\cdot, x_2))\right) dx_2 =: E(u)$, *where* $E(u)$, *the* transition energy, *is finite:* $E(u) < \infty$. *On the other hand,* $\int_{\mathbb{R}^2} \frac{1}{2}|u_{x_1}(x_1, x_2)|^2 dx = \infty$, *that is the 'length of the interface' is infinite. Moreover,* u *minimizes* $E(\cdot)$ *in* $W_S^{1,2} \cap C_c^1$, *and more generally in*

$$E^{xp} := \{z \in W_S^{1,2}(\mathbb{R}, \mathbb{R}^m) : |z(s) - a^+| + |z_s(s)| \leq K'e^{-k's}, \, \forall s \geq 0\},$$

where $K', k' > 0$ *are constants.*

6. *u is a critical point of the Jacobi functional*

$$L(z) = \int_{\mathbb{R}} \sqrt{2\mathscr{W}(z(\cdot, x_2))} \, \|z_{x_2}(\cdot, x_2)\| \, dx_2$$

in the class of 'curves' $z : \mathbb{R} \to \mathrm{E}^{\mathrm{xp}}$, $z(\cdot, \pm\infty) = e^{\pm}(\cdot)$, and moreover $L(u) = E(u)$.

Note: By Proposition 5.4 $\int_{\mathbb{R}^2} W(u)dx = \infty$.

Proof

Step 1 (Estimates Before Taking the Limit)
We begin with

$$\min J_{C_R^{\mu}}, \quad J_{C_R^{\mu}}(u) = \int_{C_R^{\mu}} \left(\frac{1}{2}|\nabla u|^2 + W(u)\right)dx, \quad C_R^{\mu} = [-\mu, \mu] \times [-R, R],$$

$$(8.61)$$

in the symmetry class, with $u(x_1, \pm R) = e^{\pm}(x_1)$, for $x_1 \in (-\mu, \mu)$, and free otherwise. We denote by $u^{R,\mu}$ a minimizer, which we can assume that it is positive and, by $\mathbf{H}_2$, that it satisfies the estimate

$$\left\| u^{R,\mu} \right\|_{C^{1,\alpha}(\overline{C}_R^{\mu}; \mathbb{R}^m)} \le C = C(M).$$

$$(8.62)$$

We refer to the beginning of Sect. 6.6 for a similar point. By considering the comparison map

$$\tilde{u}(x_1, x_2) = \left(\frac{x_2 + R}{2R}\right)e^+(x_1) - \left(\frac{x_2 - R}{2R}\right)e^-(x_1), \quad x \in C_R^{\mu},$$

we obtain the estimate

$$J_{C_R^{\mu}}(u^{R,\mu}) \le CR,$$

$$(8.63)$$

with C depending on M. Hence, by (8.62), we can take the limit along a subsequence $\mu \to \infty$, and obtain

$$u^R = \lim_{\mu \to \infty} u^{R,\mu},$$

$$(8.64a)$$

$$\left\| u^R \right\|_{C^{1,\alpha}(\overline{C}_R)} \le C(M), \quad C_R := \mathbb{R} \times [-R, R],$$

$$(8.64b)$$

$$J_{C_R}(u^R) \le CR.$$

$$(8.64c)$$

Claim 1

$$u^R(x_1, x_2) \to a^{\pm}, \quad \text{as } x_1 \to \pm\infty \text{ uniformly in } x_2. \tag{8.65}$$

This follows from (8.64c) via a contradiction argument using the uniform continuity of u^R by (8.64b) and positivity, which proceeds by constructing a sequence of disjoints discs tending to infinity, over which $|u^R(x) - a^{\pm}|$ is bounded uniformly away from zero. This clearly violates the finiteness of the energy.

Claim 2

$$\int_{\gamma}^{\gamma+L} \left(\int_{\mathbb{R}} \frac{1}{2}|\nabla u^R|^2 + W(u^R)\mathrm{d}x_1 \right)\mathrm{d}x_2 \leq C + J_{\mathbb{R}}(e^+)L, \; \forall [\gamma, \gamma+L] \subset [-R, R]. \tag{8.66}$$

To see this, we consider the comparison map in C_R^{μ},

$$\hat{u}(x_1, x_2) = \begin{cases} e^+(x_1), & \text{for } 1 \leq x_2 \leq R, \\ \left(\frac{1+x_2}{2}\right)e^+(x_1) + \left(\frac{1-x_2}{2}\right)e^-(x_1), & \text{for } -1 \leq x_2 \leq 1, \\ e^-(x_1), & \text{for } -R \leq x_2 \leq -1. \end{cases} \tag{8.67}$$

and obtain the estimate

$$J_{C_R^{\mu}}(u^{R,\mu}) \leq J_{C_R^{\mu}}(\hat{u}) \leq C + 2R J_{\mathbb{R}}(e^+), \tag{8.68}$$

for a constant C independent of μ and $R \geq 1$. Letting $\mu \to \infty$, we have

$$J_{C_R}(u^R) \leq C + 2R J_{\mathbb{R}}(e^+). \tag{8.69}$$

On the other hand

$$J_{C_R}(u^R) = J_{C_{\mathbb{R} \times [\gamma.\gamma+L]}}(u^R) + \int_{\{-R<x_2<\gamma\} \cup \{\gamma+L<x_2<R\}} \left(\frac{1}{2}|\nabla u^R|^2 + W(u^R)\right)\mathrm{d}x$$

$$\geq J_{C_{\mathbb{R} \times [\gamma.\gamma+L]}}(u^R) + \int_{\{-R<x_2<\gamma\} \cup \{\gamma+L<x_2<R\}} \left(\frac{1}{2}|u_{x_1}^R|^2 + W(u^R)\right)\mathrm{d}x$$

$$\geq J_{C_{\mathbb{R} \times [\gamma.\gamma+L]}}(u^R) + (2R - L)J_{\mathbb{R}}(e^+). \tag{8.70}$$

Inequality (8.66) follows by combining estimates (8.69) and (8.70).

Claim 3

$$\int_{\gamma}^{\gamma+L} \int_{\mathbb{R}} \frac{1}{2}|u_{x_2}^R|^2 \mathrm{d}x_1\mathrm{d}x_2 + \int_{\gamma}^{\gamma+L} \left(\int_{\mathbb{R}} \left(\frac{1}{2}|u_{x_1}^R|^2 + W(u^R)\right)\mathrm{d}x_1 - J_{\mathbb{R}}(e^+) \right)\mathrm{d}x_2 \leq C. \tag{8.71}$$

This is almost immediate from (8.66), since:

$$\int_{\gamma}^{\gamma+L}\int_{\mathbb{R}}\frac{1}{2}|u_{x_2}^R|^2 dx_1 dx_2 + J_{\mathbb{R}}(e^+)L \le \int_{\gamma}^{\gamma+L}\left(\int_{\mathbb{R}}\frac{1}{2}|u_{x_1}^R|^2 + \frac{1}{2}|u_{x_2}^R|^2 + W(u^R)dx_1\right)dx_2$$

$$= J_{\mathbb{R}\times[\gamma,\gamma+L]}(u^R) \le C + J_{\mathbb{R}}(e^+)L.$$

Step 2 (Passing to the Limit $R \to \infty$)

Minimizing connections do not intersect by [2, Proposition 2.3, p. 365]. Hence, there exists $j \in \{2, \ldots, m\}$ such that $e_j^-(0) \ne e_j^+(0)$. For notational convenience we assume that $j = 2$. By continuity, there exists $t_R \in (-R, R)$ such that

$$u_2^R(0, t_R) = \frac{1}{2}[e_2^-(0) + e_2^+(0)] \ne e_2^-(0), \; e_2^+(0). \tag{8.72}$$

We define

$$\tilde{u}^R(x_1, x_2) = u^R(x_1, x_2 + t_R) \quad \text{on } \mathbb{R} \times [-R - t_R, R - t_R]$$

so that

$$\tilde{u}_2^R(0, t_R) = \frac{1}{2}[e_2^-(0) + e_2^+(0)]. \tag{8.73}$$

We examine how some of the estimates in Step 1 above are translated in terms of $\tilde{u}^R$. Clearly, (8.64b) is unaffected:

$$\left\|\tilde{u}^R\right\|_{C^{1,\alpha}(\overline{C}_R)} \le M. \tag{8.74}$$

Fix $[\hat{\gamma}, \hat{\gamma} + L] \subset [-R - t_R, R - t_R]$ and set $\gamma_R = \hat{\gamma} + t_R$, and apply (8.66) for $[\gamma_R, \gamma_R + L]$:

$$\int_{\gamma_R}^{\gamma_R+L}\left(\int_{\mathbb{R}}\frac{1}{2}|\nabla u^R(x_1, y)|^2 + W(u^R(x_1, y))dx_1\right)dy \le C + J_{\mathbb{R}}(e^+)L. \tag{8.75}$$

Setting $y = x_2 + t_R$ we end up with

$$\int_{\hat{\gamma}}^{\hat{\gamma}+L}\left(\int_{\mathbb{R}}\frac{1}{2}|\nabla \tilde{u}^R(x_1, x_2)|^2 + W(\tilde{u}^R(x_1, x_2))dx_1\right)dx_2 \le C + J_{\mathbb{R}}(e^+)L. \tag{8.76}$$

By (8.74), we can take the limit along a subsequence $R_n \to \infty$ and obtain convergence over compacts to a solution

$$u(x) = \lim_{R_n \to \infty} \tilde{u}^{R_n}(x). \tag{8.77}$$

According to whether $\{R_n - t_{R_n}\}$ is bounded or not, we distinguish cases, and by taking further subsequences we have the following possibilities

- Case 1: u is defined on $\mathbb{R}^2$.
- Case 2: u is defined on $\mathbb{R}^2_- = \{(x_1, x_2) : x_2 \leq k, \text{ some } k > 0\}$.
- Case 3: u is defined on $\mathbb{R}^2_+ = \{(x_1, x_2) : x_2 \geq -k, \text{ some } k > 0\}$.

By (8.73), $u_2(0, 0) \neq e_2^-(0), e_2^+(0)$, and hence

$$u(x_1, x_2) \not\equiv e^-(x_1), e^+(x_1). \tag{8.78}$$

Also, by (8.76),

$$u(x_1, x_2) \not\equiv \text{Const.} \tag{8.79}$$

Step 3 (Case 1)

Statement 1 follows by Proposition 6.4. We recall the main points. First, the upper bound $\int_{B_\rho(x_0)} \left(\frac{1}{2}|\nabla u|^2 + W(u)\right)dx \leq C\rho$, $\forall B_\rho(x_0) \subset \mathbb{R}^2_r$, which follows by minimality. This in turn implies that given $c_0 > 0$, there is $d_0 > 0$, depending on c_0, such that $|u(x) - a^+| \leq c_0$, $\forall x \in \mathbb{R}^2_r$, $x_1 \geq d_0$. The proof is by a contradiction argument based on the density estimate in Theorem 5.2, and is relatively simple, since no reflection planes are involved. Finally, the exponential estimate follows by taking $c_0 > 0$ small and employing linear theory. We refer the reader to the second proof of Proposition 6.4 for a similar argument.

Statement 5

From (8.71), we obtain, as for (8.76),

$$\int_{\hat{\gamma}}^{\hat{\gamma}+L} \left(\int_{\mathbb{R}} \frac{1}{2}|\tilde{u}_{x_2}^R|^2 dx_1\right)dx_2 + \int_{\hat{\gamma}}^{\hat{\gamma}+L} \left(\int_{\mathbb{R}} \left(\frac{1}{2}|\tilde{u}_{x_1}^R|^2 + W(u^R)\right)dx_1 - J_{\mathbb{R}}(e^+)\right)dx_2$$
$$\leq C, \quad \forall [\hat{\gamma}, \hat{\gamma} + L] \subset [-R - t_R, R - t_R]. \tag{8.80}$$

Hence, taking $R \to \infty$, we obtain

$$\int_{\hat{\gamma}}^{\hat{\gamma}+L} \left(\int_{\mathbb{R}} \frac{1}{2}|u_{x_2}|^2 dx_1\right)dx_2 + \int_{\hat{\gamma}}^{\hat{\gamma}+L} \left(\int_{\mathbb{R}} \left(\frac{1}{2}|u_{x_1}|^2 + W(u)\right)dx_1 - J_{\mathbb{R}}(e^+)\right)dx_2 \leq C.$$

Finally, taking $\hat{\gamma} \to -\infty$ and $L \to +\infty$ gives

$$E(u) := \int_{\mathbb{R}} \left(\frac{1}{2}\left\|u_{x_2}(\cdot, x_2)\right\|^2 + \mathcal{W}(u(\cdot, x_2))\right)dx_2 \leq C < \infty. \tag{8.81}$$

That u minimizes $E(u)$ follows immediately by minimality for compactly supported perturbations, and by Lemma 6.2 in [4], in the class E^{xp}. That $\int_{\mathbb{R}^2} |u_{x_1}(x_1, x_2)|^2 dx = \infty$ is a simple consequence of Statement 1, which implies that given $\epsilon > 0$, there is $\mu > 0$ such that

$$\int_{-\mu}^{\mu} |u_{x_1}| dx_1 \geq (|a^+ - a^-| - \epsilon), \quad \forall x_2.$$

Statement 2
Our next claim is that either

$$\lim_{x_2 \to \pm\infty} \left\| u(\cdot, x_2) - e^{\pm}(\cdot) \right\|_{L^2(\mathbb{R}, \mathbb{R}^m)} = 0, \text{ or } \lim_{x_2 \to \pm\infty} \left\| u(\cdot, x_2) - e^{\mp}(\cdot) \right\|_{L^2(\mathbb{R}, \mathbb{R}^m)} = 0.$$

$$(8.82)$$

By the estimate

$$|u(x_1, x_2) - a^+| + |\nabla u(x_1, x_2)| \leq K e^{-kx_1}, \quad x_1 \geq 0, \tag{8.83}$$

which follows from Statement 1, and elliptic regularity ($\Delta u_{x_i} = W_{uu}(u)u_{x_i}$, for $i = 1, 2$) we deduce that

$$|u(x_1, x_2) - a^+| + |\nabla u(x_1, x_2)| + |D^2 u(x_1, x_2)| \leq K' e^{-k'x_1}, \quad x_1 \geq 0. \tag{8.84}$$

Thus, the functions $x_2 \mapsto \int_{\mathbb{R}} W(u(x_1, x_2)) dx_1$, and $x_2 \mapsto \int_{\mathbb{R}} |u_{x_i}(x_1, x_2)|^2 dx_1$, are Lipschitz continuous. Indeed, we have for instance

$$\left| \int_{\mathbb{R}} |u_{x_1}(x_1, x_2)|^2 dx_1 - \int_{\mathbb{R}} |u_{x_1}(x_1, x_2')|^2 dx_1 \right| \leq 2K' \int_{\mathbb{R}} \left| u_{x_1}(x_1, x_2) - u_{x_1}(x_1, x_2') \right| dx_1$$

$$\leq 2K' |x_2 - x_2'| \int_{\mathbb{R}} K' e^{-k'|x_1|} dx_1.$$

In particular, the function $\mathscr{W} : x_2 \mapsto J_{\mathbb{R}}(u(\cdot, x_2)) - J_{\mathbb{R}}(e^{\pm}(\cdot))$ is Lipschitz continuous, and since by Statement 5 $\int_{\mathbb{R}} \mathscr{W}(u(\cdot, x_2)) dx_2 < \infty$, we infer that

$$J_{\mathbb{R}}(u(\cdot, x_2)) \to J_{\mathbb{R}}(e^{\pm}(\cdot)), \text{ as } x_2 \to \pm\infty. \tag{8.85}$$

Now, suppose by contradiction that there exists a sequence $y_n \to +\infty$ such that $\left\| u(\cdot, y_n) - e^{\pm}(\cdot) \right\|_{L^2(\mathbb{R}, \mathbb{R}^m)} \geq \delta > 0$. According to estimate (8.84), we can extract a subsequence $y_n' \to \infty$ such that $u(\cdot, y_n')$ converges in C^1_{loc} to a function $v \in E^{xp}$. By dominated convergence we obtain $\left\| v - e^{\pm} \right\|_{L^2} \geq \delta > 0 \Rightarrow v \neq e^{\pm}$, and $J_{\mathbb{R}}(u(\cdot, y_n')) \to J_{\mathbb{R}}(v(\cdot)) > J_{\mathbb{R}}(e^{\pm}(\cdot))$, which is a contradiction. Thus, the possible limit points of $\{u(\cdot, x_2)\}$ as $x_2 \to \pm\infty$ are the two isolated points $e^{\pm}$. Since the maps $x_2 \mapsto u(\cdot, x_2) - e^{\pm}(\cdot)$ are continuous with respect to the $L^2(\mathbb{R}, \mathbb{R}^m)$ norm, we deduce that $u(\cdot, x_2)$ converges either to $e^+(\cdot)$ or to $e^-(\cdot)$, as $x_2 \to \pm\infty$. To complete the proof of (8.82) it remains to show that $u(\cdot, x_2)$ cannot have the same limit at $+\infty$ and $-\infty$. Indeed, if for instance

$$\lim_{x_2 \to \pm\infty} \left\| u(\cdot, x_2) - e^+(\cdot) \right\|_{L^2(\mathbb{R}, \mathbb{R}^m)} = 0 \tag{8.86}$$

we apply Theorem 8.1:

$$\mathscr{L}^1(B_1^1(0) \cap \{x_2 : \|u(\cdot, x_2) - e^+(\cdot)\|_{L^2} \geq \lambda\}) > 0,$$

which holds since $u \not\equiv e^-, e^+$ by (8.78), implies

$$\mathscr{L}^1(B_R^1(0) \cap \{x_2 : \|u(\cdot, x_2) - e^+(\cdot)\|_{L^2} \geq \lambda\}) \geq CR, \tag{8.87}$$

which is in conflict with (8.86). The proof of the claim is complete. Statement 2 follows from this by the hyperbolicity of $e^\pm$ (see 'Conclusion' in the proof of Corollary 8.2).

Statement 4
We begin by deriving the equipartition relation. By Statement 1, the Hamiltonian identity (3.33) holds:

$$\int_{\mathbb{R}} \left[\frac{1}{2} |u_{x_1}(x_1, x_2)|^2 + W(u((x_1, x_2))) \right] \mathrm{d}x_1 = \int_{\mathbb{R}} \frac{1}{2} |u_{x_2}(x_1, x_2)|^2 \mathrm{d}x_1 + C, \forall x_2 \in \mathbb{R}. \tag{8.88}$$

Statement 5 gives $\int_{\mathbb{R}} \left\| u_{x_2}(\cdot, x_2) \right\|^2 \mathrm{d}x_2 < \infty$, and so by the uniform continuity of $x_2 \mapsto \left\| u_{x_2}(\cdot, x_2) \right\|^2$, we conclude that

$$\lim_{x_2 \to \pm\infty} \left\| u_{x_2}(\cdot, x_2) \right\|^2 = 0. \tag{8.89}$$

Using this in (8.88), and (8.85) we obtain $C = J_{\mathbb{R}}(e^+)$. Thus,

$$\mathscr{W}(u(\cdot, x_2)) = \frac{1}{2} \int_{\mathbb{R}} |u_{x_2}(x_1, x_2)|^2 \mathrm{d}x_1, \tag{8.90}$$

and Statement 4 is established.

Statement 6
This is analogous to Corollary 2.1. Our claim is that if u_0 is a local minimizer of E and u_0 is equipartitioned, then u_0 is a critical point of L. To see this, let u_t be a perturbation of u_0 for $|t|$ small. Then, $\phi(t) := E(u_t) - L(u_t) \geq 0$, and by equipartition $\phi(0) = E(u_0) - L(u_0) = 0$. Hence, $\phi'(0) = 0$ and so $\frac{\mathrm{d}}{\mathrm{d}t}\big|_{t=0} L(u_t) = 0$.

   The proof of Theorem 8.5 is complete. What remains is to show that cases (2) and (3) are not possible.

First, we note that estimate (8.64b) is conserved by passing to the limit:

$$\|\tilde{u}\|_{C^{1,\alpha}(\{x_2 \leq k\})} \leq C(M), \tag{8.91}$$

and thus u is C^1 smooth up to the boundary $x_2 = k$. One can check that $u(x_1, k) = e^+(x_1)$. On the other hand (8.76) holds for all $[\hat{\gamma}, \hat{\gamma}+L]$ with $\hat{\gamma}+L \leq k$. From (8.76), arguing as in (8.65), we conclude that $u(x_1, x_2) \to a^\pm$ as $x_1 \to \pm\infty$, for all $x_2 \leq k$. Using this, we derive the Hamiltonian identity

$$\int_{\mathbb{R}} \left[\frac{1}{2}|u_{x_1}(x_1, x_2)|^2 + W(u((x_1, x_2))) \right] dx_1 = \int_{\mathbb{R}} \frac{1}{2}|u_{x_2}(x_1, x_2)|^2 dx_1 + C, \quad \forall x_2 \leq k. \tag{8.92}$$

We also note that the analog of (8.81) holds:

$$E(u) = \int_{-\infty}^{k} \left(\frac{1}{2} \left\| u_{x_2}(\cdot, x_2) \right\|^2 + \mathscr{W}\left(u(\cdot, x_2)\right) \right) dx_2 \leq C < \infty. \tag{8.93}$$

This ensures that we can find a sequence $\{x_2^n\}$, such that

$$\lim_{x_2^n \to -\infty} \left\| u_{x_2}(\cdot, x_2) \right\|^2 = 0 \text{ and } \lim_{x_2^n \to -\infty} \mathscr{W}(u(\cdot, x_2)) = 0. \tag{8.94}$$

Taking $x_2^n \to -\infty$ in (8.92), and utilizing (8.94), we obtain $C = J_{\mathbb{R}}(e^+)$. Returning now to (8.92) and setting $x_2 = k$, we have

$$\int_{\mathbb{R}} \left[\frac{1}{2}|u_{x_1}(x_1, k)|^2 + W(u((x_1, k))) \right] dx_1 = \int_{\mathbb{R}} \frac{1}{2}|u_{x_2}(x_1, k)|^2 dx_1 + J_{\mathbb{R}}(e^+). \tag{8.95}$$

Since $u(x_1, k) = e^+(x_1)$ it follows that $\int_{\mathbb{R}} \frac{1}{2}|u_{x_2}(x_1, k)|^2 dx_1 = 0$, that is, $u_{x_2}(x_1, k) = 0$. Now, we extend u to $\mathbb{R}^2$ by setting

$$\tilde{u}(x_1, x_2) = \begin{cases} e^+(x_1), & \text{for } x_2 \geq k, \\ u(x_1, x_2) & \text{for } x_2 \leq k. \end{cases} \tag{8.96}$$

Clearly, $\tilde{u} \in W^{1,2}_{\text{loc}}(\mathbb{R}^2; \mathbb{R}^m)$, since it is C^1 smooth. We are going to check that $\Delta\tilde{u} - W_u(\tilde{u}) = 0$ weakly in $W^{1,2}_{\text{loc}}(\mathbb{R}^2; \mathbb{R}^m)$. For this purpose consider a ball B intersecting $x_2 = k$, and denote by B^+, B^- the two parts, above and below the line. Let $\phi \in C^1_c(B; \mathbb{R}^m)$, an integration by parts gives:

$$\int_{B^+} \left(\nabla\tilde{u}\nabla\phi + W_u(\tilde{u})\phi\right) = \int_{B^+} \left(-\Delta\tilde{u}\nabla\phi + W_u(\tilde{u})\right)\phi - \int_{\mathbb{R}} \tilde{u}_{x_2}(x_1, k)\phi(x_1, k)dx_1 = 0,$$

and

$$\int_{B^-} \left(\nabla \tilde{u} \nabla \phi + W_u(\tilde{u})\phi \right) = \int_{B^-} \left(- \Delta \tilde{u} \nabla \phi + W_u(\tilde{u}) \right)\phi + \int_{\mathbb{R}} \tilde{u}_{x_2}(x_1, k)\phi(x_1, k)\mathrm{d}x_1 = 0.$$

Thus, $\int_B \left(\nabla \tilde{u} \nabla \phi + W_u(\tilde{u})\phi \right) = 0$, and $\tilde{u}$ solves (8.59) weakly. By elliptic regularity, $\tilde{u}$ is also a $C^{2,\alpha}$ classical solution. Since $W \in C^3$, we can differentiate the equation and apply the unique continuation theorem in [8, Theorem 4.2] to the linearized equation, and conclude that $\tilde{u}_{x_2} \equiv 0$. From this we deduce that $\tilde{u} \equiv e^+$, hence $u \equiv e^+$. This however contradicts (8.78). Thus case (2) is dismissed as impossible. Similarly case (3) is dismissed. The proof of the theorem is complete. $\qquad\square$

Remark 8.3 We note the analogy in establishing $\tilde{u} \equiv e^+$ via the equipartition $\mathscr{W}(u(\cdot, x_2)) = \frac{1}{2}\int_{\mathbb{R}} |u_{x_2}(x_1, x_2)|^2\mathrm{d}x_1$, for all $x_2 \in \mathbb{R}$, and establishing $u \equiv a^+$ via the equipartition $\frac{1}{2}|u_x|^2 = W(u)$, $u(\bar{x}) = a^+$.

8.6   Scholia on Chap. 8

Theorem 8.1 and Corollary 8.1 were obtained in [4] and [5], respectively. The localized version of the density estimate (Theorem 8.2) is new and so is most of the presentation in Sect. 8.3, although influenced by [5]. The results in Sect. 8.4 were obtained in [5] by a method analogous to that developed in Sect. 5.5, and adapted to the present set-up. Alessio[3] approached Theorem 8.5 via direct minimization of the normalized energy $E(u)$. Monteil and Santambrogio [12] established Theorem 8.5 via the Jacobi method.

Acknowledgements We would like to thank Mihalis Nikolouzos for useful discussions on the proof of Corollary 8.2.

References

1. Agmon, S.: Lectures on Exponential Decay of Solutions of Second Order Elliptic Equations. Math. Notes Princeton University No. 29. Princeton University Press, Princeton (1982)
2. Alama, S., Bronsard, L., Gui, C.: Stationary layered solutions in $\mathbb{R}^2$ for an Allen–Cahn system with multiple well potential. Calc. Var. **5**(4), 359–390 (1997)
3. Alessio, F.: Stationary layered solutions for a system of Allen-Cahn type equations. Indiana Univ. Math. J. **62**(5), 1535–1564 (2013)
4. Alikakos, N.D., Fusco, G.: Density estimates for vector minimizers and application. Discrete Cont. Dyn. Syst. **35**(12), 5631–5663 (2015)
5. Alikakos, N.D., Fusco, G.: Asymptotic behavior and rigidity results for symmetric solutions of the elliptic system $\Delta u = W_u(u)$. Annali della Scuola Normale Superiore di Pisa, **XV**(issue special), 809–836 (2016)
6. Alikakos, N.D., Betelú, S.I., Chen, X.: Explicit stationary solutions in multiple well dynamics and non-uniqueness of interfacial energies. Eur. J. Appl. Math. **17**, 525–556 (2006)

7. Bronsard, L., Gui, C., Schatzman, M.: A three-layered minimizer in $\mathbb{R}^2$ for a variational problem with a symmetric three-well potential. Commun. Pure. Appl. Math. **49**(7), 677–715 (1996)
8. Garofalo, N., Lin, F.H.: Monotonicity properties of variational integrals, A_p weights and unique continuation. Ind. Univ. Math. J. **35**(2), 245–268 (1986)
9. Gilbarg, D., Trudinger, N.S.: Elliptic Partial Differential Equations of Second Order. Grundlehren der Mathematischen Wissenschaften, vol. 224, revised 2nd edn. Springer, Berlin (1998)
10. Gui, C., Schatzman, M.: Symmetric quadruple phase transitions. Ind. Univ. Math. J. **57**(2), 781–836 (2008)
11. Hislop, P.D., Sigal, I.M.: Introduction to Spectral Theory with Applications to Schrödinger Operators. Applied Mathematical Sciences, vol. 113. Springer, New York (1996)
12. Monteil, A., Santambrogio, F.: Metric methods for heteroclinic connections in infinite dimensional spaces. arXiv: 1709-02117v1 (To appear)
13. Schatzman, M.: Asymmetric heteroclinic double layers. ESAIM: Control Optim. Calc. Var. **8**(A tribute to J. L. Lions), 965–1005 (electronic) (2002)

Chapter 9
Vector Minimizers in $\mathbb{R}^2$

Abstract Let $W : \mathbb{R}^m \to \mathbb{R}$ be a nonnegative potential with exactly two nondegenerate zeros $a^- \neq a^+ \in \mathbb{R}^m$. Assume that there are $N \geq 1$ distinct heteroclinic orbits connecting a^- to a^+, represented by maps $\bar{u}_1, \ldots, \bar{u}_N$ that minimize the one-dimensional energy $J_{\mathbb{R}}(u) = \int_{\mathbb{R}}(\frac{|u'|^2}{2} + W(u))ds$. Under a nondegeneracy condition on $\bar{u}_j$, $j = 1, \ldots, N$ and in two space dimensions we characterize the minimizers $u : \mathbb{R}^2 \to \mathbb{R}^m$ of the energy $J_{\Omega}(u) = \int_{\Omega}(\frac{|\nabla u|^2}{2} + W(u))dx$ that converge uniformly to $a^\pm$ as one of the coordinates converges to $\pm\infty$. We prove that a bounded minimizer $u : \mathbb{R}^2 \to \mathbb{R}^m$ is necessarily an heteroclinic connection between suitable translates $\bar{u}_-(\cdot - \eta_-)$ and $\bar{u}_+(\cdot - \eta_+)$ of some $\bar{u}_\pm \in \{\bar{u}_1, \ldots, \bar{u}_N\}$. Then, assuming $N = 2$ and denoting $\bar{u}_-, \bar{u}_+$ representatives of the two orbits connecting a^- to a^+ we give a new proof of the existence (first proved in Schatzman [40]) of a solution $u : \mathbb{R}^2 \to \mathbb{R}^m$ of $\Delta u = W_u(u)$, that connects certain translates of $\bar{u}_\pm$.

9.1 Introduction

We have seen in Theorem 2.3 that, in one space dimension ($n = 1$) and under the assumption that $W : \mathbb{R}^m \to \mathbb{R}$ is a nonnegative potential that vanishes on a finite set, there is a complete characterization of minimal solutions $u : \mathbb{R} \to \mathbb{R}^m$. In higher dimension ($n > 1$), even in the scalar case ($m = 1$), a classification of minimizers is far from being complete, in spite of many deep results that have appeared in the last 20 years, motivated by a famous conjecture of De Giorgi. De Giorgi conjectured that, for $n \leq 8$, if a solution $u : \mathbb{R}^n \to \mathbb{R}$ of

$$\Delta u = u^3 - u, \quad |u| < 1, \tag{9.1}$$

is monotone in one direction, then it is necessarily one-dimensional. That is, there exist a number $s_0 \in \mathbb{R}$ and a unit vector $v \in \mathbb{R}^n$ such that

$$u(x) = \bar{u}(s_0 + x \cdot v), \quad x \in \mathbb{R}^n, \tag{9.2}$$

© Springer Nature Switzerland AG 2018

N. D. Alikakos et al., *Elliptic Systems of Phase Transition Type*,
Progress in Nonlinear Differential Equations and Their Applications 91,
https://doi.org/10.1007/978-3-319-90572-3_9

where $\bar{u}(s) = \tanh \frac{s}{\sqrt{2}}$ is, up to translation, the unique monotone increasing solution in $\mathbb{R}$ that connects -1 to 1. The conjecture was proven true by Ghoussoub and Gui in [28] for $n = 2$, for $n = 3$ by Ambrosio and Cabré in [8], and for $4 \leq n \leq 8$ by Savin in [38] under the extra requirement that

$$\lim_{x_n \to \pm\infty} u(x_1, \ldots, x_n) = \pm 1. \tag{9.3}$$

For $n \geq 9$ a counterexample was constructed by del Pino et al. in [18]. An important relationship between De Giorgi conjecture and minimal solutions was established by Alberti et al. in [2], where they proved that solutions of (9.1) which satisfy (9.3) and are monotone in x_n are minimizers. If we consider minimizers that satisfy (9.3) and insist that the limit in (9.3) be uniform in $(x_1, \ldots, x_{n-1})$, then the restriction $n \leq 8$ can be dropped and we have [20]

$$u(x) = \bar{u}(s_0 + x_n), \quad x \in \mathbb{R}^n, \tag{9.4}$$

for every minimizer and for all $n \geq 1$.

If we drop the monotonicity requirement as well as (9.3) and simply ask about the structure of minimal solutions of (9.1), then we have from [38] that, for $n \leq 7$ a minimizer is necessarily one-dimensional. On the other hand, in [33] it is proved that, for $n \geq 8$ there exist nontrivial minimizers which are not one-dimensional. The results in [33] and [18] indicate that in higher dimensions there is a large variety of minimizers.

In the vector case $m > 1$ the situation is quite different even if, in analogy with the scalar case, we keep the assumption that $W : \mathbb{R}^m \to \mathbb{R}$, $W \geq 0$ has only two zeros, say $a^- \neq a^+ \in \mathbb{R}^m$. The problem is that now the orbit connecting a^- to a^+ is not necessarily unique: in fact, (see [7]) there may exist two or more distinct minimal orbits connecting a^- to a^+. Minimal here means that the orbits are represented by maps $\bar{u}_1, \ldots, \bar{u}_N$ that minimize the energy

$$J_{\mathbb{R}}(\bar{u}_j) = \min_u J_{\mathbb{R}}(u), \quad J_{\mathbb{R}}(u) = \int_{\mathbb{R}} \left(\frac{1}{2}|\dot{u}|^2 + W(u) \right) ds, \tag{9.5}$$

where the minimization is carried out on the set of $W^{1,2}_{\text{loc}}(\mathbb{R}; \mathbb{R}^m)$ maps that satisfy

$$\lim_{s \to \pm\infty} u(s) = a^{\pm}. \tag{9.6}$$

A consequence of the non-uniqueness of the connection is that, even for $n = 2$, as shown by Schatzman in [40] and as we prove later, there exist minimizers $u : \mathbb{R}^2 \to \mathbb{R}^m$ that satisfy

$$\lim_{y \to \pm\infty} u(x, y) = a^{\pm}, \tag{9.7}$$

uniformly in $x \in \mathbb{R}$ but, instead of (9.4), satisfy

$$\lim_{x \to \pm\infty} u(x, y) = \bar{u}_{\pm}(y - \eta^{\pm}), \tag{9.8}$$

where $\bar{u}_-$ and $\bar{u}_+$ are distinct ($\bar{u}_- \neq \bar{u}_+(\cdot - r)$ for $r \in \mathbb{R}$) minimizer of (9.5) and η_- and η_+ are real numbers. Note that, since the energy $J_{\mathbb{R}}$ is translation invariant, if $\bar{u}$ is a minimizer of (9.5), then the translate $\bar{u}(\cdot - r)$ is also a minimizer. That is, we have a one-parameter family of minimizers, all of them representing the same orbit connecting a^- to a^+. Therefore $\eta_{\pm}$ are extra unknowns of the problem needed to specify to which particular elements of the manifolds of the translates of $\bar{u}_{\pm}$ the minimizer $u : \mathbb{R}^2 \to \mathbb{R}^m$ is asymptotic to. This poses an extra difficulty compared with to the problem we have discussed in Sect. 8.5, where we considered potentials symmetric under the reflection that exchanges a^- and a^+ and proved the existence of symmetric solutions. Working in the symmetric context simplifies the problem and, in particular fixes the position of the *interface* that separates the half-spaces where u is near a^- or a^+, and automatically implies $\eta_{\pm} = 0$.

In this chapter, besides a proof of Schatzman's result, we show that if W is as before and $u : \mathbb{R}^2 \to \mathbb{R}^m$ is a minimizer that satisfies (9.7) uniformly in x, then u is of the form (9.8). Finally, we prove that, if the connection is unique, then $u(x, y) = \bar{u}(y - \eta)$ for some $\eta \in \mathbb{R}$.

9.2 Assumptions and Statements

We make the following assumptions:

H$_1$ $W : \mathbb{R}^m \to \mathbb{R}$ is a C^3 function that satisfies

$$0 = W(a^{\pm}) < W(u), \quad u \notin \{a^-, a^+\} \tag{9.9}$$

for some $a^- \neq a^+ \in \mathbb{R}^m$ that are non-degenerate in the sense that the Hessian matrix $W_{uu}(a^{\pm})$ is positive definite.

H$_2$

$$\liminf_{|u| \to +\infty} W(u) > 0.$$

H$_3$ There are exactly $N \geq 1$ distinct orbits connecting a^- to a^+ represented by minimizers $\bar{u}_1, \ldots, \bar{u}_N$ of (9.5). The connections are non-degenerate in the sense that 0 is a simple eigenvalue of the operator $T : W^{2,2}(\mathbb{R}; \mathbb{R}^m) \to L^2(\mathbb{R}; \mathbb{R}^m)$:

$$T\varphi = -\varphi'' + W_{uu}(\bar{u})\varphi, \quad \bar{u} \in \{\bar{u}_1, \ldots, \bar{u}_N\}. \tag{9.10}$$

Remark 9.1 Note that, since given $\bar{u} \in \{\bar{u}_1, \ldots, \bar{u}_N\}$, for each $r \in \mathbb{R}$, $\bar{u}(\cdot - r)$ is also a minimizer of $J_{\mathbb{R}}$ and therefore a solution of

$$u'' = W_u(u), \quad s \in \mathbb{R}, \tag{9.11}$$

by differentiating this equation with respect to r we see that 0 lies in the spectrum of T and that $\bar{u}'$ is a corresponding eigenvector.

We use the notation $\varsigma(s)$ for the sign of $s \in \mathbb{R}$. We prove

Theorem 9.1 *Assume W, $a^{\pm}$ and $\bar{u}_1, \ldots, \bar{u}_N$ satisfy* $\mathbf{H}_1$–$\mathbf{H}_3$ *and let $u : \mathbb{R}^2 \to \mathbb{R}^m$ be a minimizer that satisfies (9.7) uniformly in $x \in \mathbb{R}$.*

Then there are $\bar{u}_{\pm} \in \{\bar{u}_1, \ldots, \bar{u}_N\}$, numbers $\eta_{\pm} \in \mathbb{R}$ and constants k, K, k', $K' > 0$ such that

$$\begin{aligned}
|u(x, y) - a^{\varsigma(y)}| &\leq K e^{-k|y|}, \\
|u(x, y) - \bar{u}_{\varsigma(x)}(y - \eta_{\varsigma(x)})| &\leq K' e^{-k'|x|}.
\end{aligned} \tag{9.12}$$

Theorem 9.2 *Assume W, $a^{\pm}$ as in Theorem 9.1 and there is a unique orbit connecting a^- to a^+ or that $\bar{u}_- = \bar{u}_+ = \bar{u}$. Then there is $\eta \in \mathbb{R}$ such that*

$$u(x, y) = \bar{u}(y - \eta), \tag{9.13}$$

Theorem 9.3 *Assume $N = 2$ and let W, $a^{\pm}$ and $\bar{u}_{\pm}$ satisfy* $\mathbf{H}_1$–$\mathbf{H}_3$. *Assume that*

$$W(sz) \geq W(z), \quad \text{for } |z| \geq M, \ s \geq 1. \tag{9.14}$$

for some $M > 0$. Then there exist $u \in C^{2+\alpha}(\mathbb{R}^2; \mathbb{R}^m)$ and $\eta_{\pm} \in \mathbb{R}$ that solve

$$\Delta u - W_u(u) = 0, \tag{9.15}$$

and satisfy (9.12).

After Schatzman a new proof of Theorem 9.3 was given by Alessio and Montecchiari [3] by minimization of the normalized energy and by Monteil and Santambrogio [35] via minimization of the Jacobi functional.

Remark 9.2 There is a perfect analogy between Theorems 9.1, 9.2 and 2.3. The zeros of W, $a_1, \ldots, a_N$, correspond to the zeros $\bar{u}_1, \ldots, \bar{u}_N$ of the effective potential

$$\mathscr{W} = J_{\mathbb{R}}(u) - J_{\mathbb{R}}(\bar{u}), \quad \bar{u} \in \{\bar{u}_1, \ldots, \bar{u}_N\} \tag{9.16}$$

and, similarly to Theorem 2.3 and Theorem 9.1 says that, in two space dimensions and provided (9.7) holds uniformly in $x \in \mathbb{R}$, a minimizer is necessarily a heteroclinic connection between two of the zeros of the effective potential. If $N = 1$ Theorem 9.2 says that a minimizer is necessarily of the form (9.13) and therefore constant in $x \in \mathbb{R}$. The analogy becomes even more tight if we regard the minimizer $u : \mathbb{R}^2 \to \mathbb{R}^m$ as a map

$$\mathbb{R} \ni x \mapsto u(x, \cdot) \in \bar{v} + W^{1,2}(\mathbb{R}; \mathbb{R}^m)$$

(where $\bar{v}$ is either one of the $\bar{u}_j$, or any smooth map $\bar{v} : \mathbb{R} \to \mathbb{R}^m$ with the same asymptotic behaviour) and interpret the elliptic system (8.59) as an O.D.E. in the infinite-dimensional function space $\bar{v} + W^{1,2}(\mathbb{R}; \mathbb{R}^m)$:

$$u_{xx}(x, \cdot) = \nabla_{L^2(\mathbb{R};\mathbb{R}^m)}(J_\mathbb{R}(u(x, \cdot)) - \sigma),$$

$$\sigma = J_\mathbb{R}(\bar{u}_j), \quad j = 1, \ldots, N.$$

This equation is then analogous to the Eq. (9.11) satisfied by $\bar{u}_1, \ldots, \bar{u}_N$. To stress this analogy, differently from what we have done in Chap. 8, we reserve to $x \in \mathbb{R}$ the role of independent variable.

Equation (9.15) arises naturally in the local analysis of solutions of the reaction-diffusion equation

$$u_t = \epsilon^2 \Delta u - W_u(u), \quad x \in \Omega, \tag{9.17}$$

which for $0 < \epsilon << 1$ is a widely studied model for *phase transitions*. In this context, a^- and a^+ represent different *phases* in which a specific substance may exist. For small $\epsilon > 0$ typical solutions u^ϵ of (9.17) divide Ω as $\Omega_- \cup \Gamma \cup \Omega_+$ with $\Omega_\pm = \{u^\epsilon \approx a^\pm\}$ and Γ an *interface* of thickness $O(\epsilon)$ that separates the regions Ω_- and Ω_+ where the substance is in phase a^- or in phase a^+. Across the interface we have

$$u^\epsilon(x) \approx \bar{u}(d(x)/\epsilon),$$

where $d(x)$ is the signed distance from Γ and $\bar{u} : \mathbb{R} \to \mathbb{R}^m$ represents an heteroclinic connection between a^- and a^+. As we have already noted, in the scalar case $m = 1$ there is a unique orbit connecting a^- to a^+. In the vector case $m > 1$, as we assume in $\mathbf{H}_3$, there may exist two (or more) distinct connections between a^- and a^+ represented by minimizers $\bar{u}_-$ and $\bar{u}_+$ of problem (9.5). When this is the case, Eq. (9.17) can model more complex situations where the profile of u^ϵ across Γ changes along Γ. In two space dimensions ($n = 2$) Γ is an arc which can be divided into two sub-arcs (one with profile $\bar{u}_-$ and one with profile $\bar{u}_+$) separated by a point I. The point I can be regarded as an *interface* in the interface. It is an interesting mathematical problem to describe the structure of u^ϵ in a neighborhood of I. Equation (9.15) arises by blowing up coordinates around I as the equation that describes the first term of the asymptotic expansion of u^ϵ. If $(x, y) \in \mathbb{R}^2$ ranges in a neighborhood of $I = (0, 0)$ with x along Γ and y orthogonal to it, the transformation $(x, y) \to (x/\epsilon, y/\epsilon)$ leads to the problem of the existence of a solution $u : \mathbb{R}^2 \to \mathbb{R}^m$ of (1.2) that satisfies asymptotic conditions of the form (9.7), (9.8).

Theorem 9.1 shows that when we go from $m = 1$ to $m > 1$, even in two space dimensions and under the assumption that W has only two zeros, the set of minimizers becomes wider and qualitatively different. If W has three or more zeros, the interface Γ can exhibit multiple points like triple junctions or quadruple junctions, where three or four different regions corresponding to three or four different phases come together. At principal order the local structure of u^ϵ is again described by solutions of (1.2) with more complex asymptotic behavior (see [10–12, 30] and Theorems 6.1 and 7.1 for the symmetric case) and we can expect a new large class of minimizers.

9.3　The Proof of Theorem 9.1

Since u is a minimizer, it is a bounded solution of (9.15). This, the fact that $W \in C^3(\mathbb{R}^m; \mathbb{R})$, and elliptic theory imply that

$$\|u\|_{C^{2,\gamma}(\mathbb{R}^2;\mathbb{R}^m)} \leq M, \tag{9.18}$$

for some $M > 0$, $\gamma \in (0, 1)$. It is a standard fact that the assumption that $a^\pm$ is non-degenerate and the smoothness of W imply ($\varsigma(s) := \operatorname{sign}(s)$)

$$|\bar{u}(s) - a^{\varsigma(s)}|, |\bar{u}'(s)|, |\bar{u}''(s)| \leq \bar{K}e^{-\bar{k}|s|}, \quad \text{for } s \in \mathbb{R}, \ \bar{u} \in \{\bar{u}_1, \dots, \bar{u}_N\}, \tag{9.19}$$

for some $\bar{k}, \bar{K} > 0$.

We let $\langle f, g \rangle$ denote the standard inner product and $\|f\|$ the standard norm in $L^2(\mathbb{R}; \mathbb{R}^m)$. We use the notation $D^\alpha = \frac{\partial^{\alpha_1}}{\partial x^{\alpha_1}} \frac{\partial^{\alpha_2}}{\partial y^{\alpha_2}}, \alpha = (\alpha_1, \alpha_2)$.

Lemma 9.1 *There exist $k, K > 0$ such that, for $\alpha \in \mathbb{N}^2$, $\alpha_1 + \alpha_2 \leq 2$, it holds that*

$$|(D^\alpha(u - \bar{u}))(x, y)| \leq Ke^{-k|y|}, \quad \bar{u} = \bar{u}_1, \dots, \bar{u}_N, \tag{9.20}$$

$$\|(D^\alpha(u - \bar{u}))(x, \cdot)\| \leq \frac{K}{\sqrt{k}}, \quad x \in \mathbb{R}, \ \bar{u} = \bar{u}_1, \dots, \bar{u}_N. \tag{9.21}$$

Proof From the assumption that u satisfies (9.7) uniformly in $x \in \mathbb{R}$ it follows that

$$|u(x, y) - a^-| \leq \delta, \quad x \leq -\lambda,$$

$$|u(x, y) - a^+| \leq \delta, \quad x \geq \lambda,$$

for some $\delta \in (0, \frac{1}{2}|a^+ - a^-|)$ and $\lambda > 0$. Consequently, the minimizer u satisfies the hypothesis of Theorem 5.3 with respect to $a = a^+$ in the half-space $\{(x, y) : y \geq \lambda\}$ and with respect to $a = a^-$ in $\{(x, y) : y \leq -\lambda\}$. This and Proposition 5.2 imply

$$|u(x, y) - a^{\varsigma(y)}| \leq Ke^{-k|y|}, \tag{9.22}$$

for some constants $k, K > 0$. Therefore from (9.19) we obtain

$$|u(x, y) - \bar{u}(y)| \leq Ke^{-k|y|}, \quad (x, y) \in \mathbb{R}^2, \ \bar{u} = \bar{u}_1, \dots, \bar{u}_N. \tag{9.23}$$

This estimate and elliptic interior regularity imply (9.20). The bound (9.21) is a plain consequence of (9.20). The proof is complete. $\square$

A simple and useful consequence of the estimate (9.20) is that in the Definition 4.1 of minimality of u we can extend the class of sets Ω to include strips aligned with the y-axis: $\mathscr{R}_L(x_0) := (x_0, x_0 + L) \times \mathbb{R}$, $x_0 \in \mathbb{R}$ and $L > 0$.

Lemma 9.2 *Let u be the minimizer in Lemma 9.1. Then*

$$J_{\mathscr{R}_L(x_0)}(u) \leq J_{\mathscr{R}_L(x_0)}(v),$$

for every $v : \mathscr{R}_L(x_0) \to \mathbb{R}^m$ that satisfies

$$|v(x, y) - a^{\varsigma(y)}|, \ |(\nabla u)(x, y)| \leq Ke^{-k|y|},$$
$$v(x_0, y) = u(x_0, y), \quad v(x_0 + L, y) = u(x_0 + L, y).$$

Proof Assume there are $\eta > 0$ and $v \in u + W_0^{1,2}(\mathscr{R}_L(x_0); \mathbb{R}^m)$ such that

$$J_{\mathscr{R}_L(x_0)}(u) - J_{\mathscr{R}_L(x_0)}(v) \geq \eta. \tag{9.24}$$

For each $l > 0$ define $\tilde{v}$ by

$$\tilde{v} = \begin{cases} v, & \text{on } [x_0, x_0 + L] \times [-l, l], \\ (1 + l - y)v + (y - l)u, & \text{on } [x_0, x_0 + L] \times [l, l + 1], \\ (l + 1 + y)v - (y + l)u, & \text{on } [x_0, x_0 + L] \times [-l - 1, -l], \\ u, & \text{on } [x_0, x_0 + L] \times ([l + 1, +\infty) \cup (-\infty, -l - 1]). \end{cases}$$

From the assumptions on v and Lemma 9.1 it follows

$$J_{(x_0, x_0 + L) \times ((-l-1,-l) \cup (l,l+1))}(\tilde{v}) = \mathrm{O}(e^{-kl}).$$

This and the minimality of u imply

$$0 \geq J_{(x_0,x_0+L)\times(-l-1,l+1)}(u) - J_{(x_0,x_0+L)\times(-l-1,l+1)}(\tilde{v})$$
$$= J_{(x_0,x_0+L)\times(-l-1,l+1)}(u) - J_{(x_0,x_0+L)\times[-l,l]}(v) + \mathrm{O}(e^{-kl}). \qquad (9.25)$$

Letting $l \to +\infty$ in (9.25) yields

$$0 \geq J_{\mathscr{R}_L(x_0)}(u) - J_{\mathscr{R}_L(x_0)}(v),$$

in contradiction with (9.24). The proof is complete. $\square$

The minimality of u implies an upper bound for the energy.

Lemma 9.3 *There exists $C_0 > 0$ independent of $x_0 \in \mathbb{R}$ and $L > 0$ such that*

$$J_{\mathscr{R}_L(x_0)}(u) \leq \sigma L + C_0, \qquad (9.26)$$

where $\sigma = J_{\mathbb{R}}(\bar{u}_j)$, $j = 1, \ldots, N$.

Proof Fix $\bar{u} \in \{\bar{u}_1, \ldots, \bar{u}_N\}$, assume $L > 2$ and define a competing map v by setting

$$v(x,\cdot) = \begin{cases} u(x,\cdot), & \text{if } x \in (-\infty, x_0] \cup [x_0 + L, +\infty), \\ (1 - (x - x_0))u(x_0,\cdot) + (x - x_0)\bar{u}(\cdot), & \text{if } x \in (x_0, x_0 + 1), \\ \bar{u}, & \text{if } x \in [x_0 + 1, x_0 + L - 1], \\ (x_0 + L - x)\bar{u}(\cdot) + (1 + x - x_0 - L)u(x_0 + L, \cdot), & \\ & \text{if } x \in (x_0 + L - 1, x_0 + L). \end{cases} \qquad (9.27)$$

Using Lemma 9.1 and (9.19) one checks that the energy of v restricted to the set $((x_0, x_0 + 1) \cup (x_0 + L - 1, x_0 + L)) \times \mathbb{R}$ is bounded by a constant and (9.26) follows. The proof is complete. $\square$

Lemma 9.3 yields an upper bound for the *kinetic energy*.

Lemma 9.4 *It results*

$$\int_{\mathbb{R}} \int_{\mathbb{R}} \frac{|u_x|^2}{2} dy dx \leq C_0. \qquad (9.28)$$

$$\lim_{x \to \pm\infty} \int_{\mathbb{R}} |u_x(x, y)|^2 dy = 0. \qquad (9.29)$$

Proof The minimality of $\bar{u}_1, \ldots, \bar{u}_N$ implies $J_{\mathbb{R}}(u(x, \cdot)) - \sigma \geq 0$, for $x \in [x_0, x_0 + L]$ and therefore from (9.26) we obtain

$$\int_{x_0}^{x_0+L} \int_{\mathbb{R}} \frac{|u_x|^2}{2} dy dx \leq \int_{x_0}^{x_0+L} (J_{\mathbb{R}}(u(x, \cdot)) - \sigma) dx + \int_{x_0}^{x_0+L} \int_{\mathbb{R}} \frac{|u_x|^2}{2} dy dx \leq C_0$$

and, since this inequality is valid for all $x_0 \in \mathbb{R}$ and $L > 0$, (9.28) follows. If (9.29) does not hold, then $\int_{\mathbb{R}} |u_x(x_j, y)|^2 dy \geq \epsilon > 0$ along a sequence x_j, $j = 1, 2, \ldots$ that converges either to $-\infty$ or to $+\infty$. Lemma 9.1 implies

$$\left| \frac{d}{dx} \int_{\mathbb{R}} |u_x(x, y)|^2 dy \right| \leq 2 \|u_x(x, \cdot)\| \|u_{xx}(x, \cdot)\| \leq \frac{2K^2}{k}.$$

It follows that the map $x \mapsto \int_{\mathbb{R}} |u_x(x, y)|^2 dy$ is uniformly continuous and we have

$$\int_{\mathbb{R}} |u_x(x, y)|^2 dy \geq \frac{\epsilon}{2}, \quad x \in (x_j - \delta, x_j + \delta)$$

for some $\delta > 0$ independent of $j = 1, 2, \ldots$. By passing to a subsequence we can assume $|x_{j+1} - x_j| \geq 2\delta$ and conclude (assuming for example that $x_j \to +\infty$)

$$j \delta \epsilon \leq \int_{x_1 - \delta}^{x_j + \delta} \int_{\mathbb{R}} |u_x(x, y)|^2 dy \leq 2C_0,$$

which contradicts (9.28) for large j. This concludes the proof. $\square$

9.3.1   The Decomposition of a Map **u** Near a Translate of $\bar{u}_j$

Let $\bar{v} : \mathbb{R} \to \mathbb{R}^m$ be a smooth map with the same asymptotic behavior as the $\bar{u}_j$. Recall that $\| \cdot \|$ denotes the norm in $L^2(\mathbb{R}; \mathbb{R}^m)$.

Lemma 9.5 *Assume that* $u \in \bar{v} + L^2(\mathbb{R}, \mathbb{R}^m)$. *Then*

1. *there exist* $h \in \mathbb{R}$ *and* $\bar{u} \in \{\bar{u}_1, \ldots, \bar{u}_N\}$, *such that*

$$\|u - \bar{u}(\cdot - h)\| = \min_j \min_r \|u - \bar{u}_j(\cdot - r)\|, \tag{9.30}$$

and consequently

$$\langle u - \bar{u}(\cdot - h), \bar{u}'(\cdot - h) \rangle = 0. \tag{9.31}$$

2. *There exists* $q^0 > 0$ *such that, for* $q := \|u - \bar{u}(\cdot - h)\| \leq q^0$, h *and* $\bar{u}$ *are uniquely determined.*

3. *For $q < q^0$, h is a C^2 function of u and it results*

$$(D_u h)w = -\frac{\langle w, \bar{u}'(\cdot - h)\rangle}{\|\bar{u}'\|^2 - \langle u - \bar{u}(\cdot - h), \bar{u}''(\cdot - h)\rangle}.$$
(9.32)

Remark 9.3 If in (9.30) we replace the L^2 norm $\|\cdot\|$ with the $W^{1,2}$ norm $\|\cdot\|_1$, then the same arguments that prove Lemma 9.5, yield the analog of Lemma 9.5 for the $W^{1,2}$ norm. In particular, we have that, for $q^0 > 0$ sufficiently small, the condition

$$\min_{j} \min_{r \in \mathbb{R}} \|u - \bar{u}_j(\cdot - r)\|_1 \le p, \quad p \in (0, q^0]$$
(9.33)

implies the existence of unique h_1 and $\bar{u}$ that solve (9.30) with $\|\cdot\|_1$ instead of $\|\cdot\|$, and $\bar{u}$ does not depend on which norm is used. As expected, the difference $h - h_1$ between the solutions h and h_1 of (9.30) in the L^2 and $W^{1,2}$ sense converges to zero with p.

Lemma 9.6 *If (9.33) holds with $p \in (0, q^0]$ and $q^0 > 0$ is sufficiently small, then the solutions h and h_1 of (9.30), in the L^2 and the $W^{1,2}$ sense, respectively, satisfy*

$$|h - h_1| \le Cp.$$
(9.34)

Moreover,

$$\|u - \bar{u}(\cdot - h)\|_1 \le \bar{C}p.$$
(9.35)

For the proof of Lemmas 9.5 and 9.6 we refer to Section 2 of [40].

Lemma 9.7 *Let q^0 the constant in Lemma 9.5. Let $u \in C^2(I \times \mathbb{R}; \mathbb{R}^m)$, with $I \subset \mathbb{R}$ an open interval, be a map that, for some $k, K > 0$, satisfies*

$$|(D^\alpha(u - \bar{u}_j))(x, y)| \le Ke^{-k|y|}, \quad (x, y) \in I \times \mathbb{R}, \ \alpha \in \mathbb{N}^2, \ \alpha_1 + \alpha_2 \le 2.$$
(9.36)

Set

$$q(x) = \min_{j} \min_{r \in \mathbb{R}} \|u(x, \cdot) - \bar{u}_j(\cdot - r)\|.$$

Then, $q : I \to \mathbb{R}$ is continuous and if

$$q(x) \le q^0, \quad x \in I,$$

there exist unique $\bar{u} \in \{\bar{u}_1, \ldots, \bar{u}_N\}$ (independent of $x \in I$) and $h : I \to \mathbb{R}$ such that

$$q(x) = \|u(x, \cdot) - \bar{u}(\cdot - h(x))\|, \quad x \in I,$$
(9.37)

and $u(x, \cdot)$ *can be uniquely decomposed as*

$$\begin{cases} u(x, \cdot) = \bar{u}(\cdot - h(x)) + v(x, \cdot - h(x)) \\ \langle v(x, \cdot), \bar{u}' \rangle = 0, \quad where \end{cases} \tag{9.38}$$

$$v(x, y) := u(x, y + h(x)) - \bar{u}(y).$$

It holds, $\|h\|_{C^1(I;\mathbb{R})} \leq C$ *for some* $C > 0$ *and there exist* $k_1, K_1 > 0$ *such that*

$$|(D_y^i v)(x, y)| \leq K_1 e^{-k_1|y|}, \quad for \ (x, y) \in I \times \mathbb{R}, \ i = 0, 1, 2 \tag{9.39}$$

$$\|(D_y^i v)(x, \cdot)\| \leq \frac{K_1}{\sqrt{k_1}}, \quad for \ x \in I, \ i = 0, 1, 2. \tag{9.40}$$

Proof The continuity of $q : I \to \mathbb{R}$ is obvious. Existence and uniqueness of $\bar{u}$ and of the map $h : I \to \mathbb{R}$ that satisfies (9.37) and (9.38) follow from Lemma 9.5. From (9.36) it follows that h is bounded. To show that the same is true for h' we invoke 3. in Lemma 9.5, which implies that

$$h'(x) = -\frac{\langle u_x(x, \cdot), \bar{u}'(\cdot - h(x)) \rangle}{\|\bar{u}'\|^2 - \langle u(x, \cdot) - \bar{u}(\cdot - h(x)), \bar{u}''(\cdot - h(x)) \rangle},$$

$$|h'(x)| \leq \frac{\|u_x(x, \cdot)\| \|\bar{u}'\|}{\|\bar{u}'\|^2 - q^0 \|\bar{u}''\|} \leq C,$$

where we have used (9.36) and $(u - \bar{u}_j)_x = u_x$ that imply

$$\|u_x(x, \cdot)\| \leq \frac{K}{\sqrt{k}}, \quad for \ x \in I. \tag{9.41}$$

The estimate (9.39) follows from (9.36), the bound on h and (9.19); the estimate (9.40) follows from (9.39). The proof is complete. $\qquad\square$

We have $\|v(x, \cdot)\| = q(x)$ and, if $q(x) > 0$, we can write $v(x, \cdot) = q(x)\mathbf{n}(x, \cdot)$ with $\|\mathbf{n}(x, \cdot)\| = 1$. We call $q(x)$ the modulus and $\mathbf{n}(x, \cdot)$ the direction vector of $v(x, \cdot)$. Assuming $q(x) \leq q^0$ in I from (9.38), using also that $\langle \phi, \psi \rangle = \langle \phi(\cdot - r), \psi(\cdot - r) \rangle$, we derive the identities

$$\langle v_x(x, \cdot), \bar{u}' \rangle = \langle \mathbf{n}_x(x, \cdot), \bar{u}' \rangle = 0,$$

$$\|v_y(x, \cdot)\|^2 + \langle v(x, \cdot), v_{yy}(x, \cdot) \rangle = 0, \tag{9.42}$$

$$\langle \mathbf{n}_x(x, \cdot), \mathbf{n}(x, \cdot) \rangle = \langle \mathbf{n}_y(x, \cdot), \mathbf{n}(x, \cdot) \rangle = 0,$$

9.3.2   The Effective Potential

Lemma 9.8 *Assume $v \in L^2(\mathbb{R}; \mathbb{R}^m)$ is differentiable and $|v'| \leq C$ for some $C > 0$. Then*

$$\|v\|_{L^\infty(\mathbb{R};\mathbb{R}^m)} \leq (\tfrac{3}{2}C)^{\frac{1}{3}}\|v\|^{\frac{2}{3}}. \tag{9.43}$$

Proof The assumptions on v imply the existence of $\bar{s} \in \mathbb{R}$ such that

$$|v(\bar{s})| = V := \|v\|_{L^\infty(\mathbb{R};\mathbb{R}^m)}.$$

This and $|v'| \leq C$ imply

$$|v(s)| \geq V - C|s - \bar{s}| \quad s \in (\bar{s} - \frac{V}{C}, \bar{s} + \frac{V}{C}).$$

This and a routine computation complete the proof. $\square$

For later reference we note that if $v \in L^2(\mathbb{R}; \mathbb{R}^m)$ is as in Lemma 9.8 with $\|v\| \leq C_1$, for some $C_1 > 0$, then there is a constant $C_w > 0$ such that

$$\left| \int_{\mathbb{R}} \left(W(\bar{u} + v) - W(\bar{u}) - W_u(\bar{u}) \cdot v - \frac{1}{2}W_{uu}(\bar{u})v \cdot v \right)ds \right| \leq C_w\|v\|^{\frac{8}{3}},$$

$$\left| \int_{\mathbb{R}} \left(W_u(\bar{u} + v) \cdot v - W_u(\bar{u}) \cdot v - W_{uu}(\bar{u})v \cdot v \right)ds \right| \leq C_w\|v\|^{\frac{8}{3}}, \tag{9.44}$$

$$\left| \int_{\mathbb{R}} \left(W_{uu}(\bar{u} + v) - W_{uu}(\bar{u}) \right)v \cdot v \, ds \right| \leq C_w\|v\|^{\frac{8}{3}}.$$

From (9.43), $\|v\| \leq C_1$ and the smoothness of W we have (with $C_2 = (\tfrac{3}{2}C)^{\frac{1}{3}}C_1^{\frac{2}{3}}$)

$$\max_{|z| \leq C_2} |W_{uuu}(\bar{u} + z)| \leq \text{Const.}, \quad \bar{u} \in \{\bar{u}_1, \ldots, \bar{u}_N\}.$$

This and Taylor's formula imply that, for each $s \in \mathbb{R}$,

$$|W(\bar{u} + v) - W(\bar{u}) - W_u(\bar{u}) \cdot v - \frac{1}{2}W_{uu}(\bar{u})v \cdot v| \leq C|v|^3 \leq C\|v\|^{\frac{2}{3}}|v|^2,$$

where, for the last inequality, we have used (9.43). Then $(9.44)_1$ follows by integrating on $\mathbb{R}$. The other inequalities are proved in the same way.

If $v \in W^{1,2}(\mathbb{R}; \mathbb{R}^m)$, $v \neq 0$ we sometimes write v in the form

$$v(s) = q\mathbf{n}(s), \quad s \in \mathbb{R},$$

where $q = \|v\|$ is the L^2 norm of v, and $\mathbf{n} \in \{w \in W^{1,2}(\mathbb{R}, \mathbb{R}^m) : \|w\| = 1\}$.

For $\bar{u} \in \{\bar{u}_1, \ldots, \bar{u}_N\}$ fixed, the effective potential defined in (9.16) can be considered a function of $q \in \mathbb{R}$ and $\mathbf{n} \in W^{1,2}(\mathbb{R}, \mathbb{R}^m)$. To emphasize this point of view, we will sometimes write $\mathscr{W}(q, \mathbf{n})$ instead of $\mathscr{W}(q\mathbf{n})$. Recall that $\|v\|_1$ denotes the $W^{1,2}(\mathbb{R}, \mathbb{R}^m)$ norm of v.

Lemma 9.9 *Let $v \in W^{1,2}(\mathbb{R}; \mathbb{R}^m)$ be as in Lemma 9.8 and assume that*

$$\langle v, \bar{u}' \rangle = 0.$$

Then the constant q^0 in Lemma 9.5 can be chosen so that the effective potential $\mathscr{W}(q, \mathbf{n})$ is increasing in q for $q \in [0, q^0]$ and there is a $\mu > 0$ such that

$$\frac{\partial^2}{\partial q^2}\mathscr{W}(q, \mathbf{n}) \geq \mu(1 + \|\mathbf{n}'\|^2), \quad q \in (0, q^0], \tag{9.45}$$

and

$$\mathscr{W}(q, \mathbf{n}) \geq \frac{1}{2}\mu q^2(1 + \|\mathbf{n}'\|^2), \quad q \in (0, q^0]$$

$$\Longleftrightarrow \tag{9.46}$$

$$\mathscr{W}(v) \geq \frac{1}{2}\mu\|v\|_1^2, \quad \|v\| \in (0, q^0].$$

Moreover,

$$\left|\mathscr{W}(v) - \frac{1}{2}\langle Tv, v\rangle\right| \leq C\|v\|^{\frac{8}{3}}, \quad \|v\| \in (0, q^0], \tag{9.47}$$

where T is defined in (9.10).

Proof We have

$$\mathscr{W}(q, \mathbf{n}) = J_{\mathbb{R}}(\bar{u} + q\mathbf{n}) - J_{\mathbb{R}}(\bar{u})$$

$$= \langle \bar{u}', q\mathbf{n}'\rangle + \frac{1}{2}q^2\|\mathbf{n}'\|^2 + \int_{\mathbb{R}}\Big(W(\bar{u} + q\mathbf{n}) - W(\bar{u})\Big)ds. \tag{9.48}$$

Differentiating twice $\mathscr{W}(q, \mathbf{n})$ with respect to q, we get

$$\frac{\partial^2}{\partial q^2}\mathscr{W}(q, \mathbf{n}) = \|\mathbf{n}'\|^2 + \int_{\mathbb{R}} W_{uu}(\bar{u} + q\mathbf{n})\mathbf{n} \cdot \mathbf{n}\, ds$$

$$= \frac{\partial^2}{\partial q^2}\mathscr{W}(0, \mathbf{n}) + \int_{\mathbb{R}}\Big(W_{uu}(\bar{u} + q\mathbf{n}) - W_{uu}(\bar{u})\Big)\mathbf{n} \cdot \mathbf{n}\, ds. \tag{9.49}$$

Since $q\mathbf{n} = v$, (9.44) yields

$$\left| \int_{\mathbb{R}} \Big(W_{uu}(\bar{u} + q\mathbf{n}) - W_{uu}(\bar{u}) \Big) \mathbf{n} \cdot \mathbf{n}\, ds \right| \le C q^{\frac{2}{3}}. \tag{9.50}$$

We now observe that

$$\frac{\partial^2}{\partial q^2} \mathscr{W}(0, \mathbf{n}) = \langle T\mathbf{n}, \mathbf{n} \rangle; \tag{9.51}$$

here T is a self-adjoint operator which is positive by the minimality of $\bar{u}$. By assumption, the matrix $W_{uu}(a^{\pm})$ is positive definite and Theorem A.2 in [32] implies that the essential spectrum of T is bounded below by a positive constant $\mu_{\mathrm{e}} > 0$. Since $\bar{u}'$ is an eigenvector of T, the assumption that 0 is a simple eigenvalue of T implies that if $\mu_1 < \mu_{\mathrm{e}}$ is an eigenvalue of T, then $\mu_1 = \inf_{\langle \mathbf{n}, \bar{u}' \rangle = 0} \langle T\mathbf{n}, \mathbf{n} \rangle > 0$. From this, (9.51) and Theorem 13.31 in [37] it follows that there is $\mu_2 > 0$ such that $\frac{\partial^2}{\partial q^2} \mathscr{W}(0, \mathbf{n}) \ge \mu_2 > 0$, which together with (9.50), provided $q^0 \le (\frac{\mu_2}{2C})^{\frac{3}{2}}$, implies

$$\frac{\partial^2}{\partial q^2} \mathscr{W}(q, \mathbf{n}) \ge \frac{\mu_2}{2}, \qquad \text{for } q \in [0, q^0]. \tag{9.52}$$

To upgrade this estimate to (9.45) we use a trick from [14]. Recalling (9.49), we have

$$\frac{\partial^2}{\partial q^2} \mathscr{W}(q, \mathbf{n}) - \mu(1 + \|\mathbf{n}'\|^2)$$

$$= (1 - \mu)\Big(\|\mathbf{n}'\|^2 + \int_{\mathbb{R}} W_{uu}(\bar{u} + q\mathbf{n})\mathbf{n} \cdot \mathbf{n}\, ds \Big)$$

$$+ \mu \int_{\mathbb{R}} \Big(W_{uu}(\bar{u} + q\mathbf{n}) - I \Big) \mathbf{n} \cdot \mathbf{n}\, ds$$

$$\ge (1 - \mu)\frac{\mu_2}{2} - \mu\Big| \int_{\mathbb{R}} \Big(W_{uu}(\bar{u} + q\mathbf{n}) - W_{uu}(\bar{u}) \Big) \mathbf{n} \cdot \mathbf{n}\, ds \Big|$$

$$- \mu\Big| \int_{\mathbb{R}} \Big(W_{uu}(\bar{u}) - I \Big) \mathbf{n} \cdot \mathbf{n}\, ds \Big|$$

$$\ge (1 - \mu)\frac{\mu_2}{2} - \mu(C q^{\frac{2}{3}} + C') \ge (1 - \mu)\frac{\mu_2}{2} - 2\mu C' = 0,$$

$$\text{for } \mu = \frac{\mu_2}{\mu_2 + 4C'}, \; q \le q^0 \le \left(\frac{C'}{C} \right)^{\frac{3}{2}}.$$

where we have used (9.50) and $|W_{uu}(\bar{u}) - I| \leq C'$. This concludes the proof of (9.45). The inequality in (9.46) follows from (9.45) and $\mathscr{W}(0, \mathbf{n}) = \frac{\partial}{\partial q}\mathscr{W}(0, \mathbf{n}) = 0$, which is a consequence of the definition of $\mathscr{W}(q, \mathbf{n})$ and of the minimality of $\bar{u}$. To complete the proof, we note that (9.47) follows from (9.44) and (9.48) that, after observing $\langle \bar{u}', v' \rangle = -\langle \bar{u}'', v \rangle = -\langle W_u(\bar{u}), v \rangle$, can be rewritten as

$$
\mathscr{W}(v) = \langle \bar{u}', v' \rangle + \frac{1}{2}\|v'\|^2 + \int_{\mathbb{R}} \Big(W(\bar{u} + q\mathbf{n}) - W(\bar{u}) \Big)ds
$$
$$
= \frac{1}{2}\langle Tv, v \rangle + \int_{\mathbb{R}} \Big(W(\bar{u} + v) - W(\bar{u}) - W_u(\bar{u})v - \frac{1}{2}W_{uu}(\bar{u})v \cdot v \Big)ds
\tag{9.53}
$$

The proof is complete. $\square$

Lemma 9.9 describes the properties of the effective potential $\mathscr{W}$ in a neighborhood of one of the connections represented by $\bar{u}_1, \ldots, \bar{u}_N$. We also need a lower bound for the effective potential away from a neighborhood of the N connections. Recall that $\|\cdot\|_1$ stands for the norm in $W^{1,2}(\mathbb{R}; \mathbb{R}^m)$ and that $\sigma = J_{\mathbb{R}}(\bar{u}_j)$, $j = 1, \ldots, N$. We have (see also Corollary 3.2 in [40] and Lemma 3.6 in [26])

Lemma 9.10 *Assume that* $u \in \bar{v} + W^{1,2}(\mathbb{R}; \mathbb{R}^m)$ *satisfies*

$$
\|u - \bar{u}_j(\cdot - r)\|_1 \geq p, \quad r \in \mathbb{R}, \; j = 1, \ldots, N,
\tag{9.54}
$$

for some $p > 0$. *Then there exists* $e_p > 0$ *such that*

$$
J_{\mathbb{R}}(u) - \sigma \geq e_p.
$$

Proof If u satisfies (9.54) and has $J_{\mathbb{R}}(u) \geq 2\sigma$, we can take $e_p = \sigma$. It follows that in the proof we can assume

$$
J_{\mathbb{R}}(u) < 2\sigma.
\tag{9.55}
$$

Note that since $a^{\pm}$ are non-degenerate zeros of $W \geq 0$, there exist positive constants γ and $r_0 > 0$ such that

$$
W_{uu}(a^{\pm} + z)\zeta \cdot \zeta \geq \gamma^2 |\zeta|^2, \quad \zeta \in \mathbb{R}^m, \; |z| \leq r_0,
\tag{9.56}
$$

Set

$$
q_0 = \min\{r_0, \frac{\gamma^2}{8C_W}\},
\tag{9.57}
$$

where

$$
C_W = \max\{|W_{uuu}(a^{\pm} + z)| : |z| \leq 3r_0\}.
\tag{9.58}
$$

Given $q \in (0, q_0)$ define

$$J_z^+(q) = \min_{v \in \mathscr{V}_z^+(q)} J(v),$$

$$\mathscr{V}_z^+(q) = \{v \in W_{\text{loc}}^{1,2}((0, \tau^v); \mathbb{R}^m) : v(0) = z, |z - a^+| = q, \lim_{s \to \tau^v} v(s) = a^+\},$$

$$J^-(q) = \min_{v \in \mathscr{V}^-(q)} J(v),$$

$$\mathscr{V}^-(q) = \{v \in W_{\text{loc}}^{1,2}((0, \tau^v); \mathbb{R}^m) : |v(0) - a^+| = q, \lim_{s \to \tau^v} v(s) = a^-\},$$

$$J_0(q) = \min_{v \in \mathscr{V}_0(q)} J(v),$$

$$\mathscr{V}_0(q) = \{v \in W^{1,2}((0, \tau^v); \mathbb{R}^m) : |v(0) - a^+| = q_0, |v(\tau^v) - a^+| = q\}. \tag{9.59}$$

Observe that there exists a positive function $\psi : (0, q_0) \to \mathbb{R}$ that converges to zero with q and satisfies

$$J_z^+(q) \le \psi(q). \tag{9.60}$$

Note also that $J_{\mathbb{R}}(\bar{u}_j) = \sigma$ and the minimality of $\bar{u}_j$ imply $J^-(q) + \psi(q) \ge \sigma$ and therefore we have

$$\sigma - \psi(q) \le J^-(q). \tag{9.61}$$

Since $\mathrm{u} \in \bar{v} + W^{1,2}(\mathbb{R}; \mathbb{R}^m)$ implies

$$\lim_{s \to \pm\infty} \mathrm{u}(s) = a^\pm, \tag{9.62}$$

we can define

$$s^{\mathrm{u},+}(\rho) = \min\{s : |\mathrm{u}(t) - a^+| \le \rho, \text{ for } t \ge s\},$$
$$s^{\mathrm{u},-}(\rho) = \max\{s : |\mathrm{u}(t) - a^-| \le \rho, \text{ for } t \le s\}. \tag{9.63}$$

Since $\psi(q) \to 0$ as $q \to 0$ while $\lim_{q \to 0} J_0(q) = \bar{J}_0$, $\bar{J}_0$ a positive constant, we can fix $q = q(q_0)$ so that

$$2J_0(q(q_0)) - \psi(q(q_0)) \ge \bar{J}_0. \tag{9.64}$$

We claim that in the proof of the lemma it suffices to consider only maps that satisfy the condition

$$s^{\mathrm{u},+}(q_0) - s^{\mathrm{u},-}(q_0) \le \frac{2\sigma}{W_m(q(q_0))}, \tag{9.65}$$

where $W_m(t) = \min_{a \in \{a^-, a^+\}, |z| \geq t} W(a + z)$. To see this, set

$$\bar{s}^{u,+} = \min\{s : |u(s) - a^+| = q(q_0)\},$$
$$\bar{s}^{u,-} = \max\{s : |u(s) - a^-| = q(q_0)\},$$

and observe that the definition of $\bar{s}^{u,\pm}$ implies $|u(s) - a^\pm| > q(q_0)$, for $s \in (\bar{s}^{u,-}, \bar{s}^{u,+})$. It follows that

$$(\bar{s}^{u,+} - \bar{s}^{u,-}) W_m(q(q_0)) \leq 2\sigma. \tag{9.66}$$

Assume first that

$$|u(s) - a^-| < q_0, \quad \text{for } s \in (-\infty, \bar{s}^{u,-}),$$
$$|u(s) - a^+| < q_0, \quad \text{for } s \in (\bar{s}^{u,+}, +\infty). \tag{9.67}$$

In this case we have

$$\bar{s}^{u,-} < s^{u,-}(q_0) < s^{u,+}(q_0) < \bar{s}^{u,+},$$

which together with (9.66) implies (9.65). Now assume that (9.67) does not hold and there exists $s^* \in (\bar{s}^{u,+}, +\infty)$ such that $|u(s^*) - a_+| = q_0$ (or $s^* \in (-\infty, \bar{s}^{u,-})$ such that $|u(s^*) - a_-| = q_0$). For definiteness we consider the first possibility; the other can be treated in a similar way.

To estimate the energy of u we focus on the intervals $(-\infty, \bar{s}^{u,+})$, $(\bar{s}^{u,+}, s^{u,+}(q(q_0)))$, and $(s^{u,+}(q(q_0)), +\infty)$.

We have $J_{(-\infty, \bar{s}^{u,+})}(u) \geq J^-(q(q_0))$ and since $s^* \in (\bar{s}^{u,+}, s^{u,+}(q(q_0)))$ we also have $J_{(\bar{s}^{u,+}, s^{u,+}(q(q_0)))}(u) \geq 2J_0(q(q_0))$. This, (9.61) and (9.64) imply

$$J_{\mathbb{R}}(u) \geq J_{(-\infty, \bar{s}^{u,+})}(u) + J_{(\bar{s}^{u,+}, s^{u,+}(q(q_0)))}(u) \geq J^-(q(q_0)) + 2J_0(q(q_0))$$
$$\geq \sigma - \psi(q(q_0)) + 2J_0(q(q_0)) \geq \sigma + \bar{J}_0.$$

This completes the proof of the claim. Indeed, this computation shows that if s^* with the above properties exists, then we can take $e_p = \bar{J}_0$.

Since $J_{\mathbb{R}}$ is translation invariant, we can also restrict the analysis to the set of the maps that satisfy

$$-s^{u,-}(q(q_0))) = s^{u,+}(q(q_0))) \leq \frac{\sigma}{W_m(q(q_0))}. \tag{9.68}$$

and assume that also $\bar{u}_j$, $j = 1, \ldots, N$ satisfy (9.68). We remark that the set of maps that satisfy (9.55) and (9.65) is equibounded and equicontinuous. Indeed, (9.55) implies

$$|u(s_1) - u(s_2)| \leq \sqrt{2\sigma} |s_1 - s_2|^{\frac{1}{2}}, \tag{9.69}$$

which together with (9.65) yields

$$|u(s)| \leq M_0 := |a^-| + 3q_0 + \sqrt{2\sigma} \left(\frac{2\sigma}{W_m(q(q_0))} \right)^{\frac{1}{2}}. \tag{9.70}$$

We first prove the lemma with (9.54) replaced by

$$\|u - \bar{u}_j(\cdot - r)\| \geq p, \quad r \in \mathbb{R}, \; j = 1, \ldots, N. \tag{9.71}$$

Assume the lemma is false. Then there is a sequence $\{u_n\} \subset \bar{v} + W^{1,2}(\mathbb{R}; \mathbb{R}^m)$ that satisfies (9.62) and

$$\lim_{n \to +\infty} J_{\mathbb{R}}(u_n) = \sigma,$$

$$\|u_n - \bar{u}_j(\cdot - r)\| \geq p, \quad r \in \mathbb{R}, \; j = 1, \ldots, N. \tag{9.72}$$

Since the sequence $\{u_n\}$ is equibounded and equicontinuous, there are subsequence, still labeled $\{u_n\}$, and a continuous map $\bar{u} : \mathbb{R} \to \mathbb{R}^m$ such that

$$\lim_{n \to +\infty} u_n(s) = \bar{u}(s), \tag{9.73}$$

uniformly on compact sets. From the bound $\int_{\mathbb{R}} |u_n'|^2 < 4\sigma$ and the fact that u_n is uniformly bounded, by passing to a further subsequence if necessary, we have that u_n converges to $\bar{u}$ weakly in $W^{1,2}_{\mathrm{loc}}(\mathbb{R}; \mathbb{R}^m)$. A standard argument then shows that

$$J_{\mathbb{R}}(\bar{u}) = \sigma,$$

and therefore, by the assumption that $\bar{u}_j, \; j = 1, \ldots, N$ and their translates are the only minimizers of $J_{\mathbb{R}}$, we conclude that $\bar{u}$ coincides with $\bar{u}_j(\cdot - r)$ for some j and with $|r| \leq \lambda_0$ where λ_0 is determined by the condition that $\bar{u}$ satisfies (9.68).

From (9.19) it follows that there exist positive constants k, K such that $|\bar{u}(s) - a^+| \leq Ke^{-ks}, s > 0$. Fix a number $l > \lambda_0$ such that

$$Ke^{-kl} \leq q_0, \quad \text{and} \quad \frac{K}{C_W}e^{-kl} \leq \frac{p^2}{8}, \tag{9.74}$$

and observe that $\bar{u}$ restricted to the interval $[-l, l]$ is a minimizer of $J_{(-l,l)}(u)$ in the class of maps u that satisfy $u(\pm l) = \bar{u}(\pm l)$. From this observation it follows that

$$J_{(-l,l)}(u_n) \geq J_{(-l,l)}(\bar{u}) - Cl\delta_n, \tag{9.75}$$

where $C > 0$ is a constant and $\delta_n = \max_{\pm} |u_n(\pm l) - \bar{u}(\pm l)|$.

By the properties of u and $\bar{u}$,

$$|u_n(s) - \bar{u}(s)| \leq |u_n(s) - a^+| + |\bar{u}(s) - a^+| \leq q_0 + Ke^{-kl} \leq 2q_0, \quad \text{for } s \geq l. \tag{9.76}$$

Let us estimate the differences $J_{(-\infty,-l)}(u_n) - J_{(-\infty,-l)}(\bar{u})$ and $J_{(l,+\infty)}(u_n) - J_{(l,+\infty)}(\bar{u})$. We have, with $u_n = \bar{u} + v_n$,

$$J_{(l,+\infty)}(u_n) - J_{(l,+\infty)}(\bar{u}) = \int_l^\infty \left(\bar{u}' \cdot v_n' + \frac{1}{2}|v_n'|^2 + W(\bar{u} + v_n) - W(\bar{u}) \right) ds$$

$$= -\bar{u}'(l) \cdot v_n(l) + \int_l^\infty \left(-\bar{u}'' \cdot v_n + \frac{1}{2}|v_n'|^2 + W(\bar{u} + v_n) - W(\bar{u}) \right) ds$$

$$= -\bar{u}'(l) \cdot v_n(l) + \int_l^\infty \left(\frac{1}{2}|v_n'|^2 + W(\bar{u} + v_n) - W(\bar{u}) - W_u(\bar{u}) \cdot v_n \right) ds$$

$$\geq -2q_0 Ke^{-kl} + \int_l^\infty \left(\frac{1}{2}(|v_n'|^2 + W_{uu}(\bar{u})v_n \cdot v_n) \right.$$

$$\left. + W(\bar{u} + v_n) - W(\bar{u}) - W_u(\bar{u}) \cdot v_n - \frac{1}{2}W_{uu}(\bar{u})v_n \cdot v_n \right) ds \tag{9.77}$$

Set $I(v_n) = W(\bar{u} + v_n) - W(\bar{u}) - W_u(\bar{u}) \cdot v_n - \frac{1}{2}W_{uu}(\bar{u})v_n \cdot v_n$. Then we have

$$I(v_n) = \int_0^1 \int_0^1 \int_0^1 \rho^2 \sigma W_{uuu}(\bar{u} + \rho\sigma\tau v_n)(v_n, v_n, v_n) d\tau d\sigma d\rho$$

$$= \int_0^1 \int_0^1 \int_0^1 \rho^2 \sigma W_{uuu}(a^+ + (\bar{u} - a^+) + \rho\sigma\tau v_n)(v_n, v_n, v_n) d\tau d\sigma d\rho.$$

It follows that $|I(v_n)| \leq 2q_0 C_W |v_n|^2$, with C_W the constant in (9.58). This, (9.77), and (9.56) yield

$$J_{(l,+\infty)}(u_n) - J_{(l,+\infty)}(\bar{u})$$

$$\geq -2q_0 Ke^{-kl} + \int_l^\infty \frac{1}{2}(|v_n'|^2 + \gamma^2|v_n|^2)ds - 2q_0 C_W \int_l^\infty |v_n|^2 ds$$

$$\geq -\frac{\gamma^2}{4C_W}Ke^{-kl} + \frac{1}{4}\gamma^2 \int_l^\infty |v_n|^2 ds \tag{9.78}$$

$$\geq -\gamma^2 \frac{p^2}{32} + \frac{1}{4}\gamma^2 \int_l^\infty |v_n|^2 ds,$$

where we have used (9.57) and (9.74). From this, the analogous estimate valid in the interval $(-\infty, -l)$, and (9.75) we obtain

$$0 = \lim_{n \to +\infty} (J_\mathbb{R}(\bar{u} + v_n) - \sigma)$$

$$\geq \lim_{n \to +\infty} \left[-Cl\delta_n - \gamma^2 \frac{p^2}{16} + \frac{1}{4}\gamma^2 \left(\int_{-\infty}^{-l} |v_n|^2 ds + \int_l^\infty |v_n|^2 ds \right) \right]. \tag{9.79}$$

Since v_n converges to 0 uniformly in $[-l, l]$, for n large we have

$$\int_{-l}^l |v_n|^2 ds \leq \frac{p^2}{2}$$

and therefore from (9.71) it follows that

$$\int_{-\infty}^{-l} |v_n|^2 ds + \int_l^\infty |v_n|^2 ds \geq \frac{p^2}{2}.$$

This and (9.79) imply

$$0 = \lim_{n \to +\infty} (J_\mathbb{R}(\bar{u} + v_n) - \sigma)$$

$$\geq \lim_{n \to +\infty} \left(- Cl\delta_n - \gamma^2 \frac{p^2}{16} + \gamma^2 \frac{p^2}{8} \right) = \gamma^2 \frac{p^2}{16}. \tag{9.80}$$

This contradiction concludes the proof of the lemma when (9.54) is replaced by (9.71). To complete the proof, we note that it suffices to consider the case $p \leq 2(2 + \sqrt{2})\sqrt{\sigma} =: 2p_0$. Indeed, (9.55) implies $\|u'\| \leq 2\sqrt{\sigma}$, which together with $\|\bar{u}'_j\| \leq \sqrt{2\sigma}$ yields

$$\|u' - \bar{u}'_j(\cdot - r)\| \leq p_0, \quad r \in \mathbb{R}, \ j = 1, \ldots, N. \tag{9.81}$$

It follows that $p \geq 2p_0$ implies $\|u - \bar{u}_j(\cdot - r)\| \geq p_0$ and the existence of e_p follows from the first part of the proof.

Set

$$C_W^0 = \max\{|W_{uu}(\bar{u}_j(s) + z)| : s \in \mathbb{R}, \ |z| \leq 2p_0, \ j = 1, \ldots, N\}$$

and define $\tilde{p} = \tilde{p}(p)$ by

$$\tilde{p}(p) = \frac{p}{\sqrt{2(1 + C_W^0)}}.$$

We distinguish the following alternatives:

a) $\|u - \bar{u}_j(\cdot - r)\| \geq \tilde{p}$, for $r \in \mathbb{R}$, $j = 1, \ldots, N$;
b) there exist $\bar{r} \in \mathbb{R}$ and $\bar{u} \in \{\bar{u}_1, \ldots, \bar{u}_N\}$ such that

$$\|u - \bar{u}(\cdot - \bar{r})\| < \tilde{p}. \tag{9.82}$$

In case a) the lemma is true by the first part of the proof with $e_p = e_{\tilde{p}}$.

Case b). From (9.54) and (9.82) it follows

$$\|u' - \bar{u}'(\cdot - \bar{r})\|^2 > p^2 - \tilde{p}^2. \tag{9.83}$$

For simplicity we write $\bar{u}$ instead of $\bar{u}(\cdot - \bar{r})$ and set $v = u - \bar{u}$. Note that from (9.81), (9.82) and $\tilde{p} \leq p_0$ it follows that

$$|v(s)|^2 \leq 2 \int_{-\infty}^s |v(s)||v'(s)|ds \leq 2\|v\|\|v'\| \leq 4p_0^2.$$

We compute

$$\begin{aligned}
J(u) - \sigma &= \frac{1}{2}\|v'\|^2 + \int_{\mathbb{R}} \int_0^1 \Big(W_u(\bar{u} + \tau v) - W_u(\bar{u})\Big)v\,d\tau ds \\
&= \frac{1}{2}\|v'\|^2 + \int_{\mathbb{R}} \int_0^1 \int_0^1 \tau W_{uu}(\bar{u} + t\tau v)v \cdot v\big)d\tau dt ds.
\end{aligned} \tag{9.84}$$

Since

$$\left| \int_0^1 \int_0^1 \tau W_{uu}(\bar{u} + t\tau v)v \cdot v\big)vd\tau dt \right| \leq \frac{1}{2}C_W^0 |v|^2, \tag{9.85}$$

we have from (9.83) and (9.84)

$$J(u) - \sigma \geq \frac{1}{2}(p^2 - \tilde{p}^2) - \frac{1}{2}C_W^0 \tilde{p}^2 = \frac{1}{4}p^2. \tag{9.86}$$

This concludes the proof. $\qquad\square$

9.3.3   *Hamiltonian Identities and a Representation Formula for the Energy*

In this section we show that, by means of the Hamiltonian identities derived in Theorems 3.2 and 3.3 in Sect. 3.4, we have a special representation formula for the energy which is basic for our proof of Theorems 9.1, 9.2 and 9.3. These identities, considered also in [31] and [13], were noted (see Lemma 8.2 in [40]) but not exploited in [40]. The first identity, already considered in [29] and [4], generalizes to the present P.D.E. setting the classical theorem of conservation of mechanical energy. The other identity expresses an approximate orthogonality condition which does not have a finite-dimensional counterpart. We begin with an alternative proof of these identities which is based on the minimality of u.

Lemma 9.11 *Let $I \subset \mathbb{R}$ be an interval and assume $u : I \times \mathbb{R} \to \mathbb{R}^m$ is a minimizer that satisfies (9.20). Then there exist constants ω and $\tilde{\omega}$ such that, for $x \in I$,*

$$\int_{\mathbb{R}} \frac{1}{2} |u_x(x, y)|^2 dy = \int_{\mathbb{R}} \left(W(u(x, y)) + \frac{1}{2} |u_y(x, y)|^2 \right) dy - \sigma + \omega \qquad (9.87)$$

and

$$\int_{\mathbb{R}} u_x(x, y) \cdot u_y(x, y) dy = \tilde{\omega}, \quad \text{for } x \in I. \qquad (9.88)$$

Proof Given $[x_0, x_0 + L] \subset I$, let $g : [x_0, x_0 + L] \to [x_0, x_0 + L]$ be a continuous increasing surjection with inverse $\gamma : [x_0, x_0 + L] \to [x_0, x_0 + L]$. Define

$$v(s, y) = u(g(s), y), \quad \text{for } s \in [x_0, x_0 + L], \ y \in \mathbb{R}.$$

Then the energy $J_{\mathscr{R}_L(x_0)}(v)$ of v in the strip $\mathscr{R}_L(x_0) := (x_0, x_0 + L) \times \mathbb{R}$ is given by

$$J_{\mathscr{R}_L(x_0)}(v) = \int_{x_0}^{x_0+L} \int_{\mathbb{R}} \left(W(v(s, y)) + \frac{1}{2} (|v_s(s, y)|^2 + |v_y(s, y)|^2) \right) dyds$$

$$= \int_{x_0}^{x_0+L} \gamma'(x) \int_{\mathbb{R}} \left(W(u(x, y)) + \frac{1}{2} |u_y(x, y)|^2 \right) dydx$$

$$+ \int_{x_0}^{x_0+L} \frac{1}{\gamma'(x)} \int_{\mathbb{R}} \frac{1}{2} |u_x(x, y)|^2 dydx.$$

$$(9.89)$$

The minimality of u implies

$$J_{\mathscr{R}_L(x_0)}(v) \geq J_{\mathscr{R}_L(x_0)}(u), \qquad (9.90)$$

for all γ. Note that, since $\mathscr{R}_L(x_0)$ is unbounded, to state (9.90) we need to invoke Lemmas 9.1 and 9.2. By (9.90), if we set $\gamma(x) = x + \lambda f(x)$ with f an arbitrary C^1

function that satisfies $f(x_0) = f(x_0 + L) = 0$, we obtain

$$
\begin{aligned}
0 &= \frac{d}{d\lambda} J_{\mathscr{R}_L(x_0)}(v)\Big|_{\lambda=0} \\
&= \int_{x_0}^{x_0+L} \int_{\mathbb{R}} \left(W(u(x,y)) + \frac{1}{2}|u_y(x,y)|^2 - \frac{1}{2}|u_x(x,y)|^2 \right) f'(x)\,dy\,dx.
\end{aligned}
$$

$$(9.91)$$

Since f' is an arbitrary function with zero average and (9.91) holds for every x_0 and every $L > 0$ we obtain (9.87).

Let $g : [x_0, x_0 + L] \to \mathbb{R}$ be a C^1 function that satisfies

$$
g(x_0) = g(x_0 + L) = 0. \tag{9.92}
$$

Define

$$
v^\lambda(x,y) = u(x, y - \lambda g(x)), \qquad \text{for } x \in [x_0, x_0 + L], \ y \in \mathbb{R}, \ \lambda \in (-1, 1).
$$

Then

$$
\begin{aligned}
J_{\mathscr{R}_L(x_0)}(v^\lambda) &= \int_{x_0}^{x_0+L} \int_{\mathbb{R}} \left(W(v^\lambda) + \frac{1}{2}(|v_x^\lambda|^2 + |v_y^\lambda|^2) \right) dy\,dx \\
&= \int_{x_0}^{x_0+L} \int_{\mathbb{R}} \left(W(u) + \frac{1}{2}|u_x|^2 + (1 + \lambda^2|g'|^2)\frac{1}{2}|u_y|^2 - \lambda g' u_x \cdot u_y \right) dy\,dx.
\end{aligned}
$$

$$(9.93)$$

From (9.92) it follows that $v(x_0, \cdot) = u(x_0, \cdot)$ and $v(x_0 + L, \cdot) = u(x_0 + L, \cdot)$, which together with the minimality of u implies

$$
0 = \frac{d}{d\lambda} J_{\mathscr{R}_L(x_0)}(v^\lambda)|_{\lambda=0} = -\int_{x_0}^{x_0+L} g' \int_{\mathbb{R}} u_x \cdot u_y\,dy\,dx
$$

for all g' such that $\int_{x_0}^{x_0+L} g'\,dx = 0$. This proves (9.88) for some constant $\tilde{\omega}$. The proof is complete. $\qquad\square$

If $I = \mathbb{R}$ we have $\omega = \tilde{\omega} = 0$.

Lemma 9.12 *Assume* $u : \mathbb{R}^2 \to \mathbb{R}^m$ *is the minimizer in Theorem* 9.1. *Then*

$$
\int_{\mathbb{R}} \frac{1}{2}|u_x|^2\,dy = \int_{\mathbb{R}} \left(W(u) + \frac{1}{2}|u_y|^2 \right) dy - \sigma, \quad x \in \mathbb{R},
$$

$$
\Leftrightarrow \tag{9.94}
$$

$$\frac{1}{2}\|u_x(x,\cdot)\|^2 = J_{\mathbb{R}}(u(x,\cdot)) - \sigma, \quad x \in \mathbb{R}.$$

$$\int_{\mathbb{R}} u_x \cdot u_y \, dy = 0, \quad x \in \mathbb{R}, \tag{9.95}$$

Proof From (9.87) and (9.29) in Lemma 9.4 it follows that

$$\lim_{x \to +\infty} \int_{\mathbb{R}} \left(W(u) + \frac{1}{2}|u_y|^2 \right) dy - \sigma = -\omega \geq 0.$$

If $-\omega = |\omega| > 0$, than there exists x_ω such that

$$\int_{\mathbb{R}} \left(W(u) + \frac{1}{2}|u_y|^2 \right) dy \geq \sigma + \frac{|\omega|}{2}, \quad x \geq x_\omega.$$

and therefore (9.26) in Lemma 9.3 yields

$$(\sigma + \frac{|\omega|}{2})(x - x_\omega) \leq \int_{x_\omega}^{x} \int_{\mathbb{R}} \left(W(u) + \frac{1}{2}|u_y|^2 \right) dy \, dx$$

$$\leq J_{(x-x_\omega)\times\mathbb{R}}(u) \leq \sigma(x - x_\omega) + C_0, \quad x \geq x_\omega.$$

which is incompatible with the assumption $-\omega > 0$. This establishes (9.94). To prove (9.95), note that from (9.88) and (9.29) it follows

$$|\tilde{\omega}| = \lim_{x \to +\infty} | \int_{\mathbb{R}} u_x \cdot u_y \, dy | \leq C \lim_{x \to +\infty} \|u_x(x,\cdot)\| = 0,$$

where we have also used (9.21) and (9.19), which imply $\|u_y\| \leq C$. The proof is complete. $\qquad\square$

We can now derive a representation formula for the kinetic energy $\frac{1}{2}\int_{\mathbb{R}} |u_x|^2 dy$ of u valid in each interval I where $q(x) \leq q^0$ ensures the validity of the decomposition (9.38) for u.

Lemma 9.13 *Let $q^0 > 0$ the constant in Lemma 9.7 and let $I \subset \mathbb{R}$ an interval. Assume that u, the minimizer in Theorem 9.1, satisfies $0 < q(x) \leq q^0$ in I. Then we have*

$$h'(x) = \frac{\langle v_x(x,\cdot), v_y(x,\cdot) \rangle}{\|\bar{u}' + v_y(x,\cdot)\|^2} = \frac{q^2(x)\langle \mathbf{n}_x(x,\cdot), \mathbf{n}_y(x,\cdot) \rangle}{\|\bar{u}' + q(x)\mathbf{n}_y(x,\cdot)\|^2}, \tag{9.96}$$

and

$$\|u_x(x,\cdot)\|^2 = \|v_x(x,\cdot)\|^2 - \frac{\langle v_x(x,\cdot), v_y(x,\cdot) \rangle^2}{\|\bar{u}' + v_y(x,\cdot)\|^2}$$

$$= q_x^2(x) + q^2(x)\|\mathbf{n}_x(x,\cdot)\|^2 - q^4(x)\frac{\langle \mathbf{n}_x(x,\cdot), \mathbf{n}_y(x,\cdot) \rangle^2}{\|\bar{u}' + q(x)\mathbf{n}_y(x,\cdot)\|^2}. \tag{9.97}$$

Moreover, the map

$$(0, q(x)] \ni p \mapsto f(p, x)\|\mathbf{n}_x(x, \cdot)\|^2 := p^2\|\mathbf{n}_x(x, \cdot)\|^2 - p^4 \frac{\langle \mathbf{n}_x(x, \cdot), \mathbf{n}_y(x, \cdot)\rangle^2}{\|\bar{u}' + p\mathbf{n}_y(x, \cdot)\|^2}$$

is non-negative and nondecreasing for each fixed $x \in I$.

Proof From (9.38) we obtain

$$u_x(x, \cdot) = -h'(x)\Big(\bar{u}'(\cdot - h(x)) + v_y(x, \cdot - h(x))\Big) + v_x(x, \cdot - h(x)),$$

$$u_y(x, \cdot) = \bar{u}'(\cdot - h(x)) + v_y(x, \cdot - h(x)).$$

and therefore (9.95) in Lemma 9.12 and (9.42) imply

$$0 = \langle u_x(x, \cdot), u_y(x, \cdot)\rangle = -h'(x)(\|\bar{u}' + v_y(x, \cdot)\|^2 + \langle v_x(x, \cdot), v_y(x, \cdot)\rangle. \tag{9.98}$$

From (9.40) and (9.42) it follows that

$$\|v_y(x, \cdot)\|^2 \le \|v(x, \cdot)\|\|v_{yy}(x, \cdot)\| \le \frac{K_1}{\sqrt{k_1}}\|v(x, \cdot)\| \le \frac{K_1}{\sqrt{k_1}}q^0, \tag{9.99}$$

and in turn, since $q^0 > 0$ is small,

$$\frac{1}{2}\|\bar{u}'\| \le \|\bar{u}' + v_y(x, \cdot)\| \le \frac{3}{2}\|\bar{u}'\|. \tag{9.100}$$

Therefore (9.98) can be solved for h' and the first expression of h' in (9.96) is established. The other expression follows by (9.42), which implies $\langle v_x, v_y\rangle = \langle q_x\mathbf{n} + q\mathbf{n}_x, q\mathbf{n}_y\rangle = q^2\langle \mathbf{n}_x, \mathbf{n}_y\rangle$. A similar computation that also uses (9.96) yields (9.97).

It remains to prove the monotonicity of $f(p, \cdot)\|\mathbf{n}_x\|^2$. We can assume $\|\mathbf{n}_x\| > 0$, otherwise there is nothing to prove. We have, using also (9.99), (9.100) and $p \le q(x) \le q^0$,

$$D_p f(p, \cdot) = 2p - 4p^3 \frac{\langle \frac{\mathbf{n}_x}{\|\mathbf{n}_x\|}, \mathbf{n}_y\rangle^2}{\|\bar{u}' + p\mathbf{n}_y\|^2} + 2p^4 \frac{\langle \frac{\mathbf{n}_x}{\|\mathbf{n}_x\|}, \mathbf{n}_y\rangle^2 \langle \bar{u}' + p\mathbf{n}_y, \mathbf{n}_y\rangle}{\|\bar{u}' + p\mathbf{n}_y\|^4}$$

$$\ge p\Big(2 - p^2\|\mathbf{n}_y\|^2 \frac{16}{\|\bar{u}'\|^2} - p^3\|\mathbf{n}_y\|^3 \frac{16}{\|\bar{u}'\|^3}\Big)$$

$$\ge p\Big(2 - q^0 C^2 \frac{16}{\|\bar{u}'\|^2} - (q^0)^{\frac{3}{2}} C^3 \frac{16}{\|\bar{u}'\|^3}\Big).$$

This proves that $D_p f(p, \cdot) > 0$ for $q^0 \le \frac{\|\bar{u}'\|^2}{32C^2}$, as needed. $\qquad\square$

9.3.4   Completing the Proof of Theorem 9.1

From (9.94) and (9.29) it follows that there exists $x_0 \in \mathbb{R}$ such that

$$J_{\mathbb{R}}(u(x, \cdot)) - \sigma < \frac{e_{q^0}}{2}, \qquad x \geq x_0 \tag{9.101}$$

and Lemma 9.10 and the norm inequality $\|\cdot\| \leq \|\cdot\|_1$ imply

$$q(x) = \min_{j} \min_{r \in \mathbb{R}} \|u(x, \cdot) - \bar{u}_j(\cdot - r)\| < q^0, \qquad x \geq x_0. \tag{9.102}$$

Hence, Lemma 9.7 ensures the existence of a uniquely determined $\bar{u}_+ \in \{\bar{u}_1, \ldots, \bar{u}_N\}$ independent of $x \geq x_0$ and of a function $h : [x_0, +\infty) \to \mathbb{R}$ such that (9.38) holds with $\mathrm{u} = u$ and $\bar{u} = \bar{u}_+$.

Note that (9.102) implies $q(x) = \|v(x, \cdot)\| < q^0$ and therefore, for $x \geq x_0$, we have that $u(x, \cdot)$ remains in the *convex region* of the effective potential where (9.45) holds. We can expect that this implies exponential decay of $u(x, \cdot)$ to a translate of $\bar{u}_+$. We have indeed

Lemma 9.14 *There exist $k, C > 0$ and $x_+, \eta_+ \in \mathbb{R}$ such that*

$$q(x) = \|v(x, \cdot)\| \leq \sqrt{2} q^0 e^{-k(x - x_+)}, \qquad x \geq x_+. \tag{9.103}$$

and

$$\|u(x, \cdot) - \bar{u}_+(\cdot - \eta_+)\| \leq C(q^0)^{\frac{1}{2}} e^{-\frac{k}{2}(x - x_+)}, \qquad x \geq x_+. \tag{9.104}$$

Analogous statements hold true for the interval $(-\infty, x_-]$ for some $x_-, \eta_- \in \mathbb{R}$ and some $\bar{u}_- \in \{\bar{u}_1, \ldots, \bar{u}_N\}$.

Before giving the proof we remark on the different meaning of (9.103) and (9.104). Equation (9.103) says that, as $x \to +\infty$, $u(x, \cdot)$ converges exponentially to the manifold of the translates of $\bar{u}_+$, while (9.104) implies convergence to a specific element of that manifold.

Proof 1. There is $x_0 \in \mathbb{R}$ such that

$$\frac{d^2}{dx^2} \|v(x, \cdot)\|^2 \geq \frac{1}{2} \mu \|v(x, \cdot)\|^2, \qquad x \geq x_0, \tag{9.105}$$

where $\mu > 0$ is the constant in (9.45). To show this we begin with the elementary inequality

$$\frac{d^2}{dx^2} \|v(x, \cdot)\|^2 = \frac{d^2}{dx^2} \|u(x, \cdot) - \bar{u}_+(\cdot - h(x))\|^2$$

$$\geq 2 \left\langle \frac{d^2}{dx^2} \left(u(x, \cdot) - \bar{u}_+(\cdot - h(x)) \right), u(x, \cdot) - \bar{u}_+(\cdot - h(x)) \right\rangle. \tag{9.106}$$

From

$$\frac{d^2}{dx^2}\Big(u(x,\cdot) - \bar{u}_+(\cdot - h(x))\Big)$$

$$= u_{xx}(x,\cdot) - \bar{u}''_+(\cdot - h(x))(h'(x))^2 + \bar{u}'_+(\cdot - h(x))h''(x),$$

and (9.106), using also (9.96) (and $\langle \phi, \psi \rangle = \langle \phi(\cdot - r), \psi(\cdot - r) \rangle$), it follows that

$$\frac{d^2}{dx^2}\|v(x,\cdot)\|^2 \geq 2\langle u_{xx}(x,\cdot), u(x,\cdot) - \bar{u}_+(\cdot - h(x))\rangle$$

$$- 2\langle \bar{u}''_+, v(x,\cdot)\rangle \frac{\langle v_x(x,\cdot), v_y(x,\cdot)\rangle^2}{\|\bar{u}'_+ + v_y(x,\cdot)\|^4} = 2I_1 + 2I_2. \tag{9.107}$$

Since both u and $\bar{u}_+$ solve (9.15), we have

$$u_{xx}(x,\cdot) = W_u(u(x,\cdot)) - W_u(\bar{u}_+(\cdot - h(x))) - (u(x,\cdot) - \bar{u}_+(\cdot - h(x)))_{yy}.$$

Then, recalling the definition (9.10) of T and that $v(x,\cdot) = u(x,\cdot + h(x)) - \bar{u}_+$, after an integration by parts, we obtain

$$I_1 = \langle W_u(\bar{u}_+ + v(x,\cdot)) - W_u(\bar{u}_+) - v_{yy}(x,\cdot), v(x,\cdot)\rangle$$

$$= \langle W_u(\bar{u}_+ + v(x,\cdot)) - W_u(\bar{u}_+), v(x,\cdot)\rangle + \|v_y(x,\cdot)\|^2$$

$$= \langle W_u(\bar{u}_+ + v(x,\cdot)) - W_u(\bar{u}_+) - W_{uu}(\bar{u}_+)v(x,\cdot), v(x,\cdot)\rangle + \langle Tv(x,\cdot), v(x,\cdot)\rangle,$$

whence, in conjunction with, (9.44) and (9.47),

$$I_1 \geq 2\mathscr{W}(v(x,\cdot)) - C\|v\|^{\frac{8}{3}}, \quad x \geq x_0. \tag{9.108}$$

To estimate I_2 we note that for $q^0 > 0$ small (9.99) implies (9.100) (with $\bar{u} = \bar{u}_+$) and

$$\frac{\langle v_x(x,\cdot), v_y(x,\cdot)\rangle^2}{\|\bar{u}'_+ + v_y(x,\cdot)\|^2} \leq \frac{1}{2}\|v_x(x,\cdot)\|^2, \quad x \geq x_0. \tag{9.109}$$

Then (9.97) and (9.94) imply

$$\|v_x(x,\cdot)\|^2 \leq 4\mathscr{W}(v(x,\cdot)), \quad x \geq x_0, \tag{9.110}$$

and we obtain

$$I_2 \leq C\|v(x,\cdot)\|\mathscr{W}(v(x,\cdot)), \quad x \geq x_0.$$

From this and (9.108), using also (9.46) and $\|v(x, \cdot)\| \leq q^0$, we conclude that

$$I_1 + I_2 \geq (2 - C\|v(x, \cdot)\|)\mathscr{W}(v(x, \cdot)) - C\|v(x, \cdot)\|^{\frac{8}{3}}$$

$$\geq \frac{1}{4}\mu\|v(x, \cdot)\|^2, \quad x \geq x_0,$$

and (9.105) follows from (9.107).

2. Since by (9.102) we have $\|v(x, \cdot)\| \leq q^0$ for $x \geq x_0$, from 1. and the maximum principle we get, for every $l > 0$,

$$\|v(x, \cdot)\|^2 \leq \varphi_l(x), \quad x \in [x_0, x_0 + 2l], \tag{9.111}$$

where

$$\varphi_l(x) := (q^0)^2 \frac{\cosh\sqrt{\frac{\mu}{2}}(l - (x - x_0))}{\cosh\sqrt{\frac{\mu}{2}}l}, \quad x \in (x_0, x_0 + 2l),$$

is the solution of the problem

$$\begin{cases} \varphi'' = \frac{\mu}{2}\varphi, \quad x \in (x_0, x_0 + 2l), \\ \\ \varphi(x_0) = \varphi(x_0 + 2l) = (q^0)^2. \end{cases}$$

The estimate (9.103), with $k = \frac{1}{2}\sqrt{\frac{\mu}{2}}$ and $x_+ = x_0$, follows from (9.111) and the inequality $\varphi_l(x) \leq 2(q^0)^2 e^{-\sqrt{\frac{\mu}{2}}(x - x_0)}$, $x \in [x_0, x_0 + l]$, which holds for all $l > 0$.

3. Thus, we have that

$$|h'(x)| \leq C\|v(x, \cdot)\|^{\frac{1}{2}}, \quad x \geq x_0. \tag{9.112}$$

From (9.110) and (9.101), which implies $\mathscr{W}(v(x, \cdot)) \leq e_{q^0}$, we have

$$\|v_x(x, \cdot)\|^2 \leq 4e_{q^0}, \quad x \geq x_0.$$

Then 3. follows from (9.96), (9.99) and (9.100).

4. Point 3. and (9.103) imply

$$|h'(x)| \leq C(q^0)^{\frac{1}{2}} e^{-\frac{k}{2}(x - x_+)}, \quad x \geq x_+.$$

It follows that

$$|\eta_+ - h(x)| \leq C(q^0)^{\frac{1}{2}} e^{-\frac{k}{2}(x - x_+)}, \quad x \geq x_+, \tag{9.113}$$

for some $\eta_+ \in \mathbb{R}$. Since h is bounded there is $C > 0$ such that

$$\|\bar{u}_+(\cdot - (h(x) - \eta_+)) - \bar{u}_+\| \le C|h(x) - \eta_+|. \tag{9.114}$$

The inequality (9.104) follows from (9.114), (9.113), (9.103), and

$$\begin{aligned}
\|u(x, \cdot) &- \bar{u}_+(\cdot - \eta_+)\| \\
&\le \|u(x, \cdot) - \bar{u}_+(\cdot - h(x))\| + \|\bar{u}_+(\cdot - h(x)) - \bar{u}_+(\cdot - \eta_+)\| \\
&= \|v(x, \cdot)\| + \|\bar{u}_+(\cdot - (h(x) - \eta_+)) - \bar{u}_+\|.
\end{aligned}$$

The proof is complete. $\qquad\square$

Lemma 9.14 concludes the proof of Theorem 9.1. Indeed we have already established $(9.12)_1$ in (9.22) and that $(9.12)_2$ follows from (9.104) and Lemma 9.8.

9.4 The Proof of Theorem 9.2

From the proof of Theorem 9.1 we know that $u(x, \cdot)$ remains in a q^0 neighborhood of $\bar{u}(\cdot - \eta_-)$ in $(-\infty, x_-]$ and of $\bar{u}(\cdot - \eta_+)$ in $[x_+, +\infty)$. The problem is to analyze what happens in the interval (x_-, x_+). We prove that for $u(x, \cdot)$ is more convenient to remain near the manifold of the translates of $\bar{u}$ also in (x_-, x_+). Indeed we show that to travel away from this manifold and come back to it is more penalizing from the point of view of minimizing the energy.

In the following, for x in certain subintervals of (x_-, x_+), we use test functions of the form

$$\hat{u}(x, y) = \bar{u}(y - \hat{h}(x)) + \hat{q}(x)\mathbf{n}(x, y - \hat{h}(x)) \tag{9.115}$$

for suitable choices of the functions $\hat{q} = \hat{q}(x)$ and $\hat{h} = \hat{h}(x)$. We always take $\hat{q}(x) \le q(x) \le q^0$. Note that in (9.115) the direction vector $\mathbf{n}(x, \cdot)$ is the one associated to $v(x, \cdot) = q(x)\mathbf{n}(x, \cdot)$, with $v(x, \cdot)$ defined in the decomposition (9.38) of u.

From (9.115) it follows that

$$\int_{\mathbb{R}} |\hat{u}_x|^2 dy = (\hat{h}')^2 \|\bar{u}' + \hat{q}\mathbf{n}_y\|^2 - 2\hat{h}'\hat{q}^2\langle\mathbf{n}_x, \mathbf{n}_y\rangle + \hat{q}_x^2 + \hat{q}^2\|\mathbf{n}_x\|^2. \tag{9.116}$$

We choose the value of $\hat{h}$ that minimizes (9.116), that is,

$$\hat{h}' = \hat{q}^2\frac{\langle\mathbf{n}_x, \mathbf{n}_y\rangle}{\|\bar{u}' + \hat{q}\mathbf{n}_y\|^2}, \tag{9.117}$$

and then we get $\int_{\mathbb{R}} |\hat{u}_x|^2 dy = \hat{q}_x^2 + \hat{q}^2 \|\mathbf{n}_x\|^2 - \hat{q}^4 \frac{\langle \mathbf{n}_x, \mathbf{n}_y \rangle^2}{\|\bar{u}' + \hat{q}\mathbf{n}_y\|^2}$ Therefore, the energy density of the test map $\hat{u}$ is given by

$$\int_{\mathbb{R}} \frac{1}{2} |\hat{u}_x|^2 dy + \int_{\mathbb{R}} \left(W(\hat{u}) + \left(\frac{1}{2} |\hat{u}_y|^2 \right) dy - \sigma \right) + \sigma$$

$$= \frac{1}{2} \left(\hat{q}_x^2 + \hat{q}^2 \|\mathbf{n}_x\|^2 - \hat{q}^4 \frac{\langle \mathbf{n}_x, \mathbf{n}_y \rangle^2}{\|\bar{u}' + \hat{q}\mathbf{n}_y\|^2} \right) + \mathscr{W}(\hat{q}, \mathbf{n}) + \sigma.$$

Note that, since we do not change the direction vector $\mathbf{n}(x, \cdot)$, this expression is completely determined once we fix the function $\hat{q}$. Note also that, since we use this expression only when computing differences of energy, the constant σ can be disregarded.

Lemma 9.15 *Let $I \subset \mathbb{R}$ be an interval and assume that the minimizer $u : \mathbb{R}^2 \to \mathbb{R}^m$ satisfies*

$$q(x) \leq q^0, \quad x \in I.$$

Then the map $x \mapsto q(x)$ cannot have points of maximum in I, meaning that there are no $x_1 < x^* < x_2 \in I$ such that

$$q(x_i) < q(x^*), \quad i = 1, 2.$$

Proof Assume instead that $x_1 < x^* < x_2 \in I$ with $q(x_i) < q(x^*)$, $i = 1, 2$ exist. Since $q = q(x)$ is continuous, we can assume $q(x^*) = \max_{x \in [x_1, x_2]} q(x)$ and, by restricting the interval (x_1, x_2) if necessary, that

$$q_0 = q(x_i) < q(x) \leq q(x^*), \quad i = 1, 2, \ x \in (x_1, x_2)$$

for some $q_0 \in (0, q^0)$ that satisfies the condition $q(x^*) \leq 2q_0$.
We show that this contradicts the minimality of u by constructing a competing function $\tilde{u}$ defined as follows: in the interval $(-\infty, x_1)$ we take

$$\tilde{u}(x, \cdot) = u(x, \cdot), \quad \text{for } x \in (-\infty, x_1). \tag{9.118}$$

In the interval $[x_1, x_2]$ we take:

$$\tilde{u}(x, \cdot) = \hat{u}(x, \cdot), \quad \text{with } \hat{q}(x) = 2q_0 - q(x), \quad \text{for } x \in [x_1, x_2], \tag{9.119}$$

where $\hat{u}$ is defined in (9.115) with $\hat{q} = 2q_0 - q$ and $\hat{h}$ the solution of (9.117) with initial condition $\hat{h}(x_1) = h(x_1)$. With this definition $\tilde{u}$ is continuous at $x = x_1$. Indeed, since $\hat{q}(x_1) = 2q_0 - q(x_1) = q_0 = q(x_1)$, we have

$$\hat{u}(x_1, y) = \bar{u}(y - \hat{h}(x_1)) + \hat{q}(x_1)\mathbf{n}(x, y - \hat{h}(x_1))$$

$$= \bar{u}(y - h(x_1)) + q(x_1)\mathbf{n}(x, y - h(x_1)) = u(x_1, y).$$

For $x = x_2$ we have instead

$$\hat{u}(x_2, y) = \bar{u}(y - \hat{h}(x_2)) + \hat{q}(x_2)\mathbf{n}(x, y - \hat{h}(x_2))$$

$$= \bar{u}(y - \hat{h}(x_2)) + q(x_2)\mathbf{n}(x, y - \hat{h}(x_2))$$

$$= \bar{u}(y - h(x_2) - (\hat{h}(x_2) - h(x_2))) + q(x_2)\mathbf{n}(x, y - h(x_2) - (\hat{h}(x_2) - h(x_2)))$$

$$= u(x_2, y - (\hat{h}(x_2) - h(x_2))).$$

That is, at $x = x_2$, the function $\hat{u}(x_2, \cdot)$ coincides with the translate $u(x_2, \cdot - (\hat{h}(x_2) - h(x_2)))$ of $u(x_2, \cdot)$ where

$$\hat{h}(x_2) - h(x_2) = \int_{x_1}^{x_2} (\hat{h}' - h')dx = \int_{x_1}^{x_2} \left(\frac{(2q_0 - q)^2 \langle \mathbf{n}_x, \mathbf{n}_y \rangle}{\|\bar{u}' + (2q_0 - q)\mathbf{n}_y\|^2} - \frac{q^2 \langle \mathbf{n}_x, \mathbf{n}_y \rangle}{\|\bar{u}' + q\mathbf{n}_y\|^2} \right)dx.$$

To compensate for this translation, it is natural to complete the definition of $\tilde{u}$ by setting

$$\tilde{u}(x, \cdot) = u\left(x, \cdot - (\hat{h}(x_2) - h(x_2))(1 - \frac{x - x_2}{l})\right), \quad x \in (x_2, x_2 + l],$$

$$\tilde{u}(x, \cdot) = u(x, \cdot), \quad x \in (x_2 + l, +\infty) \tag{9.120}$$

so that $\tilde{u}(x, \cdot)$ is continuous at $x_2 + l$ and coincides with $u(x, \cdot - (\hat{h}(x_2) - h(x_2)))$ for $x = x_2$. The idea here is that, for large $l > 0$, the contribution of the interval $(x_2, x_2 + l)$ to the difference of energy between u and $\tilde{u}$ is negligible compared to the contribution of the interval (x_1, x_2). Proceeding as in the proof of Lemma 9.11 and using the identity (9.95), we obtain

$$J_{(x_2, x_2 + l) \times \mathbb{R}}(\tilde{u}) - J_{(x_2, x_2 + l) \times \mathbb{R}}(u)$$

$$= \int_{x_2}^{x_2 + l} \frac{(\hat{h}(x_2) - h(x_2))^2}{l^2} \int_{\mathbb{R}} \frac{1}{2} |u_y|^2 dy dx \leq C \frac{(\hat{h}(x_2) - h(x_2))^2}{l}, \tag{9.121}$$

where we have also used (9.21) and (9.19).

By the definition (9.118), (9.119) and (9.120) of $\tilde{u}$, in (x_1, x_2) it holds that $\hat{q}_x = (2q_0 - q)_x = -q_x$, and therefore

$$\hat{q}_x^2 = q_x^2. \tag{9.122}$$

Moreover from (9.99) and Lemma 9.13 it follows that

$$f(\hat{q}(x)) \leq f(q(x)), \quad x \in (x_1, x_2).$$

Using this, (9.122), and (9.45) in Lemma 9.9 which implies the strict monotonicity of the map $q \to \mathscr{W}(q, \mathbf{n})$, we conclude that

$$J_{(x_1,x_2)\times\mathbb{R}}(\tilde{u}) < J_{(x_1,x_2)\times\mathbb{R}}(u).$$

This inequality and (9.121), imply that, for $l > 0$ sufficiently large,

$$J_{(x_1,x_2+l)\times\mathbb{R}}(u) - J_{(x_1,x_2+l)\times\mathbb{R}}(\tilde{u})$$

$$\geq J_{(x_1,x_2)\times\mathbb{R}}(u) - J_{(x_1,x_2)\times\mathbb{R}}(\tilde{u}) - C\frac{(\hat{h}(x_2) - h(x_2))^2}{l} > 0,$$

in contradiction with the minimality of u. The proof is complete. $\square$

Remark 9.4 Later we will consider a situation where the minimizer u is defined in a bounded strip $[0, L] \times \mathbb{R}$ and satisfies a boundary condition of the form $u(L, \cdot) = \bar{u}_+(\cdot - \eta)$, where $\eta \in \mathbb{R}$ is a free parameter. In this situation the conclusion of Lemma 9.15 still applies with a simpler proof. Indeed, the competing map $\tilde{u}$ can be defined exactly as in (9.118) in the interval $[0, x_1]$ and as in (9.119) in the interval (x_1, x_2) and, since $\eta \in \mathbb{R}$ can be chosen freely, by simply setting

$$\tilde{u}(x, \cdot) = u(x, \cdot - (\hat{h}(x_2) - h(x_2))), \quad x \in [x_2, L].$$

From Lemmas 9.15 and 9.14 it follows that, under the assumption $\bar{u}_- = \bar{u}_+ = \bar{u}$, if u does not satisfy (9.13), then

$$\{x \in \mathbb{R} : q(x) > q^0\} \neq \emptyset. \tag{9.123}$$

Indeed, since $\lim_{x\to\pm\infty} q(x) = 0$ by Lemma 9.14, if $q(x) \leq q^0$ for all $x \in \mathbb{R}$, then Lemma 9.15 implies $q(x) \equiv 0$, and by (9.96), $h'(x) \equiv 0$, and we conclude that (9.13) holds.

We show that (9.123) cannot occur by constructing a map that competes energetically with u. Our construction is inspired by the proof of Lemma 2.5 (see also Lemma 3.4 in [5]). We fix a point $x^* \in \{x \in \mathbb{R} : q(x) > q^0\}$ and focus on the intervals $[\tilde{\xi}_1, \tilde{\xi}_2] \subset (\xi_1, \xi_2)$ defined by

$$\tilde{\xi}_1 = \min\{x < x^* : q(x) \geq q^0\}, \ \xi_1 = \max\{x < \tilde{\xi}_1 : q(x) \leq \tfrac{q^0}{2}\},$$

$$\tilde{\xi}_2 = \max\{x > x^* : q(x) \geq q^0\}, \ \xi_2 = \min\{x > \tilde{\xi}_2 : q(x) \leq \tfrac{q^0}{2}\}.$$

Note that

$$q(\tilde{\xi}_1) = q(\tilde{\xi}_2) = q^0,$$

$$q(\xi_1) = q(\xi_2) = \frac{q^0}{2}, \tag{9.124}$$

and also that

$$q(x) \in (\frac{q^0}{2}, q^0), \quad \text{for} \quad x \in (\xi_1, \tilde{\xi}_1) \cup (\tilde{\xi}_2, \xi_2). \tag{9.125}$$

We define the competing map $\tilde{u}$. For $x \in (-\infty, \xi_1)$ we take $\tilde{u}(x, \cdot) = u(x, \cdot)$. In $[\xi_1, \tilde{\xi}_1] \cup [\tilde{\xi}_2, \xi_2]$ we set $\tilde{u}(x, \cdot) = \hat{u}(x, \cdot)$, with $\hat{q} = \hat{q}(x)$ and $\hat{h} = \hat{h}(x)$ defined as follows. We take

$$\hat{q}(x) = q^0 - q(x), \quad x \in [\xi_1, \tilde{\xi}_1] \cup [\tilde{\xi}_2, \xi_2].$$

Note that (9.124) implies that $\hat{q}$ extends q continuously at $x = \xi_1$ and $x = \xi_2$, and moreover that $\hat{q}(x) \in [0, \frac{q^0}{2}]$ for $x \in [\xi_1, \tilde{\xi}_1] \cup [\tilde{\xi}_2, \xi_2]$. In the interval $[\xi_1, \tilde{\xi}_1]$ we let $\hat{h}$ be the solution of (9.117) with initial condition $\hat{h}(\xi_1) = h(\xi_1)$. In the interval $[\tilde{\xi}_2, \xi_2]$ again we take the solution of (9.117) with initial condition $\hat{h}(\tilde{\xi}_2) = \hat{h}(\tilde{\xi}_1)$. It remains to specify $\tilde{u}(x, \cdot)$ for $x \in (\tilde{\xi}_1, \tilde{\xi}_2) \cup [\xi_2, +\infty)$. We take

$$\tilde{u}(x, \cdot) = \bar{u}(\cdot - \hat{h}(\tilde{\xi}_1)), \quad x \in (\tilde{\xi}_1, \tilde{\xi}_2),$$

$$\tilde{u}(x, \cdot) = u\left(x, \cdot - (\hat{h}(\xi_2) - h(\xi_2))(1 - \frac{x - \xi_2}{l})\right), \quad x \in [\xi_2, \xi_2 + l],$$

$$\tilde{u}(x, \cdot) = u(x, \cdot), \quad x \in (\xi_2 + l, +\infty).$$

With these definitions, one checks that $x \mapsto \tilde{u}(x)$ is continuous and piece-wise smooth and coincides with $u(x, \cdot)$ outside $(\xi_1, \xi_2 + l)$. Arguing as in the proof of Lemma 9.15 we show that

$$J_{(\xi_2, \xi_2+l) \times \mathbb{R}}(\tilde{u}) - J_{(\xi_2, \xi_2+l) \times \mathbb{R}}(u) \leq C \frac{(\hat{h}(\xi_2) - h(\xi_2))^2}{l}, \tag{9.126}$$

and

$$J_{(\xi_1, \tilde{\xi}_1) \cup (\tilde{\xi}_2, \xi_2) \times \mathbb{R}}(\tilde{u}) < J_{(\xi_1, \tilde{\xi}_1) \cup (\tilde{\xi}_2, \xi_2) \times \mathbb{R}}(u),$$

$$\sigma(\tilde{\xi}_2 - \tilde{\xi}_1) = J_{[\tilde{\xi}_1, \tilde{\xi}_2] \times \mathbb{R}}(\tilde{u}) < J_{[\tilde{\xi}_1, \tilde{\xi}_2] \times \mathbb{R}}(u).$$

Therefore, for $l > 0$ large, we obtain $J_{(\xi_1, \xi_2+l) \times \mathbb{R}}(\tilde{u}) < J_{(\xi_1, \xi_2+l) \times \mathbb{R}}(u)$. This contradicts the minimality of u and concludes the proof of Theorem 9.2.

Remark 9.5 Under the assumptions of Theorem 9.2, by means of the weighted Hamiltonian identity introduced in Remark 3.6 (here rewritten with different notation),

$$F(u, x) = \int_{\mathbb{R}} y\left(W(u(x, y)) + \frac{1}{2}(|u_y(x, y)|^2 - |u_x(x, y))|^2\right)dy = \bar{\omega}, \tag{9.127}$$

one easily shows that

$$\eta_+ = \eta_- = \eta, \tag{9.128}$$

for some $\eta \in \mathbb{R}$. To prove (9.128) note that (9.12), (9.20) and the Lebesgue dominated convergence theorem imply that, along a sequence $x_j^\pm \to \pm\infty$, and using also (9.29), we have

$$\lim_{j \to \pm\infty} F(u, x_j^\pm) = \int_{\mathbb{R}} y\left(W(\bar{u}(y - \eta_\pm)) + \frac{1}{2}(|\bar{u}'(y - \eta_\pm)|^2\right)dy$$

$$= \int_{\mathbb{R}} (y + \eta_\pm)\left(W(\bar{u}(y)) + \frac{1}{2}(|\bar{u}'(y)|^2\right)dy = \bar{\omega}.$$

Therefore,

$$\eta_\pm = \frac{\bar{\omega} - \int_{\mathbb{R}} y\left(W(\bar{u}) + \frac{1}{2}|\bar{u}'|^2\right)dy}{\sigma}.$$

From (9.128) and (9.12) it is natural to expect the validity of Theorem 9.2, that is, $h(x) = \eta$ for all $x \in \mathbb{R}$, or equivalently, $q(x) = 0$, $x \in \mathbb{R}$. However, to deduce this from (9.128) is not straightforward. The problem is that one cannot take advantage of the equality $\lim_{x \to \pm\infty} q(x) = 0$, which follows from (9.12), by simply reducing $q(x)$ in some large interval. Indeed, reducing $q(x)$ without controlling at the same time the translation of $\bar{u}$ in the y direction may result in an increase of the energy.

9.5   Proof of Theorem 9.3

We begin with a description of the proof and discuss how we can overcome certain difficulties that hinder a straightforward application of the direct method of variational calculus. These are: loss of compactness due to translation invariance in the x and y directions and the fact that the solution we are looking for has infinite energy. To deal with these obstructions, we consider a bounded strip $\mathscr{R}_L = \{(x, y) : x \in (0, L); y \in \mathbb{R}\}$, $L > 1$ and, for each $\eta \in \mathbb{R}$, consider the problem

$$\min_{\mathscr{A}_{L,\eta}} J(u), \quad J(u) = \int_{\mathscr{R}_L} \left(W(u) + \frac{1}{2}|\nabla u|^2\right)dxdy, \tag{9.129}$$

$$\mathscr{A}_{L,\eta} = \{u \in W_{\text{loc}}^{1,2}(\mathscr{R}_L; \mathbb{R}^m) : u(0, \cdot) = \bar{u}_-, \; u(L, \cdot) = \bar{u}_+(\cdot - \eta)\}.$$

Working in a bounded strip with imposed Dirichlet conditions removes at once the difficulties mentioned above. But, on the other hand, raises the problem of understanding the relationship between minimizers $u^{L,\eta}$ of (9.129) and the solution

u we are looking for. We regard the minimization problem (9.129) as a first step where we require the minimizer to connect two given elements of the manifold of the translates of $\bar{u}_+$. We note in passing that what actually matters is the difference $\eta = \eta_+ - \eta_-$ rather than the values of η_- and η_+ separately. Indeed, a translation in the y direction reduces the problem to the case considered in (9.129).

By means of the cut-off Lemma 4.1 (see also section 2.2 in [6]) we show that the admissible set $\mathscr{A}_{L,\eta}$ in (9.129) can be restricted to maps converging to $a^\pm$ as $y \to \pm\infty$ with a well controlled rate. Then standard arguments imply that, given $L > 1$, there exist a minimizer $u^{L,\eta} \in \mathscr{A}_{L,\eta}$ of problem (9.129) for each $\eta \in \mathbb{R}$ and a map u^L that satisfies the condition

$$J(u^L) = \min_{\eta \in \mathbb{R}} J(u^{L,\eta}), \tag{9.130}$$

which we impose to determine the value of η. This yields a family of maps u^L, $L > 1$, which for large L are expected to be good approximations of a translate of the sought solution u in Theorem 9.3. Therefore, we expect that

$$u(x, y) = \lim_{L_j \to +\infty} u^{L_j}(x - l_j, y), \tag{9.131}$$

for suitable sequences L_j, l_j, $j = 1, 2, \ldots$. To show that this is indeed the case we derive precise point-wise estimates on u^L.

9.5.1 *Existence of the Minimizers $u^{L,\eta}$ and u^L*

We start by showing that, in the minimization problem (9.129), we can restrict to the subset of maps $u \in \mathscr{A}_{L,\eta}$ that satisfy

$$\|u\|_{L^\infty(\mathscr{R}_L;\mathbb{R}^m)} \leq M, \tag{9.132}$$

where $M > 0$ is the constant in (9.14). Indeed, given $u \in \mathscr{A}_{L,\eta}$, set $u_M = 0$ if $u = 0$ and $u_M = \min\{|u|, M\}u/|u|$ otherwise, and note that (9.14) implies

$$W(u_M) \leq W(u), \quad \text{a.e.}$$

while we have

$$|\nabla u_M| \leq |\nabla u|, \quad \text{a.e.}$$

since the mapping $u \mapsto u_M$ is a projection. It follows that

$$J(u) - J(u_M) = \int_{|u| \geq M} \left(W(u) - W(u_M) + \frac{1}{2}(|\nabla u|^2 - |\nabla u_M|^2) \right) \geq 0,$$

which proves the claim. We now show that in the minimization problem (9.129) we can assume that

$$J(\mathrm{u}) \le J(\tilde{u}^{L,\eta}) < +\infty, \tag{9.133}$$

for a suitable map $\tilde{u}^{L,\eta} \in \mathscr{A}_{L,\eta}$. As before, we set $\sigma = J_{\mathbb{R}}(\bar{u}_\pm)$.

Lemma 9.16 *There exist $\tilde{u}^{L,\eta} \in \mathscr{A}_{L,\eta}$ and $C_0 > 0$ such that*

$$J(\tilde{u}^{L,\eta}) \le C_0(1 + |\eta|) + \sigma L. \tag{9.134}$$

Proof For $L > 1$ and $\eta \in \mathbb{R}$, define $\tilde{u}^{L,\eta} : [0, L] \times \mathbb{R}$ by setting

$$\tilde{u}^{L,\eta}(x, y) = \begin{cases} (1 - x)\bar{u}_-(y) + x\bar{u}_+(y - \eta), & \text{for} \quad (x, y) \in [0, 1] \times \mathbb{R}, \\ \bar{u}_+(y - \eta), & \text{for} \quad (x, y) \in (1, L] \times \mathbb{R}. \end{cases} \tag{9.135}$$

Then, since both u_- and u_+ satisfy (9.19), one checks that the energy of $\tilde{u}^{L,\eta}$ in the strip $(0, 1) \times \mathbb{R}$ is the sum of a term proportional to $|\eta|$ plus a constant and (9.134) follows. The proof is complete.

Next we show that we can further restrict $\mathscr{A}_{L,\eta}$ to the set of maps that converges uniformly to $a^\pm$ as $y \to \pm\infty$. $\qquad\qquad\square$

Lemma 9.17 *In the minimization problem (9.129), the admissible set $\mathscr{A}_{L,\eta}$ can be restricted to the subset of the maps $\mathrm{u} \in \mathscr{A}_{L,\eta}$ that satisfy (9.132), (9.133) and*

$$|\mathrm{u}(x, y) - a^+| \le \frac{C_{L,\eta}}{\sqrt{y}}, \quad \text{for} \quad y \ge y_{L,\eta},$$

$$\tag{9.136}$$

$$|\mathrm{u}(x, y) - a^-| \le \frac{C_{L,\eta}}{\sqrt{-y}}, \quad \text{for} \quad y \le -y_{L,\eta}$$

for some constants $C_{L,\eta} > 0$, $y_{L,\eta} > 0$.

Proof From (9.19) we have

$$|\bar{u}_-(y) - a^+| \le \frac{r}{4}, \quad \text{for} \quad y \ge y_r,$$

$$\tag{9.137}$$

$$|\bar{u}_+(y - \eta) - a^+| \le \frac{r}{4}, \quad \text{for} \quad y \ge y_r + \eta$$

with $y_r = \frac{1}{k} \ln \frac{4\bar{K}}{r}$. Assume now $r \in (0, r_0]$, with r_0 the constant in (9.56), and define

$$Y_r := \{y \ge y_r + \max\{0, \eta\} : |\mathrm{u}(x_y, y) - a^+| \ge \frac{r}{2}, \text{ for some } x_y \in (0, L)\}.$$

Then, for $y \in Y_r$, we have

$$|\mathrm{u}(x_y, y) - \bar{u}_-(y)| \geq |\mathrm{u}(x_y, y) - a^+| - |\bar{u}_-(y) - a^+| \geq \frac{r}{4}.$$

It follows, recalling also the boundary condition $\mathrm{u}(0, \cdot) = \bar{u}_-$, that

$$\frac{r}{4} \leq |\mathrm{u}(x_y, y) - \bar{u}_-(y)| = |\mathrm{u}(x_y, y) - \mathrm{u}(0, y)| \leq L^{\frac{1}{2}} \Big(\int_0^L |\mathrm{u}_x(x, y)|^2 dx \Big)^{\frac{1}{2}},$$

and therefore $|Y_r| \frac{r^2}{16L} \leq \int_0^L \int_{Y_r} |\mathrm{u}_x(x, y)|^2 dx dy \leq 2J(\tilde{u}^{L,\eta})$, that is

$$|Y_r| \leq \frac{C'_{L,\eta}}{r^2}, \quad \text{with } C'_{L,\eta} = 32LJ(\tilde{u}^{L,\eta}).$$

It follows that there is an increasing sequence $y_{r,j}$, $j = 1, 2, \ldots$, that diverges to $+\infty$ and satisfies

$$y_{r,1} \leq y_r + \max\{0, \eta\} + \frac{C'_{L,\eta}}{r^2}, \tag{9.138}$$

$$y_{r,j} \in \mathbb{R} \setminus Y_r.$$

This and (9.137) imply that

$$|\mathrm{u}(x, y) - a^+| \leq \frac{r}{2}, \quad \text{on } \partial R_j,$$

where $R_j = (0, L) \times (y_{r,j}, y_{r,j+1})$, $j = 1, \ldots$
We can then invoke the cut-off Lemma 4.1 and conclude the existence of a map $\tilde{u} \in \mathscr{A}_{L,\eta}$ that coincides with u for $y \leq y_{r,1}$ and satisfies

$$|\tilde{u}(x, y) - a^+| \leq \frac{r}{2}, \quad \text{for } x \in [0, L], \ y \geq y_{r,1} \tag{9.139}$$

and

$$J(\tilde{u}) \leq J(\mathrm{u}),$$

with strict inequality whenever $|Y_r| > 0$. Therefore in the minimization problem (9.129) we are allowed to assume that $\mathrm{u} \in \mathscr{A}_{L,\eta}$ satisfies (9.139). By increasing the value of $C'_{L,\eta}$ if necessary, we can assume that

$$y_r + \max\{0, \eta\} \leq \frac{C'_{L,\eta}}{r^2}, \quad \text{for } r \in (0, r_0].$$

Then $y = \frac{2C'_{L,\eta}}{r^2}$ implies $y \geq y_{r,1}$ and therefore from (9.139) it follows that

$$|\mathbf{u}(x, y) - a^+| \leq \sqrt{\frac{C'_{L,\eta}}{2y}}, \quad \text{for} \quad y \geq \frac{2C'_{L,\eta}}{r_0^2}. \tag{9.140}$$

This proves $(9.136)_1$ with $C_{L,\eta} = \sqrt{\frac{C'_{L,\eta}}{2}}$ and $y_{L,\eta} = \frac{2C'_{L,\eta}}{r_0^2}$. The other inequality is proved in a similar way. $\qquad\square$

We are now in the position to prove the existence of the minimizers $u^{L,\eta}$ and u^L of problems (9.129) and (9.130).

Lemma 9.18 *There exists $u^{L,\eta} \in \mathscr{A}_{L,\eta}$ that solves problem (9.129)*

$$J(u^{L,\eta}) = \min_{\mathbf{u} \in \mathscr{A}_{L,\eta}} J(\mathbf{u}).$$

Moreover $u^{L,\eta}$ satisfies (9.132) and (9.136).

Proof By Lemma 9.16,

$$0 \leq \inf_{\mathbf{u} \in \mathscr{A}_{L,\eta}} J(\mathbf{u}) \leq J(\tilde{u}^{L,\eta}) < +\infty. \tag{9.141}$$

Let $\{\mathbf{u}_j\}_{j=1}^{\infty} \subset \mathscr{A}_{L,\eta}$ be a minimizing sequence. By Lemma 9.17 and the discussion above, we can assume that $\mathbf{u}_j$ satisfies (9.132) and (9.136). From (9.141) we have

$$\int_{\mathscr{R}_L} \frac{1}{2}|\nabla \mathbf{u}_j|^2 dxdy \leq J(\mathbf{u}_j) \leq J(\tilde{u}^{L,\eta}).$$

Hence, using also that $\|\mathbf{u}_j\|_{L^\infty(\mathscr{R}_L;\mathbb{R}^m)} \leq M$, weak compactness ensures that, possibly by passing to a subsequence,

$$\mathbf{u}_j \rightharpoonup u, \quad \text{in } W^{1,2}_{\mathrm{loc}}(\mathscr{R}_L; \mathbb{R}^m),$$

for some $u \in W^{1,2}_{\mathrm{loc}}(\mathscr{R}_L; \mathbb{R}^m)$.

By the compactness of the embedding, we can assume that $\mathbf{u}_j \to u$ strongly in $L^2_{\mathrm{loc}}(\mathscr{R}_L; \mathbb{R}^m)$ and therefore, along a further subsequence,

$$\lim_{j \to +\infty} \mathbf{u}_j(x, y) = u(x, y), \quad \text{a.e. in } \mathscr{R}_L. \tag{9.142}$$

Weak lower semi-continuity of the L^2 norm gives

$$\liminf_{j \to +\infty} \int_{\mathscr{R}_L} \frac{1}{2}|\nabla \mathbf{u}_j|^2 dxdy \geq \int_{\mathscr{R}_L} \frac{1}{2}|\nabla u|^2 dxdy,$$

and by Fatou's lemma,

$$\liminf_{j\to+\infty} \int_{\mathscr{R}_L} W(u_j)dxdy \geq \int_{\mathscr{R}_L} W(u)dxdy.$$

Moreover, (9.142) shows that u satisfies (9.132) and (9.136). It follows that we can identify the map u with the sought minimizer $u^{L,\eta}$. The proof is complete. $\qquad\square$

Since $u^{L,\eta}$ satisfies (9.132) and W and the boundary functions $\bar{u}_-$ and $\bar{u}_+(\cdot - \eta)$ are smooth, elliptic theory implies

$$\|u^{L,\eta}\|_{C^{2,\gamma}(\mathscr{R}_L;\mathbb{R}^m)} \leq C, \tag{9.143}$$

for some constant $C > 0$ and $\gamma \in (0, 1)$ independent of L and η.

Lemma 9.19 *There exist $\bar{\eta} \in \mathbb{R}$ and $u^L \in \mathscr{A}_{L,\bar{\eta}}$ such that*

$$J(u^L) = \min_{\eta} J(u^{L,\eta}) \leq C_0 + \sigma L.$$

Proof 1. There exists $\bar{y} > 0$ such that

$$(|\eta| - 2\bar{y})\frac{|a|^2}{2L} \leq J(u), \quad \text{for} \quad |\eta| \geq 2\bar{y}, \ u \in \mathscr{A}_{L,\eta}, \tag{9.144}$$

where $a = (a^+ - a^-)/2$.

Assume first $\eta \geq 0$. Since both $\bar{u}_-$ and $\bar{u}_+$ satisfy (9.19), there exists $\bar{y} > 0$ such that

$$|\bar{u}_-(y) - a^+| \leq \frac{1}{2}|a|, \quad \text{for} \quad y \geq \bar{y},$$

$$|\bar{u}_+(y - \eta) - a^-| \leq \frac{1}{2}|a|, \quad \text{for} \quad y \leq \eta - \bar{y}.$$

It follows that

$$|\bar{u}_+(y - \eta) - \bar{u}_-(y)| \geq 2|a| - |\bar{u}_+(y - \eta) - a^- - (\bar{u}_-(y) - a^+)|$$

$$\geq |a|, \quad \text{for} \quad \bar{y} \leq y \leq \eta - \bar{y}.$$

This and $u(L, y) - u(0, y) = \bar{u}_+(y - \eta) - \bar{u}_-(y)$ imply

$$|a| \leq |u(L, y) - u(0, y)| \leq L^{\frac{1}{2}}(\int_0^L |u_x(x, y)|^2 dx)^{\frac{1}{2}}, \quad y \in (\bar{y}, \eta - \bar{y})$$

and in turn

$$(\eta - 2\bar{y})|a|^2 \le L \int_{\bar{y}}^{\eta-\bar{y}} \int_0^L |u_x(x, y)|^2 dx \le 2LJ(u).$$

This and similar estimates valid for $\eta < 0$ prove 1.

2. Let u^{L,η_j}, $j = 1\ldots$ a minimizing sequence. From Lemma 9.16 we can choose a minimizing sequence that satisfies

$$\lim_{j\to+\infty} J(u^{L,\eta_j}) = \inf_{\eta} J(u^{L,\eta}) \le C_0 + \sigma L \qquad (9.145)$$

From (9.143), by passing to a subsequence, we can assume that there is a continuous function u^L such that

$$\lim_{j\to+\infty} u^{L,\eta_j} = u^L,$$

uniformly on compact sets. From (9.145) and (9.144) it follows that the sequence η_j, $j = 1, 2, \ldots$ is bounded and therefore, along a further subsequence,

$$\lim_{j\to+\infty} \eta_j = \bar{\eta}.$$

This and the uniform convergence of u^{L,η_j} to u^L imply that u^L satisfies the boundary conditions in $\mathscr{A}_{L,\bar{\eta}}$. From this point on we can proceed as in Lemma 9.18 to conclude that u^L is the sought-for minimizer. The proof is complete. $\square$

The minimizer u^L exhibited in Lemma 9.19 can be identified with $u^{L,\bar{\eta}}$. Indeed, since u^L satisfies the boundary conditions for $\eta = \bar{\eta}$ we have

$$J(u^{L,\bar{\eta}}) \le J(u^L) = \min_{\eta} J(u^{L,\eta}) \le J(u^{L,\bar{\eta}}).$$

In the following, when it is clear from the context, we simply write u instead of u^L and we do the same with other functions of L that we introduce later.

9.5.2   Basic Lemmas

In this section we prove a few lemmas that are basic for deriving estimates on u^L that are uniform in L. In Lemma 9.20 we prove that u^L decays exponentially to $a^\pm$ as $y \to \pm\infty$. In Lemma 9.21 we show that $\int_0^L \int_{\mathbb{R}} |u_x^L|^2 dx dy$ is uniformly bounded in L. This is a simple result that is essential for the ensuing analysis.

Note that u^L satisfies (9.143) and is a classical solution of (9.15). Note also that, since $\bar{\eta}$ in Lemma 9.19 depends only on L, when applied to $u = u^L$, (9.136) takes the form

$$|u(x, y) - a^+| \le \frac{C_L}{\sqrt{y}}, \qquad \text{for} \quad y \ge y_L,$$

$$|u(x, y) - a^-| \le \frac{C_L}{\sqrt{-y}}, \qquad \text{for} \quad y \le -y_L$$

$$(9.146)$$

for some constants $C_L > 0$, and $y_L > 0$.

The fact that u^L solves (9.15) implies a sharper asymptotic behavior for $y \to \pm\infty$.

Lemma 9.20 *There exist constants $k, K > 0$, independent of $L > 0$, such that $u = u^L$ satisfies*

$$|u(x, y) - a^+| \le r_0 e^{-k(y - y_L)}, \qquad \text{for} \quad y \ge y_L,$$

$$|u(x, y) - a^-| \le r_0 e^{k(y_L + y)}, \qquad \text{for} \quad y \le -y_L,$$

$$(9.147)$$

where $r_0 > 0$ is the constant in (9.56).

Moreover, for $\alpha \in \mathbb{R}^2$, $1 \le |\alpha| \le 2$, we have

$$|(D^\alpha u)(x, y)| \le K e^{-k(y - y_L)}, \quad \text{for} \quad y \ge y_L,$$

$$|(D^\alpha u)(x, y)| \le K e^{k(y + y_L)}, \quad \text{for} \quad y \le -y_L.$$

$$(9.148)$$

Proof From (9.146), (9.19) and the boundary conditions imposed on $u = u^L$ in (9.129) it follows that, by increasing $y_L > 0$ if necessary, we can assume

$$|u(x, y) - a^+| \le r_0 \quad \text{for} \quad x \in [0, L], \ y \ge y_L,$$

$$|u(0, y) - a^+| = |\bar{u}_-(y) - a^+| \le r_0 e^{-\bar{k}(y - y_L)}, \qquad \text{for} \quad y \ge y_L, \qquad (9.149)$$

$$|u(L, y) - a^+| = |\bar{u}_+(y) - a^+| \le r_0 e^{-\bar{k}(y - y_L)}, \qquad \text{for} \quad y \ge y_L.$$

Since the minimizer $u = u^L$ is a solution of (9.15), from (9.56) and (9.149)$_1$ we also have

$$\Delta|u - a^+|^2 \ge 2(\Delta u) \cdot (u - a^+) = 2W_u(u) \cdot (u - a^+)$$

$$= 2(W_u(u) - W_u(a^+)) \cdot (u - a^+) \ge \gamma^2 |u - a^+|^2,$$

$$(9.150)$$

where γ is the constant in (9.56) and we have used $W_u(a^+) = 0$. This, (9.149) and a standard comparison argument yield (9.147)$_1$, and (9.147)$_2$ follows by a similar reasoning. The estimates for the derivatives follow from (9.147) and elliptic regularity. The proof is complete. $\qquad \square$

From Lemma 9.20 it follows that, for $u = u^L$, the quantities

$$q(x) = \min_{r\in\mathbb{R},p\in\{-,+\}} \|u(x, \cdot) - \bar{u}_p(\cdot - r)\|,$$

$$\int_{\mathbb{R}} W(u(x, y))dy, \quad \int_{\mathbb{R}} |u_x(x, y)|^2 dy, \quad \int_{\mathbb{R}} |u_y(x, y)|^2 dy,$$

are well defined and continuous for $x \in [0, L]$.

Lemma 9.21 *Let u^L be as in Lemma 9.19. Then*

$$\sigma L \le J(u^L) \le \sigma L + C_0,$$

$$\int_0^L \int_{\mathbb{R}} |u_x^L|^2 dxdy \le 2C_0. \tag{9.151}$$

where σ and C_0 are the constants in Lemma 9.16.

Proof Lemma 9.17 implies $\lim_{y\to\pm\infty} u^L = a^\pm$ and therefore from $\sigma = J_{\mathbb{R}}(\pm\bar{u}_\pm)$ and the minimizing character of $\bar{u}_\pm$ it follows that

$$\int_{\mathbb{R}} \left(W(u^L(x, y)) + \frac{1}{2}|u_y^L(x, y)|^2\right)dy \ge \sigma, \quad \text{for } x \in [0, L]. \tag{9.152}$$

From this and Lemma 9.19 we obtain that

$$\frac{1}{2}\int_0^L \int_{\mathbb{R}} |u_x^L|^2 dxdy$$

$$\le \int_0^L \left(\int_{\mathbb{R}} \left(W(u^L) + \frac{1}{2}|u_y^L|^2\right)dy - \sigma\right)dx + \frac{1}{2}\int_0^L \int_{\mathbb{R}} |u_x^L|^2 dxdy \le C_0.$$

The proof is complete. $\square$

Lemma 9.22 *Assume $u = u^L$ is as in Lemma 9.19. Then u satisfies* (9.87) *and* (9.88) *for $x \in [0.L]$ and for some constants $\omega = \omega_L$ and $\tilde{\omega} = \tilde{\omega}_L$. Moreover*

$$0 \le \omega \le \frac{C_0}{L},$$

$$\tilde{\omega} = 0. \tag{9.153}$$

Proof The first part of the Lemma is proved as in Lemma 9.11. By (9.87) and (9.152),

$$\frac{1}{2}\int_{\mathbb{R}} |u_x(x, y)|^2 dy = J_{\mathbb{R}}(u(x, \cdot)) - \sigma + \omega \ge \omega, \quad \text{for } x \in [0, L]. \tag{9.154}$$

Because of the boundary value $u(0, \cdot) = \bar{u}_-$, which implies $J_{\mathbb{R}}(u(0, \cdot)) = \sigma$, we see that, for $x = 0$, (9.154) yields $\omega \geq 0$. The inequality $\omega \leq \frac{C_0}{L}$ follows by integrating (9.154) on $[0, L]$ and by Lemma 9.21. To prove that $\tilde{\omega} = 0$ it suffices to observe that $u = u^L$ is a solution of (9.130) and therefore the restriction (9.92) in the proof of Lemma 9.11 can be removed. The proof is complete. $\square$

9.5.3   Structural Properties of u^L

We are now able to derive detailed information on the structure of the minimizer $u^L \in \mathscr{A}_{L,\bar{\eta}}$ determined in Lemma 9.19. This knowledge of u^L, in particular the fact that, as we show below, the constant y_L in Lemma 9.20 is bounded independently of L, will allow us to pass to the limit in (9.131) and show that the limit map is a solution of (9.15) with the properties required in Theorem 9.3.

Set $u = u^L$. Let $p \in (0, q^0]$ be a number to be fixed later and $S_p \subset [0, L]$ be the complement of the set $\widetilde{S}_p$ defined by

$$\widetilde{S}_p := \{x \in (0, L) : \|u_x(x, \cdot)\|^2 > e_p\} \tag{9.155}$$

where e_p is the constant in Lemma 9.10. From Lemma 9.21 it follows that the measure of $\widetilde{S}_p$ is bounded independently of $L > 1$. Indeed, we have

$$|\widetilde{S}_p| e_p \leq \int_0^L \|u_x(x, \cdot)\|^2 dx \leq 2C_0,$$

and therefore

$$|\widetilde{S}_p| \leq \frac{2C_0}{e_p} \quad \text{and} \quad |S_p| \geq L - \frac{2C_0}{e_p}. \tag{9.156}$$

From Lemma 9.22 we know that $u = u^L$ satisfies (9.87) with $\omega \geq 0$, and hence

$$J(u(x, \cdot)) - \sigma \leq \frac{1}{2}\|u_x(x, \cdot)\|^2 \leq \frac{e_p}{2}, \quad x \in S_p. \tag{9.157}$$

This and Lemma 9.10 imply ($\| \cdot \|_1$ is the $W^{1,2}(\mathbb{R}; \mathbb{R}^m)$ norm)

$$\min_{p \in \{-,+\}} \min_{r \in \mathbb{R}} \|u(x, \cdot) - \bar{u}_p(\cdot - r)\|_1 < p \leq q^0. \tag{9.158}$$

It follows that $q(x) = \min_{p \in \{-,+\}} \min_{r \in \mathbb{R}} \|u(x, \cdot) - \bar{u}_p(\cdot - r)\| < p \leq q^0$ and therefore, for each $x \in S_p$, there are unique $\bar{u} \in \{\bar{u}_-, \bar{u}_+\}$ and $h(x) \in \mathbb{R}$ that allow to represent $u = u^L$ as in (9.38). If $I \subset S_p$ is an open interval, $\bar{u}$ is the same for all $x \in I$ and (from Lemma 9.7) $h(x)$ is continuously differentiable in I.

Moreover, from Lemmas 9.13 and 9.22, which yields $\tilde{\omega} = 0$, it follows that the expressions (9.96) and (9.97), respectively, are valid for $h'(x)$ and $\|u_x(x, \cdot)\|^2$ for $x \in S_p$.

If needed, we indicate that the map $\bar{u} \in \{\bar{u}_-, \bar{u}_+\}$ associated to $x \in S_p$ depends on x by using the notation $\bar{u} = \bar{u}_{\mathrm{p}_x}$ with $\mathrm{p}_x \in \{-, +\}$.

Note that (9.35) in Lemma 9.6 implies

$$\|v_y(x, \cdot)\| \le \|v(x, \cdot)\|_1 = \|u^L(x, \cdot + h(x)) - \bar{u}_{\mathrm{p}_x}\|_1 \le \bar{C}p. \tag{9.159}$$

This, provided $p > 0$ is chosen sufficiently small, implies that (9.109) holds for $x \in S_p$ and (9.97) implies

$$\|v_x(x, \cdot)\| \le 2^{\frac{1}{2}}\|u_x(x, \cdot)\|, \quad x \in S_p. \tag{9.160}$$

We also observe that, since for $x \in S_p$ we have $q(x) < p \le q^0$, we can use (9.46) and deduce from (9.157) that

$$\frac{1}{2}\mu\|v_y(x, \cdot)\|^2 \le \mathscr{W}(v) = J(u(x, \cdot)) - \sigma \le \frac{1}{2}\|u_x(x, \cdot)\|^2$$

and then

$$\|v_y(x, \cdot)\| \le \frac{1}{\mu^{\frac{1}{2}}}\|u_x(x, \cdot)\|, \quad x \in S_p. \tag{9.161}$$

Proposition 9.1 *Let $u = u^L$ the minimizer in Lemma 9.19. Then, if $p \in (0, q^0]$ is sufficiently small, it results*

1. There is a constant $C_h > 0$ independent of L such that

$$|h(x)| \le C_h, \quad x \in S_p. \tag{9.162}$$

2. There exists $y_0 > 0$, independent of L, such that

$$|u(x, y) - a^-| \le \frac{r_0}{2}, \quad x \in [0, L], \ y \le -y_0,$$

$$|u(x, y) - a^+| \le \frac{r_0}{2}, \quad x \in [0, L], \ y \ge y_0, \tag{9.163}$$

where r_0 is the constant in (9.56).

Proof $\tilde{S}_p$ is the union of a countable family of intervals: $\tilde{S}_p = \bigcup_j(\alpha_j, \beta_j)$. Therefore, for each $x \in S_p$, we have

$$|h(x)| \le \int_{S_p} |h'|dx + \sum_j |h(\beta_j) - h(\alpha_j)|. \tag{9.164}$$

Let $\lambda > 0$ be a small number to the chosen later and set $I_\lambda = \{j : \beta_j - \alpha_j \le \lambda\}$, $\widetilde{I}_\lambda = \{j : \beta_j - \alpha_j > \lambda\}$. Note that $\widetilde{I}_\lambda$ contains at most $\frac{|\widetilde{S}_p|}{\lambda} \le \frac{2C_0}{\lambda e_p}$ elements. For $j \in I_\lambda$ and $\xi \in (\alpha_j, \beta_j)$ we have

$$|u(\xi, y) - u(\alpha_j, y)| \le \int_{\alpha_j}^{\xi} |u_x(x, y)|dx \le |\xi - \alpha_j|^{\frac{1}{2}} \left(\int_{\alpha_j}^{\xi} |u_x(x, y)|^2 dx \right)^{\frac{1}{2}},$$

and therefore

$$\int_{\mathbb{R}} |u(\xi, y) - u(\alpha_j, y)|^2 dy \le |\beta_j - \alpha_j| \int_{\alpha_j}^{\beta_j} \int_{\mathbb{R}} |u_x(x, y)|^2 dy dx \le \lambda 2 C_0$$

where C_0 is the constant in Lemma 9.21. From this estimate and the fact that $\alpha_j \in S_p$, which implies

$$\|u(\alpha_j, \cdot) - \bar{u}(\cdot - h(\alpha_j))\| \le p \le q^0,$$

it follows that, if p and λ are sufficiently small, then, for each $x \in (\alpha_j, \beta_j)$, $u(x, \cdot)$ satisfies the conditions in Lemma 9.5 ensuring that $\bar{u}$ and $h(x)$ are uniquely determined and either $\bar{u} = \bar{u}_-$ or $\bar{u} = \bar{u}_+$ for every $x \in [\alpha_j, \beta_j]$. Moreover, h is a smooth function of $u(x, \cdot)$ and, from (9.32), which implies $h'(x) = (D_u h)u_x(x, \cdot)$, we obtain using also (9.159), that

$$|h'(x)| \le C \|u_x(x, \cdot)\|, \quad \text{for} \quad x \in [\alpha_j, \beta_j], \ j \in I_\lambda.$$

Therefore,

$$\sum_{j \in I_\lambda} |h(\beta_j) - h(\alpha_j)| \le \int_{\cup_{j \in I_\lambda} [\alpha_j, \beta_j]} |h'(x)| dx$$

$$\le C \int_{\cup_{j \in I_\lambda} [\alpha_j, \beta_j]} \|u_x\| dx \le C |\widetilde{S}_p|^{\frac{1}{2}} \Big(\int_0^L \|u_x\|^2 dx \Big)^{\frac{1}{2}} \le (2C_0)^{\frac{1}{2}} C |\widetilde{S}_p|^{\frac{1}{2}}.$$

$$\tag{9.165}$$

Assume now $j \in \widetilde{I}_\lambda$ and observe that, (9.19) implies that there is $\bar{y} > 0$ such that, for $p, q \in \{-, +\}$

$$|\bar{u}_p(y) - \bar{u}_q(y - r)| \ge |a|, \quad \text{for } \bar{y} \le y \le r - \bar{y}, \ \text{if } r \ge 2\bar{y},$$

$$|\bar{u}_p(y) - \bar{u}_q(y - r)| \ge |a|, \quad \text{for } r + \bar{y} \le y \le -\bar{y}, \ \text{if } r \le -2\bar{y},$$

$$\tag{9.166}$$

where as before $a = (a^+ - a^-)/2$.

Consider first the indices $j \in \widetilde{I}_\lambda$ such that $|h(\beta_j) - h(\alpha_j)| \leq 4\bar{y}$. We have

$$\sum_{j \in \widetilde{I}_\lambda, |h(\beta_j) - h(\alpha_j)| \leq 4\bar{y}} |h(\beta_j) - h(\alpha_j)| \leq 4\bar{y} \frac{|\widetilde{S}_p|}{\lambda}. \tag{9.167}$$

If $r > 4\bar{y}$ the interval $(\bar{y}, r - \bar{y})$ (if $r < -4\bar{y}$ the interval $(r + \bar{y}, -\bar{y})$) has measure larger then $\frac{|r|}{2}$. Therefore, for each $j \in \widetilde{I}_\lambda$ with $|h(\beta_j) - h(\alpha_j)| > 4\bar{y}$, there are y_j^0, y_j^1, with $y_j^1 - y_j^0 = |h(\beta_j) - h(\alpha_j)|/2$, such that

$$\begin{aligned}
|u(\beta_j, y) - u(\alpha_j, y)| &\geq |\bar{u}_{\mathrm{p}\beta_j}(y - h(\beta_j)) - \bar{u}_{\mathrm{p}\alpha_j}(y - h(\alpha_j))| \\
&\quad - |u(\beta_j, y) - \bar{u}_{\mathrm{p}\beta_j}(y - h(\beta_j))| - |u(\alpha_j, y) - \bar{u}_{\mathrm{p}\alpha_j}(y - h(\alpha_j))| \\
&= |\bar{u}_{\mathrm{p}\beta_j}(y - h(\beta_j)) - \bar{u}_{\mathrm{p}\alpha_j}(y - h(\alpha_j))| \\
&\quad - |v(\beta_j, y - h(\beta_j))| - |v(\alpha_j, y - h(\alpha_j))| \\
&\geq |\bar{u}_{\mathrm{p}\beta_j}(y - h(\beta_j)) - \bar{u}_{\mathrm{p}\alpha_j}(y - h(\alpha_j))| - 2 \cdot 2^{\frac{1}{2}} \bar{C} p \\
&\geq |a| - 2(2\bar{C})^{\frac{1}{2}} p \geq \frac{|a|}{2}, \quad \text{for } y \in (y_j^0, y_j^1).
\end{aligned} \tag{9.168}$$

where we have also used (9.166), $\|v(\alpha_j, \cdot)\|, \|v(\beta_j, \cdot)\| < p$, and (9.159), which imply

$$\|v(x, \cdot)\|_{L^\infty(\mathbb{R};\mathbb{R}^m)} \leq 2^{\frac{1}{2}} \|v(x, \cdot)\|^{\frac{1}{2}} \|v_y(x, \cdot)\|^{\frac{1}{2}} < 2^{\frac{1}{2}} \bar{C} p, \quad x \in S_p,$$

and assumed p small. Integrating (9.168) in (y_j^0, y_j^1) yields

$$\begin{aligned}
\frac{|a|}{4} |h(\beta_j) - h(\alpha_j)| &\leq \int_{y_j^0}^{y_j^1} |u(\beta_j, y) - u(\alpha_j, y)| dy \leq \int_{y_j^0}^{y_j^1} \int_{\alpha_j}^{\beta_j} |u_x| dx dy \\
&\leq \frac{1}{2^{\frac{1}{2}}} |h(\beta_j) - h(\alpha_j)|^{\frac{1}{2}} (\beta_j - \alpha_j)^{\frac{1}{2}} \left(\int_{y_j^0}^{y_j^1} \int_{\alpha_j}^{\beta_j} |u_x|^2 dx dy \right)^{\frac{1}{2}} \\
&\leq |h(\beta_j) - h(\alpha_j)|^{\frac{1}{2}} (\beta_j - \alpha_j)^{\frac{1}{2}} C_0^{\frac{1}{2}}
\end{aligned}$$

where C_0 is the constant in Lemma 9.21. Hence, $|h(\beta_j) - h(\alpha_j)| \leq \frac{16C_0}{|a|^2}(\beta_j - \alpha_j)$, and in turn

$$\begin{aligned}
\sum_{j \in \widetilde{I}_\lambda, |h(\beta_j) - h(\alpha_j)| > 4\bar{y}} &|h(\beta_j) - h(\alpha_j)| \\
&\leq \frac{16C_0}{|a|^2} \sum_{j \in \widetilde{I}_\lambda, |h(\beta_j) - h(\alpha_j)| > 4\bar{y}} (\beta_j - \alpha_j) \leq \frac{16C_0}{|a|^2} |\widetilde{S}_p|.
\end{aligned} \tag{9.169}$$

From (9.165), (9.167) and (9.169) we conclude that the sum in the right hand side of (9.164) is bounded by a constant independent of L. It remains to show that $\int_{S_p} |h'| dx \leq C$ with $C > 0$ independent of $L > 1$. This follows from (9.96), (9.159), (9.160) and (9.161), which, for $p \in (0, q^0]$ small, imply

$$|h'(x)| \leq \frac{2^{\frac{1}{2}} 4 \|u_x(x, \cdot)\|^2}{\mu^{\frac{1}{2}} \|\bar{u}'\|^2}, \qquad (9.170)$$

and from Lemma 9.21. This concludes the proof of 1.

Since $u = u^L$ satisfies (9.143) and $\bar{u}_\pm$ is bounded in $C^1(\mathbb{R}; \mathbb{R}^m)$, on the basis of Lemma 9.8, we can assume that $p > 0$ has been chosen so small that

$$\|u(x, \cdot) - \bar{u}_{p_x}(\cdot - h(x))\|_{L^\infty(\mathbb{R}; \mathbb{R}^m)} \leq C p^{\frac{2}{3}} \leq \frac{r_0}{8}, \qquad x \in S_p. \qquad (9.171)$$

Define

$$Y := \{y \geq C_h + y_{r_0} : |u(x_y, y) - a^+| \geq \frac{r_0}{2} \quad \text{for some } x_y \in (0, L)\},$$

where $y_{r_0} > 0$ is such that

$$|\bar{u}_\pm(y) - a^+| \leq \frac{r_0}{8}, \quad \text{for } y \geq y_{r_0}. \qquad (9.172)$$

Note that this and 1. imply

$$|\bar{u}_{p_x}(y - h(x)) - a^+| \leq \frac{r_0}{8}, \quad \text{for } y \geq C_h + y_{r_0}, \ x \in S_p.$$

Then (9.171) yields

$$|u(x, y) - a^+| \leq |u(x, y) - \bar{u}_{p_x}(y - h(x))| + |\bar{u}_{p_x}(y - h(x)) - a^+|$$
$$\leq \frac{r_0}{4}, \quad x \in S_p, \ y \geq C_h + y_{r_0}. \qquad (9.173)$$

This inequality shows that $y \in Y$ implies that x_y belongs to $\widetilde{S}_p$ and therefore to one of the intervals, say (α, β), that compose $\widetilde{S}_p$. This and (9.173), computed for $x = \alpha$, yield

$$|u(x_y, y) - u(\alpha, y)| \geq |u(x_y, y) - a^+| - |u(\alpha, y) - a^+| \geq \frac{r_0}{4}, \quad y \in Y.$$

It follows that

$$\frac{r_0}{4} \leq \int_\alpha^{x_y} |u_x(x, y)| dx \leq |\beta - \alpha|^{\frac{1}{2}} \left(\int_\alpha^\beta |u_x(x, y)|^2 dx \right)^{\frac{1}{2}},$$

and in turn Lemma 9.21 implies

$$|Y|\frac{r_0^2}{16} \le |\tilde{S}_p| \int_{\tilde{S}_p} \int_\alpha^\beta |u_x(x,y)|^2 dxdy \le 2C_0|\tilde{S}_p|,$$

and we have that the measure of Y is bounded independently of $L > 1$. This shows that there exists an increasing sequence $y_j \to +\infty$ such that

$$y_1 \le C_h + y_{r_0} + |Y|,$$

$$|u(x,y_j) - a^+| < \frac{r_0}{2}, \quad x \in [0,L], \quad j = 1,2,\dots$$

Therefore, using also (9.172) that implies

$$|u(L,y) - a^+| = |\bar{u}_+(y - h(L)) - a^+| \le \frac{r_0}{8}, \quad y \ge C_h + y_{r_0},$$

we can argue as in the proof of Lemma 9.17 and conclude with the help of the cut-off Lemma 4.1 (see also [6]) that the second inequality in (9.163) holds with $y_0 = y_1$. The other inequality follows by the same argument. The proof is complete. $\square$

Note that, based on Proposition 9.1, we can refine the estimates (9.147). Indeed, since in (9.163) y_0 is independent of $L > 1$, from the argument in the proof of Lemma 9.20, and the fact that $u = u^L$ is bounded it follows that there exist constants $k, K > 0$ independent of $L > 1$ and such that $u = u^L$ satisfies

$$|u(x,y) - a^+| \le Ke^{-ky}, \quad \text{for} \quad y \ge 0,$$

$$|u(x,y) - a^-| \le Ke^{ky}, \quad \text{for} \quad y \le 0. \tag{9.174}$$

This and elliptic regularity imply

$$|(D^\alpha u)(x,y)| \le Ke^{-k|y|}, \quad \text{for} \quad y \in \mathbb{R},$$

$$\|D^\alpha u(x,\cdot)\| \le \frac{K}{\sqrt{2k}}, \quad \text{for} \quad x \in [0,L], \tag{9.175}$$

for $\alpha \in \mathbb{N}^2$, $1 \le |\alpha| \le 2$.

From (9.174) we have that $u = u^L$ satisfies (9.36) in Lemma 9.7 with $\bar{u}_j$ replaced by $\bar{u}_+$. It follows that the conclusions of Lemmas 9.7 and 9.13 apply to $u = u^L$ and, in particular, the decomposition (9.38) and the expressions of $h'(x)$ and $\|u_x(x,\cdot)\|$ in (9.96) and (9.97) are valid if $q(x) = \|v(x,\cdot)\| \le q^0$, where $v = v^L$ is defined by

$$v^L(x,y) = u^L(x,y + h(x)) - \bar{u}_{p_x}(y), \quad x \in [0,L], \ y \in \mathbb{R}.$$

Next we focus on the sets

$$\Sigma_\beta := \{x \in [0, L] : q(x) \leq \beta q^0\}$$

for $\beta = 1$ and $\beta = \frac{1}{2}$ and show that $\Sigma_{\frac{1}{2}}$ has the simplest possible structure.

Proposition 9.2 *Set $u = u^L$. Then there exist $0 < l_- < l_+ < L$ such that*

$$\Sigma_{1/2} = [0, l_-] \cup [l_+, L] \tag{9.176}$$

and

$$l_+ - l_- \leq C, \tag{9.177}$$

where $C > 0$ is a constant independent of $L > 1$. For $x \in [0, l_-]$, the map $\bar{u} \in \{\bar{u}_-, \bar{u}_+\}$ in the decomposition (9.38) coincides with $\bar{u}_-$ and for $x \in [l_+, L]$ with $\bar{u}_+$. Moreover, the map $x \mapsto q(x)$ is nondecreasing in $[0, l_-]$ and nonincreasing in $[l_+, L]$.

Proof 1. Define

$$l_- := \max\{x \in \Sigma_{1/2} : \bar{u} = \bar{u}_-\},$$
$$l_+ := \min\{x \in \Sigma_{1/2} : \bar{u} = \bar{u}_+\} \tag{9.178}$$

and observe that the continuity of the map $[0, L] \ni x \mapsto q(x) \in \mathbb{R}$ and the conditions $u(0, \cdot) = \bar{u}_-, u(L, \cdot) = \bar{u}_+(\cdot - \bar{\eta})$ imply

$$0 < l_\pm < L,$$
$$q(l_-) = q(l_+) = \frac{q^0}{2}. \tag{9.179}$$

If $[l_+, L] \not\subset \Sigma_{\frac{1}{2}}$, then there exists $x^* \in (l_+, L)$ such that

$$\frac{q^0}{2} < q(x^*),$$
$$q(x) \leq q(x^*), \quad x \in [l_+, L].$$

Then there are two possibilities

(a) $\frac{q^0}{2} < q(x^*) \leq q^0$,
(b) $q^0 < q(x^*)$.

We can immediately exclude case (a) by Lemma 9.15 which, as observed in Remark 9.4, can be applied to the present situation. By arguing as in the proof of

Theorem 9.2 after Remark 9.4, we can also exclude case (b). Indeed, if $[\tilde{\xi}_1, \tilde{\xi}_2] \subset (\xi_1, \xi_2)$ are defined by

$$\tilde{\xi}_1 = \min\{x > l_+ : q(x) \geq q^0\}, \quad \xi_1 = \max\{x < \tilde{\xi}_1 : q(x) \leq \tfrac{q^0}{2}\},$$

$$\tilde{\xi}_2 = \max\{x : q(x) \geq q^0\}, \qquad \xi_2 = \min\{x > \tilde{\xi}_2 : q(x) \leq \tfrac{q^0}{2}\}.$$

Then, as in (9.124) and (9.125), we have

$$q(\tilde{\xi}_1) = q(\tilde{\xi}_2) = q^0,$$

$$q(\xi_1) = q(\xi_2) = \frac{q^0}{2}, \tag{9.180}$$

$$q(x) \in (\frac{q^0}{2}, q^0), \quad x \in (\xi_1, \tilde{\xi}_1) \cup (\tilde{\xi}_2, \xi_2).$$

Note that the definition of $\tilde{\xi}_1$ and $\tilde{\xi}_2$ implies $[l_+, \tilde{\xi}_1] \subset \Sigma_1$ and $[\tilde{\xi}_2, L] \subset \Sigma_1$, and therefore we have

$$\bar{u} = \bar{u}_+, \quad \text{for } x \in [l_+, \tilde{\xi}_1] \cup [\tilde{\xi}_2, L].$$

On the basis of these observations we can define a competing map $\tilde{u}$ by setting

$$\tilde{u}(x, \cdot) = u(x, \cdot), \quad x \in [0, \xi_1),$$

$$\tilde{u}(x, \cdot) = u(x, \cdot - (\hat{h}(\xi_2) - h(\xi_2))), \quad x \in (\xi_2, L],$$

and by defining $\tilde{u}$ in the interval $[\xi_1, \xi_2]$ exactly as in the proof of Theorem 9.2. Then arguing as in that proof we conclude that (b) is in contradiction with the minimality of u and $[l_+, L] \subset \Sigma_{1/2}$ is established.

Since the proof that $[0, l_-] \subset \Sigma_{1/2}$ is similar we obtain (9.176). We have $\bar{u} = \bar{u}_-$ in $[0, l_-]$ and $\bar{u} = \bar{u}_+$ in $[l_+, L]$ and therefore $l_- < l_+$. To prove (9.177), we observe that the definition of $\Sigma_{1/2}$ and (9.176) imply

$$q(x) = \min_{\mathrm{p}\in\{-,+\}} \min_{r\in\mathbb{R}} \|u(x, \cdot) - \bar{u}_\mathrm{p}(\cdot - r)\| > \frac{q^0}{2}, \quad x \in (l_-, l_+).$$

From this and Lemma 9.10 we obtain, using also Lemma 9.21, that

$$e_{q^0/2}(l_+ - l_-) \leq \int_{l_-}^{l_+} (J_{\mathbb{R}}(u(x, \cdot)) - \sigma)dx \leq C_0$$

and (9.177) follows with $C = \frac{C_0}{e_{q^0/2}}$. The monotonicity of the map $x \mapsto q(x)$ in the intervals $[0, l_-]$ and $[l_+, L]$ follow from Lemma 9.15. The proof is complete. $\qquad\square$

From Proposition 9.2 we know that the function $x \mapsto q(x) \leq q^0/2$ is monotone in $[0, l_-]$ and in $[l_+, L]$. Next we show that $q(x)$ converges to 0 exponentially in $[l_+, L]$ and that a corresponding statement applies to $[0, l_-]$.

Lemma 9.23 *We have*

$$q(x) \leq \frac{q^0}{2} e^{-\sqrt{\frac{\mu}{8}}(x-l_+)}, \qquad \text{for } x \in [l_+, L], \tag{9.181}$$

and

$$|h'(x)| \leq C e^{-\frac{1}{2}\sqrt{\frac{\mu}{8}}(x-l_+)}, \qquad \text{for } x \in [l_+, L], \tag{9.182}$$

where $\mu > 0$ is the constant in (9.45) and $C > 0$ is independent of $L > 1$. An analogous statement applies to the interval $[0, l_-]$.

Proof Proposition 9.2 implies $q(x) \leq q^0/2$ for $x \in [l_+, L]$ and $q(l_+) = q^0/2$, $q(L) = 0$. Therefore, we can proceed as in the proof of Lemma 9.14 and use (9.105) and the maximum principle to deduce that

$$q(x)^2 \leq \varphi(x), \quad x \in [l_+, L],$$

where

$$\varphi(x) = \left(\frac{q^0}{2}\right)^2 \frac{\sinh\sqrt{\frac{\mu}{2}}(L-x)}{\sinh\sqrt{\frac{\mu}{2}}(L-l_+)} \leq \left(\frac{q^0}{2}\right)^2 e^{-\sqrt{\frac{\mu}{2}}(x-l_+)}, \qquad \text{for } x \in [l_+, L],$$

is the solution of $\varphi'' = (\mu/2)\varphi$ with the boundary conditions $\varphi(l_+) = (q^0/2)^2$ and $\varphi(L) = 0$. This implies (9.181) and (9.182) follows as in Lemma 9.14. A similar argument applies to the interval $[0, l_-]$. The proof is complete. $\qquad\square$

9.5.4   Conclusion of the Proof of Theorem 9.3

We focus on the family of maps $\tilde{u} = \tilde{u}^L$, $L > 1$ defined via $u = u^L$, the minimizer in Lemma 9.19, by

$$\tilde{u}(x, y) := u(x - l_-, y), \quad (x, y) \in [-l_-, L - l_-] \times \mathbb{R}, \quad L > 1. \tag{9.183}$$

By (9.177) in Proposition 9.2 we can assume that, along a subsequence,

$$\lim_{L \to +\infty} (l_+ - l_-) = \ell \leq C. \tag{9.184}$$

It follows that, along a further subsequence, at least one between l_- and $L - l_+$ diverges to $+\infty$ as $L \to +\infty$. Therefore we need to consider two alternatives:

I

$$\lim_{L \to +\infty} l_- = \lim_{L \to +\infty} L - l_+ = +\infty. \tag{9.185}$$

II One of the limits in I is bounded. We will discuss the case (the other case is analogous)

$$\lim_{L \to +\infty} l_- = \ell_- < +\infty, \qquad \lim_{L \to +\infty} L - l_+ = +\infty. \tag{9.186}$$

If (9.185) prevails, then (9.143) implies that, along a subsequence, we have

$$\lim_{L \to +\infty} \tilde{u}(x, y) = u(x, y), \quad (x, y) \in \mathbb{R}^2,$$

where $u \in C^2(\mathbb{R}^2; \mathbb{R}^m)$ and the convergence is locally in $C^2(\mathbb{R}^2; \mathbb{R}^m)$. It follows that u is a solution of (9.15) and satisfies (9.174). From Lemma 9.23 we can also assume that, as $L \to +\infty$, the functions $q(\cdot - l_-)$ and $h(\cdot - l_-)$ converge pointwise in $[\ell, +\infty)$ to the functions $q^{\mathrm{u}}(\cdot)$ and $h^{\mathrm{u}}(\cdot)$ defined by

$$q^{\mathrm{u}}(x) = \|u(x, \cdot) - \bar{u}_+(\cdot - h^{\mathrm{u}}(x))\| = \min_{r \in \mathbb{R}} \|u(x, \cdot) - \bar{u}_+(\cdot - r)\|,$$

and moreover that

$$q^{\mathrm{u}}(x) \leq \frac{q^0}{2} e^{-\sqrt{\frac{\mu}{8}}(x-\ell)}, \quad x \in [\ell, +\infty),$$

$$|h^{\mathrm{u}}_+ - h^{\mathrm{u}}(x)| \leq C(1 - e^{-\frac{1}{2}\sqrt{\frac{\mu}{8}}(x-\ell)}), \quad x \in [\ell, +\infty),$$

where $h^{\mathrm{u}}_+ = \lim_{x \to +\infty} h^{\mathrm{u}}(x)$. These estimates proves $(9.12)_2$ for $x \to +\infty$ with $\eta_+ = h^{\mathrm{u}}_+$. A similar reasoning completes the proof of $(9.12)_2$ for $x \to -\infty$. Therefore, u can be identified with the map u in Theorem 9.3.

Suppose now that (9.186) holds. Proceeding as before, we show that along a subsequence, in the limit for $L \to +\infty$, $\tilde{u}$ converges to a solution $u : [-\ell, +\infty] \times \mathbb{R} \to \mathbb{R}^m$ of (9.15) and that u satisfies (9.174) and $(9.12)_2$ for $x \to +\infty$. On the other hand, we have

$$u(-\ell_-, \cdot) = \bar{u}_-(\cdot - \eta_-) \tag{9.187}$$

for some $\eta_- \in \mathbb{R}$. Moreover by Lemma 9.22, u satisfies (9.87) with $\omega = 0$ and hence $\|u_x(-\ell_-, \cdot)\| = 0$, and therefore

$$u_x(-\ell_-, y) = 0, \quad y \in \mathbb{R}.$$

This implies that u can be extended to $\mathbb{R}^2$ as a C^1 map by setting

$$u(x, \cdot) = \bar{u}_-(\cdot - \eta_-), \quad \text{for } x < -\ell_-.$$

The map u extended in this way is a weak solution of (9.15) in $\mathbb{R}^2$, and from the assumption that W is C^3 and elliptic theory it follows that u is a C^2 solution of (8.59). The extended map u trivially satisfies $(9.12)_2$ for $x \to -\infty$, and therefore also in case II the map u can be identified with the map u in Theorem 9.3. The proof is complete. $\qquad\qquad\square$

Remark 9.6 Actually, the occurrence of case II can be excluded. Indeed, on the basis of the previous discussion, (9.186) implies the existence of two solutions of (8.59) that coincide in an open set, namely u and the map v defined by

$$v(x, y) = \bar{u}_-(y - \eta_-), \quad (x, y) \in \mathbb{R}^2,$$

and as observed in [1], this contradicts the unique continuation theorem in [27].

9.6 Scholia on Chap. 9

The scalar Allen–Cahn equation

$$\Delta u = W_u(u), \quad x \in \mathbb{R}^n, \tag{9.188}$$

with $W : \mathbb{R} \to \mathbb{R}$ a C^2 potential, is a widely studied model for various physical phenomena and, in particular, for phase transitions. Classifying bounded entire solutions $u : \mathbb{R}^n \to \mathbb{R}$ and their relationship with minimal surface theory is an active and important field in P.D.E. In spite of the fact that many interesting and unexpected solutions have been discovered, see [15–17, 19, 31, 33, 34] and [43] for a review, the classification is still largely incomplete. If we restrict to solutions that are minimizers in the sense of Definition 4.1 and to the case of phase transition potentials $W \geq 0$ with two nondegenerate zeros that we assume to be ± 1, then at least in low dimension the situation is well understood. Savin [38, 39] has shown that in dimension $n \leq 7$ a minimizer $u : \mathbb{R}^n \to \mathbb{R}$ is necessarily one-dimensional, meaning that the level set of u are hyperplanes. Alberti et al. proved in [2] that a solution u of (9.188) which is strictly increasing in x_n and satisfies $|u| < 1$ and

$$\lim_{x_n \to \pm\infty} u(x', x_n) = \pm 1 \tag{9.189}$$

is a minimizer. Using this fact, Savin in [38], see also [8, 9] proved that for $n \leq 8$ such a solution is one-dimensional. For $n = 9$ there exists a minimizer that satisfies (9.189), is strictly increasing in x_n, but is not one-dimensional [18]. When we move to the vector case $m \geq 1$ the set of bounded entire solution of (9.188) is richer and harder to classify. Even if we restrict to phase transition potentials and to a symmetry context, Theorems 7.1 and 7.2 yield an endless variety of solutions and one can expect that, if the symmetry constraint is removed, families of solutions analogous to the one discovered in the scalar case will appear. The vector case is also quite different from the scalar case when we restrict to the class of minimal solutions. Since, for $m = 1$, connections exist only between neighboring zeros of $W : \mathbb{R} \to \mathbb{R}$, it is natural to conjecture that, in the scalar case, there is no minimizer $u : \mathbb{R}^n \to \mathbb{R}$ that connects three or more zeros $a_1, \ldots, a_N$ of W in the sense that $\lim_{s \to +\infty} u(s v_i) = a_i$ for some unit vector v_i, $i = 1, \ldots, N$. On the other hand, Theorem 7.1 yields for $n = 2$ the triple-junction and for $n = 3$ the quadruple-junction solution, which are minimal in the context of equivariant maps and are expected to be minimal under general compact perturbations. Another difference between the cases $m = 1$ and $m \geq 2$ is related to the fact that, in the vector case there may exist two or more distinct orbits connecting the zeros a^- and a^+ of W. When this is the case, then as discussed in Theorem 8.5 in the symmetric case and in Theorem 9.3 in the general case, already for $n = 2$ there exist minimal solutions u that share with scalar minimizers for $n \leq 7$ the property that

$$\lim_{\epsilon \to 0^+} u \left(\frac{x}{\epsilon} \right) = \chi a^- + a^+(1 - \chi),$$

with χ the characteristic function of a half-space, but at the same time are not constant on hyperplanes. The solutions in Theorems 8.5 and 9.3 also imply that, without some extra assumption, the Gibbons conjecture, namely that, for all $n \geq 1$, a minimizer such that the limit (9.189) is uniform in x' is one-dimensional [9, 20, 36], is false in the vector case even for $n = 2$. This is again related to the existence of multiple connecting orbits, a phenomenon that in the scalar case does not exist. Therefore, the question arises if, under the assumption that there is a unique orbit connecting a^- to a^+, for a minimizer the condition that the limit (9.189) is uniform in x' is sufficient for being -dimensional. Theorem 9.2 states that this is the case if $n = 2$. The proof of Theorem 9.2 does not extend to higher dimensions and the question remains open for $n \geq 3$. If we restrict to the case of potentials with only two zeros, under generic nondegeneracy assumptions, it is possible to give a complete characterization of minimizers for $n = 1$ and, under the crucial assumption that (9.189) is uniform in x', also for $n = 2$. This is done in Theorem 2.3 for $n = 1$ and in Theorem 9.1 for $n = 2$. It is an open difficult question whether the same is true without requiring uniform convergence in (9.189). Clearly the possibility of characterize vector minimizer for $n = 2$ is related to the assumption that W has only two zeros. If W has three zeros, then aside from the symmetric solutions given by Theorem 8.5 and the triple junction given by Theorem 7.1 nothing is known. In particular, there is no proof of the conjectured existence of a triple-junction solution without any assumption of symmetry. The existence of the solution u described in Theorem 8.5 was established in [1], where it was also pointed out that, in contrast to the scalar case, u is not one-dimensional. Theorem 9.3 was proved by Schatzman in [40] (see also [3, 25, 35]) and to our knowledge, the double heteroclinic solution constructed by Schatzman is the only known example of nonsymmetric vector minimizer for phase transition potentials. In relation with Theorem 9.2 we mention the work of Fazly and Ghoussoub [24] who proved, in dimension $n = 2$, one-dimensionality for vector solutions for which all components are strictly monotone in x_2 and under certain conditions on the mixed derivative of W. We remark that the class of potentials that satisfy these conditions includes some phase transition potentials. Various symmetry results, including one-dimensionality, for vector minimizers and different kind of potentials can be found in [21–23, 41, 42].

References

1. Alama, S., Bronsard, L., Gui, C.: Stationary layered solutions in $\mathbb{R}^2$ for an Allen–Cahn system with multiple well potential. Calc. Var. **5**(4), 359–390 (1997)
2. Alberti, G., Ambrosio, L., Cabré, X.: On a long-standing conjecture of E. De Giorgi: symmetry in 3D for general non linearities and a local minimality property. Acta Appl. Math. **65**, 9–33 (2001)
3. Alessio, F., Montecchiari, P.: Brake orbit solutions for semilinear elliptic systems with asymmetric double well potential. J. Fixed Point Theory Appl. **19**(1), 691–717 (2017)
4. Alikakos, N.D.: Some basic facts on the system $\Delta u - \nabla W(u) = 0$. Proc. Am. Math. Soc. **139**, 153–162 (2011)
5. Alikakos, N.D., Fusco, G.: On the connection problem for potentials with several global minima. Indiana Univ. Math. J. **57**, 1871–1906 (2008)
6. Alikakos, N.D., Fusco, G.: A maximum principle for systems with variational structure and an application to standing waves. J. Eur. Math. Soc. **17**(7), 1547–1567 (2015)
7. Alikakos, N.D., Betelú, S.I., Chen, X.: Explicit stationary solutions in multiple well dynamics and non-uniqueness of interfacial energies. Eur. J. Appl. Math. **17**, 525–556 (2006)
8. Ambrosio, L., Cabré, X.: Entire solutions of semilinear elliptic equations in $\mathbb{R}^3$ and a conjecture of De Giorgi. J. Am. Math. Soc. **13**, 725–739 (2000)
9. Barlow, M.T., Bass, R.F., Gui. C.: The Liouville property and a conjecture of De Giorgi. Commun. Pure Appl. Math. **53**, 1007–1038 (2000)
10. Bates, P.W., Fusco, G., Smyrnelis, P.: Entire solutions with six-fold junctions to elliptic gradient systems with triangle symmetry. Adv. Nonlinear Stud. **13**(1), 1–12 (2013)
11. Bronsard, L., Reitich, F.: On three-phase boundary motion and the singular limit of a vector-valued Ginzburg–Landau equation. Arch. Ration. Mech. Anal. **124**(4), 355–379 (1993)
12. Bronsard, L., Gui, C., Schatzman, M.: A three-layered minimizer in $\mathbb{R}^2$ for a variational problem with a symmetric three-well potential. Commun. Pure Appl. Math. **49**(7), 677–715 (1996)
13. Busca, J., Felmer, P.: Qualitative properties of some bounded positive solutions of scalar field equations. Calc. Var. **13**, 181–211 (2001)
14. Carr, J., Pego, B.: Metastable patterns in solutions of $u_t = \epsilon^2 u_{xx} - f(u)$. Commun. Pure Appl. Math. **42**(5), 523–576 (1989)
15. Dancer, E.: New solutions of equations in $\mathbb{R}^n$. Ann. Scuola Norm. Sup. Pisa Cl. Sci. 4 **30**(3–4), 535–563 (2002)
16. del Pino, M., Kowalczyk, M., Pacard, F., Wei, J.: The Toda system and multiple end solutions of autonomous planar elliptic problem. Adv. Math. **224**(4), 1462–1516 (2010)
17. del Pino, M., Kowalczyk, M., Pacard, F., Wei, J.: Multiple end solutions to the Allen-Cahn equation in $\mathbb{R}^2$. J. Differ. Geom. **258**(2), 458–503 (2010)

18. del Pino, M., Kowalczyk, M., Wei, J.: On De Giorgi's conjecture in dimension $N \geq 9$. Ann. Math. **174**, 1485–1569 (2011)
19. del Pino, M., Kowalczyk, M., Wei, J.: Entire solutions of the Allen-Cahn equation and complete embedded minimal surfaces of finite total curvature in $\mathbb{R}^3$. J. Differ. Geom. **93**(1), 67–131 (2013)
20. Farina, A.: Symmetry for solutions of semilinear elliptic equations in $\mathbb{R}^N$ and related conjectures. Ricerche Mat. **10**(Suppl. 48), 129–154 (1999)
21. Farina, A.: Two results on entire solutions of Ginzburg-Landau system in higher dimensions. J. Funct. Anal. **214**(2), 386–395 (2004)
22. Farina, A., Soave, N.: Monotonicity and 1-dimensional symmetry for solutions of an elliptic system arising in Bose–Einstein condensation. Arch. Ration. Mech. Anal. **213**(1), 287–326 (2014)
23. Farina, A., Sciunzi, B., Valdinoci, E.: Bernstein and De Giorgi type problems: new results via a geometric approach. Ann. Scuola Norm. Sup. Pisa Cl. Sci. **5**(7), 741–791 (2008)
24. Fazly, M., Ghoussoub, N.: De Giorgi type results for elliptic systems. Calc. Var. **47**(3), 809–823 (2013)
25. Fusco, G.: Layered solutions to the vector Allen-Cahn equation in $\mathbb{R}^2$. Minimizers and heteroclinic connections. Commun. Pure Appl. Math. **16**(5), 1807–1841 (2017)
26. Fusco, G., Gronchi, G.F., Novaga, M.: Existence of periodic orbits near heteroclinic connections (2018). http://cvgmt.sns.it/paper/3882/
27. Garofalo, N., Lin, F.H.: Unique continuation for elliptic operators: a geometric-variational approach. Commun. Pure Appl. Math. **40**(3), 347–366 (1987)
28. Ghoussoub, N., Gui, C.: On a conjecture of De Giorgi and some related problems. Math. Ann. **311**, 481–491 (1998)
29. Gui, C.: Hamiltonian identities for partial differential equations. J. Funct. Anal. **254**(4), 904–933 (2008)
30. Gui, C., Schatzman, M.: Symmetric quadruple phase transitions. Indiana Univ. Math. J. **57**(2), 781–836 (2008)
31. Gui, C., Malchiodi, A., Xu, H.: Axial symmetry of some steady state solutions to nonlinear Schrdinger equations. Proc. Am. Math. Soc. **139**, 1023–1032 (2011)
32. Henry, D.: Geometric Theory of Semilinear Parabolic Equations. Lecture Notes in Mathematics, vol. 840. Springer, Berlin (1981)
33. Liu, Y., Wang, K., Wei, J.: Global minimizers of the Allen-Cahn equation in dimension $n \geq 8$. J. Math. Pures Appl. **108**(6), 818–840 (2017)
34. Malchiodi, A.:Some new entire solutions of semilinear elliptic equations on $\mathbb{R}^n$. Adv. Math. **221**(6), 1843–1909 (2009)
35. Monteil, A., Santambrogio, F.: Metric methods for heteroclinic connections in infinite dimensional spaces. To appear. axXiv: 1602.05487v1
36. Poláčik, P.: Propagating terraces in a proof of the Gibbons conjecture and related results. J. Fixed Point Theory Appl. **19**(1), 113–128 (2017)
37. Rudin, W.: Functional Analysis. McGraw-Hill Series in Higher Mathematics. McGraw-Hill, New York (1973)
38. Savin, O.: Regularity of flat level sets in phase transitions. Ann. Math. **169**, 41–78 (2009)
39. Savin, O.: Minimal surfaces and minimizers of the Ginzburg landau energy. Cont. Math. Mech. Analysis AMS **526**, 43–58 (2010)
40. Schatzman, M.: Asymmetric heteroclinic double layers. Control Optim. Calc. Var. **8** (A tribute to J. L. Lions), 965–1005 (2002, electronic)
41. Soave, N., Tavares, H.: New existence and symmetry results for least energy positive solutions of Schrödinger systems with mixed competition and cooperation terms. J. Differ. Equ. **261**(1), 505–537 (2016)
42. Soave, N. Terracini, S.: Liouville theorems and 1-dimensional symmetry for solutions of an elliptic system modelling phase separation. Adv. Math. **279**, 29–66 (2015)
43. Wei, J.: Geometrization program of semilinear elliptic equations. AMS/IP Stud. Adv. Math. **51**, 831–857 (2012)

Appendix A
Radial Solutions of $\Delta u = c^2 u$

A.1   An Exponential Estimate

We have the following estimate for the radial solutions to

$$\begin{cases} \Delta\varphi = c^2\varphi, & x \in B_r = B(0; r), \\ \varphi = 1, & x \in \partial B_r, \end{cases} \tag{A.1}$$

with $\varphi = \varphi(x; r) = \Phi(|x|; r) = \Phi(s; r)$ (Fig. A.1).

Lemma A.1 *Let Φ as above. Then, $r \mapsto \frac{\partial\Phi}{\partial s}(r; r)$ is strictly increasing in $(0, +\infty)$ and*

$$\lim_{r \to +\infty} \frac{\partial\Phi}{\partial s}(r; r) = c. \tag{A.2}$$

Moreover, there exists a strictly increasing function $h : (0, +\infty) \to (0, +\infty)$ such that

$$\Phi(s; r) \leq e^{h(r)(s-r)}, \quad s \in [0, r], \tag{A.3}$$

and

$$\lim_{r \to \infty} h(r) = c. \tag{A.4}$$

Proof We suppress the second variable and write $\Phi(s)$, $\Phi'(s)$ instead of $\Phi(s; r)$, $\frac{\partial\Phi}{\partial s}(s; r)$, and we convert (A.1) into a Riccati equation. Set $w(s) = \Phi'(s)/\Phi(s)$.

© Springer Nature Switzerland AG 2018
N. D. Alikakos et al., *Elliptic Systems of Phase Transition Type*,
Progress in Nonlinear Differential Equations and Their Applications 91,
https://doi.org/10.1007/978-3-319-90572-3

Fig. A.1 The behaviour of Φ

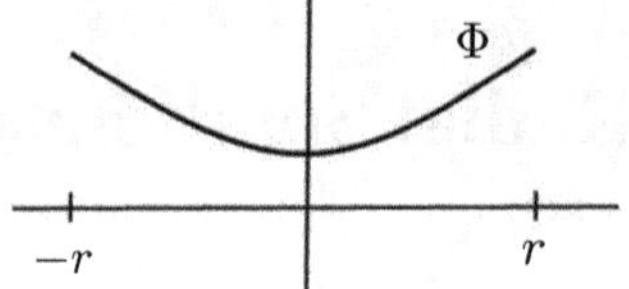

Then, (A.1) and $\Phi'(0) = 0$ imply that w solves

$$\begin{cases} w' = c^2 - \dfrac{n-1}{s} w - w^2, \\ w(0) = 0. \end{cases} \tag{A.5}$$

The right-hand side $g(s, w)$ of (A.5) has a positive root

$$z(s) = \frac{1-n}{2s} + \sqrt{\left(\frac{n-1}{2s}\right)^2 + c^2}, \tag{A.6}$$

and

$$\operatorname{sgn} g(s, w) = \operatorname{sgn}(z(s) - w). \tag{A.7}$$

It follows that

$$\lim_{s \to 0^+} z(s) = 0 \quad \text{and} \quad \lim_{s \to 0^+} z'(s) = \frac{c^2}{n-1},$$

while (A.1) implies

$$\Delta\varphi(0) = n\Phi''(0) = c^2 \Phi(0)$$

and therefore

$$w'(0) = \frac{c^2}{n}.$$

As a consequence,

$$\begin{cases} w(0) = z(0) = 0, \\ w'(0) < z'(0), \end{cases} \tag{A.8}$$

which shows that the curve $s \to w(s)$ starts below the curve $s \to z(s)$. Therefore, we conclude that

$$w(s) < z(s), \quad s \in (0, +\infty). \tag{A.9}$$

Indeed, the curves $s \to w(s)$ and $s \to z(s)$ cannot cross, since $z(s)$ is strictly increasing in $(0, +\infty)$ and $g(s, z(s)) = 0$ from (A.6). The inequalities (A.9) and (A.7) yield

$$\begin{cases} w'(s) > 0, & s \in (0, +\infty), \\ \lim_{s \to +\infty} w(s) = \lim_{s \to +\infty} z(s) = c. \end{cases} \tag{A.10}$$

From (A.10) we deduce that $\Phi'(r) = w(r)$ is strictly increasing in $(0, +\infty)$ and that (A.2) holds.

To prove (A.3), we set

$$h(r) = \frac{1}{r} \int_0^r w(s)\mathrm{d}s . \tag{A.11}$$

Then, by (A.10), h is increasing and satisfies (A.4). Let $\widetilde{\Phi}(s)$ be the right-hand side of (A.3). Then,

$$\frac{\Phi(s)}{\widetilde{\Phi}(s)} = e^{\int_r^s [w(\tau) - h(r)]\mathrm{d}\tau} \le 1, \quad s \in [0, r]. \tag{A.12}$$

Indeed, (A.11) and $h'(r) > 0$ imply that

$$\int_r^s [w(\tau) - h(r)]\mathrm{d}\tau = \int_r^s w(\tau)\mathrm{d}\tau - \frac{s - r}{r} \int_0^r w(\tau)\mathrm{d}\tau$$
$$= \int_0^s w(\tau)\mathrm{d}\tau - \frac{s}{r} \int_0^r w(\tau)\mathrm{d}\tau = s[h(s) - h(r)] \le 0, \quad s \in [0, r].$$

$\square$

Remark A.1 Let us also point out that

$$r_1 < r_2, \, t \in (0, r_1] \longrightarrow \Phi(r_1 - t; r_1) > \Phi(r_2 - t; r_2).$$

Indeed, setting

$$\theta(x) = \begin{cases} \Phi(|x| - r_2 + r_1; r_1), & \text{for } r_2 - r_1 \le |x| \le r_2, \\ \Phi(0; r_1), & \text{for } |x| \le r_2 - r_1, \end{cases}$$

we can see that θ satisfies $\Delta\theta \le c^2\theta$ weakly in B_{r_2}. Thus, by the maximum principle, $\theta(x) > \varphi(x; r_2)$, for all $x : |x| > r_2$.

Index

© Springer Nature Switzerland AG 2018

N. D. Alikakos et al., *Elliptic Systems of Phase Transition Type*,
Progress in Nonlinear Differential Equations and Their Applications 91,
https://doi.org/10.1007/978-3-319-90572-3